T0142128

Second Edition

Accident/Incident Prevention Techniques

Second Edition

Accident/Incident Prevention Techniques

Charles D. Reese

CRC Press is an imprint of the
Taylor & Francis Group, an **informa** business

CRC Press
Taylor & Francis Group
6000 Broken Sound Parkway NW, Suite 300
Boca Raton, FL 33487-2742

First issued in paperback 2017

Version Date: 2011913

ISBN 13: 978-1-138-07282-4 (pbk)
ISBN 13: 978-1-4398-5509-6 (hbk)

Library of Congress Cataloging-in-Publication Data

Reese, Charles D.
Accident/incident prevention techniques / author, Charles D. Reese. -- 2nd ed.
p. cm.
Includes bibliographical references and index.
ISBN 978-1-4398-5509-6 (hardcover : alk. paper)
1. Industrial safety. 2. Industrial hygiene. I. Title.

T55.R433 2012
363.11'7--dc23 2011035867

Visit the Taylor & Francis Web site at
http://www.taylorandfrancis.com

and the CRC Press Web site at
http://www.crcpress.com

This book is dedicated to the working men and women of this country who every day risk their health and well-being in the workplace to provide the goods and services needed by the people of the United States and other countries.

Contents

Preface

Accident/Incident Prevention Techniques, Second Edition is based on the premise that all types of businesses and industries must face the reality that accidents and incidents that result in occupational injuries and illnesses will in most cases transpire at their business or facility. The results of these events have economic, legal, and human impact on the company's bottom line. In most situations, the impact is usually negative.

With this said, anything that a corporation or company can do to prevent these events will in most instances result in positive outcomes for that company.

The causes of occupationally related accidents or incidents are most often the result of some form of energy or agent being released. Seldom is there one single cause for accidents and incidents. There are many contributing factors that combine to become the root causes of occupational accidents and incidents.

Because employers and safety and health professionals are faced with multiple causal factors for these occupational happenings, they must utilize multiple approaches to prevent the occurrence of accidents and incidents. These range from prevention program development, to behavioral approaches, acceptable best prevention techniques, sophisticated analysis methods, engineering controls, and personal protection for the workers. The approach taken by employers and safety and health professionals will include a combination of prevention approaches that best meets the needs of their unique industry or business. The approach used will be tailored to meet their needs.

Thus, *Accident/Incident Prevention Techniques* provides the plethora of techniques and tools needed to structure a prevention approach to meet the needs of corporations and companies. These techniques are those that have been found to work in the past, as well as an intermingling of the best theoretical methods. It is laced with practical examples and tools to help those responsible for occupational safety and health develop the best prevention initiative for them and their workforce.

This book is a single-source guide of techniques and approaches to prevent the occurrence of occupational injuries and illnesses.

Acknowledgments

I appreciate the courtesy extended to me by the following organizations and individuals:

National Aeronautical and Space Administration
National Institute for Occupational Safety and Health
Occupational Health and Safety Administration
United States Department of Energy
Mine Safety and Health Administration
Robert Franko
David Kline
Steve Austin

About the Author

For 30 years Dr. Charles D. Reese has been involved with occupational safety and health as an educator, manager, or consultant. In Dr. Reese's early beginnings in occupational safety and health, he held the position of industrial hygienist at the National Mine Health and Safety Academy. He later assumed the responsibility of manager for the nation's occupational trauma research initiative at the National Institute for Occupational Safety and Health's (NIOSH) Division of Safety Research. Dr. Reese has played an integral part in trying to ensure that workplace safety and health are provided for all those within the workplace. As the managing director for the Laborers' Health and Safety Fund of North America, his responsibilities were aimed at protecting the 650,000 members of the laborers' union in the United States and Canada.

Dr. Reese has developed many occupational safety and health training programs, which run the gamut from radioactive waste remediation to confined space entry. He has written numerous articles, pamphlets, and books on related safety and health issues.

Dr. Reese, professor emeritus, was a member of the graduate and undergraduate faculty at the University of Connecticut, where he taught courses on OSHA regulations, safety and health management, accident prevention techniques, industrial hygiene, ergonomics, and environmental trends and issues. As professor of environmental/occupational safety and health, he coordinated the bulk of the environmental, safety, and health efforts at the University of Connecticut. He is called upon to consult with industry on safety and health issues, and is often asked for expert consultation in legal cases.

Dr. Reese also is the principal author of the following:

Handbook of OSHA Construction Safety and Health (Second Edition)
Material Handling Systems: Designing for Safety and Health
Annotated Dictionary of Construction Safety and Health
Occupational Health and Safety Management: A Practical Approach (Second Edition)
Office Building Safety and Health
Accident/Incident Prevention Techniques (Second Edition)
The Four-Volume Set Entitled: *Handbook of Safety and Health for the Service Industry:*
Volume 1: *Industrial Safety and Health for Goods and Materials Services*
Volume 2: *Industrial Safety and Health for Infrastructure Services*
Volume 3: *Industrial Safety and Health for Administrative Services*
Volume 4: *Industrial Safety and Health for People-Oriented Services*

1 Introduction

The objectives of Healthy People 2020 have not changed but will experience some modifications. The 2020 objectives for work-related injuries will be to

- Reduce death from work-related injuries
- Reduce nonfatal work-related injuries
- Reduce work-related assaults
- Reduce work-related homicides
- Reduce occupational needle-stick injuries among hospital-based health-care workers

Although progress was made regarding occupation injuries and trauma deaths, new data and goal projections are not available for 2020 at this time. The 2010 goals for injuries, trauma deaths, and needle-stick injuries achieved approximately 50% of the goals while assaults goals did not make progress. The following paragraphs set out the baselines for 2010.

Healthy People 2010 Objectives from the U.S. Department of Health and Human Services (DHHS) has made the facts available relevant to occupational injuries and illnesses. Every five seconds a worker is injured. Every ten seconds a worker is temporarily or permanently disabled. Each day, an average of 137 persons die from work-related diseases, and an additional 17 die from workplace injuries on the job. Each year, about 70 youths under 18 years of age die from injuries at work and 70,000 require treatment in a hospital emergency room. In 1996, an estimated 11,000 workers were disabled each day due to work-related injuries. That same year, the National Safety Council estimated that on-the-job injuries cost society $121 billion, including lost wages, lost productivity, administrative expenses, health care, and other costs (National Safety Council Injury Facts, 2010). A study published in July 1997 reports that the 1992 combined U.S. economic burden for occupational illnesses and injuries was an estimated $171 billion (U.S. Department of Commerce, www.commerce.gov, 2001).

A number of data systems and estimates exist to describe the nature and magnitude of occupational injuries and illnesses, all of which have advantages as well as limitations. In 1996, information from death certificates and other administrative records indicated that at least 6,112 workers died from work-related injuries (Bureau of Labor Statistics, www.bls.gov, 2000). No reporting system for national occupational chronic disease and mortality currently exists in this country. Therefore, scientists and policymakers must rely on estimates to understand the magnitude of occupational disease generated from a number of data sources and published epidemiologic (or population-based) studies. Estimates generated from these sources are generally thought to underestimate the true extent of occupational disease, but the scientific community recognizes these estimates as the best available information. Such compilations indicate that an estimated 50,000 to 70,000 workers die each year from work-related diseases. A further discussion on occupational disease is found in Chapter 23.

Current data collection systems are not sufficient to monitor disparities in health-related occupational injuries and illnesses. Efforts will be made over the coming decade to improve surveillance systems and data points that may allow evaluation of health disparities for work-related illnesses, injuries, and deaths. Data from the National Institute for Occupational Safety and Health's (NIOSH) National Traumatic Occupational Fatalities Surveillance System (NTOF), based on death certificates from across the United States, demonstrate a general decrease in occupational mortality over the 15-year period from 1980 to 1994. However, the number and rates of fatal injuries from 1990 through 1994 remained relatively stable (at over 5,000 deaths per year and about 4.4 deaths per 100,000 workers). Motor vehicle-related fatalities at work, the leading cause of death for U.S. workers since 1980, accounted for 23 percent of deaths during the 15-year period. Workplace homicides became the second leading cause of death in 1990, surpassing machine-related deaths. While the rankings of individual industry divisions have varied over the years, the largest number of deaths is consistently found in construction, transportation, public utilities, and manufacturing, while those industries with the highest fatality rates per 100,000 workers are mining, agriculture/forestry/fishing, and construction. Data from the Bureau of Labor Statistics (BLS), Department of Labor, indicate that, for nonfatal injuries and illnesses, incidence rates have been relatively stable since 1980. The rate in 1980 was 8.7 per 100,000 workers and 8.4 per 100,000 workers in 1994. Incidence varied between a low of 7.7 per 100,000 workers (1982) and a high of 8.9 per 100,000 workers (1992) over the 15-year period of 1980 to 1994.

The toll of workplace injuries and illnesses continues to harm our country. Six million workers in the United States are exposed to workplace hazards ranging from falls from elevations to exposures to lead. The hazards vary depending upon the type of industry (e.g., manufacturing) and the types of work being performed by workers (e.g., welding).

The consequences of occupational accidents or incidents have resulted in pain and suffering, equipment damage, exposure of the public to hazards, lost production capacity, and liability. Needless to say, these occupationally related accidents or incidents have a direct impact on profit, which is commonly called the "bottom line."

WHY INJURY PREVENTION?

There are very real advantages for those trying to address injury prevention, advantages that do not exist when trying to address illnesses prevention; these include

- Injuries occur in real-time with no latency period (an immediate sequence of events).
- Accident or incident outcomes are readily observable (must only reconstruct a few minutes or hours).
- Root or basic causes are more clearly identified.
- It is easy to detect cause-and-effect relationships.
- Injuries are not difficult to diagnose.
- Injuries are highly preventable.

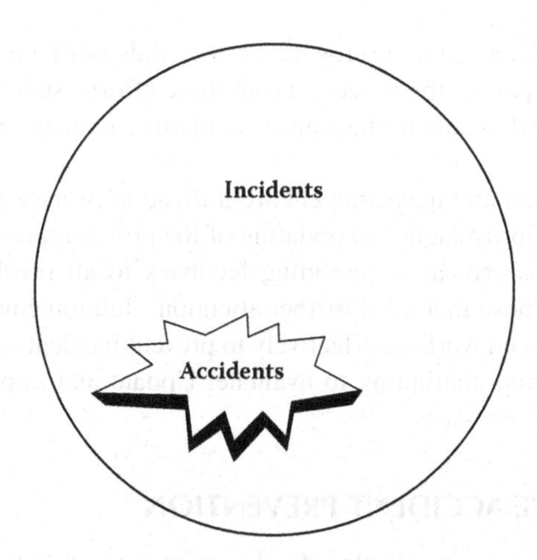

FIGURE 1.1 Incidents.

ACCIDENTS OR INCIDENTS

Debate over use of the term "accidents" versus "incidents" has been a long and continuing one. Although these terms are used as virtually interchangeable in the context of this book, you should be aware of the distinction between the two. *Accidents* are usually defined as unexpected, unplanned, and uncontrollable events or mishaps. These undesired events result in personal injury, or property damage, or equipment failure, or some combination thereof. An *incident* is all of the previous as well as adverse production effects. Accidents are a subset of incidents (see Figure 1.1).

This definition of an accident undermines the basic philosophy of this book, that we can control these types of events or mishaps. This is why we spend time identifying hazards and determining risks with the probability that a hazard will result in an accident with definable consequences. Thus, striving for a safe workplace, where the associated risks are judged to be acceptable, is a goal of safety. This will result in freedom from circumstances that can cause injury or death to workers, and damage to or loss of equipment or property. This is essentially a definition of safety.

The approach to this is that we can control factors that are the causative agents of accidents. We can prevent accidents by using the tools provided within these pages.

ACCIDENT/INCIDENT PREVENTION PROCESS

An accident/incident prevention process is as simplistic as it sounds. Within the text of this book the elements are provided for such a process. On the people side of this process, employers must commit themselves and all facets of management. The leadership for this process comes from management. This does not mean that the workforce should not be part of the process. The involvement of employees must be incorporated into the process in order to elicit a commitment and buy-in to the prevention effort. This is an important motivational factor for employees.

Another major area is the identification, ranking, and controlling of risk. This part of the process is paramount when faced with serious hazards that can be catastrophic events. Analysis of events and hazards using such tools as root cause analysis helps provide the solution part of the process. From these efforts, standard or safe operating procedures as well as safe performance standards can be developed to standardize safe production.

Constant evaluation and measurement are utilized to oversee progress regarding prevention so that improvement and updating of the process can occur.

Communication is critical in providing feedback to all involved on areas that have improved and those that need further attention. Reinforcement should be used to convey what has been working effectively to prevent incidents and implement any corrective action while continuing to evaluate. Update and improve the accident/incident prevention process.

COMPREHENSIVE ACCIDENT PREVENTION

Accident prevention is very complex due to interactions that transpire within the workplace (see Figure 1.2). These interactions are between

1. Workers
2. Management
3. Equipment and machines
4. Environment

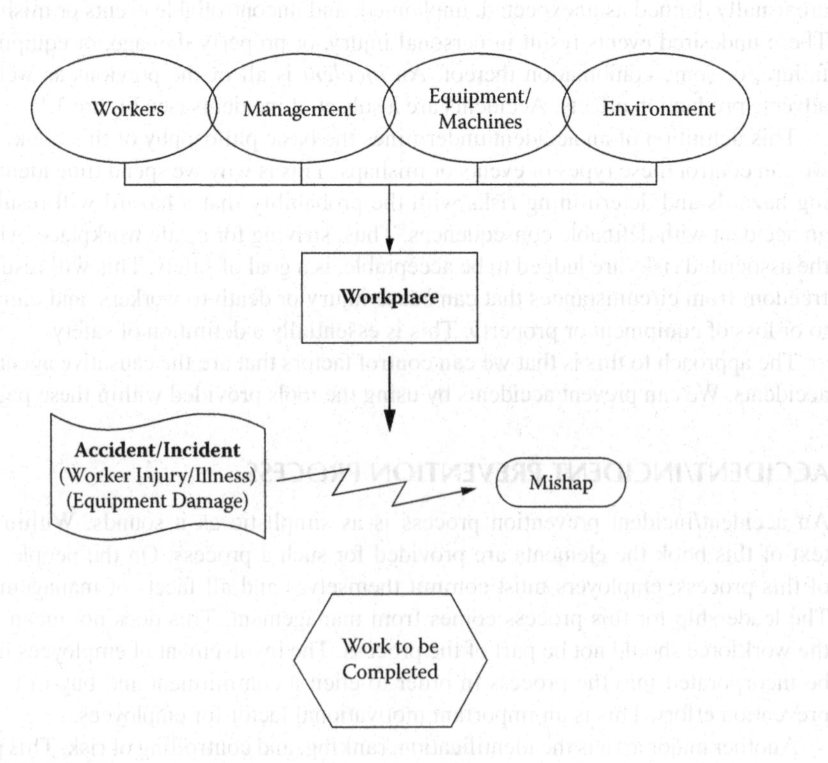

FIGURE 1.2 Interactive factors in accident prevention.

The interaction between workers, management, equipment and machinery, and the workplace environment separately are complex enough by themselves. More complexity transpires when they are blended together to become a workplace. But physical environment is not the only environment impinging on the accident prevention efforts within companies. The social environment is also an interactive factor, and one that encompasses our lives at work and beyond. Government entities establish rules and regulations and leave their mark on the workplace. Other entities in the social arena provide pressures in the workplace, such as unions, family, peer pressure, friends, associates, etc. This extends the interactions that must be attended to in order to successfully prevent accidents.

ACCIDENT PREVENTION

Prevention of occupationally related accidents/incidents is the law. The Occupational Safety and Health Act of 1970 (OSHAct) requires employers to provide a workplace free from hazards that could cause serious harm or death. Beyond that, it makes good business sense to prevent accident/incidents. More and more companies have come to realize that the OSHAct is a helpmate, not a hindrance, to their accident/incident prevention initiative. The Occupational Safety and Health Administration (OSHA) sets the foundation and assumes the role of law enforcer, allowing the employer to not be viewed as the bad guy to his or her employees. Employers can deflect responsibility to OSHA.

As business competition has increased, loss control has been seen as a logical place to curtail costs, especially direct losses from equipment damage, medical costs, and workers' compensation premiums. Prevention of accidents results in real, observable savings. Safety experts approximate the hidden cost of accidents as being conservatively five to ten times the direct cost incurred. Hidden costs include lost production, retraining, and supervisor's time lost, just to name a few.

ACCIDENT PREVENTION BENEFITS

You can expect many benefits from preventing occupational accidents/incidents. Some of the benefits you might expect are

- Reduced industrial insurance premium costs
- Reduced indirect costs of accidents
- Fewer compliance inspections and penalties
- Avoidance of adverse publicity from deaths or major accidents
- Reduced litigation and legal settlements
- Lower employee payroll deductions for industrial insurance
- Reduced pain and suffering by injured workers
- Reduced long-term or permanent disability cases
- Increased potential for retrospective rating refunds
- Increased acceptance of bids (more jobs)
- Improved morale and loyalty from individual workers
- Increased productivity from workers
- Increased pride in company personnel

Although this not an all-inclusive list, it certainly provides a snapshot of why you should undertake a loss control effort for occupationally related incidents.

PREVENTING OCCUPATIONAL ACCIDENTS/INCIDENTS

Throughout the years since the advent of OSHA's regulatory controls, some companies have been more successful than others at preventing worker-related accidents/incidents. These companies have placed occupational safety and health on an equal footing with production. They have developed programs, used accident-prevention techniques, and applied the principles of motivation to safety. In doing so, they often have devised their own unique approach, specific to their industry. One example of this is Dupont's SAFE program. The successes these companies have experienced have made them the shining examples of how safety can become a part of the cultural and philosophical approach of businesses.

Each industry is somewhat different. What works well for one company may not duplicate or replicate at other companies without tailoring to fit the unique needs of the latter. Uniqueness is often used as an excuse to not take action. Excuses include, "Our industry is different," or "It won't work in our industry." Generally, these are seen as lame excuses and a way of saying that they just don't want to try to make safety a priority. The principles of preventing accidents are applicable to any industry setting if one desires to make accident prevention an integral part of doing business.

NOTHING NEW IN PREVENTION

There is nothing earthshakingly new in the accident/incident prevention arena. Basic accident/incident prevention techniques espoused in this book have been used and modified over decades. Few, if any, dramatically new approaches have been devised using more modern techniques. Thus, accident investigation has always been accident investigation, no matter the words used. However, the tools and processes utilized for accident prevention have experienced modification and evolution as accident prevention has become an integral part of the loss control initiatives of companies.

HOW MUCH PREVENTION?

Workplaces have hazards that present a risk of injury or illness from the dangers that exist. At times, the hazards cannot be removed and the dangers exist and can result in an accident. Risk is the probability of an accident occurring. The amount of risk that you deem acceptable will do much to define the extent of your injury prevention effort. Risk related to safety and health is often a judgment call. But, even a judgment call can be quantified if you develop criteria and place value upon them. Fine (1973) has provided a mathematical model for conducting a risk assessment that results in a numerical value that can be used to compare potential risks from accidents.

Fine's first component is a risk score that compiles numerical values related to consequences, exposure, and probabilities. To calculate a risk score, extract the rating values from the Justification Formula Rating Summary Sheet found in Table 1.1.

TABLE 1.1
Justification Formula Rating Summary Sheet

Factor	Classification	Rating
Consequences	Catastrophe; numerous fatalities; damage over $1,000,000; major disruption of activities	100
Most probable result of the potential accident:	Multiple fatalities; damage $500,000 to $1,000,000	50
	Fatality, damage $100,000 to $500,000	25
	Extremely serious injury (amputation, permanent disability); damage $1000 to $100,000	15
	Disabling injury; damage up to $1000	5
	Minor cuts, bruises, bumps; minor damage	1
Exposure	*Hazard-event occurs:*	
The frequency of occurrence of the hazard event:	Continuously (or many times daily)	10
	Frequently (approximately once daily)	6
	Occasionally (from once per week to once per month)	3
	Usually (from once per month to once per year)	2
	Rarely (it has been known to occur)	1
	Remotely possible (not known to have occurred)	0.5
Probability	*Complete accident sequence:*	
Likelihood that accident sequence will follow to completion:	Is the most likely and expected result if the hazard-event takes place	10
	Is quite possible, not unusual, has an even 50/50 chance	6
	Would be an unusual sequence or coincidence	3
	Would be a remotely possible coincidence	1
	Has never happened after many years of exposure, but is conceivably possible	0.5
	Practically impossible sequence (has never happened)	0.1
Cost Factor	Over $50,000	10
Estimated dollar cost of proposed corrective action:	$25,000 to $50,000	6
	$10,000 to $25,000	4
	$1,000 to $10,000	3
	$100 to $1,000	2
	$25 to $100	1
	Under $25	0.5
Degree of Correction	Hazard positively eliminated, 100%	1
Degree to which hazard will be reduced:	Hazard reduced at least 75%	2
	Hazard reduced 50% to 75%	3
	Hazard reduced by 25% to 50%	4
	Slight effect on hazard (less than 25%)	6

The values derived are placed in the Risk Score Formula. Then you would calculate a risk score.

$$\text{Risk Score Formula} = \text{Consequences} \times \text{Exposure} \times \text{Probability}$$

Once you have a risk score, you should compare it to the following criteria to determine the need for action. If the score is between

- 18 and 85, the hazard should be eliminated without delay, but the situation is not an emergency.
- 90 and 200, this is urgent and requires attention as soon as possible.
- 270 and 1,500, this score requires immediate corrective action. All activity should be discontinued until the hazard is removed or reduced.

A risk score will not necessarily garner support for the removal or reduction of a potential hazard. The questions asked are, "How much will it cost?" and "How much hazard reduction will be derived from fixing the dangerous situation?" Fine (1973) went beyond the risk score and developed a Justification Formula. The formula includes factors for cost and the degree of correction. The resulting rating values for the cost and degree of correction factor can be extracted from Table 1.1. The Justification Formula is as follows:

$$\text{Justification Formula} = \frac{\text{Consequences} \times \text{Exposure} \times \text{Probability (Risk Score)}}{\text{Cost Factor} \times \text{Degree of Correction}}$$

For a Critical Justification Rating greater than 10, fixing the hazard is justified. For a score of less than 10, fixing the hazard is not justified.

Any time you can quantify something, you have a better chance of proving your point or gaining acceptance for it. Most managers and others pay more attention to numbers that support a case for risk reduction than to rhetoric.

Thus, when discussing risks related to occupational safety and health, being able to determine a value that equates to the degree of risk gives each company a measure for deciding how much risk it is willing to assume. This is best accomplished using a quantitative approach as discussed previously.

RISK CONTROL

The key to risk control is to prevent exposure of those who could be at risk. In the workplace it is not possible to have no exposure if anything is going to get done. It is important to limit the potential exposure or amount of exposure. The basic principles of protection from radiation exposure provide the foundation for risk control. The three elements of exposure control are distance, time, and shielding. Distance provides the best mechanism to prevent exposure. Distance can be physical distance or remote distance where robotics can provide the distance and limit the exposure. Time is an exposure limiter and a mechanism that allows employers to spread exposure over several workers. The time part of the work cycle that allows for minimal risk exposure to transpire is during a shift when fewer employees are present to be exposed. Often, second and third shifts are times when fewer workers are present. Shielding is frequently considered the least acceptable approach to risk control. Barriers or personal protective equipment should be the risk control of last resort. With occupational safety and health, other approaches to risk control are often employed.

Many ways to control hazards have been used over the years but usually these can be broken down into five primary approaches. The preferred ways to do this are through engineering controls, awareness devices, predetermined safe work practices, and administrative controls. When these controls are not feasible or do not provide sufficient protection, an alternative or supplementary method of protection is to provide workers with personal protective equipment (PPE) and the know-how to use it properly.

Engineering Controls

When a hazard is identified in the workplace, every effort should be made to eliminate it so that employees are not harmed. Elimination may be accomplished by designing or redesigning a piece of equipment or process. This could be the installation of a guard on a piece of machinery that prevents workers from contacting the hazard. The hazard can be engineered out of the operation. Another way to reduce or control the hazard is to isolate the process, such as in the manufacture of vinyl chloride used to make such items as plastic milk bottles, where the entire process becomes a closed circuit. This will result in no one being exposed to vinyl chloride gas, which is known to cause cancer. Thus, any physical controls that are put in place are considered the best approach from an engineering perspective. Keep in mind that you are a consumer of products. Thus, at times you can leverage the manufacturer to implement safeguards or safety devices on products that you are looking to purchase. Let your vendor do the engineering for you or do not purchase their product. This may not always be a viable option. To summarize the engineering controls that can be used, the following may be considered:

- Substitution
- Elimination
- Ventilation
- Isolation
- Process or design change

Awareness Devices

Awareness devices are linked to the senses. They are warning devices that can be heard and seen. They act as alerts to workers but create no type of physical barrier. They are found in most workplaces and carry with them a moderate degree of effectiveness. Such devices include

- Backup alarms
- Warning signals both audible and visual
- Warning signs

Work Practices

Work practices concern the ways in which a job task or activity is done. This may mean that you create a specific procedure for completing the task or job. It may also mean that you implement special training for a job or task. It also presupposes that you might require inspection of the equipment or machinery prior to beginning work or when a failure has occurred. An inspection should be done prior to restarting the process or task. It may also require that you require a lockout/tagout procedure be used to create a zero potential energy release.

Administrative Controls

A second approach is to control the hazard through administrative directives. This may be accomplished by rotating workers, which allows you to limit their exposure, or having workers only work in areas when no hazards exist during that part of their shift. This applies particularly to chemical exposures and repetitive activities that could result in ergonomically related incidents. Examples of administrative controls are

- Requiring specific training and education
- Scheduling off-shift work
- Worker rotation

Management controls are needed to express the company's view of hazards and their response to hazards that have been detected. The entire program must be directed and supported through the management controls. If management does not have a systematic and set procedure for addressing the control of hazards in place, the reporting/identifying of hazards is a waste of time and dollars. This goes back to the policies and directives and the holding of those responsible accountable by providing them with the resources (budget) for correcting and controlling hazards. Some aspects of management controls include

- Policies
- Directives
- Responsibilities (line and staff)
- Vigor and example
- Accountability
- Budget

The attempt to identify worksite hazards and address them should be an integral part of your management approach. If the hazards are not addressed in a timely fashion, they will not be identified or reported. If dollars become the main reason for not fixing or controlling hazards, you will lose the motivation of your workforce to identify or report them.

Personal Protective Equipment

Personal Protective Equipment (PPE) includes a variety of devices and garments to protect workers from injuries. You can find PPE designed to protect the eyes, face, head, ears, feet, hands and arms, and the whole body. PPE includes such items as goggles, face shields, safety glasses, hard hats, safety shoes, gloves, vests, earplugs, earmuffs, and suits for full-body protection.

When employees must be present and engineering or administrative controls are not feasible, it will be essential to use PPE as an interim control and not a final solution. For example, safety glasses may be required in the work area. Too often, PPE usage is considered the last thing to do in the scheme of hazard control. PPE can provide added protection to the employee even when the hazard is being controlled by other means. However, there are drawbacks to the use of PPE, including

- Hazard still looms
- Protection dependent on worker using PPE
- PPE may interfere with performing task and productivity
- Requires supervision
- Is an ongoing expense

Personal protective equipment includes the following:

- Eye and face protection (29 CFR 1910.133)
- Respiratory protection (29 CFR 1910.134)
- Head protection (29 CFR 1910.135)
- Foot and leg protection (29 CFR 1910.136)
- Electrical protective equipment (29 CFR 1910.137)
- Hand protection (29 CFR 1910.138)
- Respiratory protection from tuberculosis (29 CFR 1910.139)

Ranking Hazard Controls

In determining which hazard control procedures have the best chance of being effective, it is useful to have some sort of a ranking of them along a continuum. The five hazard controls that were espoused in the earlier part of this chapter are ranked in Figure 1.3. This should assist you in determining which control, if you have a choice of more than one, would be most effective for your purposes. The ranking goes from most effective to least effective.

STRUCTURING ACCIDENT/INCIDENT PREVENTION

Structuring should begin with a written safety and health program. There is a need to assess your accident/incident history and develop a good accident and injury

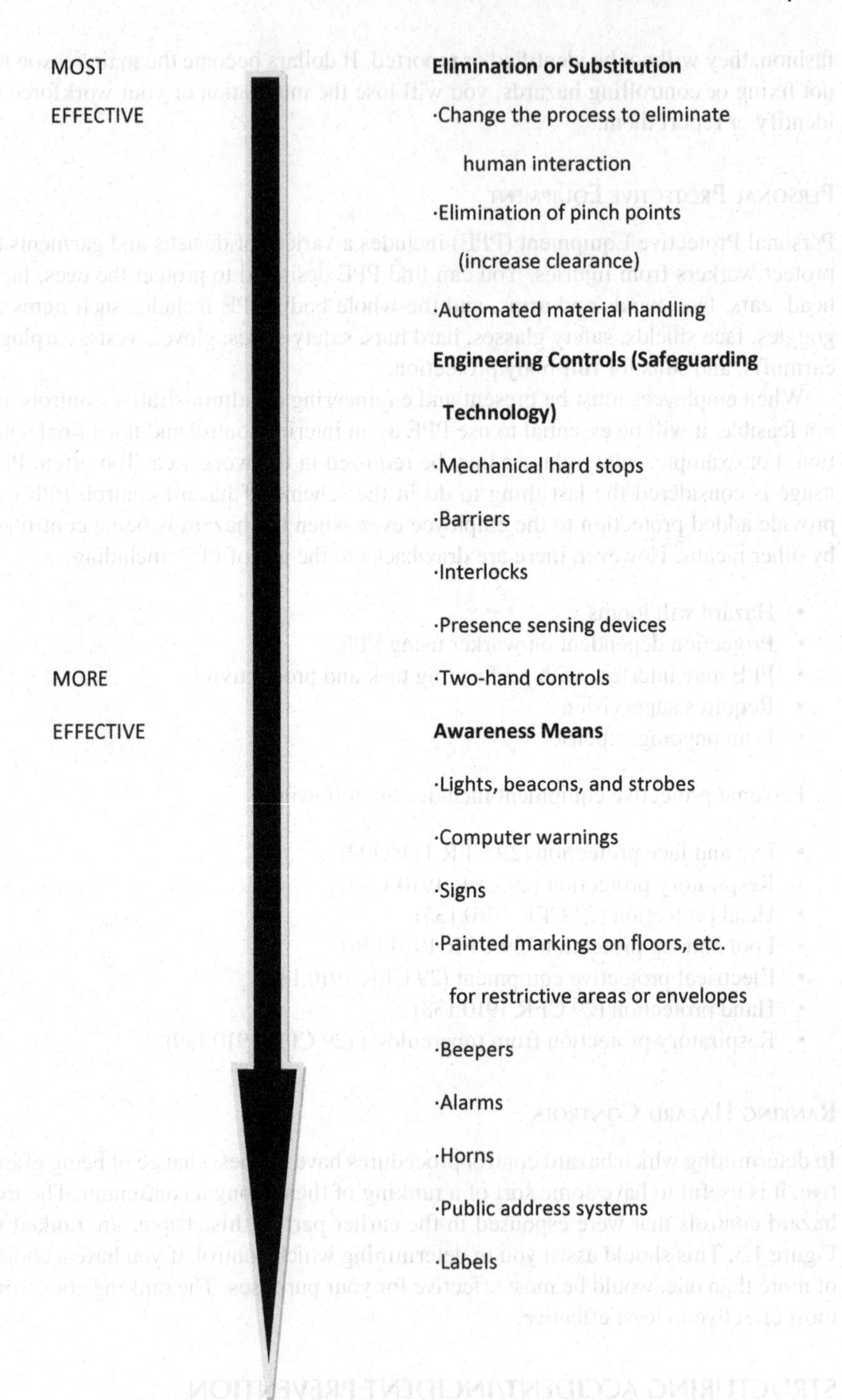
MOST
EFFECTIVE
Elimination or Substitution
·Change the process to eliminate
human interaction
·Elimination of pinch points
(increase clearance)
·Automated material handling
Engineering Controls (Safeguarding
Technology)
·Mechanical hard stops
·Barriers
·Interlocks
·Presence sensing devices
MORE
·Two-hand controls
EFFECTIVE
Awareness Means
·Lights, beacons, and strobes
·Computer warnings
·Signs
·Painted markings on floors, etc.
for restrictive areas or envelopes
·Beepers
·Alarms
·Horns
·Public address systems
·Labels

FIGURE 1.3 Ranking the effectiveness of hazard controls.

LESSER

EFFECTIVE

Training and Procedures

(Administrative Controls)

·Training

·Job rotation

·Off-shift scheduling of work

·Safe job procedures

·Safety equipment inspections/audits

·Lockout/tagout

LEAST

EFFECTIVE

Personal Protective Equipment

·Safety eyewear (face shield, etc)

·Hearing protection (ear plugs or muffs)

·Fireproof clothing

·Gloves

·Safety shoes

·Respirators

·Whole body protection (Tyveks, etc.)

FIGURE 1.3 (continued)

reporting and investigation procedure. Once you have sufficient data to identify your hazards, determine the potential interventions that will prevent the occurrence of injuries. The interventions, chosen for implementation, are the accident prevention processes and techniques you feel will help prevent mishaps. These interventions become a part of your accident prevention program. They may be processes or applications of accident prevention techniques such as job safety observations or actual physical changes (engineering) to equipment or facilities. Your safety and health program should culminate with follow-up evaluation procedures to determine the effectiveness of your loss control program.

KNOW WITH WHOM YOU ARE DEALING

You must understand the culture of the work environment before you can intelligently implement any type of program aimed at the workforce. You need to know and understand the values, attitudes, and perceptions held by the workforce relevant to job safety and health. This will afford the opportunity to place into action a motivational plan relevant to preventing accidents and incidents in your unique workplace.

DETERMINING THE CAUSE OF ACCIDENTS

The causes of the specific types of accidents/incidents that have occurred within your workplace must be assessed. The approach you wish to use in the assessment and analysis process depends greatly on your familiarity with and the types of occurrences that have transpired in your workplace. Analyses covered in this book are root cause, causal factor, change, and barrier analysis.

This is not a finite listing, as other, more complex approaches such as Management Oversight and Risk Tree Analysis (MORT), failure mode and effect analysis, and the aspects of system safety engineering are not covered within the scope of this book. These approaches can be found in many other books and articles.

ACCIDENT PREVENTION TECHNIQUES

Accident prevention techniques, such as job hazard analysis, safe operating procedures, and job safety observations, have their application to specific jobs within the workplace. In general, they aim to address and identify existing or potential work-related hazards. These types of accident prevention techniques are emphasized in this book. Most books addressing accident prevention techniques usually emphasize specific job hazards and their prevention, such as fire safety and machine guarding. This particular book places only minor emphasis on specific job hazards.

REFERENCES

Fine, W. Mathematical evaluations for controlling hazards, in J. Widner, (Ed.). *Selected Readings in safety*. Macon, GA: Academy Press, 1973.

Petersen, D. *Techniques of safety management: A systems approach* (3rd ed.). Goshen, NY: Aloray Inc., 1989.

Reese, C. D. and J. V. Eidson. *Handbook of OSHA construction safety & health.* Boca Raton, FL: CRC Press/Lewis Publishers, 1999.

United States Department of Labor, Occupational Safety and Health Administration, Office of Training and Education. OSHA Voluntary Compliance Outreach Program: Instructors Reference Manual. Des Plaines, IL: U.S. Department of Labor, 1993.

United States Department of Labor, National Mine Health and Safety Academy. Accident Prevention Techniques. Beckley, WV: U.S. Department of Labor, 1984.

United States Department of Labor, Mine Safety and Health Administration. Accident Prevention, Safety Manual No. 4, Beckley, WV: U.S. Department of Labor, revised 1990.

2 Safety and Health Programs

INTRODUCTION

The need for safety and health programs in the workplace has been an area of controversy for some time. Many companies feel that written safety and health programs are just more paperwork, a deterrent to productivity, and nothing more than another bureaucratic way of mandating safety and health on the job. But over a period of years, data and information have been mounting in support of the need to develop and implement written safety and health programs.

To effectively manage safety and health, a company must pay attention to some critical factors. These factors are the essence in managing safety and health on worksites. The questions that need to be answered regarding managing safety and health are

- What is the policy of management regarding safety and health at the company's workplace?
- What are the safety and health goals for the company?
- Who is responsible for occupational safety and health?
- How are supervisors and employees held accountable for job safety and health?
- What are the safety and health rules for this type of industry?
- What are the consequences of not following the safety rules?
- Are there set procedures for addressing safety and health at the worksite?

A written safety and health program is of primary importance in addressing these items. Have you ever wondered how your company is doing in comparison with a company without a safety and health professional and a viable safety and health program? Well, wonder no more….

In research conducted by the Lincoln Nebraska Safety Council in 1981, the following conclusions were based on a comparison of responses from a survey of 143 national companies (see Table 2.1). All conclusions have a 95 percent confidence level or more. Table 2.1 is an abstraction of conclusions from this study.

It seems apparent from the previous research that in order to have an effective safety program, at a minimum, an employer must

- Have a demonstrated commitment to job safety and health
- Commit budgetary resources

TABLE 2.1
Effectiveness of Safety and Health Programs Findings

Fact	Statement	Findings
1.	Do not have separate budget for safety.	43% more accidents
2.	No training for new hires.	52% more accidents
3.	No outside sources for safety training.	59% more accidents
4.	No specific training for supervisors.	62% more accidents
5.	Do not conduct safety inspections.	40% more accidents
6.	No written safety program compared with companies that have written programs.	106% more accidents
7.	Those using canned programs not self-generated.	43% more accidents
8.	No written safety program.	130% more accidents
9.	No employee safety committees.	74% more accidents
10.	No membership in professional safety organizations.	64% more accidents
11.	No established system to recognize safety accomplishments.	81% more accidents
12.	Did not document/review accident reports, and reviewers did not have safety as part of their job responsibility.	122% more accidents
13.	Did not hold supervisor accountable for safety through merit salary reviews.	39% more accidents
14.	Top management did not actively promote safety awareness.	470% more accidents

- Train new personnel
- Ensure that supervisors are trained
- Have a written safety and health program
- Hold supervisors accountable for safety and health
- Respond to safety complaints and investigate accidents
- Conduct safety audits

Other refinements can always be part of the safety and health program, which will help in reducing workplace injuries and illnesses. They include more worker involvement (e.g., joint labor/management committees), incentive or recognition programs, getting outside help from a consultant or safety association, and setting safety and health goals.

A decrease in occupational incidents that result in injury, illness, or damage to property is enough reason to develop and implement a written safety and health program.

REASONS FOR A COMPREHENSIVE SAFETY PROGRAM

The three major considerations involved in the development of a safety program are

1. Humanitarian:
    - Safe operation of workplaces is a moral obligation imposed by modern society. This obligation includes consideration for loss of life, human pain and suffering, family suffering and hardships, etc.

2. Legal obligation:
   - Federal and state governments have laws charging the employer with the responsibility for safe working conditions and adequate supervision of work practices. Employers are also responsible for paying the costs incurred for injuries suffered by their employees during their work activities.
3. Economic:
   - Prevention costs less than accidents. This fact has been proven consistently by the experience of thousands of industrial operations. The direct cost is represented by medical care, compensation, etc. The indirect cost of four to ten times the direct cost must be calculated, as well as the loss of wages to employees and the reflection of these losses on the entire community.

All three of these are good reasons to have a health and safety program. It is also important that these programs be formalized in writing, as a written program sets the foundation and provides a consistent approach to occupational health and safety for the company. There are other logical reasons for a written safety and health program. Some of them are

- It provides standard directions, policies, and procedures for all company personnel.
- It states specifics regarding safety and health and clarifies misconceptions.
- It delineates the goals and objectives regarding workplace safety and health.
- It forces the company to actually define its view of safety and health.
- It sets out in black and white the rules and procedures for safety and health that everyone in the company must follow.
- It is a plan that shows how all aspects of the company's safety and health initiative work together.
- It is a primary tool of communications of the standards set by the company regarding safety and health.

SAFETY AND HEALTH MANAGEMENT PROCESS

The old adage that says that failure to plan is planning to fail is very appropriate when discussing the need to have a safety and health management process in place for each workplace. The term "process" infers action, meaning to implement and organize an approach. Such an approach must set forth a roadmap to follow, as well as the rules of the road that must be followed in order to attain a safe and healthy work environment. This organized approach to occupational safety and health must contain all the components to accomplish a process that will facilitate safety and health at the workplace. The process will include

- The management component
- A commitment of a personal nature as well as resources
- The policies and procedures to be implemented

- The technical issues that relate to prevention of accidents/incidents, commonly called hazard recognition and control
- Communication of the features of the process so all will be aware of expectations
- Assuring the development of a workforce that buys in to the process, partially accomplished by effective training
- Actually evaluating the process to determine effectiveness
- Provide feedback to allow for change and reinforcement for what is working
- The beginning and revision of the components of process must start the process again with a rebirth of the process

The need to have such an organized approach to occupational safety and health as an integral element in the development of a safety and health initiative and program is paramount to a working safety and health management approach for each company and workplace.

BUILDING A SAFETY AND HEALTH PROGRAM

The length of such a written plan is not as important as the content. It should be tailored to the company's needs and the health and safety of its workforce. It could be a few pages or a multiple page document. However, it is suggested that you adhere as much as possible to the KISS principle (Keep It Simple Stupid). To ensure a successful safety program, three conditions must exist: (1) management leadership, (2) safe working conditions, and (3) safe work habits by all employees. The employers must

- Let the employees know that the employer is interested in safety on the job by consistently enforcing and reinforcing safety regulations.
- Provide a safe working place for all employees; it pays dividends.
- Be familiar with federal and state laws applying to your operation.
- Investigate and report all OSHA recordable accidents and injuries. This information may be useful in determining areas where more work is needed to prevent such accidents in the future.
- Make training and information available to the employees, especially in such areas as first aid, equipment operation, and common safety policies.
- Develop a prescribed set of safety rules to follow, and see that all employees are aware of the rules.

The basic premise of this book is that all employers should establish a workplace safety and health program to assist them in compliance with OSHA standards and the General Duty Clause of the Occupational Safety and Health Act of 1970 (OSHAct) (Section 5(a)(1)). Each employer should set up a safety and health program to manage workplace safety and health to reduce injuries, illnesses, and fatalities through a systematic approach to safety and health. The program should be appropriate to conditions in the workplace, such as the hazards to which employees are exposed and the number of employees there. The primary guidelines for employers to develop an organized safety and health program are

- Employers are advised and encouraged to institute and maintain in their establishments a program that provides systematic policies, procedures, and practices that are adequate to recognize and protect their employees from occupational safety and health hazards.
- An effective program includes provisions for the systematic identification, evaluation, and prevention or control of general workplace hazards, specific job hazards, and potential hazards that may arise from foreseeable conditions.
- Although compliance with the law, including specific OSHA standards, is an important objective, an effective program looks beyond the specific requirements of law to address all hazards. This effectively will seek to prevent injuries and illnesses, whether or not compliance is at issue.
- The extent to which the program is described in writing is less important than how effective it is in practice. As the size of a worksite or the complexity of a hazardous operation increases, however, the need for written guidance increases to ensure clear communications of policies and priorities and consistent and fair application of rules.

The primary elements that should be addressed within this program are management leadership and employee participation, hazard identification and assessment, hazard prevention and control, information and training, and evaluation of program effectiveness.

Management Commitment and Employee Involvement

Management commitment and employee involvement are complementary. Management commitment provides the motivating force and the resources for organizing and controlling activities within an organization. In an effective program, management regards workers' safety and health as a fundamental value of the organization and applies its commitment to safety and health protection with as much vigor as to other organizational purposes. Employee involvement provides the means through which workers develop and express their own commitment to safety and health protection, for themselves and for their fellow workers.

Management must state clearly a worksite policy on safe and healthful work and working conditions, so that all personnel with responsibility at the site, and personnel at other locations with responsibility for the site, understand the priority of safety and health protection in relation to other organizational values.

Management must establish and communicate a clear goal for the safety and health program and objectives for meeting that goal, so that all members of the organization understand the results desired and the measures planned for achieving them.

There must be visible top management involvement in implementing the program, in order to make it known that management's commitment is serious. In turn, employees must be encouraged to be involved in the structure and operation of the program and in decisions that affect their safety and health; their commitment, insight, and energy will help achieve the safety and health program's goal and objectives.

Management should assign and communicate responsibility for all aspects of the program. Managers, supervisors, and employees in all parts of the organization must know what performance is expected of them. Adequate authority and resources must be provided to responsible parties so that assigned responsibilities can be met. Managers, supervisors, and employees must be held accountable for meeting their responsibilities to assure that essential tasks are performed. It is essential that managers understand their safety and health responsibilities, as described. Having a clear understanding will provide managers with vision and direction to effectively carry out those responsibilities.

By reviewing program operations at least annually, management can evaluate their success in meeting the goal and objectives. Deficiencies can be identified within the program and objectives can be revised, once again targeting the goal of effective safety and health protection.

Management commitment and leadership design and develop the policy statement. It should be signed by the top person in your company. Safety and health goals and objectives are also included to assist you with establishing workplace goals and objectives that demonstrate your company's commitment to safety. An enforcement policy is provided to outline disciplinary procedures for violations of your company's safety and health program. This enforcement policy should be communicated to everyone in the company.

Establish the program responsibilities of managers, supervisors, and employees for safety and health in the workplace and hold them accountable for carrying out those responsibilities. Provide managers, supervisors, and employees with the authority, access to relevant information, training, and resources they need to carry out their safety and health responsibilities. Identify at least one manager, supervisor, or employee to receive and respond to reports about workplace safety and health conditions. Where appropriate, he or she should initiate corrective action.

The safety and health program should contain the following to demonstrate management commitment and leadership:

- Policy statement with written goals is established, issued, and communicated to employees.
- Program is revised annually.
- Participation in safety meetings, inspections; agenda item in meetings.
- Commitment of resources is adequate.
- Safety rules and procedures are incorporated into jobsite operations.
- Management observes safety rules.

Assignment of responsibility identifies the responsibilities of management officials, supervisors, and employees. Emphasis on responsibility to safety and health is more credible if everyone is held accountable for their safety and health performance as related to established safety and health goals. The assignment of responsibility should include the following aspects:

- Safety designee is on-site, knowledgeable, and accountable.
- Supervisors' (including foremen) safety and health responsibilities are understood.
- Employees are aware of and adhere to safety rules.

The employer must provide employees with opportunities for participation in establishing, implementing, and evaluating the program. The employer must regularly communicate with employees about workplace safety and health matters. She or he should provide employees with access to information relevant to the program. In turn, there should be established processes for employees to become involved in hazard identification and assessment, prioritizing hazards, training, and program evaluation. Ways should be established for employees to promptly report job-related fatalities, injuries, illnesses, incidents, and hazards. There should also be an easy-to-use system to make recommendations about appropriate means to control these hazards—then promptly respond to such reports and recommendations.

Do not discourage employees from making reports and recommendations about fatalities, injuries, illnesses, incidents, or hazards in the workplace, or from otherwise participating in the workplace safety and health program.

Hazard Identification and Assessment

The employer must systematically identify and assess hazards to which employees are exposed and assess compliance with the General Duty Clause and OSHA standards. The employer must conduct inspections of the workplace; review safety and health information; evaluate new equipment, materials, and processes for hazards before they are introduced into the workplace; and assess the severity of identified hazards and rank those hazards that cannot be corrected immediately according to their severity.

Identification of hazards includes those items that can assist you in identifying workplace hazards and determining what corrective action is necessary to control them. Actions include worksite safety inspections, accident investigations, meetings of safety and health committees, and project safety meetings. To accomplish the identification of hazards, the following should be reviewed:

- Periodic site safety inspections involve supervisors
- Preventive controls in place (personal protective equipment (PPE), maintenance, engineering controls)
- Action taken to address hazards
- Establish safety committee, where appropriate
- Document technical references available
- Enforcement procedures implemented by management

The employer must carry out an initial assessment and then reassess as often thereafter as necessary to ensure compliance. Reassessment should be done at least every two years. When safety and health information or a change in workplace conditions indicates that a new or increased hazard may be present, the employer should also conduct a reassessment. The employer should investigate each work-related death, serious injury or illness, or incident (near-miss) having the potential to cause death or serious physical harm. The employer should keep records of the hazards identified, their assessment, and the actions the employer has taken, or plans to take, to control those hazards. These will be positives if OSHA ever has to inspect the workplace. It shows a good-faith effort and commitment to safety and health.

Worksite analysis involves a variety of worksite examinations to identify not only existing hazards, but also conditions and operations where changes might occur and create hazards. Being unaware of a hazard, which stems from failure to examine the worksite, is a sure sign that safety and health policies or practices are inadequate. Effective management actively analyzes the work and worksite to anticipate and prevent harmful occurrences. Worksite analysis is intended to ensure that all hazards are identified. This can be accomplished by

- Conducting comprehensive baseline worksite surveys for safety and health and periodically doing a comprehensive updated survey
- Analyzing planned and new facilities, processes, materials, and equipment
- Performing routine job hazard analyses

Providing for regular site safety and health inspection, so that new or previously missed hazards and failures in hazard controls are identified, is critical to worksite analysis. Employee insight and experience in safety and health protection should also be utilized, where employee concerns may be addressed. A reliable system for employees is to be provided and encouraged, without fear of reprisal, to notify management personnel about conditions that appear hazardous and to receive timely and appropriate responses.

All accidents and "near-miss" incidents should be investigated to determine causes, and means for their prevention identified. Analysis of injury and illness trends over time should be undertaken so that patterns with common causes can be identified and prevented.

Hazard Prevention and Control

The requirements of the General Duty Clause and OSHA standards must be met. If it is not possible for the employer to comply immediately, the employer must develop a plan for coming into compliance as promptly as possible; this includes setting priorities and deadlines and tracking progress in controlling hazards. *Note:* Any hazard identified by the employer's hazard identification and assessment process that is covered by an OSHA standard or the General Duty Clause must be controlled, as required by that standard or that clause, as appropriate. Control means to reduce exposure to hazards in accordance with the General Duty Clause or OSHA standards, including providing appropriate supplemental or interim protection, as necessary, to exposed employees. Prevention and elimination are the best forms of control.

Hazard prevention and controls are triggered by a determination that a hazard or potential hazard exists. Where feasible, hazards are prevented by effective design of the jobsite or job. Where it is not feasible to eliminate them, they are controlled to prevent unsafe and unhealthful exposure. Elimination of controls should be accomplished in a timely manner once a hazard or potential hazard is recognized.

So that all current and potential hazards, however detected, are corrected or controlled in a timely manner, procedures must be established using the following measures:

- Engineering techniques where feasible and appropriate
- Procedures for safe work that are understood and followed by all affected parties, as a result of training, positive reinforcement, correction of unsafe performance, and, if necessary, enforcement through a clearly communicated disciplinary system
- Provision of personal protective equipment
- Administrative controls, such as reducing the duration of exposure

Facility and equipment maintenance should be provided so that hazardous breakdown is prevented. Plan and prepare for emergencies and conduct training and drills as needed so that the response of all parties to emergencies becomes "second nature." Establish a medical program that includes first aid available on-site, with physician and emergency medical care nearby, to minimize harm if any injury or illness does occur.

Information and Training

The employer must ensure that each employee is provided information and training in the safety and health program. Each employee exposed to a hazard must be provided information and training in that hazard. *Note:* Some OSHA standards impose additional, more specific requirements for information and training. This rule does not displace those requirements.

Safety and health information means the establishment's fatality, injury, and illness history; OSHA 200 logs; workers' compensation claims; nurses' logs; the results of any medical screening or surveillance; employee safety and health complaints and reports; environmental and biological exposure data; information from prior workplace safety and health inspections; materials safety data sheets (MSDSs); the results of employee symptom surveys; safety manuals and health and safety warnings provided to the employer by equipment manufacturers and chemical suppliers; information about occupational safety and health provided to the employer by trade associations or professional safety or health organizations; and the results of prior accident and incident investigations at the workplace.

The employer must provide general information and training on the following subjects:

1. The nature of the hazards to which the employee is exposed and how to recognize them.
2. What is being done to control these hazards?
3. What protective measures the employee must follow to prevent or minimize exposure to these hazards.
4. The provisions of applicable standards.

The employer must provide specific information and training:

1. New employees must be informed and properly trained prior to their initial assignment to a job involving exposure to a hazard.
2. The employer is not required to provide initial information and training for which the employer can demonstrate that the employee has already been adequately trained.
3. The employer must provide periodic information and training as often as necessary to ensure that employees are adequately informed and trained; and to ensure that safety and health information and changes in workplace conditions, such as when a new or increased hazard exists, are communicated.

Safety and health training addresses the safety and health responsibilities of all personnel concerned with the site, whether salaried or hourly. It is often most effective when incorporated into other training about performance requirements and job practices. The complexity of the training depends on the size and complexity of the worksite, and on the nature of the hazards and potential hazards at the site. It is essential that all employees understand the hazards to which they may be exposed and how to prevent harm to themselves and others from exposure to such hazards. Employees should be convinced to accept and follow established safety and health procedures.

Supervisors will more likely carry out their safety and health responsibilities effectively when they understand the responsibilities and the reasons for them. Responsibilities include:

1. Analyzing the work under their supervision to identify unrecognized potential hazards
2. Maintaining physical protections in their work areas
3. Reinforcing employee training on the nature of potential hazards in their work and on needed protective measures, through continual performance feedback and, if necessary, through enforcement of safe work practices

The employer must provide all employees who have program responsibilities with the information and training necessary to carry out their safety and health responsibilities.

Evaluation of Program Effectiveness

The employer's basic obligation is to evaluate the safety and health program to ensure that it is effective and appropriate to workplace conditions. The employer must evaluate the program as often as necessary to guarantee its effectiveness. The employer must also revise the program in a timely manner to correct deficiencies identified by the program evaluation.

Multi-Employer Workplaces

Multi-employer workplace means one where there is a host employer and at least one contract employer. A host employer is an employer who controls conditions at a multi-employer worksite. The host employer's responsibilities are to

1. Provide information about hazards, controls, rules on safety and health, and emergency procedures to all employers at the workplace.
2. Make certain that safety and health responsibilities are assigned as appropriate to other employers at the workplace.

In turn, the responsibilities of a contract employer are to

1. Ensure that the host employer is aware of the hazards associated with the contract employer's work and what the contract employer is doing to address them.
2. Advise the host employer of any previously unidentified hazards that the contract employer identifies at the workplace.

A contract employer is an employer who performs work for a host employer at the host employer's workplace. A contract employer does not include an employer who provides incidental services that do not influence the workplace safety and health program, or whose employees are only incidentally exposed to hazards at the host employer's workplace (e.g., food and drink services, delivery services, or other supply services).

CHARACTERISTICS OF AN OCCUPATIONAL SAFETY AND HEALTH PROGRAM

A review of research on successful safety and health programs reveals that a number of factors comprise these programs. Strong management commitment to health and safety and frequent, close contacts between workers, supervisors, and management on health and safety are the two most dominant factors in good health and safety programs. Other relevant factors include workforce stability, stringent housekeeping, training emphasizing early indoctrination and follow-up instruction, and special adaptation of conventional health and safety practices to enhance their suitability to the workplace.

Factors Affecting Safety and Health

1. Management factors:
   a. Management commitment, as reflected by management involvement in aspects of the health and safety program in a formal way, and as reflected in the amount of employer resources committed to the company's health and safety program
   b. Management adherence to principles of good management in the utilization of resources (i.e., people, machinery, and materials), supervision of employees and production planning, and monitoring
   c. Designated health and safety personnel reporting directly to top management

2. Motivational factors:
   a. Humanistic approach to interacting with employees
   b. High levels of employee/supervisor contact
   c. Efficient production planning
3. Hazard control factors:
   a. Effort to improve the workplace
   b. Continuing development of the employees
   c. Clean working environment
   d. Regular, frequent inspections
4. Illness and injury investigations and recordkeeping factors:
   a. Investigation of all incidents of illness and injury as well as non-lost-time accidents
   b. Recording of all first aid cases

Components of a Safety and Health Program

A listing of the components that comprise a successful health and safety program are as follows:

- Health and safety program management
- Inspections and job observations
- Illness and injury investigations
- Task analysis
- Training
- Personal protection
- Communication and promotion of health and safety
- Personal perception
- Off-the-job health and safety

This is only a representative list; it could be either expanded or consolidated, depending on the unique needs of the company or contractor. Health and safety programs should be tailored to meet individual requirements.

Evaluative Questions Regarding a Safety and Health Program

This section breaks down these components into subparts or subelements using a questioning approach as the mechanism to draw attention to the intricacies of a health and safety program. These breakouts and questions, which should guide your evaluation and development of a successful health and safety program, are as follows:

1. Health and safety program management:
   a. Is there a safety policy signed and dated by top management?
   b. Is there someone responsible for health and safety?
   c. Does a health and safety manual or handbook exist?
   d. Is there a set time devoted to health and safety during management meetings?

  e. Are employees encouraged to participate in the safety and health program?
  f. Are there health and safety rules and regulations for employees and specific jobs?
  g. Is there a discipline policy for disobeying safety and health rules?
  h. Are health and safety rules enforced and violators disciplined?
  i. Who is held accountable for health and safety?
  j. Are there special goals for health and safety?
2. Inspections/job observations:
  a. Inspections:
    – Are health and safety inspections conducted?
    – How often are inspections conducted?
    – Are unsafe conditions or hazards found and corrected immediately?
    – Is equipment inspected?
    – Is equipment maintained?
    – Is there a preventive maintenance program?
    – Are housekeeping inspections conducted?
    – Is housekeeping maintained?
    – Does monitoring occur for health hazards?
    – Are written inspection reports completed?
    – Are inspection reports disseminated?
  b. Job observations:
    – Are job observations done?
    – Who does job observations?
    – Do job observations result in new work practices, workplace design, training, retraining, and task analysis including Job Safety Analysis (JSA) or Safe Operating Procedure (SOP)?
    – Are job observations done for punitive reasons?
3. Illness and injury investigations:
  a. Are employees encouraged to report hazards and accidents?
  b. Are all incidents involving illness or injury investigated?
  c. Have written reports been generated for all incidents?
  d. Are preventative recommendations being implemented?
  e. Do employees review incident reports?
  f. Are incident data analyzed to determine illness or injury trends and damage experience?
4. Task analysis:
  a. Do inspections, job observations, and incident investigations result in a task analysis?
  b. Do task analyses result in changes in work practices or workplace design?
  c. Does task analysis facilitate the development of JSAs or SOPs?
  d. Does task analysis result in new training or retraining?
5. Training:
  a. Do all employees receive health and safety training?
  b. Do employees receive site-specific training?
  c. Are employees given job-specific or task-specific training?
  d. Is training well planned and organized?

e. Does both classroom and On-the-Job Training (OJT) or Job Instructional Training (JIT) occur?
f. Is refresher training conducted?
g. Do management and supervisors receive health and safety training?
h. Are training records maintained?

6. Personal protection:
 a. Does the work require personal protective equipment (PPE)?
 b. Is the proper PPE available?
 c. Have employees been trained in the use of PPE?
 d. Do you need a respirator program (29 CFR 1910.134)?
 e. Should you follow the requirements for handling hazardous waste (29 CFR 1910.120)?
 f. Are the rules and use of PPE enforced?
7. Communication and promotion of health and safety:
 a. Communication:
 – Is health and safety visible?
 – Are company and contractors' health and safety goals communicated?
 – Are health and safety meetings held (i.e., tool box)?
 – Do health and safety talks convey relevant information?
 – Are personal health and safety contacts made?
 – Are bulletin boards used to communicate health and safety issues?
 – Do those responsible for health and safety request feedback?
 – Are health and safety suggestions given consideration or used?
 – Are supervisors interested in health and safety?
 b. Promotion:
 – Is there an award/incentive program tied to safety and health?
 – Are health and safety exhibits or posters used?
 – Are paycheck stuffers used?
8. Personal perception:
 a. Do company and contractors extend considerable effort to ensure an effective health and safety program?
 b. Do supervisors support and enforce all aspects of the health and safety program?
 c. Do most employees insist on doing all tasks in a safe and healthy manner?
9. Off-the-job health and safety:
 a. Is off-the-job health and safety promoted as part of the total health and safety program?
 b. Do the company and contractors provide a wellness/fitness program?
 c. Do the company and contractors foster and encourage healthier lifestyles?
 d. Do the company and contractors support an employee assistance program?

A sample of a written occupational safety and health program, which includes examples of the major program elements and components, is found in Appendix A.

TOOLS FOR A SAFETY AND HEALTH PROGRAM ASSESSMENT

There are three basic methods for assessing safety and health program effectiveness. This discussion explains each of them. It also provides more detailed information on how to use these tools to evaluate each element and subsidiary component of a safety and health program. The three basic methods for assessing safety and health program effectiveness are

1. Checking documentation of activity
2. Interviewing employees at all levels for knowledge, awareness, and perceptions
3. Reviewing site conditions and, where hazards are found, finding the weaknesses in management systems that allowed the hazards to occur or to be "uncontrolled"

Some elements of the safety and health program are best assessed using one of these methods. Others lend themselves to assessment by two or all three methods.

Documentation

Checking documentation is a standard audit technique. It is particularly useful for understanding whether the tracking of hazards to correction is effective. It can also be used to determine the quality of certain activities, such as self-inspections or routine hazard analysis.

Inspection records can tell the evaluator whether serious hazards are being found, or whether the same hazards are being found repeatedly. If serious hazards are not being found and accidents keep occurring, there may be a need to train inspectors to look for different hazards. If the same hazards are being found repeatedly, the problem may be more complicated. Perhaps the hazards are not being corrected. If so, this would suggest a tracking problem or a problem in accountability for hazard correction.

If certain hazards recur repeatedly after being corrected, someone is not taking responsibility for keeping those hazards under control. Either the responsibility is not clear, or those who are responsible are not being held accountable.

Employee Interviews

Talking to randomly selected employees at all levels will provide a good indication of the quality of employee training and of employee perceptions of the program. If safety and health training is effective, employees will be able to tell you about the hazards they work with and how they protect themselves and others by keeping those hazards controlled. Every employee should also be able to say precisely what he or she is expected to do as part of the program. And all employees should know where to go and the route to follow in an emergency.

Employee perceptions can provide other useful information. An employee's opinion of how easy it is to report a hazard and get a response will tell you a lot about how well the hazard reporting system is working. If employees indicate that the system for enforcing safety and health rules and safe work practices is inconsistent or confusing, you will know that the system needs improvement.

Interviews should not be limited to hourly employees. Much can be learned from talking with first-line supervisors. It is also helpful to query line managers about their understanding of the safety and health responsibilities for which they are responsible.

Site Conditions and Root Causes of Hazards

Examining the conditions of the workplace can reveal existing hazards; it can also provide information about the breakdown of management systems meant to prevent or control hazards.

Looking at conditions and practices is a well-established technique for assessing the effectiveness or implementation of safety and health programs. For example, let us say that in areas where PPE is required, you see large and understandable signs communicating this requirement and all employees, with no exceptions, must wear equipment properly. You have obtained valuable visual evidence that the PPE program is working.

Another way to obtain information about safety and health program management is through root analysis of observed hazards. This approach is much like the most sophisticated accident investigation techniques, in which many contributing factors are located and corrected or controlled.

When evaluating each part of a worksite's safety and health program, use one or more of the above methods as appropriate. The remainder of this discussion identifies the components found in each element of a quality safety and health program and describes useful ways to assess these components.

ASSESSING THE KEY COMPONENTS OF LEADERSHIP, PARTICIPATION, AND LINE ACCOUNTABILITY

Is there a worksite policy on safe and healthful working conditions? If there is a written policy, does it clearly declare the priority of worker safety and health over other organizational values, such as production? When asked, can employees at all levels express the worksite policy on worker safety and health? If the policy is written, can hourly employees tell you where they have seen it? Can employees at all levels explain the priority of worker safety and health over other organizational values, as the policy intends?

Have injuries occurred because employees at any level did not understand the importance of safety precautions in relation to other organizational values, such as production?

Goal and Objectives for Worker Safety and Health

If there is a written goal for the safety and health program, is it updated annually? If there are written objectives, such as an annual plan to reach that goal, are they clearly stated? If managers and supervisors have written objectives, do these documents include objectives for the safety and health program?

Do managers and supervisors have a clear idea of their objectives for worker safety and health? Do hourly employees understand the current objectives of the safety and health program?

Visibility of Top Management Leadership

Are there one or more written programs involving top-level management in safety and health activities? For example, top management can receive and sign off on inspection reports either after each inspection or in a quarterly summary. These reports can then be posted for employees to see. Top management can provide "open-door" times each week or each month for employees to come in to discuss safety and health concerns. Top management can reward the best safety suggestions each month or at other specified intervals.

Can hourly employees describe how management officials are involved in safety and health activities? Do hourly employees perceive that managers and supervisors follow safety and health rules and work practices, such as wearing appropriate personal protective equipment?

When employees are found not wearing required personal protective equipment or not following safe work practices, have any of them said that managers or supervisors also did not follow these rules?

Employee Participation

Are there one or more written programs that provide for employee participation in decisions affecting their safety and health? Is there documentation of these activities, for example, employee inspection reports, and minutes of joint employee–management or employee–committee meetings? Is there written documentation of any management response to employee safety and health program activities? Does the documentation indicate that employee safety and health activities are meaningful and substantive? Are there written guarantees of employee protection from harassment resulting from safety and health program involvement?

Are employees aware of ways they can participate in decisions affecting their safety and health? Do employees appear to take pride in the achievements of the worksite safety and health program? Are employees comfortable answering questions about safety and health programs and conditions at the workplace? Do employees feel they have the support of management for their safety and health activities?

Assignment of Responsibility

Are responsibilities written out so that they can be clearly understood? Do employees understand their own responsibilities and those of others? Are hazards caused in part because no one was assigned the responsibility to control or prevent them? Are hazards allowed to exist in part because someone in management did not have the clear responsibility to hold a lower-level manager or supervisor accountable for carrying out assigned responsibilities?

Adequate Authority and Resources

Do safety staff members, or any other personnel with responsibilities for ensuring safe operation of production equipment, have the authority to shut down that equipment or to order maintenance or parts? Do employees talk about not being able to get safety or health improvements because of cost? Do employees mention the need for more safety or health personnel or expert consultants? Do recognized hazards go uncorrected because of a lack of authority or resources? Do hazards go unrecognized because greater expertise is needed to diagnose them?

Budgeting for Safety and Health

Safety and health must have a separate and distinct budget with dollars allocated for the various responsibilities, mandates, and requirements placed upon the occupational safety and health initiative in order to be considered an important management function.

Without an extensive history of cost related to occupational safety and health, the development of a budget for safety and health can be a rather inexact undertaking. Without such a history of past budgets or spending on occupational safety and health, budgeting becomes somewhat of a guesstimate. This is not to suggest that a reasonable and logical budget cannot be formulated. However, the budget will require much more effort, research, and justification.

Few seminars or classes exist regarding budgeting for occupational safety and health; the only book that gives the topic a better than adequate coverage is *Safety and Health Management Planning* by James P. Kohn and Theodore S. Ferry. It seems safe to say that most safety and health professional have had little or no training or experience related to budgeting for occupational safety and health unless they have had hands-on experience from having to develop budgets. This is a function of doing while you learn and probably resulted in some painful and time-consuming lessons.

One section in a book cannot provide comprehensive coverage of this topic. This is especially true because a budget must be specific for the function of the company and the resources available. Budgets are usually considered a roadmap or planning document for the completion of the assigned tasks and responsibilities based on the resources allocated by the company and is never all that it should be or a safety ad professional would want it to be.

Budget Items

There are a whole host of budget items that may need to be budgeted for regarding occupational safety and health. Table 2.2 is not comprehensive but is a representation of potential budget items.

TABLE 2.2
Listing of Potential OSH Budget Items

- Personnel cost:
  - Salaries
  - Fringe benefits
  - Social security
  - Health insurance
  - Workers' compensation
  - Retirement
  - Disability insurance
  - Unemployment benefits
  - Skill levels
  - Training
  - Profession meeting
  - Travel
- Equipment – expendable
- Equipment – nonexpendable:
  - Purchase
  - Rental
  - Repair agreements
  - Calibration or certification cost
- Travel
- Administrative cost:
  - Office supplies
  - Computers
  - Computer software
  - Computer hardware
  - Office equipment
  - Repair of office equipment
  - Ergonomic furniture
  - Storage cabinets
  - Other office furniture
  - Telecommunication services (Internet, etc.)
  - Cell phones
  - Postage
- Compliance cost:
  - Program development
  - Personal protective equipment
  - Medical examinations
  - Supplies (first aid, etc.)
  - Professional services (e.g., industrial hygienist)
  - Consultants (e.g., engineers for redesign of safeguards)
- Contracts (e.g., hazardous waste disposal)
- Facilities:
  - Utilities
  - Repairs and upgrades

continued

TABLE 2.2 (continued)
Listing of Potential OSH Budget Items

- Maintenance
- Liability insurance
- Budgeting for hiring new personnel
- Budgeting for long-term or multiyear projects
- Allowance for unforeseen emergencies
- Cost for OSHA citations and violations

Budget Approach

While it would be great to be able to predict the future and plan for events and items as though they would certainly happen, it is unlikely that the use of the budget as a precise document to follow will occur. There are many factors that are not under the control of the person responsible for occupational safety and health.

Thus, developers of an occupational safety and health budget must hitch their proverbial wagon to as many real to life safety and health issues as possible. This means that the best approach is to tie as much of the expenditures to compliance with regulatory requirements as humanly possible. Another approach is to show how the use of dollars for intervention and prevention of potential cost (a cost-avoidance strategy) can demonstrate savings.

Developing a budget that identifies specific items to be completed for a specific cost with proper justification provides the resources to complete the agreed-upon safety and health task. A budget should be broken down into several identifiable categories, such as

- Workplace health issues
- Workplace safety issues
- Safety and health management issues
- Environmental safety issues
- Product safety issues

The Compliance Factor

The OSHA database regarding the average cost of having citations can be used as a lever to get management's attention related to the investment in safety and health being requested in the budget. It is a real eye-opener for management when safety and health professionals can say that the cost of compliance is $500 for bloodborne pathogens and the average violation produces a fine of $1,000 per citation. Also, Occupational Safety and Health Administration (OSHA) professionals can research the most frequent violation for their particular industry sector. In many cases, the most frequent violation is the requirement for a program and training under the Hazard Communication Standard. Using compliance is, by far, one of the best ways to justify expenditures. Some of the regulations require programs, some medical examinations, and some training for compliance. An OSHA violation or citation strikes fear in most employers and is an excellent lever to use to justify a budget request. Other such levers include

- Cost of accidents and incidents
- Medical cost
- Workers' compensation cost
- Real dollar savings
- Loss potential from not performing the action item

The Written Budget

When it is time to develop a written budget, it is important to identify operating cost, potential benefits such as direct benefits (e.g., reduced labor cost, lower accidents, reduced insurance cost, or productivity gains). Indirect benefits should also be considered in light of quality improvements—reduced scrap; less rework; reduced product liability, exposure, or product recall expenses; improved corporate image; or increased market share. At times, indirect benefits are improved employee morale, which reduces absenteeism and turnover or increases teamwork and ownership. The potential reduction in the numbers of compliance penalties can be a benefit.

The safety and health professional who develops a responsible budget that looks for ways to control cost and maintain or improve the company's bottom line will be view as a part of the team and not just a necessary evil.

The safety and health professional can look for ways to control cost by sharing resources, using his or her available resources to the fullest, working on projects with other departments, lending expertise when possible, using mail order to cut costs, using a cost-bid service to obtain more for dollars spent, using in-house engineering expertise when possible, volunteering the use of his or her staff expertise when possible, using consultants or contractors for temporary or short-term projects or when a cost savings can be attained, performing as much as possible with his or her own staff, being productive, and taking low-cost action by being a good shopper.

There is always a degree of inaccuracy in any budget. Prevent cost overruns if at all possible, as overruns almost always have detrimental effects. May sure that you use the most qualified person to develop the occupational safety and health budget. Remember that there will always be budget expenditures and issues beyond your control, as actual cost may not be the same as planned cost. Link budget elements to accomplishments, milestones, needs, and compliance. Provide as much substance as possible to your budget. Quantified justifications have more substance than qualified opinions. Numbers are more understandable to upper management than reducing severity rates. Use all your experience and all your tools to develop a workable and reasonable budget.

Accountability of Managers, Supervisors, and Hourly Employees

Do performance evaluations for all line managers and supervisors include specific criteria relating to safety and health protection? Is there documented evidence of employees at all levels being held accountable for safety and health responsibilities, including safe work practices? Is accountability accomplished through either performance evaluations affecting pay or promotions, or through disciplinary action?

When you ask employees what happens to people who violate safety and health rules or safe work practices, do they indicate that rule breakers are clearly and consistently held accountable? Do hourly employees indicate that supervisors and managers genuinely care about meeting safety and health responsibilities? When asked what happens when rules are broken, do hourly employees complain that supervisors and managers do not follow rules and are never disciplined for infractions?

Are hazards occurring because employees, supervisors, or managers are not being held accountable for their safety and health responsibilities? Are identified hazards not being corrected because persons assigned that responsibility are not being held accountable?

Evaluation of Contractor Programs

Are there written policies for on-site contractors? Are contractors' rates and safety and health programs reviewed before selection? Do contracts require contractors to follow site safety and health rules? Are there means for removing a contractor who violates the rules?

Do employees describe hazardous conditions created by contract employees? Are employees comfortable with reporting hazards created by contractors? Do contract employees feel they are covered by the same, or the same quality, safety and health program as regular site employees?

Do areas where contractors are working appear to be in the same, better, or worse condition as areas where regular site employees are working? Does the working relationship between site and contract employees appear cordial?

ASSESSING THE KEY COMPONENTS OF WORKSITE ANALYSIS

Comprehensive Surveys, Change Analyses, and Routine Hazard Analyses

Are there documents that provide comprehensive analysis of all potential safety and health hazards of the worksite? Are there documents that provide both the analysis of potential safety and health hazards for each new facility, equipment, material, or process and the means for eliminating or controlling such hazards? Does documentation exist outlining the step-by-step analysis of hazards in each part of each job, so that you can clearly discern the evolution of decisions on safe work procedures? If complicated processes exist, with a potential for catastrophic impact from an accident but low probability of such accident (as in nuclear power or chemical production), are there documents analyzing the potential hazards in each part of the process and the means to prevent or control them? If there are processes with a potential for catastrophic impact from an accident but low probability of an accident, have analyses such as "fault tree" or "what if?" been documented to ensure sufficient backup systems for worker protection in the event of multiple control failures?

Do employees complain that new facilities, equipment, materials, or processes are hazardous? Do any employees say they have been involved in job safety analysis or process review and are satisfied with the results? Does the safety and health staff indicate ignorance of existing or potential hazards at the worksite? Does the occupational nurse, doctor, or other healthcare provider understand the potential occupational diseases and health effects in this worksite?

Have hazards appeared in areas where management did not realize there was the potential? Where workers have faithfully followed job procedures, have accidents or near-misses occurred because of hidden hazards? Have hazards been discovered in the design of new facilities, equipment, materials, and processes after use had begun? Have accidents or near-misses occurred when two or more failures in the hazard control system occurred at the same time, surprising everyone?

Regular Site Safety and Health Inspections

If inspection reports are written, do they show that inspections are done on a regular basis? Do the hazards found indicate a good ability to recognize hazards typical of this industry? Are hazards found during inspections tracked to complete correction? What is the relationship between hazards uncovered during inspections and those implicated in injuries or illness?

Do employees indicate that they see inspections being conducted, and that these inspections appear thorough? Are the hazards discovered during accident investigations ones that should have been recognized and corrected by the regular inspection process?

Employee Reports of Hazards

Is the system for written reports being used frequently? Are valid hazards that have been reported by employees tracked to complete correction? Are the responses timely and adequate?

Do employees know who to contact and what to do if they see something they believe to be hazardous to themselves or co-workers? Do employees think that responses to their reports of hazards are timely and adequate? Do employees say that sometimes when they report a hazard, they hear nothing further about it? Do any employees say that they or other workers are being harassed, officially or otherwise, for reporting hazards?

Are hazards ever found where employees could have reasonably been expected to have recognized and reported them? When hazards are found, is there evidence that employees complained repeatedly but to no avail?

Accident and Near-Miss Investigations

Do accident investigation reports show a thorough analysis of causes, rather than a tendency to automatically blame the injured employee? Are near-misses (property damage or close calls) investigated using the same techniques as accident investigations? Are hazards identified as contributing to accidents or near-misses tracked to correction?

Do employees understand and accept the results of accident and near-miss investigations? Do employees mention a tendency on management's part to blame the injured employee? Do employees believe that all hazards contributing to accidents are corrected or controlled?

Are accidents sometimes caused, at least in part, by factors that might also have contributed to previous near-misses that were not investigated or accidents that were too superficially investigated?

Injury and Illness Pattern Analysis

In addition to the required OSHA log, are careful records kept of first aid injuries and illnesses that might not immediately appear to be work related? Is there any periodic, written analysis of the patterns of near-misses, injuries, and illnesses that may show, over time, previously unrecognized connections between incidents that indicate unrecognized hazards need correction or control?

Looking at the OSHA 200 log and, where applicable, first aid logs, are there patterns of illness or injury that should have been analyzed for previously undetected hazards? Is there an occupational nurse or doctor on the worksite? Are employees suffering from ordinary illnesses encouraged to see a nearby healthcare provider? Are the records of those visits analyzed for clusters of illness that might be work related?

Do employees mention illnesses or injuries that seem work related to them, but that have not been analyzed for previously undetected hazards?

ASSESSING THE KEY COMPONENTS OF HAZARD PREVENTION AND CONTROL

Appropriate Use of Engineering Controls, Work Practices, Personal Protective Equipment, and Administrative Controls

If there are documented comprehensive surveys, are they accompanied by a plan for systematic prevention or control of hazards found? If there is a written plan, does it show that the best method of hazard protection was chosen? Are there written safe work procedures? If respirators are used, is there a written respirator program?

Do employees say they have been trained in and have ready access to reliable, safe work procedures? Do employees say they have difficulty accomplishing their work because of unwieldy controls meant to protect them? Do employees ever mention personal protective equipment, work procedures, or engineering controls as interfering with their ability to work safely? Do employees who use PPE understand why they use it and how to maintain it? Do employees who use PPE indicate that the rules for PPE use are consistently and fairly enforced? Do employees indicate that safe work procedures are fairly and consistently enforced?

Are controls meant to protect workers actually putting them at risk or not providing enough protection? Are employees engaging in unsafe practices or creating unsafe conditions because rules and work practices are not fairly and consistently enforced? Are employees in areas designated for PPE wearing it properly, with no exceptions? Are hazards that could feasibly be controlled through improved design inadequately controlled by other means?

Facility and Equipment Preventive Maintenance

Is there a preventive maintenance schedule that provides for timely maintenance of the facilities and equipment? Is there a written or computerized record of performed maintenance that shows the schedule has been followed? Do maintenance request records show a pattern of certain facilities or equipment needing repair or breaking down before maintenance was a major cause?

Do employees mention difficulty with improperly functioning equipment? Is safety maintenance scheduled or actually performed? Do any accident/incident investigations list facility or equipment breakdowns and equipment or facilities in poor repair? Do maintenance employees believe that the preventive maintenance system is working well? Do employees believe that hazard controls needing maintenance are properly cared for? Is poor maintenance a frequent source of hazards? Are hazard controls in good working order? Does equipment appear to be in good working order?

Establishing a Medical Program

Are good, clear records kept of medical testing and assistance? Do employees say that test results were explained to them? Do employees feel that more first aid or cardiopulmonary resuscitation- (CPR-) trained personnel should be available? Are employees satisfied with the medical arrangements provided at the site or elsewhere? Does the occupational healthcare provider understand the potential hazards of the worksite, so that occupational illness symptoms can be recognized?

Have further injuries, or worsening of injuries, occurred because proper medical assistance (including trained first aid and CPR providers) was not readily available? Have occupational illnesses possibly gone undetected because no one with occupational health specialty training reviewed employee symptoms as part of the medical program?

Emergency Planning and Preparation

Are there clearly written procedures for every likely emergency, with clear evacuation routes, assembly points, and emergency telephone numbers? When asked about any kind of likely emergency, can employees tell you exactly what they are supposed to do and where they are supposed to go?

Have hazards occurred during actual or simulated emergencies due to confusion about what to do? In larger worksites, are emergency evacuation routes clearly marked? Are emergency telephone numbers and fire alarms in prominent, easy-to-find locations?

ASSESSING THE KEY COMPONENTS OF SAFETY AND HEALTH TRAINING

ENSURING THAT ALL EMPLOYEES UNDERSTAND HAZARDS

Does the written training program include complete training for every employee in emergency procedures and in all potential hazards to which employees may be exposed? Do training records show that every employee received the planned training? Do the written evaluations of training indicate that the training was successful, and that the employees learned what was intended?

Can employees tell you what hazards they are exposed to, why those hazards are a threat, and how they can help protect themselves and others? If PPE is used, can employees explain why they use it and how to use and maintain it properly? Do employees feel that health and safety training is adequate?

Have employees been hurt or made ill by hazards of which they were completely unaware, or whose dangers they did not understand, or from which they did not know how to protect themselves? Have employees or rescue workers ever been endangered by employees not knowing what to do or where to go in a given emergency situation? Are there hazards in the workplace that exist, at least in part, because one or more employees have not received adequate hazard control training? Are there any instances of employees not wearing required PPE properly because they have not received proper training, or because they simply do not want to and the requirement is not enforced?

ENSURING THAT SUPERVISORS UNDERSTAND THEIR RESPONSIBILITIES

Do training records indicate that all supervisors have been trained in their responsibilities to analyze work under their supervision for unrecognized hazards? Do they apply practices to maintain physical protections, to reinforce employee training through performance feedback, and, where necessary, enforce safe work procedures and safety and health rules?

Are supervisors aware of their responsibilities? Do employees confirm that supervisors are carrying out these duties? Has a supervisor's lack of understanding of safety and health responsibilities played a part in creating hazardous activities or conditions?

ENSURING THAT MANAGERS UNDERSTAND THEIR SAFETY AND HEALTH RESPONSIBILITIES

Do training plans for managers include training in safety and health responsibilities? Do records indicate that all line managers have received this training? Do employees indicate that managers know and carry out their safety and health responsibilities? Has an incomplete or inaccurate understanding by management of its safety and health responsibilities played a part in the creation of hazardous activities or conditions?

SUMMARY

The key to a successful and efficient evaluation is to combine elements when using each technique. First, review the documentation available relating to each element. Then walk through the worksite to observe how effectively what is on paper appears to be implemented. While walking around, interview employees to verify that what you read and what you saw reflect the state of the safety and health program.

Effective safety and health program evaluation is a dynamic process. If you see or hear about aspects of the program not covered in your document review, ask to receive the documents, if any, relating to these aspects. If the documents included program elements not visible during your walk around the site or not known to employees, probe further. Utilizing this cross-checking technique should result in an effective, comprehensive evaluation of the worksite's safety and health program.

REFERENCES

Allison, W. W. *Profitable risk control.* Des Plaines, IL: American Society of Safety Engineers, 1986.

Kohn, J. P. and Ferry, T. S. *Safety and health management planning.* Rockville, MD: Government Institutes, 1999.

Reese, C. D. *Mine safety and health for small surface sand/gravel/stone operations: A guide for operators and miners.* Storrs, CT: University of Connecticut Press, 1997.

Reese, C. D. and J. V. Eidson. *Handbook of OSHA construction safety & health.* Boca Raton, FL: CRC Press/Lewis Publishers, 1999.

United States Department of Labor, Occupational Safety and Health Administration. Field Inspection Reference Manual (FIRM) (OSHA Instruction CPL 2.103). Washington, D.C.: U.S. Department of Labor, September 26, 1994.

United States Department of Labor, Occupational Safety and Health Administration. Citation Policy for Paperwork and Written Program Requirement Violations (OSHA Instruction CPL 2.111). Washington, D.C.: U.S. Department of Labor, November 27, 1995.

United States Department of Labor, Occupational Safety and Health Administration. *Federal Register: Safety and Health Program Management Guidelines* (Vol. 54, No. 16). pp. 3904–3916. Washington, D.C.: U.S. Department of Labor, January 26, 1989.

United States Department of Labor, Occupational Safety and Health Administration. Office of Construction and Maritime Compliance Assistance (OSHA Instruction STD 3-1.1). Washington, D.C.: U.S. Department of Labor, June 22, 1987.

3 Accident/Incident Investigation

INTRODUCTION

The material in this chapter is a combination of information obtained and developed by the Occupational Safety and Health Administration (OSHA) and the Mine Health and Safety Administration based on their experience in conducting accident investigations over a period of many years.

Millions of accidents/incidents occur throughout the United States every year. The inability of people, equipment, supplies, or surroundings to behave or react as expected causes most of these accidents/incidents. Accident/incident investigations determine how and why each incident has occurred. By using the information gained, a similar or perhaps more disastrous accident may be prevented. Accident/incident investigations should be conducted with accident prevention in mind. The mission is one of fact finding. Investigations are not to find fault.

An accident, by definition, is any unplanned event that results in personal injury or in property damage. When the personal injury requires little or no treatment, or is minor, it is often called a first aid case. If it results in a fatality or in a permanent total, permanent partial, or temporary total (lost-time) disability, it is serious. Likewise, if property damage results, the event may be minor or serious. All accidents should be investigated regardless of the extent of injury or damage.

Accidents are part of a broad group of events that adversely affect the completion of a task. These events are defined as incidents. With this said, the most commonly used term for accidents and incidents is "accident," which will be used to refer to both accidents and incidents because the basic precepts are applicable to both.

PURPOSE OF ACCIDENT INVESTIGATIONS

An important element of a safety and health program is accident investigation. Although it may seem to be too little too late, accident investigations serve to correct the problems that contribute to an accident and will reveal accident causes that might otherwise remain uncorrected.

The main purpose of conducting an accident investigation is to prevent a recurrence of the same or a similar event. It is important to investigate all accidents regardless of the extent of injury or damage. The kinds of accidents that should be investigated and reported include

1. Disabling injury accidents
2. Nondisabling injury accidents that require medical treatment
3. Circumstances that have contributed to acute or chronic occupational illness
4. Noninjury property damage accidents that exceed a normally expected operating cost
5. Near-accidents with a potential for serious injury or property damage

ACCIDENT PREVENTION

When accidents are not reported, their causes usually go uncorrected. This allows the chance for the same accident to happen again. Every accident, if properly investigated, serves as a learning experience for the people involved. However, investigations should not become a mechanical routine. They should strive to establish what happened, why it happened, and what must be done to prevent a recurrence. An accident investigation must be conducted to learn the facts, not to place blame.

In any accident investigation, consider the aspect of multiple causation. The contributing factors surrounding an accident, as well as the unsafe acts and unsafe conditions, should be considered. If only the unsafe acts and conditions are considered when investigating an accident, little will be accomplished toward any accident prevention effort because the "root causes" still remain. This leaves the possibility for an accident to recur. The "root causes" are items such as management policies and decisions, and the personal and environmental factors that could prevent accidents when corrected.

Accidents are usually complex and are the result of multiple causes. A detailed analysis of an accident will normally reveal three cause levels: basic, indirect, and direct. At the lowest level, an accident results only when a person or object receives the release of an amount of energy or exposure to hazardous material that cannot be absorbed safely. This energy or hazardous material is the *direct* cause of the accident. The second causal areas are usually the result of one or more unsafe acts or unsafe conditions, or both. Unsafe acts and conditions are the *indirect* causes or symptoms. In turn, indirect causes are usually traceable to poor management policies and decisions, or to personal or environmental factors. These are the *basic* causes.

Despite their complexity, most accidents are preventable by eliminating one or more causes. Accident investigations determine not only what happened, but also how and why. The information gained from these investigations can prevent the recurrence of similar or perhaps more disastrous accidents. Accident investigators are interested in each event as well as in the sequence of events that led to the accident. The accident type is also important to the investigator. The recurrence of accidents of a particular type or those with common causes show areas needing special accident prevention emphasis.

It is important to have some mechanism in place to investigate accidents/incidents in order to determine the basis of cause-and-effect relationships. These types of relationships can only be determined when you actively investigate all accidents and incidents that result in injuries, illnesses, or damage to property, equipment, or machinery.

Accident investigation becomes more effective when all levels of management—particularly top management—take a personal interest in controlling accidents.

Management adds a contribution when it actively supports accident investigations. It is normally the responsibility of line supervisors to investigate all accidents; and in cases where there is serious injury or equipment damage, other personnel such as department managers, might also have an investigation team become involved.

Once the types of accidents/incidents occurring have been determined, prevention and intervention activities can be undertaken to ensure that no recurrences will transpire. Even if the company is not experiencing large numbers of accidents/incidents, it still needs to implement activities that actively search for, identify, and correct the risk from hazards on jobsites. Reasons to investigate accidents/incidents include

- To know and understand what happened
- To gather information and data for present and future use
- To deter cause and effect
- To provide answers for the effectiveness of intervention and prevention approaches
- To document the circumstances for legal and workers' compensation issues
- To become a vital component of your safety and health program

If only a few accidents/incidents are occurring, investigations might want to move down one step to examine near-misses and first aid related cases. It is only a matter of luck or timing that separates the near-miss or first aid event from being a serious, recordable, or reportable event. The truth is that the company probably has been lucky by seconds or inches. (That is, a second later and it would have hit someone or an inch more and it would have cut off a finger.) Truly, it pays dividends to take the time to investigate accidents/incidents occurring in the workplace.

REPORTING ACCIDENTS

When accidents are not reported, their causes usually go uncorrected, allowing the chance for the same accident to result again. Every accident, if properly investigated, serves as a learning experience to the people involved. The investigation should avoid becoming a mechanical routine. It should strive to establish what happened, why it happened, and what must be done to prevent a recurrence. An accident investigation must be conducted to find out the facts and not to place blame.

The first step in an effective investigation is the prompt reporting of accidents. The company cannot respond to accidents, evaluate their potential, and investigate them if they are not reported when they happen. Prompt reporting is the key to effective accident investigations. Hiding small accidents does not help prevent the serious accidents that kill people, put the company out of business, and take away jobs. If workers do not report accidents to the supervisor, they are stealing part of the supervisor's authority to manage his or her job (see Table 3.1).

The results-oriented supervisor recognizes that the real value of investigation can only be achieved when his or her workers report every problem, incident, or accident that they know of. To promote conscientious reporting, it may be helpful to know

TABLE 3.1
Sound Reasons for Reporting All Accidents

1. You learn nothing from unreported accidents.
2. Accident causes go uncorrected.
3. Infection and injury aggravations can result.
4. Failure to report injuries tends to spread and become accepted practice.

some of the reasons why workers fail or avoid reporting accidents. There are usually reasons that workers espouse for not reporting accidents. They include

- Fear of discipline
- Concern about the company's record
- Concern for the company's reputation
- Fear of medical treatment
- Dislike of medical personnel
- Desire to keep personal record clear
- Avoidance of red tape
- Desire to prevent work interruptions
- Concern about the attitudes of others
- Poor understanding of the importance of reporting accidents

How can we combat these reporting problems?

- React in a more positive way.
- Indoctrinate workers on the importance of reporting all accidents.
- Make sure everyone knows what kinds of accidents should be reported.
- Give more attention to prevention and control.
- Recognize individual performance.
- Develop the value of reporting.
- Show disapproval of injuries neglected and not reported.
- Demonstrate belief by action.
- Do not make mountains out of molehills.

Let the worker know that you appreciate him or her for reporting accidents promptly. Inquire about his or her knowledge of the accident. Do not interrogate or "grill" the worker. Stress the value of knowing about problems while they are still small. Focus on accident prevention and loss control. Emphasize compliance with practices, rules, protective equipment, and promptly commend good performance. Pay attention to the positive things workers do and give sincere, meaningful recognition where and when it is deserved. Use compliments as often as you use warnings. Use group and personal meetings to point out and pass on knowledge gained from past accidents. Give an accident example as an important part of every job instruction. Show that the organization believes in what it says by taking corrective action promptly. The company can always do something right at the moment, even if permanent correction requires time to develop new methods, buy new equipment, or modify the building.

ORGANIZING AND ASSIGNING RESPONSIBILITIES

Organizing a division of responsibility toward accident investigations should include a broad range of experience, responsibility, and authority. Although more serious accidents or equipment damage should be investigated primarily by a person with more authority, upper-level management personnel, department supervisors, and first-line supervisors should still play an active role in accident investigations.

Specific accident investigation responsibilities should be assigned to various people. Accidents involving serious injury or major equipment damage should be investigated by a committee. The committee should involve all levels of management as well as a range of technical competencies. Most accidents that are potentially serious or have the potential for major equipment damage should be investigated by the department's supervisor in conjunction with their immediate managers. Minor injuries and minor equipment damage accidents should normally be investigated by the responsible first-line supervisor.

SUPERVISORS AND ACCIDENT INVESTIGATION

Line supervisors should be involved in accident investigations, basically because they are normally the people in direct contact with the worker and understand their problems, personalities, and capabilities. Involving supervisors increases their responsibilities toward the accident prevention effort (see Figure 3.1). Supervisors are normally responsible for training, so becoming involved in investigations will make the supervisor more aware of what causes accidents as well as ways to prevent a recurrence.

Let's take a look at the reasons why the line supervisor should conduct the investigation. An accident happens because there is a problem somewhere, and the problem exists because

1. A deficiency is not known.
2. The risk involved in the deficiency is thought to be less than it really is.
3. Despite the deficiency, someone without authority decided to go ahead.
4. Someone with authority decides that the cost is more than the money that is available, or that the cost outweighs the benefit of correcting the deficiency.

When taking enough time, anyone can find out what the exact problem is. For every problem there is an answer—a single best way to correct the problem—and for several reasons the supervisor is the best person to put this all together.

1. The supervisor has a personal interest in protecting his or her workforce.
2. The supervisor knows the most about the workers and conditions.
3. The supervisor knows best how to get information.
4. The supervisor will take the action anyway.

FIGURE 3.1 Line supervisors make valuable contributions to accident investigations. (*Source:* Courtesy of the Mine Safety and Health Administration.)

Actions the supervisor can take to encourage workers to report minor injury accidents include

1. Instruct the employees on the importance of reporting such accidents.
2. Do not discourage the employees from reporting minor injuries.
3. Periodically discuss failure to report minor injuries.
4. Show disapproval of neglected injuries.

INVESTIGATIONS BENEFIT THE SUPERVISOR

There are other benefits a line supervisor can gain from performing the accident investigation, including

- An increase in production time
- Evidence of the supervisor's concern for the workers
- A reduction in operating costs
- A demonstration of the supervisor's control

The supervisor who gets right into an investigation and takes immediate action to prevent other accidents shows workers that he or she really cares for their physical welfare and their effectiveness in doing their job. The supervisor who lets others do his or her work makes a poor impression on the workers.

One common effect of an accident is that it interrupts work. Effective investigations prevent repeated accidents and control work interruptions and inefficiencies.

The cost of delays, injuries, property damage, and insurance all add to the cost of the production. Control of costs is one way of judging a supervisor's job performance.

Persistent efforts to solve problems reflect the image of a capable supervisor who is in control of things. Actions speak louder than words. People like to work for a supervisor who takes control of any situation, and who shows he or she is capable of dealing with any problem.

PREPLANNING AN ACCIDENT INVESTIGATION

Before an accident occurs, you should contact emergency services—medical, paramedical, fire department, police, utilities, bomb disposal, and poison control—to determine their procedures for response. Operation and locations of emergency shutdown switches of equipment must be identified unless it must be operating to prevent further damage or loss. Be prepared to isolate sources of secondary accidents and injuries. As well, provide for interruption of business for sufficient periods to permit thorough investigation.

Upon discovery of an accident, management personnel should be notified commensurate with the classification of the accident and pertinent regulatory agencies as required by public laws. Institute security and implement investigative procedures using appropriate tools and equipment.

Upon arrival at the accident scene or site, the first priority should be to rescue the injured or endangered in such a manner that injury is not inflicted or compounded. Then reduce risk of further injury or damage to control loss. Identify elements involved in the accident to initiate reporting and investigation of the accident.

After the emergency is under control, plan interfaces with other investigators prescribed by public laws; ensure that the scene is systematically preserved and evidence is not destroyed. Be prepared to provide special briefings for investigators, specialized assistants, and the press.

Part of the preparation is having an accident investigation kit already available so that the investigator does not waste valuable time in procuring the items and tools needed to conduct an effective accident investigation. Table 3.2 contains a list of many of the items and tools that would be appropriate for an accident investigation kit. Of course, this list is not all-inclusive.

When urgent data collection is completed, have sources of technical expertise identified for examination and testing of evidence. Make sure that the company has communicated and interfaced with suppliers of equipment and materials on product liability.

When work at the accident site is finished, have a plan for temporary storage of evidence that might be required later for additional examination. Determine when and where final disposal of evidence, damaged equipment, and materials should be. Make plans for the follow-on phases of the investigation. Conduct a planned inspection of the workplace or equipment before allowing full, normal activity to resume.

INVESTIGATION PROCESS

Accidents should be investigated as soon as possible after they happen. The more time that passes before questioning personnel and witnesses involved, the greater the risk of not getting an accurate story of what happened and why. The injured

TABLE 3.2
Items and Tools for an Accident Investigation Kit

Accident report form	"Keep Out" placards
Adjustable wrench	Knife
Air velocity meter	Large envelopes (9 × 12 inches)
Attachable tags and labels	Magnifying glass
Barrier tape	Metric conversion chart
Calculator	Nail and screws (assorted)
Camera (35 mm)	Note pads (lined)
Camera (digital)	Orange reflective vest
Camera (video)	Pens and pencils
Chemical gloves	Pliers
Chemical goggles	Protractor
Chemical sampling containers	Rain suit
CO and CO_2 detection meter	Respirator (assorted cartridges)
Compass	Rope (nylon)
Duct and masking tape	Ruler
Ear protection	Rubber boots
Electrical tester	Safety glasses
Emergency flares	Scissors
Explosive level meter	Scotch tape
Face shield	Screwdrivers
Film for all cameras	Small collapsible shovel
First aid kit	Small paint brush
Flashlight	Socket set
Grid (square) paper	String/twine
Hammer	Synthetic smoke
Hand-held audio recorder	Tape measure
Ice pick or awl	Thermometer
Indelible pen or marker	Tyvek suit
Investigator's notebook	Zip-lock bags

person's version of an accident should be obtained as soon as practical before his or her memory distorts what really happened. There are two circumstances when it is proper to postpone questioning of injured persons:

1. Do not question a person if doing so delays medical treatment.
2. Do not question a person who is extremely upset or in pain.

There are various investigative procedures that should be followed during an investigation. The following procedures should be followed to gain maximum results:

1. The investigator should possess familiarity with the equipment, operation, or process involved that is likely to bring about accidents.
2. The scope of the investigation and the size of the investigation team should be determined depending on the nature of the accident, its magnitude, and its technical complexity.

3. To obtain accurate facts, the investigator must get to the scene as promptly as possible.
4. The theory of multiple causation must be considered when investigating accidents.
5. The real purpose is to gather facts that led to the accident, and not to focus emphasis on placing blame.

The investigator will need to maintain a notebook that will include observation, data, and information pertaining to the investigation as it proceeds through its many steps and phases. The investigator's notebook should, at a minimum, contain the following:

1. At the accident scene, it is the investigator's job to gather all information:
   a. Field notes are the continuous written record of the investigation.
   b. Notes aid in the investigation, interviews, follow-up investigations, and if called to testify in court.
   c. Notes are the basic source of information for the written report.
2. The notebook itself:
   a. The inside cover should have information as to the owner: name, address (office, not home), and telephone number.
   b. It should be carried at all times.
   c. It should be large enough to allow the recording of substantial information on each page, as well as diagrams. Approximately 4 × 6 inches is considered best.
3. Basic essentials to include in field notes:
   a. Who?
      - Who was involved?
      - Who discovered the incident?
      - Who saw, or heard, something of importance?
      - Who was interviewed?
      - Who marked the evidence?
   b. What?
      - What happened?
      - What do witnesses know?
      - What evidence was obtained?
      - What action was taken?
      - What other agencies were notified?
      - What time did the incident occur?
      - What time did you arrive?
   c. Where?
      - Where did the incident occur?
      - Where were the victims?
      - Where were the witnesses?
      - Where was evidence marked?
      - Where was the individual suspected of causing the incident?

d. When?
 - When was the victim last seen?
 - When was the suspect last seen?
 - When did help arrive?
e. How?
 - How was the offense committed?
 - How did the incident occur?
 - How were the tools or weapons obtained?
 - How did you get your information?
f. With what?
 - With what trade or profession are tools or evidence associated?
 - With what other incident(s) is this one related?
g. Why?
 - Why did the incident occur?
 - Why was it reported?
 - Why were witnesses reluctant?
 - Why was there a delay?
h. With whom?
 - With whom was the victim seen?
 - With whom are the witnesses connected?
i. How much?
 - How much damage was done?
 - How much property is missing?

The actual procedures used in a particular investigation depend on the nature and results of the accident. The employer must determine the scope of the investigation based on the amount of resources available for the investigation. The following procedure should (or might) be followed, depending on the accident:

Conducting the Investigation

Determine who will conduct the investigation. This may be a single individual or a team approach, depending on the availability of qualified personnel. No matter which procedure is used, someone should be assigned responsibility. Some effort should be expended in briefing everyone involved regarding what is known about the accident and providing the company with written programs and procedures. Other information might include photographs, maps, any changes since the accident, events prior to the accident, and a list of any witnesses.

Visit the Site

As soon as feasibly possible, conduct a site visit. The site should have been secured. Do not disturb the location of anything on the site unless hazards or an emergency situation exists. Someone may need to make sketches or take photographs of the site. Some guidelines for sketches and photographs can be found in Table 3.3 and Table 3.4, respectively.

TABLE 3.3
Sketching Checklist

By eliminating irrelevant details and adding measurements, you can often sketch a scene more clearly than you can photograph it. The following points will make sketching for accident maps easy without sacrificing accuracy.

1. Use graph paper. Let each square represent a fixed distance such as a foot, and write the scale at the bottom of the sketch.
2. Use a strip of graph paper to measure diagonals on the sketch.
3. Locate each important object with a rough outline.
4. Label large objects inside their outline. Label small objects outside their outline with an arrow to the object; the arrow should just touch the object.
5. For maps with a lot of detail, use a sketch log. Use double letters to identify reference points and single letters to identify items of evidence.
6. Indicate distances of movable objects from at least two fixed points. Logs for detailed maps have columns for measurement data.
7. Include a north arrow in each sketch.
8. Mark camera positions by a letter inside a circle. Later, the appropriate letter should be used on each print.
9. Identify the sketches with a label, data box, or on the back with date, time, place, description, who did the sketch, etc., just as you would a photograph.

TABLE 3.4
General Uses of Photographs in Accident Investigations

1. Orientation to the scene of an accident
2. Record of the detail of injury and damage
3. Record of relative positions of large numbers of items or damaged fragments
3. Depiction of witnesses' views of the scene
4. Evidence of improper assembly or use of equipment, materials, and structures
5. Detail of marks, spills, instructional aids, signs, etc.
6. Records of disassembly of parts for analysis by examination
7. Evidence of deterioration, abuse, and lack of proper maintenance
8. Location of parts, or other evidence, overlooked during early stages of the investigation
9. Depiction of the sequence of failure of machines
10. Analysis worksheets
11. Illustration for follow-on training

Note: A photograph log should be maintained in order to document what was photographed and the sequence of photos (see Table 3.5).

As the site is viewed, some items, which should be recorded by position, include

1. The dead and injured people
2. Machines, vehicles, and other types of equipment involved in or affected by the accident

TABLE 3.5
Photograph Log

Accident Investigation Photography Log

Film Roll Identification Number: _______ Date: _______

Film Type, Speed, & Size: _______ Accident #: _______

Photo #	Subject	Location	Lens/Speed	Angle	Comments
1.					
2.					
3.					
4.					
5.					
6.					
7.					
8.					
9.					
10.					

3. Parts broken off or detached from equipment and materials
4. Objects that were broken, damaged, or struck during or as a result of the accident
5. Gouges, scratches, dent, paint smears, and rubber skid marks, etc. on surfaces
6. Tracks or similar traces of movements
7. Defects or irregularities in surfaces
8. Accumulations of, or stains from, fluids, whether existing before the accident or spilled as a result of the accident
9. Spilled or contaminated materials
10. Areas of debris
11. Sources of distractions or adverse environmental conditions
12. Safety devices and equipment

While some of these are not actually parts but are very useful elements of evidence, some parts that should be examined to see if they should be extracted from the accident site for analysis or even subjected to a technical examination are as follows:

1. Components of equipment, materials, and parts of structures that are fractured, distorted, scarred, chafed, or ruptured
2. Parts suspected of internal failure or subjected to sudden stoppage or abnormal stress as a consequence of the accident
3. Parts suspected of improper assembly or mating
4. Parts suspected of deficient material in fabrication, improper tempering or heat treatment, or improper bonding
5. Parts or components that seem faulty in workmanship or deficient in design, or seem to have inadequate interface with other equipment and materials

6. Parts improperly mounted or inadequately supported, such as lines, tubing, fittings, wiring and controls, and those subjected to cyclical operation, vibration, or reactive movement of pressure or tension
7. Parts requiring or showing evidence of need for lubrication or surface conditioning
8. Controls and position or operation indicators
9. Parts that are power sources, such as engines, motors, pumps, transformers, and relays
10. Substitute or modified component parts
11. Foreign objects and parts that seem different in smell, color, shape, size, or location
12. Fluid spills and stains, as well as parts that show signs of leakage

The site visit will set the tone for the remainder of the investigation. It will probably generate as many questions as are answered by the site visit. However, the site visit and its documentation are necessary in order to effectively analyze the causes of the accident as well as being an aid in the completion of your final report. Once all the physical evidence has been gathered, it is time to interview any witnesses or victims.

Interviewing

Interviewer

1. Definition of interview: the art and science of questioning or talking with someone, especially to get information or to gain understanding.
2. Before the interview:
   a. Be prepared. Have some knowledge of the situation.
      i. Gather information about people, circumstances, and the situation.
      ii. Is there anything similar or different from other interview circumstances?
      iii. What about the people involved? We see people differently. Avoid preconceptions.
      iv. Know the location of where the interviewer is going. Get directions.
   b. Know what you want to know (plan):
      i. Specifics: who, what, when, where, how.
      ii. Have a list of questions to ask.
      iii. Ask open-ended questions to get responses other than yes or no.
   c. Practice interviewing:
      i. Listen and learn from voice and questions (use a tape recorder).
      ii. Talk to other interviewers or help as needed.
      iii. Be brief, clear, and to the point.
   d. Personal:
      i. Most difficult interview—when people are forced to talk to interviewers.
      ii. Key (be patient with yourself and have confidence in yourself as the interviewer).

iii. Appearance (appropriate to the situation).
iv. Be on time.

3. The interview:
   a. Beginning:
      i. Develop personal (respect, understanding) relationship.
      ii. Note taking (shorthand, abbreviations).
      iii. Be very observant:
         – Verbal
         – Nonverbal (eye movement, body position)
   b. Middle:
      i. Check it out (everything, anything).
      ii. Ask yourself:
         – What am I doing? Am I getting what I want?
         – Do the facts agree? If as the interviewer you do not understand, ask questions, find out.
         – "Probing" for information (examples: "Tell me more about that" or "Anything else?").
         – Set the pace for the interview. If pace is fast, slow down. Take a break; walk around.
         – Ask (who, what, when, where, how). "Why" often elicits an emotional response—opinions, not information.
   c. The end:
      i. Review the outcome of what has been accomplished. Tell the person interviewed. Go over notes.
      ii. Allows the interviewer to act on information missed.
      iii. Find out if there are other sources of information.
      iv. When the interviewer has all your information: Finish by standing up, making contact (thank you), be quiet, and leave.
4. After the interview:
   a. Record immediately any impressions.
   b. Separate opinions from facts.
   c. Go over the notes (i.e., reread, fill in).
   d. Add anything additional.
   e. Jot down added questions and comments.
   f. Make decisions about future interviews.
   g. Take a break.
   h. Review positives and negatives.
5. Techniques:
   a. Prompt (through making comments on what has been said). Example: "Yes, and what about…?"
   b. Nonverbal cues (silence, pauses).
   c. Be pleasant and respond to people.
   d. Be permissive (make allowances for people being different).
   e. Keep pressure or coercion off yourself (do not hassle the interviewee).
   f. Be yourself (use this information to fit you).

Accurate Interviews

When possible, the victims and witnesses should be interviewed at the scene of the accident. There are several reason why this is the best procedure to follow:

1. The average employee has difficulty explaining with words only, how an accident happened. At the scene, the employee can point out things. This results in a better understanding of what happened and why.
2. Employees are less likely to stray from the facts when they are telling about the occurrence at the scene of the accident. When not at the scene, some employees are inclined to exaggerate the role of environmental factors as contributors to the accident.
3. Supervisors can better evaluate the part that environmental hazards played in the accident when they see the work areas with their own eyes.
4. Often, supervisors can get a better idea of what needs to be done to prevent recurrence when they have the opportunity to size up the actual situation.

When starting the interview, follow these key steps:

- Remind employees of the investigation's purpose.
- Ask employees to give their complete version of the accident.
- Ask them questions to fill in the gaps.
- Check your understanding of the accident.
- Discuss how to prevent this type of accidents recurrence.

Expanding the Interview for Details

During the interview, the investigator should introduce control questions to ensure the accuracy of statistical data as well as permit subsequent evaluation of the reliability of information supplied by the witness. The control questions should include

1. Time and location of the accident.
2. Environment: weather, lighting, temperature, noise, distractions, concealments. Include pre-contact, contact, and post-contact time periods by specific questions.
3. Positions of people, equipment, material, and their relationships to pre-contact, contact, and post-contact events. Include the position of the witness being interviewed.
4. Other witnesses, known as well as those unidentified by name, and their positions.
5. If anything was moved, repositioned, turned off or on, or taken from the scene (including the injured) during pre-contact, contact, and post-contact phases.
6. Observations of the response of emergency teams and supervisory personnel and their actions at the scene.
7. What attracted the witness' attention to the accident?
8. What would the witness do to prevent getting involved in a similar accident under similar circumstances?

At times it is useful to have the victim and other persons being interviewed reenact the accident. This may provide some answers and may enhance recall of some details previously not remembered. Reenactment is not necessary in all accident investigations.

Evaluating the Witness' State

The investigator needs to analyze the witness as well as his or her testimony during the course of the interview. He should consider and try to determine, by observation or direct questions, the witness' state with respect to

1. Delay between the time of occurrence of the accident and interview. People forget 50 percent to 80 percent of the detail in just a 24-hour period.
2. Contact with principals in the accident. People who feel they have faults may give misleading information to witnesses, such information later being remembered as observation rather than hearsay.
3. Contact with other witnesses with domineering personalities or perceived expertise that distorts the witness' memory.
4. Contact with news media or other investigators who induce inaccuracies through leading questions and alleged "facts" or opinions they present for substantiation.
5. Signs of shock, amnesia, or other psychological disturbance resulting from the accident. Unpleasant experiences are frequently blanked from people's memory.
6. General health and use of medication, drugs, or alcohol, all of which affect perception, attention, and memory.
7. Personal problems in any of the areas of the relationship of risk to stress. Self-induced stresses cause preoccupation and inattention, which may be masked by detailed testimony.
8. Personal interests or other needs to end the interview, evade responsibility, or divert an investigator's inquiry from a particular subject area.
9. Personality types with resultant tendencies.
10. Apparent rationalizations. Are parts of the witness' story entirely too smooth and complete in their first statement? The more intelligent a witness is, the stronger their tendency to mull events over in their mind, and become aware of gaps and supplement knowledge with assumptions from an honest desire to give coherent information.

The investigator should interview each victim and witness. Those who were present before the accident and those who arrived at the site shortly after the accident should be interviewed. Keep accurate records of each interview. Use a tape recorder if desired and if approved by the person being interviewed.

Summarizing the Interviewing Process

After interviewing all witnesses, the team should analyze each witness' statement. They may wish to re-interview one or more witnesses to confirm or clarify key points. While there may be inconsistencies in witnesses' statements, investigators should assemble the available testimony into a logical order. Analyze this information along with data from the accident site.

Not all people react in the same manner to a particular stimulus. For example, a witness within close proximity to the accident may have an entirely different story from one who saw it at a distance. Some witnesses may also change their stories after they have discussed it with others. The reason for the change may be additional clues.

A witness who has had a traumatic experience may not be able to recall the details of the accident. A witness who has a vested interest in the results of the investigation may offer biased testimony. Finally, eyesight, hearing, reaction time, and the general condition of each witness may affect his or her powers of observation. A witness may omit entire sequences because of a failure to observe them or because their importance was not realized.

In general, experienced personnel should conduct interviews. If possible, the team assigned to this task should include an individual with a legal background. In conducting interviews, the team should

1. Appoint a speaker for the group.
2. Get preliminary statements as soon as possible from all witnesses.
3. Locate the position of each witness on a master chart (including the direction of view).
4. Arrange for a convenient time and place to talk to each witness.
5. Explain the purpose of the investigation (accident prevention) and put each witness at ease.
6. Listen; let each witness speak freely, and be courteous and considerate.
7. Take notes without distracting the witness. Use a tape recorder only with the consent of the witness.
8. Use sketches and diagrams to help the witness.
9. Emphasize areas of direct observation. Label hearsay accordingly.
10. Be sincere and do not argue with the witness.
11. Record the exact words used by the witness to describe each observation. Do not put words into a witness' mouth.
12. Word each question carefully and be sure the witness understands.
13. Identify the qualifications of each witness (name, address, occupation, years of experience, etc.).
14. Supply each witness with a copy of his or her statement(s). Signed statements are desirable.

HANDLING THE EVIDENCE

The gathering of evidence is a fact-finding process that provides the data to be analyzed, and prevention decisions can be based upon the findings. The investigator's evidence usually comes from four sources: people, locations, equipment/parts/accessories, and paper documentation. The evidence sought should identify what was normal, what occurred abnormally, what time it occurred, how it happened, and what the results of the incident were.

At times, specific analytical techniques and tests must be performed on chemicals, parts, and equipment in order to verify suspicions, such as that a faulty or damaged part existed. This will require a laboratory test to support your other evidence.

Evidence should provide reasons why the accident occurred, the sequence of events leading up to the accident, and a possible or plausible event sequence. When you have collected all your evidence, you will want to use one or more analysis techniques discussed in Chapters 7 through 11. Once all facets are analyzed, then appropriate interventions and prevention activities can be put forward and a final report on the incident generated.

DEVELOPING ACCIDENT INVESTIGATION FORMS

The development of standard accident investigation forms, including accident-specific forms, will allow the investigator(s) to make comparisons between accidents and injuries that are similar or caused by the same event, because evidence or data have been collected in a systematic and consistent fashion (see Table 3.6). Examples of various accident investigation forms can be found in Appendix B. These forms should be tailored to meet the investigators' or company's needs, or designed to fit their own specifications. The completed accident investigation forms will assist in the final accident report.

FINAL REPORT OF INVESTIGATION

As noted earlier, an accident investigation is not complete until a report is prepared and submitted to the proper authorities. Special report forms are available in many cases. Other instances may require a more extended report. Such reports are often very elaborate and may include a cover page, a title page, an abstract, a table of contents, a commentary or narrative portion, a discussion of probable causes, and a section on conclusions and recommendations. The following outline has been found especially useful in developing the information to be included in the formal report:

1. Introduction:
   a. Purpose of the investigation
   b. Scope of the investigation
   c. Who is the investigator(s)
2. Description of the accident:
   a. Where and when the accident occurred
   b. Who and what were involved
   c. Operating personnel and other witnesses
   d. Account of the accident (what happened?):
      i. Sequence of events
      ii. Extent of damage
      iii. Accident type
      iv. Agency or source (of energy or hazardous material)
3. How was the accident investigation approached?
4. How was the evidence collected?

TABLE 3.6
Accident Investigation Report Form

ACCIDENT NUMBER: ______________________
COMPANY ______________ ADDRESS ______________
DEPARTMENT OR LOCATION OF ACCIDENT IF DIFFERENT THAN ABOVE:
____

WHO WAS INJURED, ILL, OR DIED?
NAME OF INJURED ______________ SOCIAL SECURITY NUMBER ______________
SEX ______ AGE ______ DATE OF BIRTH ______ DATE OF ACCIDENT ______
HOME ADDRESS ______________ TELEPHONE # ______________
EMPLOYEE'S USUAL OCCUPATION ______________________
OCCUPATION AT TIME OF ACCIDENT ______________________
LENGTH OF EMPLOYMENT ______ TIME IN OCCUP. AT TIME OF ACCIDENT
EMPLOYMENT CATEGORY ______________________ (i.e., full time)
TYPE OF INJURY ______________ PART OF BODY ______________
SEVERITY OF THE INJURY ______________________
NAMES OF OTHERS INJURED IN SAME ACCIDENT
____
____
____

WHEN DID THE ACCIDENT OCCUR?
DATE OF ACCIDENT ______________ TIME OF ACCIDENT ______________
SHIFT ______________________
WHERE DID THE ACCIDENT OCCUR?
LOCATION OF ACCIDENT ______________________
ON EMPLOYER'S PREMISE?
ACTIVITY AT TIME OF ACCIDENT ______________________
SUPERVISOR IN CHARGE ______________________
WHAT HAPPENED OR CAUSED THE ACCIDENT?
DESCRIBE HOW THE ACCIDENT OCCURRED
____
____
____

DESCRIBE THE ACCIDENT SEQUENCE
____
____
____
____

CAUSAL FACTOR
____
____

HOW CAN IT BE PREVENTED FROM OCCURRING AGAIN?
CORRECTIVE ACTIONS
____
____
____
____

PREPARED BY ______________________
TITLE ______________________
DEPARTMENT ______________________
SIGNATURE ______________________ DATE ______________

5. Analysis of evidence and methodology used:
   a. Direct causes (energy sources, hazardous materials)
   b. Indirect causes (unsafe acts and conditions)
   c. Basic causes (management policies, personal or environmental factors)
6. Discussion of findings, outcomes, and conclusions
7. Delineation of recommendations and interventions
8. Follow-up to solutions
9. Appendices:
   a. Accident investigation forms
   b. Documentation
   c. Analyses results
   d. Interviewee's statements
10. Signature of all those responsible for the investigation

FOLLOW-UP

Once the accident investigation process is complete and the final report done, a review process should be in place, with management involvement, to determine the actions to take regarding the findings and recommendations from the final report. A process should be installed, or formulated, to ensure the implementation of interventions and preventive activities.

Once the implementation process is in place and active, its overall effectiveness must be determined. You will want to know how the accident investigation process and outcome impact your safety and health program, and this allows you to assess the need for change in your safety and health effort.

SUMMARY

Accident investigations are an integral part of any good safety and health initiative or program. Investigations should be performed for all accidents, injuries, illnesses, property damage, and near-miss incidents. It should be a formalized part of a company's safety and health commitment.

REFERENCES

Reese, C. D. and J. V. Eidson. *Handbook of OSHA construction safety & health.* Boca Raton, FL: CRC Press/Lewis Publishers, 1999.

United States Department of Labor, Occupational Safety and Health Administration, Office of Training and Education. OSHA Voluntary Compliance Outreach Program: Instructors Reference Manual. Des Plaines, IL: U.S. Department of Labor, 1993.

United States Department of Labor, National Mine Health and Safety Academy. Accident Prevention Techniques: Accident Investigation. Beckley, WV: U.S. Department of Labor, 1984.

United State Department of Labor, Mines Safety and Health Administration. Accident Investigation (Safety Manual No. 10). Beckley, WV: U.S. Department of Labor, Revised 1990.

4 Hazard Recognition and Avoidance

HAZARD IDENTIFICATION

Potential safety hazards come from a large number of sources. Each poses it own unique danger and also varies greatly in the degree of risk that it poses as well as the type of energy that it can release when not prevented or controlled. Table 4.1 provides a list of a wide range of equipment, tools, sources, etc. that can cause safety hazards to exist.

EMPHASIS ON HAZARDS

The emphasis in this chapter is on safety hazards. Hazards are defined as a source of danger that could result in a chance event such as an accident. A danger itself is a potential exposure or a liability to injury, pain, or loss. Not all hazards and dangers are the same. Exposure to hazards may be dangerous but this depends on the amount of risk that accompanies it. The risk of water contained by a dam is different from being caught in a small boat in rapidly flowing water. Risk is the possibility of loss or injury or the degree of the possibility of such loss. Accidents do not occur if a hazard does not exist that presents a danger to those working around it. If the potential exposure is high, there is a greater risk that an undesired event will occur. An accident is an unplanned or undesirable event whose outcome is normally a trauma. Trauma is the injury to living tissue caused by some outside or extrinsic agent. Trauma is caused by an agent, force, or mechanism impinging on the human body.

The emphasis area regarding safety hazards will be to identify the hazard and its danger, and to suggest ways to remove, intervene, or mitigate its risk for the purpose of preventing accidents that result from the errant uncontrolled release of energy that normally results in trauma to those who have the exposure to that hazard.

ACCIDENT CAUSES

Experts who study accidents often do a "breakdown" or analysis of the causes. They analyze them at three different levels:

1. Direct causes (unplanned release of energy and/or hazardous material)
2. Indirect causes (unsafe acts and unsafe conditions)
3. Basic causes (management safety policies and decisions, and personal factors)

TABLE 4.1
Potential Sources of Safety Hazard

Acids	Housekeeping/waste
Abrasives	Ladders
Biohazards	Lasers
Bloodborne pathogens	Lifting
Blasting	Lighting
Caustics	Loads
Chains	Machines
Chemicals	Materials
Compressed gas cylinders	Mists
Conveyors	Noise
Cranes	Platforms
Confined spaces	Personal protective equipment
Derricks	Power sources
Electrical equipment	Power tools
Elevators and manlifts	Pressure vessels
Emergencies	Radiation
Environmental factors	Rigging
Excavations	Respirators
Explosives	Scaffolds
Falls	Slings
Fibers	Solvents
Fires	Stairways
Flammables	Storage facilities
Forklifts	Stored materials
Fumes	Transportation equipment
Generators	Transportation vehicles
Gases	Trucks
Hand tools	Unsafe conditions
Hazardous chemical processes	Unsafe act
Hazardous waste	Ventilation
Heavy equipment	Walkways and roadways
High voltage	Walls and floor openings
Hoists	Warning devices
Hoses	Welding and cutting
Hot items	Wire ropes
Hot processes	Working surfaces

Direct Causes

Most accidents are caused by the unplanned or unwanted release of large amounts of energy, or of hazardous materials. In a breakdown of accident causes, the direct cause is the energy or hazardous material released at the time of the accident. Accident investigators are interested in finding out what the direct cause of an accident is because this information can be used to help prevent other accidents, or to reduce the injuries associated with them.

Energy is classified in one of two ways: it is either potential or kinetic energy. *Potential energy* is defined as stored energy, such as a rock on the top of a hill. There are usually two components to potential energy: the weight of the object and its height. The rock resting at the bottom of the hill has little potential energy as compared to the one at the top of the hill.

The other classification is kinetic energy, which is best described as energy motion. *Kinetic energy* depends on the mass of the object. Mass is the amount of matter making up an object, such as an elephant has more matter than a mouse, and therefore more mass. The weight of an object is a factor of the mass of an object and the pull of gravity on that object. Kinetic energy is a function of an object's mass and its speed of movement (or velocity). A bullet thrown at you has the same mass as one shot at you, but the difference is in the velocity and there is not any disagreement as to which has the most kinetic energy or potential to cause the greatest damage or harm.

Energy comes in many forms, and each has its own unique potential for release of energy and the ensuing form of danger and potential for an accident and the trauma that it can cause. The forms of energy are pressure, biological, chemical, electrical, thermal, light, mechanical, and nuclear. Examples of each form of energy are found in Table 4.2.

If the direct cause is known, then equipment, materials, and facilities can be redesigned to make them safer; personal protection can be provided to reduce injuries; and workers can be trained to be aware of hazardous situations so that they can protect themselves against them. The remainder of this book is directed toward managing, preventing, and controlling hazards that occur within the nation's industries.

HAZARD ANALYSIS

Hazard analysis is a technique used to examine the workplace for hazards with the potential to cause accidents. Hazard identification, as envisioned in this section, is a worker-oriented process. The workers are trained in hazard identification and asked to recognize and report hazards for evaluation and assessment. Management is not as close to the actual work being performed as are those performing the work. Even supervisors can use extra pairs of eyes looking for areas of concern.

Workers may already have hazard concerns and have often devised ways to mitigate the hazards, thus preventing injuries and accidents. This type of information is invaluable when removing and reducing workplace hazards.

This approach to hazard identification does not require that someone with special training conduct it. It can usually be accomplished using a short fill-in-the-blank questionnaire. This hazard identification technique works well where management is open and genuinely concerned about the safety and health of its workforce. The most time-consuming portion of this process is analyzing the assessment and response regarding potential hazards identified. Empowering workers to identify hazards, make recommendations on abatement of the hazards, and then suggest how management can respond to these potential hazards is essential. Only three responses are required:

1. Identify the hazard(s).
2. Explain how the hazard(s) could be abated.
3. Suggest what the company could do. Use a form similar to the one found in Table 4.3.

TABLE 4.2
Forms of Energy and Examples of Their Sources

Pressure energy	Light energy
Pressurized vessel	Intense light
Caisson work	Lasers
Explosives	Infrared sources
Noise	Microwaves
Compressed gases	Sun
Steam source	Ultraviolet light
Liquefied gases	Welding
Air under pressure	RF fields
Diving	Radio frequency
Confined spaces	
Biological energy	**Nuclear energy**
Allergens	Alpha particles
Biotoxins	Beta particles
Pathogens	High-energy nuclear particles
Poisonous plants	Neutrons
	Gamma rays
	X-rays
Chemical energy	**Thermal (heat) energy**
Corrosive materials:	Chemical reactions
Flammable/combustible materials	Combustible materials
Toxic chemicals	Cryogenic materials
Compressed gases	Fire
Carcinogens	Flames
Confined spaces	Flammable materials
Oxidizing materials	Friction
Reactive materials:	Hot processes
Poisonous chemicals and gases	Hot surfaces
Explosives	Molten metals
Acids and Bases	Steam
Oxygen deficiency atmosphere	Solar
Fuels	Weather phenomena
Dusts or powders	Welding
Electrical energy	
Capacitors:	
Transformers	
Energized circuits:	
Power lines	
Batteries	
Exposed conductors	
Static electricity	
Lightning	

TABLE 4.3
Hazard Identification Form

HAZARD IDENTIFICATION FORM

Worker's Name (Optional): ____________________ Date: __________
Jobsite: ____________________ Job Titles: __________

1. Describe the hazard that exists.

2. What are your recommendations for reducing or removing the hazard?

3. What suggestions do you have for management for handling the hazard?

4. Manager's or supervisor's response to hazard concern identified.

***NOTE:* Use a Separate Form for Each Hazard Identified.**

The information obtained from the hazard identification process provides the foundation for making decisions upon which jobs should be altered in order for the worker to perform the work safer and expeditiously. Also, this process allows workers to become more involved in their own destiny. For some time, involvement has been recognized as a key motivator of people. This is also a positive mechanism for fostering labor/management cooperation.

It is important to remember that a worker may perceive something as a hazard, when in fact it may not be a true one; the risk may not match the ranking that the worker places on it. Also, even if hazards do exist, you need to prioritize them. Companies need to prioritize them according to the ones that can be handled quickly, the ones that may take time, or those that will cost money above their budget. If the correction will cause a large capital expense and the risk is real but does not exhibit an extreme danger to life and health, companies might need to wait until next year's budget cycle. An example of this would be when workers complain of a smell and dust created by a chemical process. If the dust is not above accepted exposure limits and the smell is not overwhelming, then the company may elect to install a new ventilation system—but not until the next year because of budgetary constraints. The use of PPE until the hazard can be removed would be required.

Hazard identification is a process controlled by management. Management must assess the outcome of the hazard identification process and determine if immediate action is necessary or if, in fact, there is an actual hazard involved. When the company does not view a reported hazard as an actual hazard, it is critical to the ongoing process to inform the workers that the company does not view it as a true hazard and explain why. This will ensure the continued cooperation of workers in hazard identification.

The expected benefits are a decrease in the incidence of injuries, a decrease in lost workdays and absenteeism, a decrease in workers' compensation costs, increased productivity, and better cooperation and communication. The baseline for determining the benefit of hazard identification can be formulated from existing company data on occupational injuries and illnesses, workers' compensation, attendance, profit, and production.

WORKSITE HAZARD ANALYSIS

Worksite analysis is the process of identifying hazards related to a project, process, or activities at the worksite. Identify hazards before determining how to protect employees. In performing worksite analyses, consider not only hazards that currently exist, but also hazards that could occur because of changes in operations or procedures or because of other factors, such as concurrent work activities. First, perform hazard analyses of all activities and projects prior to the start of work, determine the hazards involved with each phase of the project, and perform regular safety and health site inspections. Second, require supervisors and employees to inspect their work areas prior to the start of each shift or new activity, investigate accidents and near-misses, and analyze trends in accident and injury data.

When performing a worksite analysis, all hazards should be identified. This means conducting comprehensive baseline worksite surveys for safety and health and periodic comprehensive updated surveys. Companies must analyze planned and new facilities, processes, materials, and equipment, as well as perform routine job hazard analyses. This also means that regular site safety and health inspections should be conducted so that new or previously missed hazards and failures in hazard controls are identified.

A hazard reporting/response program should be developed to utilize employees' insight and experience in safety and health protection. Employees' concerns should be addressed, and a reliable system should be provided whereby employees, without fear of reprisal, can notify management personnel about conditions that appear hazardous. These notifications should receive timely and appropriate responses, and employees should be encouraged to use this system.

Another way to maintain a worksite hazard analysis is to investigate accidents and near-miss incidents so that their causes, and means for their prevention, are identified. By analyzing injury and illness trends over a period of time, patterns with common causes can be identified and prevented.

Each company may require different types of hazard analyses, depending on the company's role, the size and complexity of the worksite, and the nature of associated hazards. The company may choose to use a project hazard analysis, a phase hazard analysis, or job safety assessment.

A project hazard analysis (preliminary hazard analysis) should be performed for each project prior to the start of work and should provide a basis for the project-specific safety and health plan. The project hazard analysis should identify the following:

1. The anticipated phases of the project
2. The types of hazards likely associated with each anticipated phase
3. The control measures necessary to protect site workers from the identified hazards
4. Those phases and specific operations of the project for which activities or related protective measures must be designed, supervised, approved, or inspected by a registered professional engineer or competent person

5. Phases and specific operations of the project that will require further analyses are those that have a complexity of potential hazards or unusual activities involved; those where there may be uncertainty concerning the worksite's condition at the time of project; or those where there is concern for methods that will be used to complete a phase or operation

A phase hazard analysis may be performed for those phases of the project for which the project hazard analysis has identified the need for further analysis, and for those phases of the project for which methods or site conditions have changed since the project hazard analysis was completed. The phase hazard analysis is performed prior to the start of work on that phase of the project and is expanded, based on the results of the project hazard analysis, to provide a more thorough evaluation of related work activities and site conditions. As appropriate, the phase hazard analysis should include

1. Identification of the specific work operations or procedures
2. An evaluation of the hazards associated with the specific chemicals, equipment, materials, and procedures used or present during the performance of that phase of work
3. An evaluation of how safety and health has impacted any changes in the schedule; and how work procedures or site conditions that have occurred since performance of the project hazard analysis need to be completed
4. Identification of specific control measures necessary to protect workers from the identified hazards
5. Identification of specific operations for which protective measures or procedures must be designed, supervised, approved, or inspected by a registered professional engineer or competent person

A job safety assessment or analysis should be performed at the start of any task or operation. The designated competent or authorized person should evaluate the task or operation to identify potential hazards and determine the necessary controls. This assessment should focus on actual worksite conditions or procedures that differ from or were not anticipated in the related project or phase hazard analysis. In addition, the authorized person should ensure that each employee involved in the task or operation is aware of the hazards related to the task or operation and of the measures or procedures that workers and visitors must use to protect themselves. Note: The job safety assessment is not intended to be a formal, documented analysis, but instead is more of a quick check of actual site conditions and a review of planned procedures and precautions. A more detailed explanation of job safety analysis is provided in Chapter 12.

TRAINING ON HAZARD IDENTIFICATION

Supervisors and workers must be trained to both identify hazards in order to prevent accidents, and to identify existing and potential hazards that may prevail in the workplace. When looking at specific jobs, identify the hazards by breaking down the job into a step-by-step sequence and identifying potential hazards associated with each step. Consider the following:

1. Is there a danger of striking against, being struck by, or otherwise making injurious contact with an object? (Example: Can tools or materials be dropped from overhead, striking workers below?)
2. Can the employee be caught in, on, or between objects? (Example: an unguarded V-belt, gears, or reciprocating machinery.)
3. Can the employee slip, trip, or fall on the same level, or to another level? (Example: slipping in an oil-changing area of a garage, tripping on material left on stairways, or falling from a scaffold.)
4. Can the employee strain him- or herself by pushing, pulling, or lifting? (Example: pushing a load into place or pulling a load on a hand truck.)
5. Does the environment have hazardous toxic gas, vapors, mist, fumes, dust, heat, or ionizing or nonionizing radiation? (Example: Arc welding on galvanized sheet metal produces toxic fumes and nonionizing radiation.)

During training, practice identifying all hazards or potential hazards. With identification of the hazards, take steps to prevent the accidents or incidents from occurring. If the hazards are known, it is easier to develop interventions that mitigate the risk potential. These interventions may be in the form of safe operating procedures. Training workers to identify potential or real hazards as quickly as possible definitely will reduce the number of accidents/incidents that could occur. Once hazards are identified, the company must have a system of reporting these hazards that is accompanied by real-time response by supervision and management.

WORKSITE HAZARD IDENTIFICATION

Employers must identify the workplace hazards before they can determine how to protect their employees. In performing worksite hazard identification, employers must consider not only hazards that currently exist in the workplace, but also those hazards that could occur because of changes in operations or procedures, or because of other factors such as concurrent work activities. For this element, employers should

1. Perform hazard identification of all worksites prior to the start of work.
2. Perform regular safety and health inspections.
3. Require supervisors and employees to inspect their workspace prior to the start of each work shift or new activity.
4. Investigate accidents and near-misses.
5. Analyze trends in accident and injury data.

So that all hazards are identified, conduct comprehensive baseline worksite surveys for safety and health, and perform periodic comprehensive updated surveys. Analyze planned and new facilities, processes, materials, and equipment. Perform routine job hazard analyses.

Provide for regular site safety and health inspections to identify new or previously missed hazards and failures in hazard controls. So that employee insight and experience in safety and health protection can be utilized and employee concerns can be addressed, provide a reliable system wherein employees, without fear of reprisal, can notify management personnel about conditions that appear hazardous and receive timely and appropriate responses; and encourage employees to use the system. Provide for investigation of accidents and near-miss incidents to identify both their causes and means for their prevention. Analyze injury and illness trends over time so that patterns with common causes can be identified and prevented.

RANKING HAZARDS

Once a number of hazards have been identified, a mechanism for ranking the hazards needs to exist. The employer can devise his own or use someone else's approach. An approach is provided in this section that can be tailored and revised to meet the company's needs.

The main purposes of ranking hazards are to determine which hazards should be addressed first and provide a method to prioritize this. The first step is to determine the probability of the hazarding occurring. This can be combined with its severity or the outcomes of its occurrence. To have some numeric values to compare, start with Probability (P) multiplied by Outcomes/Severity (OS); the product becomes the factor of potential exposure and the outcome of exposure. These estimates can come from the hazard analysis and past history from accident records.

For occupational exposure purposes, the values in Table 4.4 can be used for Probability; for Outcome/Severity, Table 4.5 can be used to determine an OS value.

TABLE 4.4
Probability of Exposure Values

Frequency (Exposure)	P Value
Frequently (daily)	5
Likely (weekly)	4
Occasional (monthly)	3
Seldom (yearly)	2
Unlikely (every 5 years)	1

TABLE 4.5
Severity and Outcome of Exposure Values

Outcome	Severity	OS Value
Catastrophic	Deaths and multiple employees in hospital	5
Critical	Severe disabling injury or illness	4
Moderate	Medical treatment	3
Minor	First aid required	2
Negligible	No injury or illness	1

TABLE 4.6
Resultant Effect of Impact (REI) Values

REI	Worker	Property	Productivity	Economic
Very High – 5	Death	Total loss	Down longer than a week	≥$10,000
High – 4	Critical or severe	Major repairs	Down less than a week	<$10,000
Medium – 3	Medical treatment	Minor repairs to operate	Down less than 2 days	<$1,000
Low – 2	First aid	Cosmetic repairs	Down for a shift	A few hundred $
Very Low – 1	No injury or illness	No damage	No downtime	No monetary loss

Second, determine the impacts on the workplace and those workers present. In the work environment, the impact on a worker is the most important. From an occupational perspective the impact upon machinery and equipment is of lesser importance while the least important is the impact upon productivity and economics (monetary lose). Place a numeric weighting IP value for each impact; the four factors in order of most important to least important should be weighted as follows:

- Worker (W) – 4
- Property (Prop) – 3
- Productivity (Prod) – 2
- Economic (Eco) – 1

The above scale constitutes the weighting factor (IWF) for each impact (i.e., the multiplier for the resultant effect of each impact). The resultant effect of the impact value (REI) is determined using Table 4.6.

An REI value for each impact should be determined and multiplied by the IP value for that impact. The product of each of the four is added together to determine IMPACT (IMP). The IMP is then added to the product of P and OS. Thus, the Risk Assessment Value (RAV) allows for ranking the hazard. The higher the score, the higher the risk. The highest risk assessment value would be 70 (Highest Risk) to a low of 11 (No Risk). Ranking risk is not an exact science but it is an approximation of the potential risk. A summary formula for the RAV is

$$RAV = P \times OS + IWF_W \times REI_W + IWF_{Prop} \times REI_{Prop} + IWF_{Prod} \times REI_{Prod} + IWF_{Eco} \times REI_{Eco}$$

HAZARD AND COST AVOIDANCE

At times, even using the preceding risk assessment and cost justification models (refer to Chapter 1) does not find acceptance when trying to convince others of the need to fix or invest heavily in safety and health in the workplace. In the past, the use of the cost avoidance has usually resulted in support of my proposal.

For example, consider a worker who received a back injury from lifting 150 lb, a task that he performs two times a week. For a cost of $10,000, the injury would never occur again. Would the cost to fix the situation be worth the prevention of an injury? First of all, what employer would allow an employee to lift such a heavy load? Simply allowing this practice is a risk in itself. With that said, let's consider the options.

If a risk/cost formula is applied, it would determine that the risk assessment factor is not very high, principally because the lift is only performed twice a week and not regularly. Also, using a justification factor, it should definitely be fixed or removed. The analysis sends a mixed message to management. So the case to have the problem fixed may not be very strong in this situation. A cost avoidance approach may be a better avenue. First, let us assume that a worker gets injured. What are the costs incurred?

- Lost-time: Normal back strains require 7 to 15 days of bed rest for the injured. Let's say an employee makes $10.00 per hour × 40-hour work week = **$400.00** per week.
- Replacement worker for the injured = **$400.00** per week.
- Environmental, Health and Safety (EH&S) department doing the incident investigation: Average wage $15.00 per hour × 8 hours of investigation, assessment, and interviewing = **$120.00.** (Note that hourly wages have been grossly understated.)
- Employees/supervisor involved in the investigation and interviewing process: Worker/witness at $10.00 per hour × 1 hour = **$10.00** and supervisor at average pay of $13.00 per hour × 2 hours = **$26.00.**
- Finance personnel to do workers' compensation forms and insurance forms. It takes about 3 to 4 hours of work to collect the data. Assume the finance employee makes a salary of $10.00 per hour × 4 hours = **$40.00.**
- Take into account loss of production or equipment damage. Process shut down for 2 hours, resulting in two pieces of equipment not manufactured, each worth $2,000 × 2 = **$4,000.00.**
- Also, do not forget the injured worker who now has to start therapy. Normally a 2-week process for strains. This will cost close to, or more than, **$5,000.00,** depending on medical services needed.
- The company will need to retrain employees on proper lifting and back safety and again, looking at wages of employees' training time, lost production for training, and the EH&S person's time to do the training, this cost could be well over **$5,000.00.**
- And this does not include the rise in premiums for your workers' compensation assessed by the company's insurance carrier or Workers' Compensation Commissioner.

This is well over the cost of $10,000.00 to prevent the accident from occurring again and ensuring the safety of the employees who can return home and enjoy life.

The company might have other costs that have been missed, but suffice it to say that this is a way of opening eyes to the cost if the company does not take action. Another take on this is that the employee can file a complaint with OSHA and the company now has more help than they ever wanted from a government agency. This may greatly increase the cost from violations or citations. There are a lot of considerations to look at when looking at the safety of the employees or the environment. Using varied approaches to identify and analyze hazard-caused incidents or the possibility of incidents can certainly improve the company's safety and health approach.

HAZARD CONTROL

All identified hazardous conditions should be eliminated or controlled immediately. Where this is not possible, interim control measures should be implemented immediately to protect workers, warning signs must be posted at the location of the hazard, all affected employees must be informed of the location of the hazard and of the required interim controls, and permanent control measures must be implemented as soon as possible.

If a supervisor or foreman is not sure how to correct an identified hazard or is not sure if a specific condition presents a hazard, that supervisor or foreman should seek technical assistance from the designated competent person, safety and health officer, or technical authority.

TECHNIQUES OF HAZARD CONTROL

JOB SAFETY ASSESSMENT

Prior to the start of any task or operation, the designated competent or company-authorized person should evaluate the task or operation to identify potential hazards and to determine the necessary controls. This assessment should focus on actual worksite conditions or procedures that differ from, were not anticipated, or were not related to other hazard analyses. In addition, the competent person should ensure that each employee involved in the task or operation is aware of the hazards related to the task, or operation, and of the measures or procedures to use for protection. Note that the job safety assessment is not intended to be a formal, documented analysis, but instead is more of a quick check of actual site conditions and a review of planned procedures and precautions.

CONTROLS

Controls come in all forms, from engineering devices and administrative policy to personal protective equipment (PPE). The best controls can be placed on equipment prior to involving people and, thus, either preclude or guard the workforce against hazards. Administrative controls rely on individuals to follow policies, guidelines, and procedures in order to control hazards and exposure to hazards. But, as we all know, this certainly provides no guarantees that the protective policies and procedures will be adhered to unless effective supervision and enforcement exist. Again, this relies on the company having a strong commitment to occupational safety and health. The use of PPE will not control hazards unless individuals who are exposed to the hazards are wearing the appropriate PPE. The use of PPE is usually considered the control of last resort because it has always been difficult for companies to be sure that exposed individuals are indeed wearing the required PPE.

Accident Reporting

All incidents and accidents resulting in injury or causing illness to employees, and all events (or near-miss accidents), should be reported in order to

1. Establish a written record of factors that caused injuries or illnesses or that caused incidents such as near misses that could have possibly have resulted in an injury or illnesses and did not result in bodily harm, property damage, or vehicle damage can be used to establish cause and effect relationships for prevention purposes
2. Maintain a capability to promptly investigate incidents and events in order to initiate and support corrective or preventive action
3. Provide statistical information for use in analyzing all phases of incidents and events
4. Provide the means for complying with the reporting requirements for occupational injuries and illnesses

Your incident reporting system requirements should apply to all incidents involving company employees, on-site vendors, contractor employees, and visitors that result in (or might have resulted in) personal injury, illness, or property or vehicle damage.

Injuries and illnesses that require reporting include those injuries and illnesses occurring on the job that result in any of the following: lost work time, restrictions in performing job duties, requirement for first aid or outside medical attention, permanent physical bodily damages, or death. Examples of reportable injuries and illnesses include, but are not limited to, heat exhaustion from working in hot environments, strained back muscles from moving equipment, acid burns on fingers, etc.

Other incidents that require reporting include those incidents occurring on the job that result in any of the following: injury or illness, damage to a vehicle, fire or explosion, property damage valued at more than $100, or chemical releases requiring evacuation of, at least, the immediate area where a chemical spill has occurred.

Examples of nonreportable injuries and illnesses include small paper cuts, common colds, and small bruises not resulting in work restrictions or requiring first aid or medical attention.

Events (near-misses) and other incidents (near-misses) that, strictly by chance, do not result in actual or observable injury, illness, death, or property damage are required to be reported. The information obtained from such reporting can be extremely useful in identifying and mitigating problems before they result in actual personal or property damage. Examples of near-miss incidents that require reporting include the falling of a compressed gas cylinder; overexposures to chemical, biological, or physical agents (not resulting in an immediately observable manifestation of illness or injury); and slipping and falling on a wet surface without injury.

Incident Reporting Procedures

The following procedures should be followed by all employees in order to effectively report occupational injuries and illnesses and other incidents or events. Serious injury or illness posing a life-threatening situation should be reported immediately to the local emergency response medical services (i.e., call 9-1-1).

Injuries and illnesses should be reported by the injured employee to his or her supervisor in person or by phone as soon after any life-threatening situation has been addressed. If the injured employee is unable to report immediately, then the incident should be reported as soon as possible.

Upon notification of an occupational injury or illness, the supervisor should complete a formal incident/accident report and, if possible, send it with the injured employee to the medical professional involved. The incident/accident report form must be completed and forwarded to the company medical department even if the employee receives medical treatment at the hospital and/or from a private physician.

Incidents not involving injury or illness but resulting in property damage must also be reported within 24 hours of the incident. In case of a fire or explosion that cannot be controlled by one person, a vehicular accident resulting in injury of more than $500 worth of damage, or a chemical release requiring a building evacuation, the involved party must immediately report the incident to the emergency response services in the area (i.e., 9-1-1, police, fire, etc.)

All near-miss incidents also must be reported on the Incident/Accident Report Form within 24 hours of occurrence. In place of indicating the result of the incident (i.e., actual personal or property damage), the reporting person should indicate the avoided injury or damage.

Events, hazardous working conditions or situations, and incidents involving contractor personnel must be reported to the supervisor or safety professional immediately.

The Safety Department will complete and maintain occupational injury and illness records. The OSHA 301 "Injury and Illness Incident Report," or equivalent, must be completed within 7 days of the occurrence of an injury at the worksite and the OSHA 301 must be retained for 5 years. Also, the OSHA 300 "Log of Work-Related Injuries and Illnesses" must be completed within 7 days when a recordable injury or illness occurs, and maintained for 5 years. The OSHA 300A "Summary of Work-Related Injuries and Illnesses" must be posted yearly from February 1 to April 30.

Training

To ensure that all employees understand the incident reporting requirements and are aware of their own and others' responsibilities, annual training sessions must be held with all employees to review procedures and responsibilities. New employee orientation training should include information on incident reporting and procedures.

FIGURE 4.1 Hazard identification helps determine risk and danger.

Program Audits

The effectiveness of a program can only be accomplished if the program is implemented. Therefore, periodic reviews and audits shall be conducted to confirm that all employees are familiar with the incident reporting requirements.

Identification of hazards and controlling hazards within the workplace are the responsibilities of the employer and the company's management. Because employers are in control of the workplace, they have the right to set and enforce their own occupational safety and health requirements.

SUMMARY

It is important that all hazards are identified, and that an assessment is made of the potential risk from the hazard. This allows for the determination of the *real* danger that exists. If a high degree of risk and danger exists, then efforts must be undertaken to alleviate or mitigate the potential danger (see Figure 4.1).

REFERENCES

Petersen, D. *Techniques of safety management: A systems approach (3rd ed.).* Goshen, NY: Aloray Inc., 1989.

Reese, C. D. and J. V. Eidson. *Handbook of OSHA construction safety & health (second ed.).* Boca Raton, FL: CRC Press/ Taylor & Francis, 2006.

Reese, C. D. *Occupational health and safety management: A practical approach (second edition).* Boca Raton, FL: Lewis Publishers, 2009.

United States Department of Energy, Office of Nuclear Energy. Root Cause Analysis Guidance Document. Washington, D.C.: U.S. Department of Energy, February 1992.

United States Department of Labor, National Mine Health and Safety Academy. Accident Prevention Techniques. Beckley, WV: U.S. Department of Labor, 1984.

United States Department of Labor, Mine Safety and Health Administration. Accident Prevention, Safety Manual No. 4, Beckley, WV: U.S. Department of Labor, revised 1990.

United States Department of Labor, Occupational Safety and Health Administration, Office of Training and Education. OSHA Voluntary Compliance Outreach Program: Instructors Reference Manual. Des Plaines, IL: U.S. Department of Labor, 1993.

United States Department of Labor, Mine Health and Safety Administration. Hazard Recognition and Avoidance: Training Manual (MSHA 0105). Beckley, WV: U.S. Department of Labor, revised May 1996.

5 Accountability and Responsibility

INTRODUCTION

Everyone is responsible for accident prevention. This statement also means that no one in particular is accountable or responsible. Three entities must be accountable and responsible for accident prevention. Each one has distinct responsibilities for which they should be held accountable. These three entities are

1. The person who by background or experience has been assigned responsibility and therefore assigned accountability to ensure that the company's safety and health program is adhered to.
2. The supervisor, who models the company's safety personality, and is the liaison between management and the worker relevant to implementation of safety, and must be held both responsible and accountable for safety and health in his or her workplace.
3. Employees, who are responsible to abide by the company's rules and policies and are accountable for their own behavior—safe or unsafe.

Each of the aforementioned entities must understand both their responsibilities and accountabilities regarding the safety and health policies and procedures of the company.

THE SAFETY AND HEALTH PROFESSIONAL

The individual who has been assigned the ultimate responsibility for occupational safety and health is the safety and health professional. This may be an individual with academic training in safety and health, or an individual who has both experience and an understanding of the specific hazards that exist in the company's workplace(s). This individual may be called the safety and health director, coordinator, or person. No matter the title, his or her responsibilities are varied and wide ranging. Some possible performance expectations may be as follows:

1. Establishing programs for detecting, correcting, or controlling hazardous conditions, toxic environments, and health hazards
2. Ensuring that proper safeguards and personal protective equipment are available, properly maintained, and properly used
3. Establishing safety procedures for employees, plant design, plant layout, vendors, outside contractors, and visitors

4. Establishing safety procedures for purchasing and installation of new equipment and for purchasing and safe storage of hazardous materials
5. Maintaining an accident recording system to measure the organization's safety performance
6. Staying abreast of, and advising management on, the current federal, state, and local laws, codes, and standards related to safety and health in the workplace
7. Carrying out the company's safety and health obligations as required by law and/or union contract
8. Conducting investigations of accidents, near-misses, and property damage, and preparing reports with recommended corrective action
9. Conducting safety and health training for all levels of management, newly hired. and current employees
10. Assisting in the formation of both management and union/management safety and health committee (department heads and superintendents) and attending monthly departmental safety and health committee meetings
11. Keeping informed on the latest developments in the field of safety and health, such as personal protective equipment, new safety standards, workers' compensation legislation, new literature pertaining to safety and health, as well as attending safety and health seminars and conventions
12. Maintaining liaison with national, state, and local occupational safety and health organizations and taking an active role in the activities of such groups
13. Accompanying the Occupational Safety and Health Administration (OSHA) compliance officers during plant inspections and insurance safety and health professionals on audits and plant surveys; the safety and health professional reviews reports that relate to these activities and, with management, initiates action for necessary corrections
14. Distributing the organization's statement of policy as outlined in its organizational manual

If some facets of the safety and health effort are not going well, this individual will usually be held accountable although the authority to rectify the existing problem may not exist. Usually the safety and health person is a staff position, which seldom allows him or her to interfere in any way with the line function of production. Without some authority to impact the line function when necessary, the safety and health professional has little clout as to worksite implementation of the company's safety and health effort. Accountability must go beyond the safety and health professional. Dan Petersen, a noted safety expert, espouses that what is desired is "safe production."

THE LINE SUPERVISOR

The individual with the most impact on workplace safety and health must be the line supervisor. Almost everything that is communicated to the workers comes from the supervisor. The first-line supervisor sets the tone for his or her workplace and is the role model for the company by conveying, implementing, supporting, and enforcing all the company's policies and procedures from production to occupational safety and health.

Just think of all that the first-line supervisor does:

1. Oversees production
2. Hires new employees
3. Reports on probationary employees
4. Trains new employees
5. Conducts job safety observations
6. Holds safety meetings
7. Coaches employees on the job
8. Controls quality and quantity
9. Stops a job in progress
10. Takes unsafe tools out of production
11. Investigates accidents
12. Inspects the work area
13. Corrects unsafe conditions and unsafe acts
14. Recommends promotions or demotions
15. Transfers employees in and out of the work area
16. Grants pay increases
17. Issues warnings and administers discipline
18. Suspends and discharges
19. Prepares work schedules
20. Delegates work to others
21. Prepares vacation schedules
22. Grants leaves of absence
23. Lays off others for lack of work
24. Processes grievances
25. Authorizes maintenance and repairs
26. Makes suggestions for improvement
27. Discusses problems with management
28. Reduces waste
29. Prepares budgets
30. Approves expenditures
31. Fosters employee morale
32. Motivates workers
33. Reduces turnover

Is it any wonder that the success or failure of the safety and health program depends on the first-line supervisor? Certainly everyone would acknowledge that the first-line supervisor is responsible for safety and health within his or her work area. But seldom is he or she evaluated on his or her safety and health performance in the same manner as his or her production performance. Until such time that each supervisor is held as accountable for safety and health as production, with equal consequences for poor safety and health performance as for poor production performance, safety and health will never be a priority with supervisors. The value that the supervisor places on workplace safety and health will always be far less than the value placed on production. Using a separate evaluation form for each supervisor's safety and health performance, which compares the safety and health performance records of one supervisor with another, may make a significant difference (see Table 5.1).

USING THE SUPERVISOR EVALUATION FORM

Numbers can be placed in each blank. Each number may stand for a different set of values, depending on the section being scored.

Value	Section 1	Section 2	Section 3
1	0 days	None	Excellent
2	1–2 days	Some	Good
3	3–5 days	Many	Fair
4	6–10 days to count	Too numerous	Poor
5	>10 days	Dangerous	Unacceptable

You can devise your own form, one that better meets the needs of your company. Also, based on the condition of your company regarding supervisor safety and health performance, you may want a more stringent scoring system or a less stringent scoring system to start your evaluation of supervisors. Until supervisors' safety and health performance has some value attached to it that impacts retention of their job or bonuses, one cannot expect them to take safety and health as a serious component of their job.

THE WORKER

Although workers do not have the control that management has over their workplace, they are still responsible for complying with the company's safety and health policies and procedures. Some of the commonly accepted worker responsibilities are to

1. Comply with OSHA regulations and standards.
2. Not remove, displace, or interfere with the use of any safeguards.
3. Comply with the employer's safety and health rules and policies.
4. Report any hazardous conditions to supervisor or employer.
5. Report any job-related injuries and illnesses to supervisor or employer.
6. Report near-miss incidents to supervisor or employer.
7. Cooperate with the OSHA inspector during inspections when requested to do so.
8. Report to work on time.
9. Wear suitable work clothes.
10. Observe good personal hygiene.
11. Never sleep, gamble, horseplay, fight, steal, bring fireworks or firearms on the job, or face grounds for immediate dismissal.

TABLE 5.1
Supervisor Safety Performance Evaluation Form

MONTHLY OR QUARTERLY REVIEW SUPERVISOR SAFETY EVALUATION FORM

Supervisor name: ______________________ Date: __________

Department: ______________________

Jobsite: ______________________

Supervisor's supervisor: ______________________

Section 1:

_____ Number of consecutive days without a reportable injury: __________

_____ Number of first aid cases: __________

_____ Number of reportable injuries: __________

_____ Number of injuries with restricted work days: __________

_____ Number of injuries with lost work days: __________

_____ Total number of days away from work: __________

Section 2:

Worksite Evaluation:

_____ Housekeeping

_____ Workers wearing PPE

_____ Unsafe conditions

_____ Unsafe acts

_____ Equipment in disrepair

_____ Faulty or defective tools

_____ Safety devices or grounds removed

Section 3:

Supervisor's Safety Performance:

_____ Sets a good example

_____ Complies with OSHA standards

_____ Promptly corrects hazards

_____ Holds toolbox talks or safety meetings

_____ Displays a safety attitude

_____ Enforces safety

_____ Provides feedback to workers on safety performance

_____ Reinforces safe work procedures

_____ Recognizes safe workers

Score for month or quarter __________

Gain (+) or loss (–) __________

Ranking of the supervisor __________

Recommendations:

12. Never use or be under the influence of alcohol, narcotics, or other drugs or intoxicants on the job.
13. Wear PPE as prescribed for each task.
14. Maintain order, as housekeeping is everyone's responsibility.
15. Observe "Danger," "Warning," "Caution," and "No Smoking" signs and notices.
16. Use and handle equipment, material, and safety devices with care.

17. Never leave discharged fire extinguishers in the work areas.
18. Never expose oneself to dangerous conditions or actions.
19. Never operate any equipment for which one has not been trained.
20. Participate in all safety and health training provided.
21. Attend all safety, health, and toolbox meetings (mandatory).

If a worker elects not to follow rules or to perform work in an unsafe manner, then there should be consequences that are strictly enforced by the company and its representatives. Failure to do so negates the authority of the safety and health polices and procedures.

Each company should have a discipline policy that is progressive and stringently enforced. A policy, at its simplest, is presented as follows:

1. Verbal warning—first offense
2. Written warning—second offense
3. Dismissal—third offense

Note: The company reserves the right to immediately dismiss a worker when he or she flagrantly endangers other workers, causes an incident, or causes imminent danger to themselves or others.

Workers must be held accountable for poor safety and health performance.

As can be seen in this chapter, each of these three entities—the company's designated safety person(s), supervisors, and employees—have responsibility for safety in some form or fashion and should be held strictly accountable for their safety performance.

REFERENCES

Reese, C. D. and J. V. Eidson. *Handbook of OSHA construction safety & health.* Boca Raton, FL: CRC Press/Lewis Publishers. 1999.

United States Department of Labor, Occupational Safety and Health Administration, Office of Training and Education. OSHA Voluntary Compliance Outreach Program: Instructors Reference Manual. Des Plaines, IL: U.S. Department of Labor, 1993.

United States Department of Labor, Mine Safety and Health Administration. Accident Prevention, Safety Manual No. 4. Beckley, WV: U.S. Department of Labor, revised 1990.

Petersen, D. *Techniques of safety management.* Boston, MA: Aloray Press, 1989.

6 Motivating Safety and Health

INTRODUCTION

The people side of safety and health is little different from the people side related to any other aspect of life. It is difficult to understand the values and attitudes that workers bring with them to the workplace. Actually, there is little that can be done regarding the values and attitudes that workers bring to the workplace. We can only see their outward behavior regarding safety. Workers' safety performances should be observable and measurable. We can and do work with the behaviors that individuals exhibit.

Once we are able to obtain safe worker performance by encouraging a change in behaviors, then, little by little, attitudes toward safety will begin to change. Eventually, safe work performance will become a value for workers.

PLANNING THE MOTIVATIONAL APPROACH

Just like any other endeavor in life, if we do not plan for it, we are seldom successful in accomplishing what we want. Employers trying to motivate individuals to behave in a manner that facilitates a safe and healthy work environment require planning and an organized approach. The very same approaches that have been used for decades to motivate individuals to perform can be applied to achieving safe work performance.

All employees within the workplace need to know what the safety and health performance expectations are. Everyone likes to have a clear explanation of what is expected of them. Employees have a basic need to be involved in decisions affecting themselves, and in solving safety and health problems within their workplace. Research has shown that employee involvement is critical to the success of any motivation approach.

Most individuals need to have consequences that will occur if they fail to perform in a safe and healthy manner. There is no reason to comply with safety and health requirements unless failure to do so has consequences. Certainly, workers do not want to become injured or ill from something within their workplace. But other factors affect the value of safety and health on the job. Workers may be fired if they do not get the job done at all costs, or their peer group may bring pressure to bear if they overlook safety requirements. In either case, workers must decide whether the consequence of not getting the job done or the consequence of peer group pressure is more negative than the chore of abiding by safe and healthy work practices. The severity of the consequences and the follow-through (enforcement) will determine the value placed upon job safety and health by all workers. Granted, this is a very negative view but it is also a very realistic view if you honestly consider it.

Although consequences are strong determinants of job safety and health performance, seldom can you achieve performance solely through a negative factor. All of us must have a path or direction that provides us with an ultimate outcome. The strongest positive motivators are goals. Suffice it to say that almost anything worth anything has been achieved by us because we set a goal to accomplish it. Most sports that we play have a purpose, a "goal," or an ultimate outcome. Thus, to have an effective safety and health initiative, the company must have specific safety and health goals. Once a goal is achieved, then other goals must be formulated and worked toward. Care must be taken to ensure that these goals are realistic and achievable. Safety and health goals can have the same positive outcomes that production goals have.

When goals are set, or safety and health performance monitored, it is imperative that workers receive feedback on the progress they are making toward those goals or the quality of their safety and health performance. It is always best if the feedback is of a positive nature. It is wiser to provide positive feedback on small gains than no feedback at all, and negative feedback is better than no feedback at all. The use of negatives should be the route of last resort and not standard practice whenever possible. All of us need to know where we are in relation to what is expected of us.

A kind of feedback that is instantaneous is the development of a self-monitoring system that allows workers to track their own progress toward goals and outcomes. Most of us feel more involved and in control when we are able to keep score ourselves. A feeling of control over one's destiny is always more comforting, and self-monitoring systems provide the mechanism to do this to some extent.

A self-monitoring system is a type of reinforcement process that lets workers know when they are performing in the desired fashion. But, apart from this, most of us desire to receive recognition of some sort when we are doing something the right way. Workers need to hear from someone else that they are performing work in a safe and healthy manner, and that it is recognized and appreciated. This is reinforcement. If someone does some task in a safe manner and you call attention to that safe performance, then it is more likely that the worker will repeat the task in the same safe manner. It is important for workers to have that personal touch that supervisors or others can give when complimenting, rewarding, and reinforcing a worker's safe performance. Safe performance is reinforced by something as a simple pat on the back. It does not have to be an award or reward.

In combination with the previously mentioned plan to motivate workers, a system of rewards or an incentive program has its place if all the other components of an accident prevention program are in place (i.e., written safety and health program, training, and accident investigation). If we reward individuals for high production with bonuses, then why not reward safety and health. Care must be taken that reward systems do not end up with workers not reporting hazards or injuries. Peer pressure can cause similar behavior if a group or team award exists. Because every work environment is unique and each work population is unique, what works in one workplace may not work in all of them. Let the workers tell you what is of value to them and how such a program might work best. Some individuals will suggest money but that is a one-time award and quickly forgotten. Others prefer a gift certificate for merchandise or a meal that can be used at their discretion. When workers receive a reward, like a gift certificate, and then use it, the gift certificate reinforces twice the reason they received it. If the certificate procures merchandise, then every time they use or see the merchandise, it reinforces why it was received and hopefully strengthens behavior toward continuing to work in a safe and healthy manner. Incentive programs should be changed when they lose their desired effectiveness; they certainly are not the end-all answer to accident prevention or motivation of workers.

The components of a motivational program must be planned, evaluated, implemented, reviewed, and revised in order to successfully motivate safe work performance. How you use these principles and design them to meet your needs will determine your success in eliciting safe performance from your workforce.

SUPERVISORS

The basic philosophy espoused in this book is that the success or failure of a safety and health program lies with the first-line supervisor, who is the key person in the process of safety communication, safety training, safety enforcement, creating and supporting a safety culture, and in the commitment to safety and health programs at the employee level. All safety and health information comes to employees through that person. Likewise, all the motivation is applied by the supervisor. If the supervisor is not trained in the principles of motivation and how to present such a factor with positive reinforcement, safe performance recognition, and providing feedback, then surely an effort to motivate safe and healthy work behaviors will fail. A supervisor cannot be expected to fully support a motivational plan without the full support and commitment of middle and top management.

Managers and supervisors achieve results through the efforts of other people. While planning, directing, and controlling receive most of their attention, motivating employees is a critical part of their jobs. Employee motivation is an ongoing process. It requires a continuing commitment, an objective view of the supervisor's management style, and an understanding of the effect of their behavior on their subordinates. By and large, employees want the same things that managers want. Among other things, they want respect, a sense of accomplishment, and personal development.

In today's multi-ethnic workplace, managers and employees bring a variety of personal experiences to the job. Failure to show sensitivity to the feelings of subordinates can result in misunderstanding, resentment, anxiety, or cause communication gaps, wasted time, unnecessary work, lost productivity, poor morale, and other negative effects. Generally, a good rule for guidance is the Golden Rule (i.e., do unto others as you would have them do unto you), which is as effective in the workplace as anywhere else.

To improve the chances for motivating workers, supervisors must keep employees informed about what is going on in the organization: People like to feel that they are trusted with information as it becomes available. Make expectations clear, be honest, and stick to your decisions.

Make time available to workers so you can meet, unhurried, and without interruptions; and actively listen to what people say. It is imperative to know the workers and by all means find out if they have career aspirations of which you are not aware. It is important for supervisors to listen to the concerns and complaints of workers.

Supervisors must provide as much flexibility in decision making as possible, especially in areas that directly affect workers. This can increase their personal commitment to the organization and their feeling of control over their jobs. Include workers in the goal-setting process to give them more of a stake in helping to accomplish the goals established. Let the workers know how their tasks relate to organizational goals, so that they can feel that their efforts are significant and that they are making a contribution.

Supervisors must provide feedback often. The workers want to hear from the supervisor, whether the comments are positive or negative; letting people know how they are doing is most effective immediately following their performance. For motivational purposes, it is not wise to use formal appraisal times for feedback.

Workers expect that their supervisor will help them reach their highest potential career goals by offering them advice and guidance. Recognize and reward workers even if it is only approving a request, getting a raise, receiving a reward, or obtaining a promotion when they deserve it. Whether the supervisors realize it or not, workers are working for them and they will determine the supervisor's own success.

Supervisors must show the same respect for workers that they desire for themselves. Manners and courtesy is a two-way street. Supervisor credibility is based to a large extent on the respect he or she shows and receives. Be as considerate of workers as you would be of those individuals closest to you. They are a family away from home for most supervisors. Do not reprimand employees in front of others; admonishing employees in the presence of subordinates, peers, or colleagues is almost certain to arouse ill will.

Supervisors are role models. If they do not want to follow the rules or want to take special privileges, then they should not be supervisors. Being a supervisor requires some self-sacrifice. What the supervisors give is what they get. If supervisors want workers motivated to work in a safe and healthy manner, then they must also be motivated to act in a safe and healthy manner before them.

SELF-MOTIVATED WORKERS

It is important to remember that little can replace motivation on a one-to-one basis. Obviously, this is time consuming and a supervisor cannot spend all of his or her time motivating a single worker. However, supervisors can develop motivational skills and find the best of opportunities to use them.

Every worker is different and each comes to work with a different attitude toward work and their own safety at work. It will take a concerted, if not impossible, effort to elicit safe behavior from every worker. But, if a comprehensive approach to accident prevention and motivation is used, the opportunity for success with the majority will be greatly enhanced.

A company can go to great extremes to motivate safety and health performance, but if attention to some fundamental characteristics of people is not made, the company will not be successful in developing a good motivational environment. Some of the fundamental principles that should be considered when working with people are

1. Individuals view themselves as very special. Thus, praise, respect, responsibility, delegated authority, promotion, recognition, bonuses, and raises add to their feelings of high self-esteem and should be considered when structuring a motivational situation.
2. Instead of criticism, use positive approaches and ask for corrected behavior. Individuals usually react in a positive manner when this approach is used.
3. Verbally attacking (disciplining) individuals tends to elicit a very defensive response (even a mouse, when cornered, will fight back in defense of itself). Therefore, it is better to give praise in public and, when necessary, criticize in private.
4. Individuals are unique and, given the proper environment, will astound you with their accomplishments and creativity (even those individuals who you consider to be noncreative).
5. Each individual has unique skills for learning and understanding as well as different levels of motivation. This is why each worker must be nurtured and motivated based upon his or her unique aptitude.

Each worker has unique needs. Some need recognition, some need money, and some need to be left alone. However, the needs of an individual are very strong motivators and must be considered when devising a motivation initiative.

Remember that the final outcome lies with the employees; they will decide whether or not to perform safely. But if the supervisor has done his or her part, workers will never be able to hold the former responsible for the decisions the latter have chosen to make.

There will be people who elect to work unsafely even though the environment may be very motivational to the majority. Pay closer attention to the motivational environment, and work at making it better. But, with those who still fail to perform safely, there should be consequences and discipline administered quickly and fairly. If there is no one enforcing the speed limit, then who will abide by it? Either enforce the rules or lose the effectiveness of your motivational effort.

CHANGING BEHAVIOR

At times, unsafe and unhealthy behaviors can occur through a myriad of causes. Some of the reasons that aberrant behavior occurs are

- Cognitive overload
- Cognitive under-load/boredom
- Habit intrusion
- Lapse of memory/recall
- Lack of spatial orientation
- Mindset/preconceived idea
- Tunnel vision or lack of “big picture”
- Unawareness
- Wrong assumptions made
- Reflective/instinctive action

- Thinking and actions not coordinated
- Insufficient degree of attention applied
- Shortcuts evoked to complete job
- Complacency/lack of perceived need for concern
- Confusion
- Misdiagnosis
- Fear of failure/consequences
- Tired/fatigued

As with worker safety and health where a high risk can result from noncompliance with requirements, each of the human behavior factors should be considered prior to contemplating how and what behaviors need to change.

Changing unsafe to safe behavior is an important part of a company's motivational initiative. This concept has often been viewed as manipulation, when in fact it is an attempt to elicit behavior that will be beneficial to the worker and is not done for retribution, degradation, or punishment of workers. The process is not one of evaluating an individual but a process of identifying a behavior that can be understood and measured and can lead to positive change. By careful observation and monitoring, the extinction of the aberrant behavior (unsafe or unhealthy work practices) can be obtained.

This can be accomplished by counting or charting the behavior and providing feedback on performance as well as reinforcing the desired behavior. Success in getting rid of unsafe and unhealthy behaviors is something that can be evaluated by conducting a survey/walk-around and actually counting the number of undesirable behaviors witnessed. Actually, a similar survey could be done prior to starting the behavior modification effort. An example might be to count the number of workers wearing safety eyewear before the motivational effort and then count, after you sense the behavior has changed, to determine the impact that the effort has had.

BEHAVIOR-BASED SAFETY

Behavior-based safety (BBS) is an approach to safety that focuses on workers' behavior as the cause of most work-related injuries and illnesses. Promoters of BBS programs maintain that 80 percent to 96 percent of workplace incidents are caused by workers' unsafe behavior.

In fact, many safety professionals and researchers have viewed these figures as resulting in fault finding rather than fact finding, as well as causing workers to underreport accident/incident events to prevent themselves from becoming the target of scrutiny, enforcement, discipline, or reprisal.

The major complaint with this concept is that it overlooks the importance of, or contribution to, preventing accidents that unsafe conditions attribute to the numbers of accidents/incidents occurring. When addressing accident/incident prevention, it has never been my personal opinion or contention that addressing only unsafe acts was the cure-all in the accident prevention puzzle.

BBS today cannot be viewed as the panacea or end-all solution for the prevention of accidents and incidents, but rather only as one tool in the arsenal of tools, and does not supplant a complete and organized overall approach in addressing the occupational safety and health issues of today. There must be all the components discussed in this book in place, such as training, safety and health program, accident/incident analysis, safety engineering, controls, interventions, etc. It is only then that BBS can become an integral part of the whole occupational safety and health initiative.

It has always been a goal to get all employees to be motivated to perform their tasks in a safe and healthy manner, but this goal is only achievable when all safeguards are in place, all feasible protections are provided, all hazards are eliminated or controlled, safety and health is managed effectively, etc., and the workplace has been a structure to protect the workforce as best as possible. Only then can the application of behavioral approaches be implemented to elicit changed behavioral pattern and attitudes toward the standard practice of all the workforce to self and groups taking responsibility and becoming involved with the prevention process by adhering to policies, safe procedures/practices, and rules with regard to working and performance of all aspects of their jobs in a safe manner without any thought of circumvention of standards of practice or best practices.

BBS is not a quick fix to safety and health issues. Every BBS approach must be designed to fit the needs and culture of the organization or business. It is based on the notion that safety and health are a shared responsibility and not just a personal matter. It is a way the employers provide the tools to optimize safety performance in the employees' unique work environment by developing methods to measure successes regarding safety performance in accomplishments, rather than using the traditional failure rates.

BBS is a process used to identify at-risk behaviors that are likely to cause injury to workers and is dependent upon the involvement of workers in this process so that they become a willing participant and buy in to the concepts and purpose of BBS. They will be asked to observe each other and their co-workers in order to determine if decreases in at-risk behavior have resulted in a reduction in these unsafe behaviors.

This is a very simplistic description of BBS. Although there are an infinite number of variations to BBS programs, they all share common characteristics.

There are four steps to the BBS process, as follows.

Identifying Critical Behaviors

In this first step of behavior safety, an analysis needs to be completed that makes safe behavior more probable and at-risk behavior less probable. This is to be done by a steering committee composed of operational personnel. This includes front-line individuals such as supervisors, experienced operators, and other interested workers. They will need to use resources at their disposal to identify the major contributors to incidents, accidents, injuries, near-misses, and property damage. This group is often called a steering committee and is the first step in gaining support for behavior-based programs. They will identify a cadre of at-risk behaviors that will act as the foundation for this approach. By consulting the workers doing the jobs or tasks, a more complete set of the types of behaviors that contribute to incidents can be of help in developing a data sheet that can be used by trained observers to assess the number of safe behaviors occurring.

Gathering Data

This step is completed by trained observers who are, in most cases, fellow workers. These observers are not intent upon finding fault or blaming workers for their safety behaviors, but to document the rate at which workers perform tasks in a safe manner or in an at-risk fashion. The observations usually take 10 to 15 minutes. The data sheets developed by the steering committee should be the guide and specify the expected behaviors. These observations are strictly conducted under the conditions that no names are used and no blame is placed. These observations are best when peer-to-peer observations are performed and feedback can be given immediately. Observation data is entered into a database for analysis and problem solving. This approach builds a sense of ownership in this type of safety program.

Two-Way Feedback

The observer can provide the person performing the task with immediate feedback on at-risk behaviors and provide reinforcement on safe work performance. If the observer asks the worker why he used at-risk behaviors to perform a task, the observer can learn if there are roadblocks that prevent the worker from performing the task in a safe manner. The observer may find that protective equipment is not available or is no longer usable, which then becomes a follow-up item for the steering committee or management.

This process is founded on the premise that for every accident there are hundreds or sometimes thousands of at-risk behaviors. When at-risk behaviors decrease, the likelihood of injuries also decreases. A successful approach does not look for blame, but provides two-way feedback that promotes the idea that the worker is indeed the solution.

Any positive reinforcement that can be provided will help to strengthen the safe performance and cause it to become the norm.

Continuous Improvement

Using the comment data will allow site personnel to target areas of improvement and demonstrate to workers that their input is critical and an important component of the program. If at-risk behaviors are occurring in certain areas, then this is an accident waiting to happen. If there are barriers to safe performance, then continuous improvement is deterred. The most common barriers to safe performance are

- *Hazard recognition:* If workers did not realize that they were performing an at-risk behavior, then they could never perform the task in a safe manner.
- *Business systems:* The at-risk behavior was the result of an organization system that was unreliable due to inefficiency. If this occurs, workers will avoid using the system and will find a way around it.
- *Disagreement on safe practices:* There can be legitimate disagreement as to what constitutes safe performance and this needs to be reviewed and addressed in some manner. This is best accomplished by working toward an agreed-upon consensus.

- *Culture:* The way that it was always done may be at odds with what is a safe practice. It is hard to teach "old dogs new tricks."
- *Inappropriate rewards:* Rewards for achieving production may be at odds with safety and reinforce that at-risk short-cuts are more of a benefit than safe performance.
- *Facilities and equipment:* Outdated facilities or processes, or rigged, missing, or damaged equipment, may cause workers to act in an unsafe manner.
- *Personal factors:* This is when the at-risk behavior results from personal characteristics of the worker that results in him or her deliberately taking risks or refusing to work safely as a result of factors such as fatigue, medication, stress, or illness.
- *Personal choice:* A worker with adequate skills, knowledge, and resources chooses to work at risk to save, time, effort, or something similar.

A successful approach is able to remove these barriers by observing and talking with employees and must not, in any manner, imply that the workers are the problem. It is important that the workers are viewed as the solution.

The ideas that consequences control behavior are the foundation (conceptually) to BBS. Thus, the majority of behaviors rely on applying previous experience of consequences (both negative and positive) as the reinforcing factor. A picture of an amputated finger visually portrays the consequence of at-risk behavior, and this reminder of a negative consequence may be enough to cause a worker to alter his or her behavior prior to a similar incident.

Although the previous barriers are addressed, there are no guarantees that behavior-based safety will work in your situation; however, the principles are applicable to any situation when designed and implemented to meet your needs.

In summary, BBS is based on the general principles that behavior causes the majority of accidents, but this does not excuse employers from providing a safely engineered workplace with all controls in place to prevent the occurrence of incidents. Second, accountability inspires behavior and accountability facilitates accomplishments. Third, feedback that fosters good communications is the key to continuous improvement, and excellence in safety must be established as the underlying culture desired in the organization or company.

These premises are driven by the following strategies:

- Obtaining objective evidence of at-risk behaviors
- Defining barriers to safe behavior
- Teaching ways to substitute safe for at-risk practices
- Holding employees accountable to improve their safety-related behaviors and help others do the same
- Demonstrating the effectiveness by measurement that garners continued management support

BBS is utilized to increase safety awareness and to decrease accidents and incidents by focusing on identification and elimination of unsafe behaviors. Workers are trained to conduct safety observations and give guidance on specific behaviors while collecting the information in a readily available format for providing immediate feedback. Observations are structured to have a minimal impact on the workload and the data is shared with the entire workforce.

For such an approach to be successful, it requires a good organizational safety culture and people participation and involvement. Because the real-time safety analysis is an integral part of BBS, using operational personnel involvement to identify hazards and risks is key to effective behavior-based safety; a behavior-based process allows an organization or company to create and maintain a positive safety culture that continually reinforces safe behaviors over unsafe behaviors and ultimately results in a reduction of risk.

The organization's or company's purpose must be to continuously improve, with the ultimate goal being a workplace that is free of injuries and illnesses; while attitudes are not addressed directly, it is the deep-seated intention over time to have employees accept safety as a value.

FACTORS AFFECTING MOTIVATION

Peer pressure is a very powerful motivator and can be either rewarding or punishing. Peer groups who are doing just enough to get by tend to draw or attract lesser-motivated individuals. The lesser-motivated individuals tend to identify with the peer group and are governed more by them than are the highly motivated individuals. Normally, highly motivated individuals do not succumb as easily to peer pressure.

Social and family pressures have a role in motivating individuals, but most people sense that peer groups are the prime motivators in the workplace. Thus, to motivate an individual who is under the influence of peers, one must spend time trying to change the peer group behavior. This is the only way to achieve the motivation of an individual within that group.

Although motivating workers is not a straightforward task, one finding over the years is that supervisors must be willing to listen to people and accept and implement their ideas. This is often one of the highest-powered motivators. When you are willing to do this, the workers' jobs become more interesting and challenging to them, and they are more inclined to perform in a positive manner.

When delegating responsibility, it is important to remember that once individuals know what their responsibility is, let them alone to complete the task. Because they have played such an important part in the decision making, their best motivator will be the desire to see the end product.

Do not arbitrarily assign people to a task; rather, ask them, if possible, what they want to do and what they think they can accomplish. It is well founded that there is more than one way to accomplish most jobs. When working on an assignment, workers should feel that they can take a chance on doing their job a different way, a way that they feel will be more effective, as long as it does not endanger anyone's safety.

It is more important that the person who will be doing the task feels that he or she has had an important part in making the decisions concerning that task. Many times it has been found that suggestions given by others are just as viable, if not more so, than those given by an owner or supervisor. At times there may be mistakes made. Nevertheless, it is still more important that those involved in a task be part of the decision-making process (mistakes can normally be corrected without much difficulty!).

So, expect some mistakes. If people are willing to do something, they will be more exposed to the risk of making a mistake. Remember that only those who are doing nothing will never make a mistake. These are tried and proven techniques to motivate people. If your approach is not well thought out and planned, it will have no long-lasting effect and could merely be thought of as a band-aid remedy. You will probably have to use many approaches but you may not have thought about them as motivational factors. Seldom do we organize our motivational approach in such a manner that these techniques are viewed as a part of a complete motivational package.

One major factor in motivating people is training. The basic roots of training include the structuring of behavior, developing the ability to perform, and instilling a sense of purpose. Training is essential and should be provided in order to achieve goals and objectives. Goals and objectives result in such things as learning to drive a piece of machinery, being able to work safely, teaching a sport, learning to communicate with others, etc.

Most training is behavioral or learning objective driven. This could be viewed as the major goal in the motivational process. To achieve goals, you need to let workers know clearly what they are expected to do and precisely how you want them to accomplish their tasks. There should be an expected level of attainment, whether it be mastery or just to gain a certain amount of knowledge. To reinforce the expected skills, you should allow workers time to practice what they have learned.

Perfect results should not be expected the first time; you should be pleased with small, progressive steps. These steps should be evaluated on a regular basis, and constant feedback should be provided concerning their progress. Try dwelling on the positives instead of the failures, and guide the workers until they are able to perform a skill or pass an examination. In the training process, the true reinforcer and reward for an individual or group is being recognized for their ability to perform a skill, pass an examination, or complete a course. Individuals who are trained will perform at a higher level than those who are learning as they go. Training improves efficiency, safety, and the performance of workers.

Rewards are often short-lived motivators. They are usually not the results of performing an activity, but appear some time after the task is accomplished. Most of us are rethinking our concepts of rewards. It seems that accountability for performance and greater participation in making decisions are more critical than our previous reward structure. Rewards are often not viewed as rewards by those who receive them if they do not have value to that person. Types of incentives or rewards that could be given include

- Gift certificates for dinners or merchandise
- A "title" for those employees who deserve (titles cost a company nothing)
- Employee-of-the-Month plaques
- Special parking spaces
- Personal days off (can be used at worker's discretion with approval)
- Compensatory hours
- Money; it is good for one time and that time only

- Bonds, gold, or silver; they are usually kept and thus a constant reminder
- Items with the company logo or worker's name
- Special commemorative items or pins
- Tangible items (products)
- One of a kind or limited edition items (belt buckles, knives, etc.)

Reinforcing the desired behavior is motivational. Any merit reward that increases the likelihood of a person repeating the desired behavior is valuable. Be sure you use reinforcement (i.e., reward or commendation); it is always something perceived as positive, not negative.

The threat of something happening to us is neither a positive nor negative reinforcer. It is only after something has actually occurred that, then and only then, it becomes positive or negative (such as an occupational injury or illnesses). Therefore, how do we motivate someone to respond to a desired behavior when there is no reinforcing experience to help motivate them? To accomplish this, we must state the reason and importance of the request and compliment the person when they give the proper response.

To address a unique issue or a specific problem, the use of special emphasis programs is an appropriate motivational tool. These programs define a problem or issue, use symbolism (drawings or slogans) to elicit or trigger an appropriate response, indicate a true interest in the special area of emphasis, and use a structured approach to address the special issue. It is used to cause focused attention on a specific area. An example would be a special emphasis program on back injury prevention.

Some organizations try to use contests as motivational tools but, generally, they are found unsatisfactory. In many instances, too many negative response factors come into play. Contests are a type of competition and not everyone likes to compete. Many times there is only one winner in the contest and therefore many receive no type of reward even though they have worked very hard and to the best of their ability. This can affect an individual's status and, at times, even their morale. If you do decide that a contest is good for your particular situation, design the contest so that it will involve all individuals, foster status and pride, involve group competition, and involve management. Be sure that the same individual does not win all the time; this can be very discouraging to the other participants. If varied skill levels exist, use handicaps. Make sure winning and losing are distributed and give prizes to first-, second-, and third-place winners.

Gimmicks and gadgets used as motivational tools are novel or unconventional ideas, gifts, or devices that call attention to a desired response and maintain motivation in an unusual manner. Some examples are

1. Utilizing other gadgets such as knives, belt buckles, caps, pens, key chains, patches, rings, tee shirts, or trophies
2. Special chits for lunch or free coffee for wearing safety glasses
3. Having a weight lifter demonstrate proper lifting

Many of the motivation tools presented here are short-term motivators and should only be used to supplement an existing comprehensive program. They should not be the entire program.

SAFETY CULTURE

Defining Safety Culture

Safety culture is a concept defined at group level or higher; it refers to the shared values among all the group or company, corporation, or organization members. Safety culture is concerned with formal safety issues in an organization and is closely related, but not restricted, to the management and supervisory system. Safety culture emphasizes the contribution from everyone at every level of the organization. The safety culture of the business entity has an impact on all members of the workforce's behavior at work. Safety culture is usually reflected in the relationship between the reward system and safety performance. A positive safety culture is indicative of an organization's willingness to develop, change, and learn from errors, incidents, and accidents. Safety culture is ingrained, enduring, stable, and very resistant to change.

In summary, safety culture is the enduring values and priority placed on workers and management by everyone within the organization. It refers to the extent to which individuals and groups will commit to personal responsibility for safety; act to preserve, enhance, and communicate safety concerns; strive to actively learn, adapt, and modify (both individuals and the organization) behavior based on lessons learned from mistakes; and be rewarded in a manner consistent with these values.

Developing or Changing a Safety Culture

Without exception, development or change starts with management's commitment by actions, not words. Words have no substance, but actions and behaviors have substance. During this process, actions and development must be maintained by a continuous, constant, and consistent effort. It must be worked at, a new plan must exist, a new approach must be fostered, and support must be gained from all involved.

During development and change, individuals are not merely encouraged to work toward change; they must take action when it is needed. Inaction in the face of safety problems must be unacceptable. This will result in pressure from all facets both peers and leaders to achieve a change of culture. There is no room in the culture of safety for those who uselessly a point finger and say, "Safety is not my responsibility" and file a report and turn their back on helping to fix it.

These changes are contingent on leaders who are committed to change and open to sharing safety information. Open communication without fear of reprisal and knowledge that action will be taken when reported safety issues are tendered is a vital key to change.

Senior management must support and commit to culture change by demonstrating commitment to safety by providing resources to achieve results. The message must be consistent and sustained because it takes a long time to develop or change a safety culture.

Describing a Safety Culture

Safety culture is the value put upon workplace safety and health by an organization, corporation, or company. The message being communicated and espoused throughout the place of employment is the outward manifestation of the culture. If the safety culture is not a positive one, then all the other communication attempts are only "window dressing."

Safety culture is best described using an example of a worker wearing his or her protective safety eyewear without being told to do so. This behavior is an example of the worker's attitude regarding the wearing of personal protective equipment. This is an outward act that demonstrates the value held by the organization related to the true value the company places on safety and health.

Thus, the value the organization places on safety and health is demonstrated by the attitudes that not only the workers have, but also the attitudes that supervisors and managers hold related to safety and health issues. It is even more revealing of a positive safety and health culture when supervisors and managers in the same work environment are seen to be wearing protective eyewear. The behavior of management indicates that they do not view themselves above the values espoused by the company.

However, most of the time, safety culture is not so clear-cut. For example, one company that was surveyed espoused the values of accident prevention outwardly. Management seemed to support all that was right in protecting the workforce. Actually, the real value held by workers especially and management by acceptance was that getting injured was just part of doing the job. Getting injured was acceptable and expected by those performing the work. Although the company voiced a positive safety and health message, the reality was that the consensus in that workplace was that injury is just an acceptable part of doing business. This is an example of a communicated message that was truly not representative of the actual safety culture of the organization.

The workplace culture is not something that can be categorized into certain specific types of culture. Culture is what everyone in the workplace believes about the company, themselves, and safety. These opinions, assumptions, values, perceptions, stereotypes, rituals, leadership, and stories all mesh together to form the culture that translates into policies, procedures, and accident/incidents. There are many factors invisible from the surface, the taboos, assumptions, and norms that are never written down. These are the true forces behind outward safety behavior that reflect the real value of safety within the organization. No one espouses these deeply buried parts of the culture, but everyone knows what they are.

Many factors impact the culture of the workplace, including

- Intelligence and job knowledge
- Emotions and emotional illness
- Individual motivation toward safe work
- Physical characteristics and handicaps
- Family situations
- Peer groups
- The company itself
- Existing society and its values
- Consequence of the work itself

Because everyone at the workplace knows what the culture is, it is necessary to be very observant and listen prior to trying to communicate. It does not take long to discover the culture. It is seen in the safety, productivity, quality, and discipline of the work and the workforce. The culture often stymies efforts to communicate even when the outcomes are good. It is the sovereign duty of all to maintain the old culture as it is. Thus, the process of changing the culture is time consuming and quite complex.

It is bad enough that the workforce has its own culture, but this is compounded by the company's culture that impacts, and at times is diametrically opposed to, the culture of the workforce. The goal would be to meld the two cultures together in a viable and productive relationship.

Industries and companies that have labor unions seldom are able to perfectly integrate safety into their culture even though safety should be in everyone's best interest. Unions often hold safety hostage as a bargaining chip, tool, or weapon against management. Thus, the culture does not see safety as a level playing field on which everyone can participate equally. But in the same light, management often fights the battles related to safety and health regulations that were instituted to provide safety and health for its workforce and is intent on fighting compliance.

Thus, the culture that exists is an amalgamation of the culture of labor and the company. At times there is reasonable compatibility and the two parts form a culture that strives for effective safety while at other times they are so opposite that no one benefits from their inability to merge the cultures into a useful entity where effective and rational communications occur.

The only way the culture can be changed is when an urgent need to change is the motivator. This could be a number of occupational deaths, a catastrophe, or a series of occupational injuries or illnesses that forge partnerships in troubled times. Second, the resources must be available and the ability to change must be present. Third, a roadmap or plan must be developed and agreed upon by all parties involved in order to transition to the new culture. Most of us are reticent to change. Thus, gaining consensus for change is going to be difficult. It will be the leadership's responsibility to decide that there is a need for change, communicating it to the organization, getting consensus of all parties, and directing the implementation.

If this discussion of culture seems ominous and overwhelming, then you have some feeling of how difficult it is to communicate if the culture is not receptive to your message. This makes it imperative to understand the culture that you are trying to communicate within. As for safety, if the culture is antisafety, then your communication regarding safety will fall on deaf ears. So in order to communicate safety in your workplace, you must first understand the culture and then begin to change it. This is a communicate-or-lose situation, which is not the best of all worlds. Suffice it to say that in most cases you will be able to accomplish some degree of safety communications based in a large part on the perception of how you really feel about safety and health at your workplace.

Positive Safety Culture

An organization, corporation, or company with a positive safety culture is one that gives appropriate priority to safety and realizes that occupational safety and health must be managed like other areas of the business. Safety culture is more than just avoiding accidents or even reducing the number of accidents, although these are most likely the outcome or measurement of a positive or successful safety culture.

A positive safety culture is doing the right thing at the right time in response to normal or emergency operation. The quality and effectiveness of training will play an important role in determining the attitudes and performance toward safe production or performance. These attitudes toward safety are, in a large part, a mirror image of the culture set up by the company or corporation.

The key to achieving a positive safety culture is recognizing that all accidents are preventable by following correct procedures and established best practices, and maintaining constant awareness and thinking about safety.

Trying to Improve Safety on a Continuous Basis

It is unusual for different types of accidents to occur within an organization, and most that continue to occur after the unsafe condition has been removed are the result of unsafe acts or behaviors. The unsafe actions are usually the result of error, violations of best practices, or disregard for established rules and can be avoided. Most often, those making such errors are aware of their mistakes. They take short-cuts even though they have received training, but fall prey to a culture that allows workers to take calculated risks.

The goal is to have effectively trained workers who are motivated to self-regulate themselves and take personal responsibility to work safely using the safest practices where these have become an integrated part of the workers' value system and company culture.

Assessing Safety Culture

You might ask yourself the following questions regarding the safety culture at a company's place of business or operation:

- Is health and safety a top priority in the organization?
- Is safe work performance reinforced, recognized, and rewarded?
- Are business decisions made with safety and health being given a major consideration?
- Is there a requirement that safe and healthy attitudes and behaviors are expected of all employees, including management?
- Do employees feel comfortable to voice their concerns to management regarding any hazards or safety issues about the workplace?
- Does peer pressure act as a positive or a negative in support of a positive safety culture?
- Do production deadlines cause safe and healthy work practices to be overlooked?

If the safety culture does not coincide or integrate well with the business approach of the organization, corporation, or company, the indication is that a safety culture does not or only marginally exists.

Summary

A positive safety culture exhibits the following characteristics:

- The importance of committed leadership from top to bottom
- A clear set of explained expectations for line management
- The involvement of all employees
- Effective communication must exist
- A commonly understood and agreed set of safety and health goals
- A learning organization that is responsive to change
- A zeal for and attention to detail regarding safety and health
- A questioning attitude and problem-solving environment—not fault finding
- Trust permeates the organization
- Decision making reflects safety first

The failure to consider and address the safety culture can prove very costly.

VISUAL MOTIVATORS

Visuals, such as posters and bulletin boards, can be used as motivational tools. They are beneficial in that they serve as a constant reminder of the desired goal you are trying to reach. Bulletin board posters need to be changed often and kept updated. You can even get your employees who have graphic or artistic talents to develop posters that you can have reproduced and displayed. See Figure 6.1 for an example. This way, the employees get some recognition and you do not have the cost of commercially made posters. Videos are another excellent way of motivating individuals or groups. They can clearly reinforce a message using visual and auditory senses, and the message may actually show this behavior in a similar or the actual setting.

When giving a talk, the use of visuals normally increases the effectiveness of that presentation. By using personalized information, written literature, and statistics, it tends to more readily hold the participants' attention.

NONFINANCIAL INCENTIVES

Nonfinancial incentives can be such things as the use of praise, knowledge of results (output), competition, experience of progress, experience of achievement, or granting a request. Some of the most powerful motivators are achievement, recognition, a person's work or task, responsibility, and growth potential.

These motivators can become functional by giving workers more control, while at the same time holding them accountable. You can make them more accountable by making them responsible for a discrete outcome, or allowing them additional authority. You will also be more successful if you personally keep the people you are trying to motivate informed instead of through someone else. Challenge them with more difficult tasks and allow them to become specialized in a certain area.

Communications including interpersonal relations, employers or supervisors, promotion or recognition, work conditions, and status are some external factors

DON'T WEAR
LOOSE CLOTHES
AROUND
MOVING
MACHINERY

FIGURE 6.1 Safety and health poster. (*Source:* Courtesy of the National Mine Health and Safety Academy.)

that motivate people. The personal or internal motivators are those things that give more freedom of choice of activity, freedom from criticism, freedom to adjust the work environment, choice of peers, fewer status factors, less supervisor or employee conflict, and more opportunity to be oneself.

SUMMARY

It is evident that we spend a large portion of our lives either motivating ourselves or trying to motivate someone else. Hopefully you have gained some insight into ways that you can be more effective at motivating yourself and others. In summary, some of the key traits believed to be critical to understanding how to motivate people are

1. People are self-motivated.
2. What people do seems logical and rational to them.
3. People are influenced by what is expected of them.
4. To each individual, the most important person is oneself.

5. People support what they create or are involved in.
6. Conflict is natural (normal) and can be used positively.
7. People prefer to keep things the way they are rather than make a change.
8. People are underutilized.

With these thoughts in mind, people can be motivated by

1. Allowing them involvement and participation
2. Delegating to them responsibility with authority
3. Effectively communicating with them
4. Demonstrating concern and assisting them with training, counseling, and coaching
5. Being a good role model to them
6. Having high expectations of them
7. Providing rewards and promotions based on their achievements

Workers need to know what is expected of them, what happens if they do not perform, and what the cost, rewards, outcomes, or consequences will be for not meeting expectations. We are motivated by what we think the consequences of our actions will be. Those consequences should be immediate, certain, and nonrewarding if we expect them to be motivational.

Practically all our motivational attempts are geared toward peers or employees and this is accomplished through the employer, fellow workers, or supervisors. Supervisors are, among other things, a funnel that directs all materials and information to those who need to be motivated. The motivator also directs or carries out the vast majority of learning. Everything that motivates employees is applied by them and, obviously, their role is crucial.

People have many abilities and talents that they are unaware of or just do not use. As a motivator, the supervisor is responsible for bringing out those hidden abilities and talents and channeling them toward the goals, outcomes, behaviors, and objectives desired. If this is done ethically, responsibly, and honestly, as discussed in previous sections, it will give workers a new sense of enthusiasm and self-esteem.

Motivation takes a lot of nurturing and caring—for both the workers involved and the goals to be attained. Many organizations say that people are their most important asset but fail to exhibit that principle by the manner in which they treat their employees.

Motivation is not something that can be scheduled for a Thursday at 2:00 p.m. It is a process that requires continuing commitment and the ability to have an objective view of self and others. You must also have an understanding of your effect on others.

The essence of motivation is to find meaning in what you are doing. Motivation is the predisposition of doing something in order to satisfy a need. In real life, most people rarely have just one need; they have several needs at any one given time and are, consequently, moved to do something about them. Unfortunately, if they have too many needs facing them at any one time, they may become indecisive, highly aggressive, negative, or even irrational.

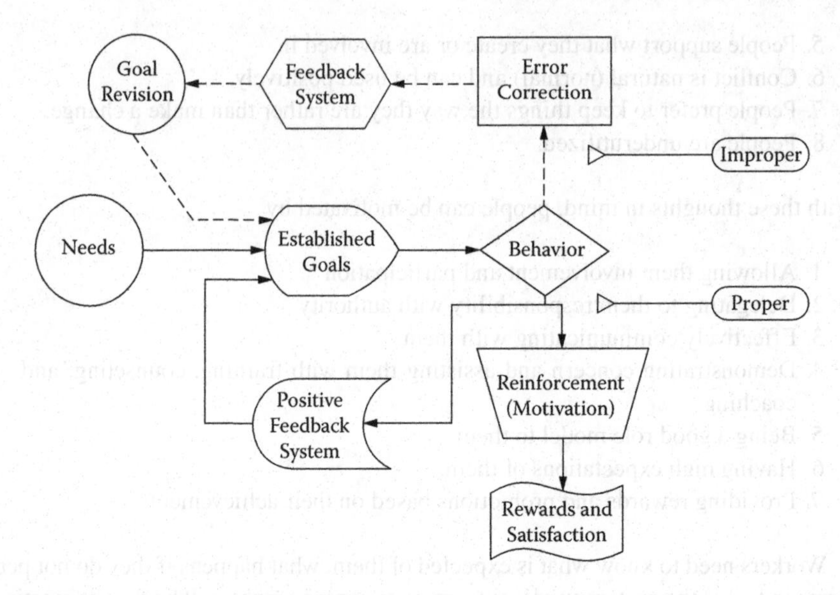

FIGURE 6.2 Flowchart for applying the principles of motivation.

Motivation is internal and can be stimulated by leadership and incentives. However, unless you know something about the needs, desires, and drives of the other person, your leadership and incentives may be completely ineffective. When people's tasks or jobs do not permit them to satisfy their own personal needs, they are less likely to work as hard at accomplishing the task you have chosen for them.

It seems safe to say that people do things well if they are excited about their assigned tasks. When their external environment ensures that their own needs, wants, and desires will be met, it further enhances their desire to do a good job. You, as a leader and motivator, are also responsible for helping others meet the demands of their world to the level of their capabilities. When each of these aspects is being fulfilled, you will have an excellent motivational situation.

The application of good motivational techniques and factors with your existing safety and health programming and accident prevention initiative will enhance the possibilities for success in decreasing accidents and injuries in your workplace (see Figure 6.2).

REFERENCES

Blake, R. R. and J. Srygley. Principles of behavior for sound management. *Training and Development Journal* (October, 1979): pp. 26–28.

Blanchard, K. How to get better feedback. *Success* (June, 1991): p. 6.

Brown, P. L. and R. J. Presbie. *Behavior modification in business, industry and government.* Paltz, NY: Behavior Improvement Associates, Inc., 1976.

Federal Aviation Administration, *System safety handbook, chapter 3.* Washington D.C.: Federal Aviation Administration, 2000.

Geller, E. S. *The psychology of safety handbook.* Boca Raton, FL: CRC Press/Lewis Publishers, 2001.

Herzberg, F. One more time: How do you motivate employees? *Harvard Business Review* (January-February, 1968): pp. 53–62.

Maslow, A. H. *Motivation and personality*. New York: Harper and Brothers, 1954.

Reese, C. D. and J. V. Eidson. *Handbook of OSHA construction safety & health*. Boca Raton, FL: CRC Press/Lewis Publishers, 1999.

Swartz, G. (Ed.). *Safety culture and effective safety management.* Itasca, IL: National Safety Council, 2000.

Weisinger, H. and N. Lobsewz. *Nobody's perfect: How to give criticism and get results*. New York: Stratford Press, 1981.

Wieneke, R. E., et al., Los Alamos National Laboratory, *Success in Behavior-Based Safety at Los Alamos National Laboratory's Plutonium Facility.* Los Alamos, NM: 2002.

7 Accident/Incident Analysis

INTRODUCTION

The material for this chapter comes from information derived from the Mine Safety and Health Administration's training manual on Hazard Recognition and Avoidance. This is a simple analysis tool that sets the stage for more detailed examination of accidents and incidents.

Many workplaces have high accident/incidence and severity rates because they are hazardous. Hazards are dangerous situations or conditions that can lead to accidents. The more hazards present, the greater the chance that there will be accidents. Unless safety procedures are followed, there will be a direct relationship between the number of hazards in the workplace and the number of accidents that will occur there.

As in most industries, people work together with machines in an environment that causes employees to face hazards that can lead to injury, disability, or even death. To prevent industrial accidents, the people, machines, and other factors that can cause accidents, including the energy associated with them, must be controlled. This can be done through education and training, good safety engineering, and enforcement.

Many accidents can be prevented. One study has shown that 88 percent were caused by human failure (unsafe acts), 10 percent were caused by mechanical failure (unsafe conditions), and only 2 percent were beyond human control (Heinrich, 1931).

If workers are aware of what the hazards are, and what can be done to get rid of them, many accidents can be prevented. For a situation to be called an accident, it must have certain characteristics. The definition of an accident is, as used here, any unplanned event that results in personal injury or property damage. The personal injury may be considered minor when it requires no treatment or only first aid.

The personal injury is considered serious if it results in a fatality, or in a permanent, partial, or temporary total disability (lost-time injuries). Property damage may also be minor or serious. For an event to be called an accident, it must have the following characteristics:

1. It must be unplanned.
2. Personal injury or property damage (or both) must result.

BREAKDOWN OF CAUSES

Experts who study accidents often do a "breakdown" or analysis of the causes. They analyze them at three different levels:

1. Direct causes (unplanned release of energy or hazardous material)
2. Indirect causes (unsafe acts and unsafe conditions)

3. Basic causes (management safety policies and decisions, and personal factors) (*See diagram at the beginning of this chapter.*)

Direct Cause

Most accidents are caused by the unplanned or unwanted release of large amounts of energy, or of hazardous materials. In a breakdown of accident causes, the direct cause is the energy or hazardous material released at the time of the accident. Accident investigators are interested in finding out what the direct cause of an accident is because this information can be used to help prevent other accidents, or to reduce the injuries associated with them.

Energy is classified in one of two ways. It is either potential or kinetic energy. *Potential energy* is defined as stored energy such as a rock on the top of a hill. There are usually two components to potential energy. They are the weight of the object and its height. The rock resting at the bottom of the hill has little potential energy as compared to the one at the top of the hill. Some examples of potential energy are depicted in Table 7.1.

The other classification is *kinetic energy,* which is best described as energy motion. Kinetic energy depends on the mass of the object. Mass is the amount of matter making up an object, such as an elephant has more matter than a mouse, and therefore more mass. The weight of an object is a factor of the mass of an object and the pull of gravity on it. Kinetic energy is a function of an object's mass and its speed of movement or velocity. A bullet thrown at you has the same mass as one shot at you, but the difference lies in the velocity—and there is no disagreement as to which has more kinetic energy or potential to cause the greatest damage or harm. Some examples of kinetic energy are given in Table 7.2.

Energy comes in many forms, and each has its own unique potential for release of energy and it ensuing form of danger and potential for an accident and the trauma that it can cause. The forms of energy are pressure, biological, chemical, electrical, thermal, light, mechanical, and nuclear. Examples of each form of energy are found in Table 7.3.

If the direct cause is known, then equipment, materials, and facilities can be redesigned to make them safer; personal protection can be provided to reduce injuries;

TABLE 7.1
Examples of Potential Energy

Compressed gases	Hand or power tool
Object at rest	Liquefied gas
Effort to move an object	Dust
Spring-loaded objects	Unfallen tree
Electrically charged component	Radiation source
Idling vehicle	Chemical source
Disengaged equipment	Biological organism
Flowable material	

TABLE 7.2
Examples of Kinetic Energy

Examples of Kinetic Energy
Operating tools or equipment
Moving conveyors
Flow of materials
Running machines
Falling objects
Running equipment
Lifting a heavy object
Moving dust
Moving vehicles or heavy equipment
Tree falling
Release of energy from radiation, chemical, or biological sources
Pinch area from moving objects
Running power tools
Energy transfer devices such as pulleys, belts, gears, shears, edgers

and workers can be trained to be aware of hazardous situations so that they can protect themselves against them.

Indirect Causes

Now that direct causes have been discussed, let us move one step down from there and discuss indirect causes, or "symptoms," that may be considered contributing factors. In most cases, the release of excessive amounts of energy or hazardous materials is caused by unsafe acts or unsafe conditions. Put another way: Unsafe acts and unsafe conditions trigger the release of large amounts of energy or hazardous materials, which directly cause the accident. As you may have noticed, this chapter refers to indirect causes as symptoms or contributing factors. That is because unsafe acts and unsafe conditions do not themselves cause accidents. These are just symptoms or indicators of such things as poor management policy, inadequate controls, lack of knowledge, insufficient knowledge of existing hazards, or other personal factors. Table 7.3 provides some examples of unsafe acts and conditions.

Basic Causes

While we often think of hazardous acts and conditions as the basic causes of accidents, actually they are symptoms of failure on another level. Unsafe acts and unsafe conditions can usually be traced to three basic causes: poor management policies and decisions, personal factors, and the physical facility design.

The first category, *management safety policies and decisions,* includes such things as management's intent (relative to safety); production and safety goals; staffing procedures; use of records; assignment of responsibility, authority, and accountability; employee selection; training, placement, direction, and supervision; communication

TABLE 7.3
Direct and Indirect Causes of Accidents

ACCIDENT CAUSE LEVELS	
Direct Causes of Accidents	**Indirect Causes of Accidents**
Energy Sources	**Unsafe Acts (May equal 95% of all accidents)**
Mechanical:	Failing to use personal protective equipment
Machinery	Failing to warn co-workers or to secure equipment
Tools	Engaging in horseplay
Moving objects	Lifting improperly
Strain (self)	Loading or placing equipment or supplies improperly
Compressed gas	Making safety devices inoperable
Explosives	Operating equipment at improper speeds
Electrical:	Operating equipment without authority
Uninsulated conductors	Servicing equipment in motion
High-voltage sources	Taking an improper working position
Chemical:	Using alcoholic beverages
Acids	Using drugs
Bases	Using defective equipment
Fuels	Using equipment improperly
Reactive materials	
Thermal:	**Unsafe Conditions (May equal 5% of all accidents)**
Flammable	Congestion of workplace
Nonflammable	Defective tools, equipment, or supplies
Radiation:	Excessive noise
Noise	Fire and explosion hazards
Lasers	Hazardous atmospheric conditions:
Microwaves	Gases
X-rays	Dusts
Radioactive materials	Fumes
Hazardous Materials	Vapors
Compressed or liquefied gas:	Inadequate supports or guards
Flames	Inadequate warning systems
Hot surfaces	Poor housekeeping
Corrosive material	Poor illumination
Flammable material:	Poor ventilation
Solid	Radiation exposure
Liquid	
Gas	
Oxidizing material	
Poison	
Radioactive material	
Etiological agent	
Dust	
Explosive	

procedures; inspection procedures; equipment, supplies, and facility design; purchasing; maintenance; standard and emergency job procedures; and housekeeping.

The second category, *personal factors,* includes motivation, ability, knowledge, training, safety awareness, assignments, performance, physical and mental state, reaction time, and personal care.

The third category, the *actual physical facility design*, includes the unsafe procedures being used, the unsafe projections, and geological and climatic conditions (see Table 7.4).

Cause Breakdown Review

Briefly, let us review the three levels of accident causes. As was discussed, the direct cause of an accident is an unplanned release of energy or hazardous material, or both. The indirect causes, which are also referred to as "symptoms," are known to trigger the release of energy or hazardous materials. We put indirect causes in two groups: unsafe acts and unsafe conditions. Some examples of unsafe acts are using defective equipment, making safety equipment inoperable, and horseplay. Some examples of unsafe conditions are defective equipment, poor lighting, excessive noise, and poor housekeeping.

We said that although unsafe acts and conditions appear to be the basic causes of accidents, they can actually be traced to such things as management safety policies and decisions, and personal factors. These things (management policies and decisions, personal and environmental factors) are the basic causes of accidents.

Some examples of management safety policies and decisions that may lead to accidents include failure to inform employees of preset safety goals; employee selection, training, placement, and supervision; and inspection procedures. Some of the personal factors that can cause accidents are motivation, ability, knowledge, and safety awareness.

When beginning to analyze the causes of accidents or incidents, use the Accident Report Form found in Table 7.5. It will act as a guide in identifying the various causes in each category and provides space for making recommendations, which will help prevent the recurrence of this type of causal factor or accident.

MISHAP PROBABILITY

Each of the preceding causes, when taken as a total package, provides guidance in determining the probability of a mishap occurring. Refer to Chapter 4 regarding ranking hazards.

SUMMARY

As has been discussed, accidents occur frequently in the work environment. This is because many hazards still exist in the workplace, providing conditions within a workplace that can lead to accidents. Many of these accidents, however, can be prevented.

First, we can prevent them by knowing what caused the accidents to happen in the first place. Accident investigators perform that task. The accident investigator determines the direct, indirect, and basic causes of an accident. But prevention does

TABLE 7.4
Basic Causes of Accidents

Management Safety Policies and Decisions

Health and safety policy is not:
- In writing
- Signed by top management
- Distributed to each employee
- Reviewed periodically

Health and safety procedures do not provide for:
- A written manual
- Safety meetings
- Adequate housekeeping
- Preventive maintenance
- Safety inspections
- Accident investigations
- Job safety analysis
- Medical surveillance
- Reports

Health and safety is not considered in the procurement of:
- Supplies
- Equipment
- Services

Inadequate personnel practices regarding:
- Employee selection
- Communications
- Training
- Assigned responsibility
- Assignment
- Job observation
- Accountability

Physical Factors

Behavior factors:
- Accident repeater
- Risk taking
- Lack of hazard awareness

Experience factors:
- Insufficient knowledge
- Accident record
- Inadequate skills
- Unsafe practices

Physical factors:
- Size
- Strength
- Stamina

Mental factors:
- Emotional instability
- Alcoholism
- Depression
- Drug use

Motivational factors:
- Needs
- Capabilities

Attitude factors:
- People
- Company
- Job

Environmental Factors

Unsafe facility designs:
- Mechanical layout
- Electrical systems
- Hydraulic systems
- Air conditioning
- Access ways
- Material handling
- Illumination
- Noise

Unsafe operating procedures
- Normal
- Emergency

Unsafe projections:
- Physical plant
- Equipment
- Supplies
- Procedures

Unsafe location factors:
- Geographical area
- Terrain
- Surroundings
- Access roads
- Weather

TABLE 7.5
Accident Report Form

DEPARTMENT:	EMPLOYEE NAME:	LOCATION OF ACCIDENT:
DATE OF ACCIDENT:	EMPLOYEE AGE:	ACCIDENT TYPE:
TIME OF ACCIDENT:	EMPLOYEE OCCUPATION:	ACCIDENT CLASSIFICATION:

DESCRIPTION OF ACCIDENT/INCIDENT: ____________________

Item Personal Injury Involved: ____________________
Itemize Property Damage Involved: ____________________
Item Tools/Equipment Involved: ____________________

CAUSES	DIRECT		INDIRECT		BASIC	
	Energy Source	Hazardous Materials	Unsafe Acts	Unsafe Conditions	Inadequate Policy and/or Decisions	Environmental and/or Personal Factors

RECOMMENDATIONS	DIRECT LEVEL		INDIRECT LEVEL		BASIC LEVEL	

Foreman's Signature: ____________ Department Head's Signature: ____________

Source: Courtesy of Mine Safety and Health Administration.

not stop there. The next step is putting that knowledge to work. If hazards can be recognized and identified, then the workforce can help prevent accidents.

Once we know what practices might cause a back injury, for instance, we can learn proper lifting procedures that can help eliminate this kind of accident.

REFERENCES

Reese, C. D. and J. V. Eidson. *Handbook of OSHA construction safety & health.* Boca Raton, FL: CRC Press/Lewis Publishers, 1999.

United States Department of Labor, Mine Safety and Health Administration. Hazard Recognition and Avoidance: Training Manual (MSHA 0105). Beckley, WV: U.S. Department of Labor, revised May 1996.

United States Department of Labor, Mine Safety and Health Administration. Accident Prevention (Safety Manual No. 4). Beckley, WV: U.S. Department of Labor, revised 1990.

United States Department of Labor, National Mine Health and Safety Academy. *Accident Prevention Techniques: Accident/Incident Analysis.* Beckley, WV: U.S. Department of Labor, 1984.

8 Root Cause Analysis

INTRODUCTION

A root cause analysis is not a search for the obvious but an in-depth look at the basic or underlying causes of occupational accidents or incidents. The following should be considered when performing analyses:

- Chart events in chronological order, developing an events and causal factors chart as initial facts become available.
- Stress aspects of the accident that may be causal factors.
- Establish accurate, complete, and substantive information that can be used to support the analysis and determine the causal factors of the accident.
- Stress aspects of the accident that may be the foundation for judgments of needs and future preventive measures.
- Resolve matters of speculation and disputed facts through investigative team discussions.
- Document methodologies used in analysis: Use several techniques to explore various components of an accident.
- Qualify facts and subsequent analysis that cannot be determined with relative certainty.
- Conduct preliminary analyses: Use results to guide additional collection of evidence.
- Analyze relationships of event causes.
- Clearly identify all causal factors.
- Examine management systems as potential causal factors.
- Consider the use of analytic software to assist in evidence analysis.

A root cause analysis is only the beginning and a fraction of the analysis process and should not be considered the sole approach to an analysis of an accident.

The basic reason for investigating and reporting the causes of occurrences is to enable the identification of corrective actions adequate to prevent recurrence and thereby protect the health and safety of the public; the workers; the equipment, machinery, and facility; and the environment. Every root cause investigation and reporting process should include five phases. While there may be some overlap between phases, every effort should be made to keep them separate and distinct. The phases of a root cause analysis are

- Phase I: Data collection
- Phase II: Assessment

- Phase III: Corrective actions
- Phase IV: Inform
- Phase V: Follow-up

The objective of investigating and reporting the cause of occurrences is to enable the identification of corrective actions adequate to prevent recurrence and thereby protect the health and safety of the public, the workers, and the environment. Programs can then be improved and managed more efficiently and safely.

The investigation process is used to gain an understanding of the occurrence, its causes, and what corrective actions are necessary to prevent recurrence. The line of reasoning in the investigation process is (1) outline what happened step by step; (2) begin with the occurrence and identify the problem (condition, situation, or action that was not wanted and not planned); (3) determine what program element was supposed to have prevented this occurrence (was it lacking or did it fail?); and (4) investigate the reasons why this situation was permitted to exist.

This line of reasoning will explain why the occurrence was not prevented and what corrective actions will be most effective. This reasoning should be kept in mind during the entire root cause process. Effective corrective-action programs include the following:

- Management emphasis on the identification and correction of problems that can affect human and equipment performance, including assigning qualified personnel to effectively evaluate equipment and human performance problems, implementing corrective actions, and following up to verify corrective actions are effective
- Developments of administrative procedures that describe the process, identify resources, and assign responsibility
- Development of a working environment that requires accountability for correction of impediments to error-free task performance and reliable equipment performance
- Development of a working environment that encourages voluntary reporting of deficiencies, errors, and omissions
- Training programs for individuals found at fault in root cause analysis
- Training of personnel and managers to recognize and report occurrences, including early identification of significant and generic problems
- Development of programs to ensure prompt investigation following an occurrence or identification of declining trends in performance to determine root causes and corrective actions
- Adoption of a classification and trending mechanism that identifies those factors which continue to cause problems with generic implications

A description of each of the five phases is provided to clarify the purpose of each phase of the root cause analysis.

PHASE I: DATA COLLECTION

It is important to begin the data collection phase of root cause analysis immediately following the occurrence identification to ensure that data are not lost. (Without compromising safety or recovery, data should be collected even during an occurrence.) The information that should be collected consists of conditions before, during, and after the occurrence; personnel involvement (including actions taken); environmental factors; and other information having relevance to the occurrence, condition, or problem. For serious cases, photographing the area of the occurrence from several views may be useful in analyzing information developed during the investigation. Every effort should be made to preserve physical evidence such as failed components, ruptured gaskets, burned leads, blown fuses, spilled fluids, and partially completed work orders and procedures. This should be done despite operational pressures to restore equipment to service. Occurrence participants and other knowledgeable individuals should be identified.

Once all the data associated with this occurrence have been collected, the data should be verified to ensure accuracy. The investigation may be enhanced if some physical evidence is retained. Establishing a quarantine area or the tagging and segregation of pieces and material should be performed for failed equipment or components.

The basic need is to determine the direct, contributing, and root causes so that effective corrective actions can be taken that will prevent recurrence. Some areas to consider when determining what information is needed include

- Activities related to the occurrence
- Initial or recurring problems
- Hardware (equipment) or software (programmatic-type issues) associated with the occurrence
- Recent administrative program or equipment changes
- Physical environment or circumstances

Some methods of gathering information include

- Conducting interviews and collecting statements—interviews must be fact-finding and not fault-finding. Preparing questions before the interview is essential to ensure that all necessary information is obtained.
- Interviews should be conducted, preferably in person, with those people who are most familiar with the problem. Individual statements could be obtained if time or the number of personnel involved makes interviewing impractical. Interviews can be documented using any format desired by the interviewer. Consider conducting a "walk-through" as part of this interview if time permits.

Although preparing for the interview is important, it should not delay prompt contact with participants and witnesses. The first interview may consist solely of hearing their narrative. A second, more-detailed interview can be arranged, if needed. The interviewer should always consider the interviewee's objectivity and frame of reference.

Undertake to interview other personnel who have performed the job in the past. Consider using a walk-through as part of the interview.

Reviewing records: Review relevant documents or portions of documents as necessary and reference their use in support of the root cause analysis. Record appropriate dates and times associated with the occurrence on the documents reviewed. Examples of documents include the following:

- Operating logs
- Correspondence
- Inspection and surveillance records
- Maintenance records
- Meeting minutes
- Computer process data
- Procedures and instructions
- Vendor manuals
- Drawings and specifications
- Functional retest specification and results
- Equipment history records
- Design basis information
- Related quality control evaluation reports
- Operational safety requirements
- Safety performance measurement system/occurrence reporting and processing system (SPMS/ORPS) reports
- Trend charts and graphs
- Facility parameter readings
- Sample analysis and results (chemistry, radiological, air, etc.)
- Work orders

By acquiring related information, there is additional information that an evaluator should consider when analyzing the causes, including the following:

1. Evaluating the need for laboratory tests, such as destructive/nondestructive failure analysis
2. Viewing the physical layout of the system, component, or work area; developing layout sketches of the area; and taking photographs to better understand the condition
3. Determining if operating experience information exists for similar events at other facilities
4. Reviewing equipment supplier and manufacturer records to determine if correspondence has been received addressing this problem

PHASE II: ASSESSMENT

The assessment phase includes analyzing the data to identify the causal factors, summarizing the findings, and categorizing the findings according to cause. Any root cause analysis method may be used that includes the following steps:

1. *Identify the problem.* Remember that actuation of a protective system constitutes the occurrence but is not the real problem; the unwanted, unplanned condition or action that resulted in actuation is the problem to be solved. An example of this would be when dust in the air actuates a false fire alarm. In this case, the occurrence is the actuation of an engineered safety feature. The smoke detector and alarm functioned as intended; the problem to be solved is the dust in the air, not the false fire alarm.

   Another example is when an operator follows a defective procedure and causes an occurrence. The real problem is the defective procedure; the operator has not committed an error. However, if the operator had been correctly trained to perform the task and, therefore, could reasonably have been expected to detect the defect in the procedure, then a personnel problem may also exist.
2. *Determine the significance of the problem.* Were the consequences severe? Could they be next time? How likely is recurrence? Is the occurrence symptomatic of poor attitude, a safety culture problem, or other widespread program deficiency? Base the level of effort of the subsequent steps of your assessment on the estimation of the level of significance.
3. *Identify the causes.* Identify the causes (conditions or actions) immediately preceding and surrounding the problem (i.e., the reason the problem occurred).
4. *Identify the reasons why the causes in the preceding step existed.* By working back to the root cause, the fundamental reason that, if corrected, will prevent recurrence of this and similar occurrences throughout the facility and other facilities under control of the organization. This root cause is the stopping point in the assessment of causal factors. It is the place where, with appropriate corrective action, the problem will be eliminated and will not recur.

Root Cause Analysis Methods

The most common root cause analysis methods are

1. *Events and Causal Factor Analysis* identifies the time sequence of a series of tasks or actions and the surrounding conditions leading to an occurrence as well as determines the causal factors. (See Chapter 9.)
2. *Change Analysis* is used when the problem is obscure. It is a systematic process that is generally used for a single occurrence and focuses on elements that were planned and unplanned changes in the system and determines their significance as causal factors in an accident. (See Chapter 10.)
3. *Barrier Analysis* is a systematic process that can be used to identify physical, administrative, and procedural barriers or controls that should have protected persons, property, and the environment from unwanted energy and prevented the occurrence. (See Chapter 11.)
4. *Management Oversight and Risk Tree (MORT) Analysis* is used to identify inadequacies in barriers and controls, specific barrier and support functions, and management functions. It identifies specific factors relating to an occurrence and identifies the management factors that permitted these

risk factors to exist. MORT (or Mini-MORT) is used to prevent oversight in the identification of causal factors. It lists on the left side of the tree specific factors relating to the occurrence; and on the right side of the tree, it lists the management deficiencies that permit specific risk factors to exist. Management factors support each of the specific barrier and control factors. Included is a set of questions to be asked for each of the barrier and control factors on the tree. As such, they are useful in preventing oversight and ensuring that all potential causal factors are considered. It is especially useful when there is a shortage of experts of whom to ask the right questions. However, because each management oversight factor may apply to specific barrier and control factors, the direct linkage or relationship is not shown but is left to the analyst. For this reason, Events and Causal Factor Analysis and MORT Analysis should be used together for serious occurrences: one to show the relationship, the other to prevent oversight. A number of condensed versions of MORT, called Mini-MORT, have been produced. For a major occurrence justifying a comprehensive investigation, a full MORT analysis could be performed while Mini-MORT would be used for most other occurrences.

5. *Human Performance Evaluation* identifies factors that influence task performance. The focus of this analysis method is on operability, work environment, and management factors. Man-machine interface studies are frequently done to improve performance. This takes precedence over disciplinary measures. Its focus is on operability and work environment, rather than training of operators to compensate for bad conditions. Human Performance Evaluations may be used to analyze most occurrences, since many conditions and situations leading to an occurrence have ultimately originated from some task performance problem results from management planning, scheduling, task assignment, maintenance, and/or inspections. Training in ergonomics and human factors is needed to perform adequate Human Performance Evaluations, especially in man-machine interface situations.

   Human Performance Evaluations may be used to analyze most occurrences, since many conditions and situations leading to an occurrence have ultimately originated from some task performance problem results from management planning, scheduling, task assignment, maintenance, and/or inspections. Training in ergonomics and human factors is needed to perform adequate Human Performance Evaluations, especially in man-machine interface situations.

6. *Kepner-Tregoe Problem Solving and Decision Making* provides a systematic framework for gathering, organizing, and evaluating information and applies to all phases of the occurrence investigation process. Its focus on each phase helps keep them separate and distinct. The root cause phase is similar to change analysis. Kepner-Tregoe is used when a comprehensive analysis is needed for all phases of the occurrence investigation process. Its strength lies in providing an efficient, systematic framework for gathering, organizing, and evaluating information and consists of four basic steps:

a. *Situation appraisal* to identify concerns, set priorities, and plan the next steps.
b. *Problem analysis* to precisely describe the problem, identify and evaluate the causes, and confirm the true cause (this step is similar to change analysis).
c. *Decision analysis* to clarify purpose, evaluate alternatives, assess the risks of each option, and make a final decision.
d. *Potential problem analysis* to identify safety degradation that might be introduced by the corrective action, identify the likely causes of those problems, take preventive action, and plan contingent action (this final step provides assurance that the safety of no other system is degraded by changes introduced by proposed corrective actions).

These four steps cover all phases of the occurrence investigation process. Thus, Kepner-Tregoe can be used for more than causal factor analysis. Separate worksheets (provided by Kepner-Tregoe) provide a specific focus on each of the four basic steps and consist of step-by-step procedures to aid in the analyses. This systematic approach prevents overlooking any aspect of concern. As formal Kepner-Tregoe training is needed for those using this method, a further description is not included in this book.

The use of different methods to conduct root cause analysis has been widely accepted. Certain methods are used for different circumstances (see Figure 8.1 and Table 8.1).

Root cause categories have been selected with the intention of addressing all problems that could arise in conducting industrial operations. Elements necessary to perform any task are equipment and material, procedures (instructions), and personnel. Design and training determine the quality and effectiveness of the equipment and personnel. These five elements must be managed; therefore, management is also a necessary element. External phenomena, beyond operational control, serves as a seventh cause element. Whenever there is an accident or incident, one of the seven program elements was inadequate to prevent the occurrence. (Note that a direct, contributing, or root cause can occur anywhere in the cause factor chain; that is, a root cause can be an operator error, while a management problem can be a direct cause, depending on the nature of the occurrence.) These seven root causal categories are further subdivided into a total of thirty-two subcategories as is illustrated and used in Causal Factor Analysis (see Table 8.2). The direct cause, contributing causes, and root cause are all derived from these subcategories. The major cause categories are

- Equipment/Material Problem
- Procedure Problem
- Personnel Error
- Design Problem
- Training Deficiency
- Management Problem
- External Phenomenon

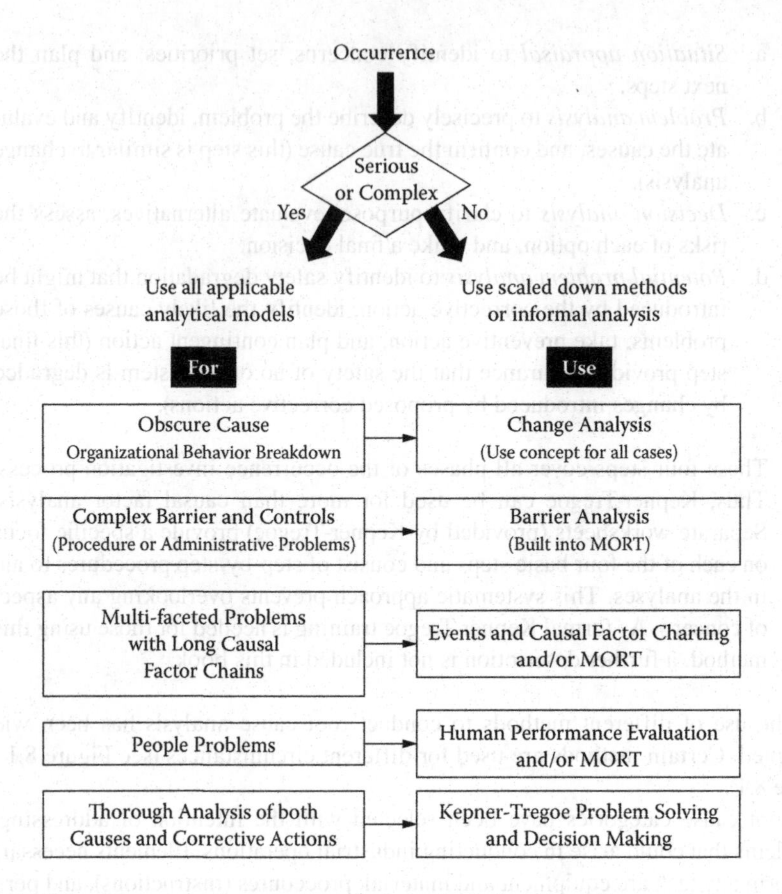

FIGURE 8.1 Summary of root cause analysis methods flowchart. (*Source:* Courtesy of the United States Department of Energy.)

PHASE III: CORRECTIVE ACTIONS

Implementing effective corrective actions for each cause reduces the probability that a problem will recur and improves reliability and safety. Root cause analysis enables improvement of reliability and safety by selecting and implementing effective corrective actions. First, identify the corrective action for each cause, and then apply the following criteria to the corrective actions to ensure they are viable. If the corrective actions are not viable, reevaluate the solutions.

1. Will the corrective action prevent recurrence?
2. Is the corrective action feasible?
3. Does the corrective action allow meeting the primary objectives or mission?
4. Does the corrective action introduce new risks? Are the assumed risks clearly stated? (The safety of other systems must not be degraded by the proposed corrective action.)
5. Were the immediate actions taken appropriate and effective?

TABLE 8.1
Summary of Root Cause Analysis Methods

Method	When to Use	Advantages	Disadvantages	Remarks
Events and Causal Factor Analysis	Use for multi-faceted problems with long or complex causal factor chain	Provides visual display of analysis process Identifies probable contributors to the condition	Time consuming and requires familiarity with process to be effective	Requires a broad perspective of the event to identify unrelated problems Helps to identify where deviations from acceptable methods occurred
Change Analysis	Use when cause is obscure; especially useful in evaluating equipment failures	Simple six-step process	Limited value because of the danger of accepting wrong, "obvious" answer	A singular problem technique that can be used in support of a larger investigation All root causes may not be identified
Barrier Analysis	Use to identify barrier and equipment failures and procedural or administrative problems	Provides systematic approach	Requires familiarity with process to be effective	This process is based on the MORT Hazard/ Target Concept
MORT/ Mini-MORT	Use when there is a shortage of experts to ask the right questions and whenever the problem is a recurring one; helpful in solving programmatic problems	Can be used with limited prior training Provides a list of questions for specific control and management factors	May only identify area of cause, not specific causes	If this process fails to identify problem areas, seek additional help or use cause-and-effect analysis
Human Performance Evaluations (HPEs)	Use whenever people have been identified as being involved in the problem cause	Thorough analysis	None if process is closely followed	Requires HPE training
Kepner-Tregoe	Use for major concerns where all aspects need thorough analysis.	Highly structured approach that focuses on all aspects of the occurrence and problem resolution	More comprehensive than may be needed	Requires Kepner-Tregoe training

Source: Courtesy of the United States Department of Energy.

TABLE 8.2
Cause Code Categories

Cause Codes

1. Equipment/Material Problem:
   a. Defective or failed part
   b. Defective or failed material
   c. Defective weld, braze, or soldered joint
   d. Error by manufacturer in shipping or marking
   e. Electrical or instrument noise
   f. Contamination
2. Procedure Problem:
   a. Defective or inadequate procedure
   b. Lack of procedure
3. Personnel Error:
   a. Inadequate work environment
   b. Inattention to detail
   c. Violation of requirement or procedure
   d. Verbal communication problem
   e. Other human error
4. Design Problem:
   a. Inadequate man-machine interface
   b. Inadequate or defective design
   c. Error in equipment or material selection
   d. Drawing, specification, or data errors
5. Training Deficiency:
   a. No training provided
   b. Insufficient practice or hands-on experience
   c. Inadequate content
   d. Insufficient refresher training
   e. Inadequate presentation or materials
6. Management Problem:
   a. Inadequate administrative control
   b. Work organization/planning deficiency
   c. Inadequate supervision
   d. Improper resource allocation
   e. Policy not adequately defined, disseminated, or enforced
   f. Other management problem
7. External Phenomenon:
   a. Weather or ambient condition
   b. Power failure or transient
   c. External fire or explosion
   d. Theft, tampering, sabotage, or vandalism

Source: Courtesy of the United States Department of Energy.

A systems approach, such as Kepner-Tregoe, should be used in determining appropriate corrective actions. It should consider not only the impact that actions will have on preventing recurrence, but also the potential that the corrective actions may actually degrade some other aspect of nuclear safety. At the same time, the impact that corrective actions will have on other facilities and their operations must be considered. The proposed corrective actions must be compatible with facility commitments and other obligations. In addition, those affected by or responsible for any part of the corrective actions, including management, should be involved in the process. Proposed corrective actions should be reviewed to ensure that the above criteria have been met, and should be prioritized based on importance, scheduled (a change in priority or schedule should be approved by management), entered into a commitment tracking system, and implemented in a timely manner. A complete corrective action program should be based not only on specific causes of occurrences, but also on items such as lessons learned from other facilities, appraisals, and employee suggestions.

A successful corrective action program requires that management be involved at the appropriate level, be willing to take responsibility, and allocate adequate resources for corrective actions. Additional specific questions and considerations in developing and implementing corrective actions include

- Do the corrective actions address all the causes?
- Will the corrective actions cause detrimental effects?
- What are the consequences of implementing the corrective actions?
- What are the consequences of not implementing the corrective actions?
- What is the cost of implementing the corrective actions (capital costs, operations, and maintenance costs)?
- Will training be required as part of the implementation?
- In what time frame can the corrective actions reasonably be implemented?
- What resources are required for successful development of the corrective actions?
- What resources are required for successful implementation and continued effectiveness of the corrective actions?
- What impact will the development and implementation of corrective actions have on other workgroups?
- Is the implementation of corrective actions measurable?

PHASE IV: INFORM

Effectively preventing recurrences requires the distribution of these reports (especially the lessons learned) to all personnel who might benefit. Methods and procedures for identifying personnel who have an interest are essential to effective communications. Also included is a discussion and explanation of the results of the analysis, including corrective actions, with management and personnel involved in the occurrence. In addition, consideration should be given to providing information of interest to other facilities. Finally, an internal self-appraisal report identifying management and control system defects should be presented to management for more serious occurrences. Consideration should then be given to directly sharing the details of root cause information with similar facilities where significant or long-standing problems may also exist.

PHASE V: FOLLOW-UP

Follow-up includes determining if corrective actions have been effective in resolving problems. First, the corrective actions should be tracked to ensure that they have been properly implemented and are functioning as intended. Second, a periodically structured review of the corrective action tracking system, normal process and change control system, and occurrence tracking system should be conducted to ensure that past corrective actions have been effectively handled. The recurrence of the same or similar events must be identified and analyzed. If an occurrence recurs, the original occurrence should be reevaluated to determine why corrective actions were not effective. Also, the new occurrence should be investigated using change analysis. The process change control system should be evaluated to determine what improvements are needed to keep up with changing conditions. Early indications of deteriorating conditions can be obtained from tracking and trend analyses of occurrence information.

SUMMARY

Determining facts related to any accident is the key to an accurate and effective analysis. Remember to

- Begin defining facts early in the collection of evidence.
- Develop an accident chronology (e.g., events and causal factors chart) while collecting evidence.
- Set aside preconceived notions and speculation.
- Allow discovery of facts to guide the investigative process.
- Consider all information for relevance and possible causation.
- Continually review facts to verify accuracy and relevance.
- Retain all information gathered, even that which is removed from the accident chronology.
- Establish a clear description of the accident.

Select the one (most) direct cause and the root (basic) cause (the one for which corrective action will prevent recurrence and have the greatest, most widespread effect). In cause selection, focus on programmatic and system deficiencies and avoid simple excuses such as blaming the employee. Note that the root (basic) cause must be an explanation (the why) of the direct cause, not a repeat of the direct cause. In addition, a cause description is not just a repeat of the category code description; it is a description specific to the occurrence. Also, up to three (contributing or indirect) causes may be selected. Describe the corrective actions selected to prevent recurrence, including the reason why they were selected, and how they will prevent recurrence. Collect additional information as necessary.

REFERENCES

Chiu, Chong. A Comprehensive Course in Root Cause Analysis and Corrective Action for Nuclear Power Plants (Workshop Manual). San Juan Capistrano: Failure Prevention Inc., 1988.

Gano, D. L. Root cause and how to find it. *Nuclear News*, August 1987.

Nertney, R.J., J. D. Cornelison, and W. A. Trost. Root Cause Analysis of Performance Indicators, (WP-21). System Safety Development Center, Idaho Falls, ID: EG&G Idaho, Inc., 1989.

United States Department of Energy, Office of Nuclear Energy. Root Cause Analysis Guidance Document. Washington, D.C.: U.S. Department of Energy, February 1992.

United States Department of Energy. Occurrence Reporting and Processing of Operations Information, (DOE Order 5000.3A). Washington, D.C.: U.S. Department of Energy, May 30, 1990.

United States Department of Energy. User's Manual, Occurrence Reporting and Processing System (ORPS), (Draft, DOE/ID-10319). Idaho Falls, ID: EG&G Idaho, Inc., 1991.

United States Department of Energy. Accident/Incident Investigation Manual, (SSDC 27, DOE/SSDC 76-45/27), 2nd ed. Washington, D.C.: U.S. Department of Energy, November 1985.

9 Causal Factor Analysis

INTRODUCTION

Accidents are rarely simple and almost never result from a single cause. They may develop from a sequence of events involving performance errors, changes in procedures, oversights, and omissions. Events and conditions must be identified and examined in order to find the cause of the accident and a way to prevent that accident and similar accidents from recurring. To prevent the recurrence of accidents, one must identify the accident's causal factors. The higher the level in the management and oversight chain in which the root cause is found, the more diffused the problem can be.

Causal factor analysis is a form of root cause analysis that aids in the development of evidence by collecting information and putting the information in the logical sequence so that it can be easily examined. This will lead to the causal factors of the accident and then to the development of new methods in order to help eliminate hazards or causes of that accident or similar accidents and prevent their recurrence in the future. By creating an event in the causal factor tree (chain or chart) as can be seen in Figure 9.1, multiple causes can be visually illustrated, as well as a visual relationship between the direct and contributing causes can be identified. By adding all harmful energy, events, exceeded/failed barriers, and people/objects affected by harmful energy that produced the undesired outcome, event causal charting also visually delineates the interactions and relationships of all involved groups or individuals. By using causal factor analysis, one can develop an event causal chain to examine the accident in a step-by-step manner by looking at the events, conditions, and causal factors chronologically, in order to prevent future accidents. Worksheets can also be used to list the applicable causes and sub-causes for each category of findings. These worksheets, such as those found in Appendix C, can then be transferred to a summary worksheet for easy viewing.

DEFINITION

Causal factor analysis is used when there are multiple problems with a long causal factor chain of events. A causal factor chain is a sequence of events that shows, step by step, the events that took place in order for the accident to occur. Causal factor analysis puts all the necessary and sufficient events and causal factors for an accident in a logical, chronological sequence. It analyzes the accident and evaluates evidence during an investigation. It is also used to help prevent similar accidents in the future and to validate the accuracy of pre-accidental system analysis. It is used to help

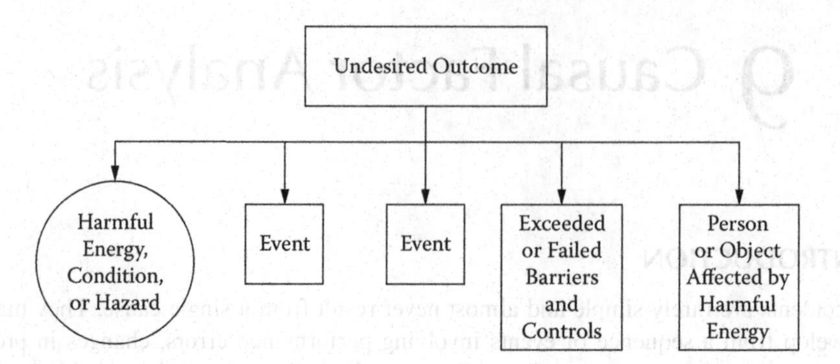

FIGURE 9.1 An event and causal factor tree (chart). (*Source:* Courtesy of the National Aeronautical and Space Administration.)

identify an accident's causal factors, which, in turn once identified, can be fixed to eliminate future accidents of the same or similar nature.

On the downside, causal factor analysis is time consuming and requires the investigator to be familiar with the process for it to be effective. As can be seen later in this chapter, the accident scene may need to be revisited a number of times and areas that are not directly related to the accident may need to viewed, in order to have a complete event and causal factor chain. It requires a broad perspective of the accident in order to identify any hidden problems that would have caused the accident.

DETERMINING CAUSAL FACTORS

The collection of facts should begin immediately after an accident occurs. Start with an accident site walk-through, interviews, and actual physical material collection. This will increase the accuracy of the information that is collected and help eliminate any uncertainty or vagueness. It will also help when it comes time to put together the event's causal factor chain. At this stage, the details will prove necessary. It is important to have as much information as possible in order to have an ideal accident investigation, including the events and the conditions at the time.

Direct Causes

Direct causes are the basic contributing factors that directly cause the accident to occur. For example, it might be determined that the immediate events or conditions leading up to an accident might be traced to the explosion of a pressurized vessel. The direct cause should be stated in one sentence. Then briefly explain what happened. A direct cause would be the release of energy or hazardous material. (Examples of the release of energy and examples of hazardous material can be found in Chapter 7.) Another example of a direct cause statement would be: "The electrician made contact between the metal rod and the exposed 220-volt wire." State the cause as simply as humanly possible.

Identifying the direct cause of an accident is optional because it is not necessary to complete a causal factor analysis. However, the direct cause should be identified when it facilitates an understanding of why the accident occurred, and it is useful in developing lessons learned from the accident. The direct cause is also used when writing up the accident report and filling out an accident report form.

Contributing Causes

Contributing causes are events or conditions that alone do not cause the accident but increase the probability of the accident occurring. In other words, they are not sufficient to cause the accident to occur by themselves but are needed for the accident to occur. Contributing causes can be unsafe acts or unsafe conditions. (Examples of unsafe acts and unsafe conditions can be found in Chapter 7.) An example of a contributing cause statement would be: "The painter failed to attach fall protection to a solid, secure object able to withstand the force of 5,000 pounds as required by OSHA regulations."

Root (Basic) Causes

Root causes are those that, if corrected, would prevent the accident from recurring or similar accidents from happening. They may surround or include several contributing causes. They are a higher order of causes that address multiple problems rather than focusing on the single direct cause. An example would be: "Management failed to implement the principles and core functions of a safety and health program." It is management's responsibility to ensure that the workplace has an effective safety and health program and that the workplace is safe for employees to work (see Figure 9.2).

EVENT FACTOR CHAINS (CHARTING) AND EXPLANATION

Events

Events are occurrences that take place in order for a task to be completed. In the case of causal factor analysis, events describe a single occurrence, and are short, to the point, and contain detail. The level of detail depends on the task that is performed. For example, "The plane descended 200 feet," not "The plane lost altitude and the pilot went by proper procedures to regain altitude." It is up to you, as the investigator, to establish that level of detail.

Events should be used in a causal factor analysis only if they pertain to the accident. If the event is not pertinent to the accident, exclude it from the analysis. For example, do not include: "The worker called her husband on her lunch break." However, do include: "The worker lifted the 100-pound block over her head, which led to her hurting her back."

Each event should precede the last event in an understandable sequence; this means that they should be chronological. If something does not appear to make sense or something is missing, then investigate those areas where information is lacking. This provides a better understanding of why the accident happened. It is important because of the need to develop a detailed step-by-step chart showing each event in a sequential manner, when trying to eliminate the causal factors of the accident.

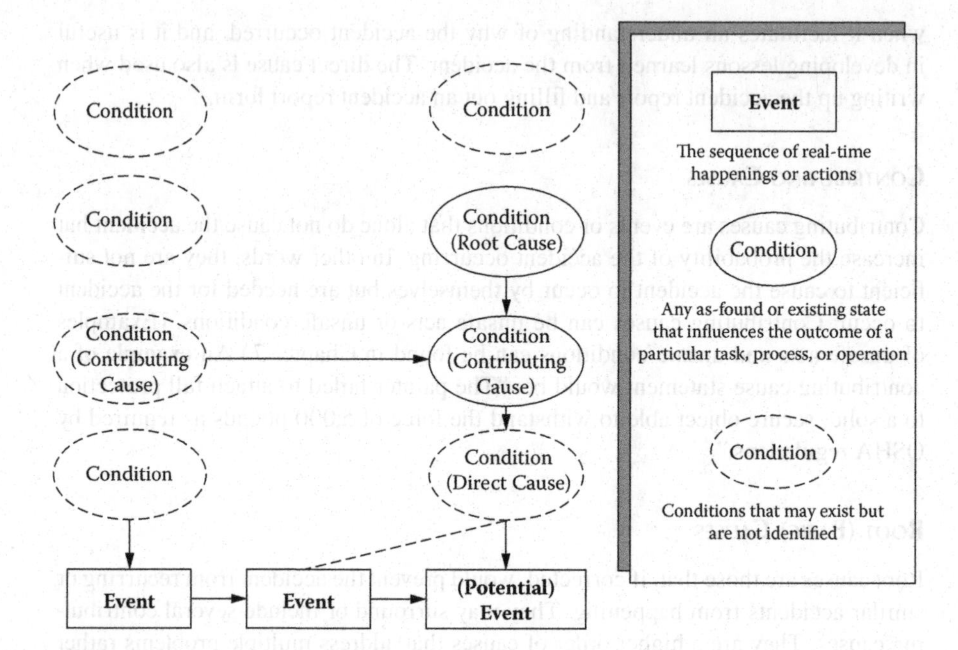

FIGURE 9.2 Causal factor relationships. (*Source:* Courtesy of the United States Department of Energy.)

CONDITIONS

Conditions are the states or circumstances surrounding the accident, rather than the happenings. They are inactive elements that increase the possibility for the accident to occur. An example of this would be: "The ground is wet." Just like events, when possible conditions should be described, quantified, and posted with time and date. Always record the conditions immediately preceding the event. An example of this would be that the worker did not know how to stop the backhoe from rolling down the hill, and the worker lost control of the backhoe.

CONTRIBUTING FACTORS

Contributing factors can be events or conditions that alone do not cause the accident to occur. They, along with conditions and events, increase the probability of the accident occurring. They "set up" the accident to occur. For example, "The road is icy." This contributing factor can set up a number of accidents, for example, a slip and fall; or a motor vehicular accident, either hitting another motor vehicle, hitting a stationary object, or hitting a person.

SECONDARY EVENT

Secondary events are events that are not directly related to the accident. However, they are like contributing factors, in that they are indirectly related to the accident, and they help set up the accident. An example would be not locking the gates to an electrical substation and kids wandering onto the site and hurting themselves.

BENEFITS OF EVENTS AND CAUSAL FACTORS CHARTING

The benefits of events and causal factors charting include

- Illustrating and validating the sequence of events leading to the accident and the conditions affecting these events
- Showing the relationship of immediately relevant events and condition to those that are associated but less apparent—portraying the relationships of organizations and individuals involved in the accident
- Directing the progression of additional data collection and analysis by identifying information gaps
- Linking facts and causal factors to organization issues and management systems
- Validating the results of other analytic techniques
- Providing a structured method for collecting, organizing, and integrating collected evidence
- Conveying the possibility of multiple causes
- Providing an ongoing method of organizing and presenting data to facilitate communication among the investigators
- Clearly presenting information regarding the accident that can be used to guide report writing
- Providing an effective visual aid that summarizes key information regarding the accident and its causes in the investigation report.

USING CAUSAL FACTOR ANALYSIS (EVENT FACTOR CHAINS OR CHARTING)

Constructing the Chart

Constructing the events and causal factors chart should begin immediately. However, the initial chart will be only a skeleton of the final product. Many events and condition will be discovered in a short amount of time; therefore, the chart should be updated almost daily throughout the investigative data collection phase. Keeping the chart up-to-date helps ensure that the investigation proceeds smoothly, that gaps in information are identified, and that the investigators have a clear representation of the accident chronology for use in evidence collection and witness interviewing.

The investigator and analyst can construct an event and causal factors chart using either a manual or computerized method. Accident investigation teams often use both techniques during the course of the investigation, developing the initial chart manually and then transferring the resulting data into a computer program.

The manual method may employ removable adhesive notes to chronologically depict events and the conditions affecting the events. The chart is generally constructed on a large conference room wall on many sheets of poster paper. Accident events and conditions are recorded on removable tack or adhesive notes and affixed sequentially to the wall in a conference room or command center. Because the exact chronology of the information is not yet known, using removable notes allows investigators to easily change the sequence of this information and to add information as it becomes available. Different-colored notes can be used to distinguish between events and conditions in this initial manual construction of the events and causal factors chart.

If the information becomes too difficult to manipulate manually, the data can be entered into a computerized analysis program. Using specialized analytical software, investigators can produce an events and causal factors graphic, as well as other analytical trees or accident models.

Whether using a manual or computerized approach, the process begins by chronologically constructing, from left to right, the primary chain of events that led to the accident. Secondary and miscellaneous events are then added to the events and causal factors chart, inserted where appropriate in a line above the primary sequence line. Conditions that affect either the primary or secondary events are then placed above or below the events.

Charting Symbols

Events should be placed in a rectangle and connected by arrows. The Primary Event Chain should be connected by bold arrows and directed to the accident. Conditions should be placed in ovals and connected by dashed arrows. Any event or condition that is presumptive, meaning the investigative team is not sure whether or not it has any effect on the accident, or cannot prove that it has occurred, or is related to the accident, should placed in dashed rectangles (events) or dashed ovals (conditions) and connected by dashed arrows. Secondary events, like primary events, should be placed in a rectangle and connected by arrows. However, when charting the secondary events, they must be on a different level, either above or below the Primary Event Line; this will help reduce confusion on which events are primary and which are secondary. An example of an Event Factor Chain is shown in Figure 9.3. A narrative that relates to the accident and Figure 9.3 follows.

> A large 2,400-volt fan system blew a fuse. The electrician obtained a fuse from the storeroom, tagged out the switch, and replaced the fuse. The system would not work, so the electrician bypassed a safety interlock and used a meter to check the fuse. A large fireball erupted, causing burns that required hospitalization and 50 lost workdays.
>
> This was classified as an off-normal, personnel safety occurrence (in-patient hospitalization). However, because this was a near fatality and because there existed a potential for significant programmatic impact, the investigation used formal Cause and Effects Analysis with charting to identify all the contributing conditions and any weaknesses in programmatic or operational control. A condensed version of the working chart is given in Figure 9.3. The significant findings are given below.

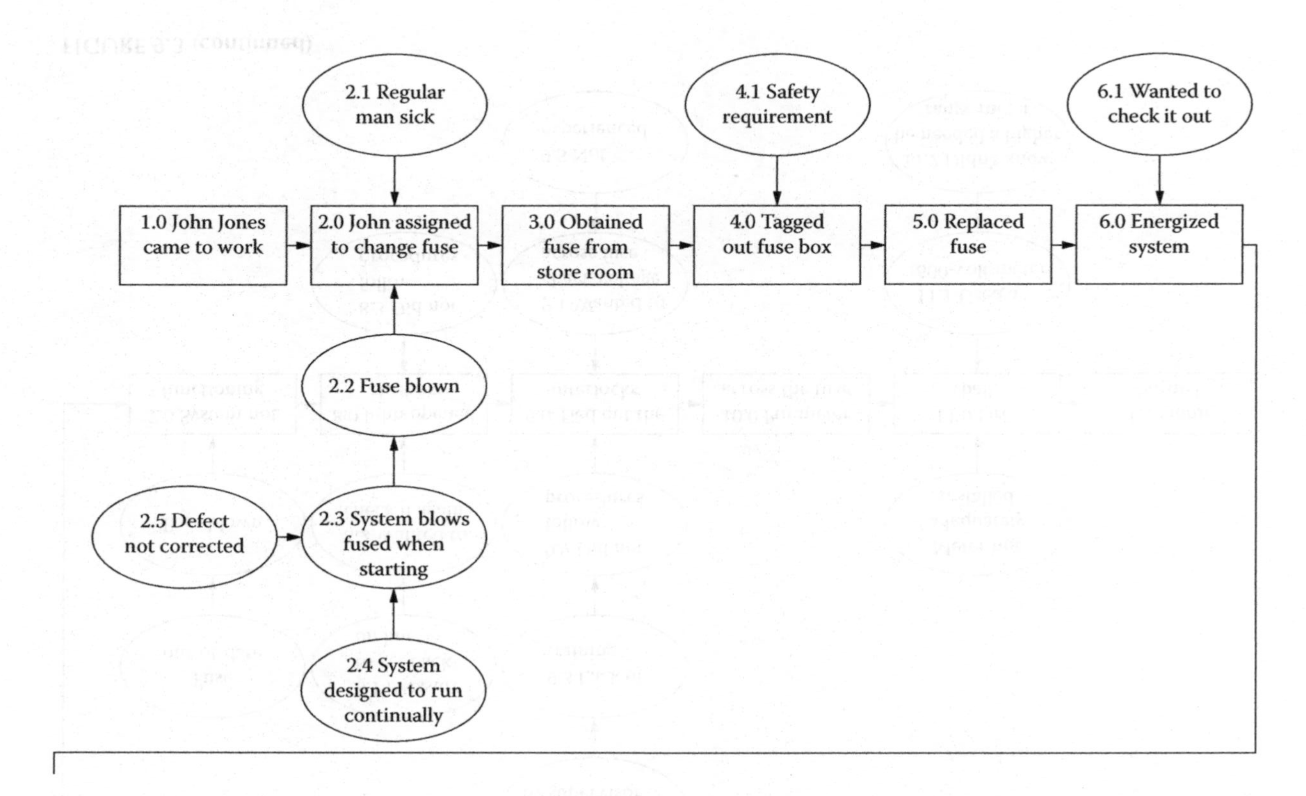

FIGURE 9.3 Events and causal factors chart. (*Source:* Courtesy of the United States Department of Energy.)

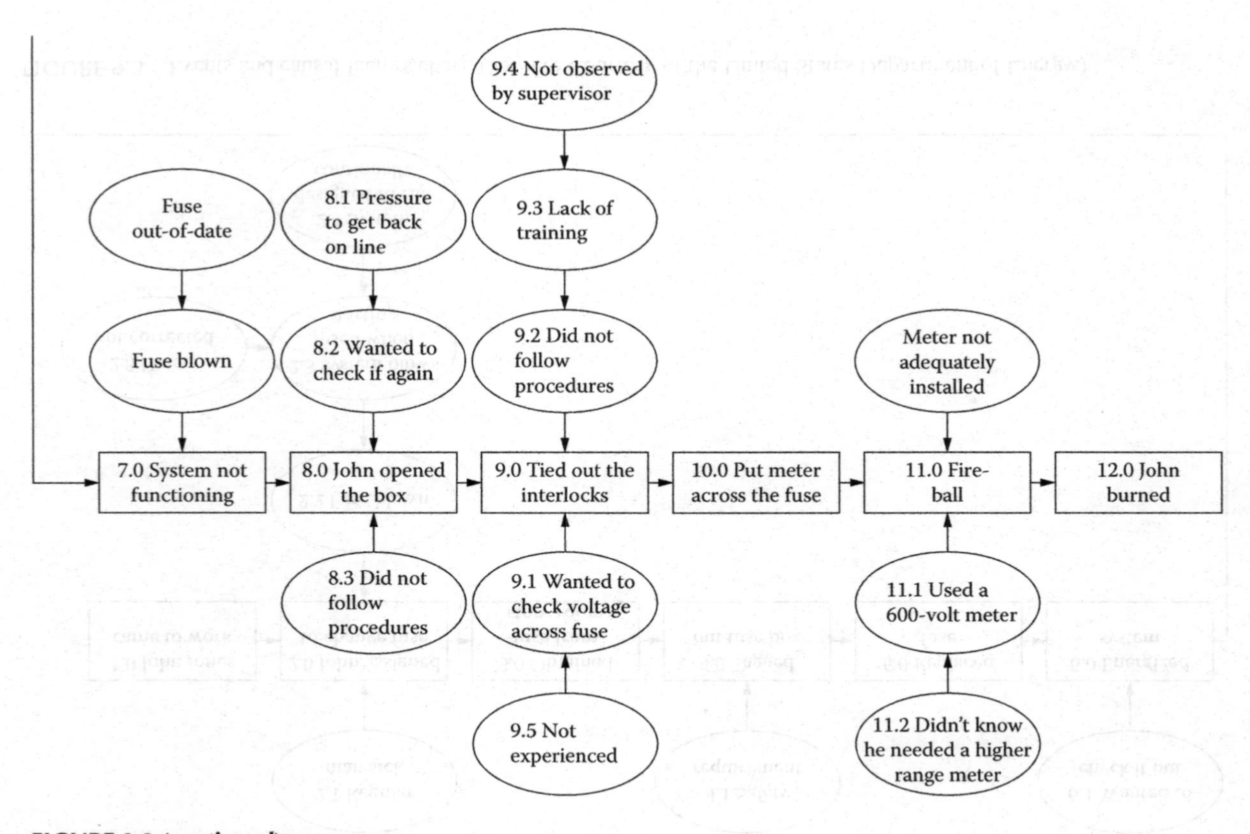

FIGURE 9.3 (continued)

Findings included

- The regular electrician was sick so a substitute who was not trained on high voltage was used (Cause Code 5A, No Training Provided).
- The substitute did not follow procedures. The substitute tied out the interlocks and used the wrong meter (Cause Code 3C, Violation of Requirement or Procedure).
- The fuse obtained from the storeroom was outdated and was no good (Cause Code 1A, Defective or Failed Part).
- The large fan was not designed for cycling (frequent startups) and had been regularly blowing fuses (Cause Code 4B, Inadequate or Defective Design).
- The supervisor knew the substitute was inexperienced but did not observe the substitute or give any special assistance (Cause Code 6C, Inadequate Supervision).
- Known defects had not been corrected (Cause Code 6A, Inadequate Administrative Control).

To correct these conditions, the following recommendations were made:

- Investigate and repair the system so that it does not blow fuses.
- Train supervisors to ensure that the worker is qualified for that task.
- Provide high-voltage training as needed.
- Evaluate management response to safety problems and operation of malfunctioning equipment.

As a result of the potential significance of this occurrence, a formal, detailed root cause analysis was performed. A high level of effort was expended but the effort was justified due to the consequences of a repeat occurrence.

When preparing to develop a Causal Factor Analysis chain as stated earlier, begin by collecting all possible facts as soon as possible. It will not be possible to collect the Events, Conditions, etc. chronologically, so investigators will need to revisit the chain over and over again and place them in order. When it is seen that something is missing or does not make sense, the need to reinvestigate that area and make sense of it should be undertaken. It is a good idea to use a charting method that can be edited easily, in order to finish the chart appropriately. For instance, use magnets on a metal background or "Post-It-Notes" on a large piece of paper, and the Events and/or Conditions can be easily moved without any or little confusion when new facts arise. If a chart is being drawn, then every time a new fact arises, drawing up a new chart creates confusion and messy work. This would also be very time consuming and not very cost effective.

As stated earlier, a Causal Factor Chain is used to trace the sequence of events surrounding an accident, including the Conditions that exist for the accident to occur. First, an examination of the event immediately preceding the accident should be performed to determine if it is the significant event. To evaluate the significance of the event, ask the following question: "If this event had not occurred, would the accident have happened?" If the answer is "no," then the event should be excluded because it is not relevant to the accident. If the answer is "yes," then the event stays because it is significant. This question should be asked for each of the events you have shown.

The Significant Events and the Events and Conditions that allowed the Significant Events to occur are the accident's Causal Factors. Once all the significant events have been identified and all the Events and Conditions that do not belong in the analysis have been excluded, then an examination of the Events and Conditions preceding the significant event should be examined by asking a series of questions.

These questions should include

- Why did the event happen?
- What Events and Conditions led to the occurrence of the event?
- What went wrong to allow the Event to occur?
- Why did these Conditions exist?
- How did these Conditions originate?
- Who had responsibility for these Conditions?
- Are there any relationships between what went wrong in this event chain and other Events and Conditions in the accident sequence?
- Is the significant event linked to other Events or Conditions that may indicate a more general or larger deficiency?

This questioning should then be repeated for each and every event and condition, where applicable, and reviewed by others to ensure that nothing has been overlooked. Once everyone has agreed on all the results of the questioning, then your Causal Factor Analysis is complete.

CAUSAL FACTOR WORKSHEETS

Accident causes can be broken down into seven different "categories": equipment/material problems, procedural problems, personnel error, design problems, training deficiencies, management problems, and external phenomenon. These categories are further broken down into subcategories to further explain the cause of the accident. The causal categories and cause codes can be found in Chapter 8, Table 8.2. These causal categories are used in a causal factor worksheet found in Appendix C. Each worksheet has a matrix that is used to list and identify the direct cause, the root cause, and any contributing causes the accident may have. As can be seen, each subcategory is found within the matrix and there is enough room to list four (I through IV in Appendix C) causes for each of the subcategories.

These worksheets are broken down by first determining if that particular cause was applicable or nonapplicable; then the subcategories are to be used to distinguish what was the direct cause, the contributing cause, and/or the root cause. There is also space to write the cause descriptions and any recommended corrective actions that can be taken.

There are seven worksheets for major cause categories, and there is an additional worksheet, found at the end of the seven, to put all causes into summary form creating one easy-to-look-at form. It contains one direct cause, one root cause, and up to three contributing causes, their descriptions, and any corrective actions. The contributing causes, which should be listed and have corrective actions, are those that can result in the most effective benefit when corrected. Although only three contributing causes can be reported, and their corrective actions reported, corrective actions should be made on all identified causes.

SUMMARY

Causal Factor Analysis is a useful tool when trying to understand why an accident occurred. It is used when there are multiple problems and the Event Factor Chain is long or complex. After investigating an accident and collecting all the information surrounding an accident, an event chain technique can be used to help understand and visually depict why the accident occurred. This is done by asking a series of questions, eliminating all events and conditions that are not significant, and keeping those that are significant. Once all the events and conditions that belong to the accident have been determined, another examination can be performed on the events and conditions that remain. The second questioning process can help in understanding why the accident occurred and in finding out ways to help prevent this accident and similar accidents from recurring in the future. Answers should then be reviewed by others to gain more insight, and reassessments should be made where applicable. Conclusions should then be drawn.

Worksheets can also be used to help identify the direct cause, the root cause, and any contributing causes of the accident. The results from the worksheets are then placed on a summary worksheet along with the cause descriptions and any recommended corrective actions. From this point on, appropriate changes can be made to help prevent that accident or any similar accident from occurring in the future.

REFERENCES

Buys, J. R. and J. L. Clark. *Events and Causal Factors Charting,* System Safety Development Center. *(SSDC-14).* Idaho Falls, ID: EG&G Idaho, Inc., August 1978.

United States Department of Energy. Office of Nuclear Energy. *Root Cause Analysis Guidance Document.* Washington, D.C.: U.S. Department of Energy, February 1992.

10 Change Analysis

INTRODUCTION

As its name implies, this technique emphasizes change. To solve a problem, one must look for deviation from the norm. Change analysis follows a logical sequence. It is based on the principle of differences and considers problems to result from some unanticipated change. An analysis of the change is employed to determine causes using the following steps:

1. Define the problem. (What happened?)
2. Establish the norm. (What should have happened?)
3. Identify, locate, and describe the change. (What, where, when, to what extent?)
4. Specify what was and what was not affected.
5. Identify the distinctive features of the change.
6. List the possible causes.
7. Select the most likely causes.

WHEN TO USE CHANGE ANALYSIS

Change analysis is used when the problem is obscure. It is a systematic process that is generally used for a single occurrence and focuses on elements that have changed. It compares the previous trouble-free activity with the occurrence to identify differences. These differences are subsequently evaluated to determine how they contributed to the occurrence.

HOW TO USE CHANGE ANALYSIS

Change analysis looks at a problem by analyzing the deviation between what is expected and what actually happened. The evaluator essentially asks what differences occurred to make the outcome of this task or activity different from all the other times this task or activity was successfully completed. This technique consists of asking the questions: What? When? Where? Who? How? Answering these questions should lead toward answering the root cause question: Why? Primary and secondary questions included within each category provide the prompting necessary to thoroughly answer the overall question. Some of the questions may not be applicable to any of the existing conditions. Some amount of redundancy exists in the

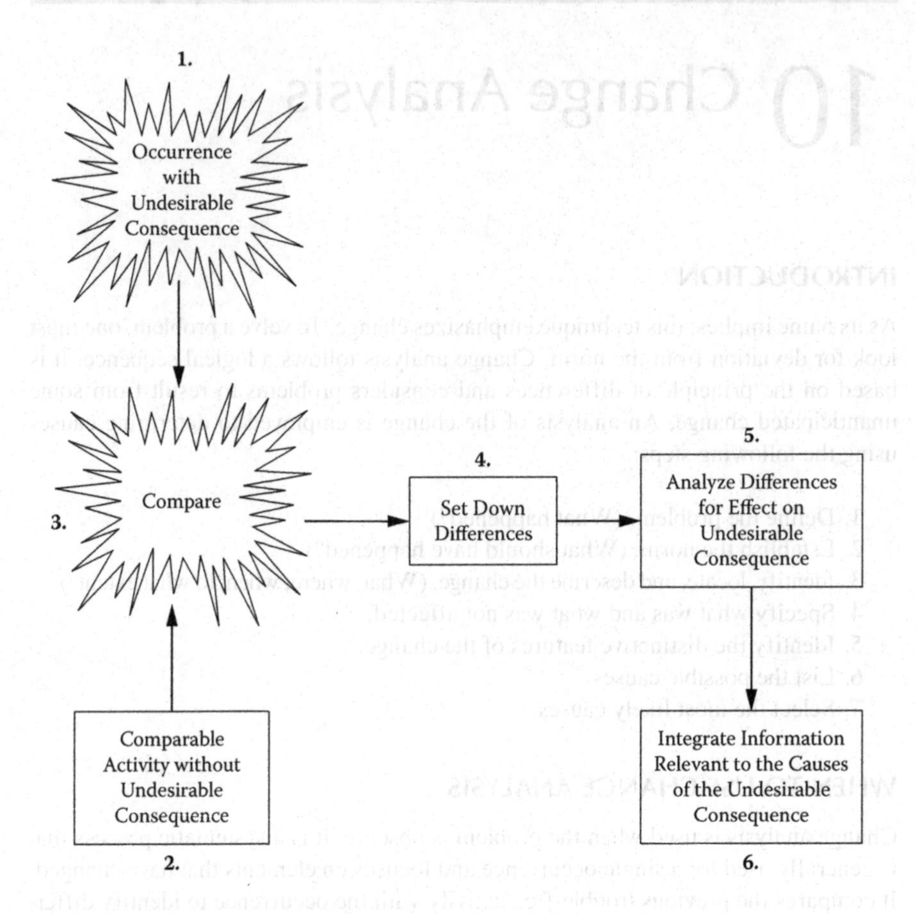

FIGURE 10.1 Six steps involved in change analysis. (*Source:* Courtesy of the United States Department of Energy.)

questions to ensure that all items are addressed. Several key elements for addressing any change in the standard operating process include

- Consider the event containing the undesirable consequences.
- Consider a comparable activity that did not have the undesirable consequences.
- Compare the condition containing the undesirable consequences with the reference activity.
- Set down all known differences whether or not they appear to be relevant.
- Analyze the differences for their effects in producing the undesirable consequences. This must be done with careful attention to detail, ensuring that obscure and indirect relationships are identified (e.g., a change in color or finish may change the heat transfer parameters and consequently affect system temperature).
- Integrate information into the investigative process relevant to the causes of, or the contributors to, the undesirable consequences.

TABLE 10.1
Change Analysis Worksheet

Change Factor	Difference/Change	Effect	Questions to Answer
What (Conditions, occurrence, activity, equipment)			
When (occurred, identified, plant status, schedule)			
Where (physical location, environmental conditions)			
How (work practice, omission, extraneous action, out-of-sequence procedure)			
Who (personnel involved, training, qualification, supervision)			

Source: Courtesy of the United States Department of Energy.

Change analysis is a good technique to use whenever the causes of the condition are obscure, you do not know where to start, or you suspect a change may have contributed to the condition. Not recognizing the compounding of change (e.g., a change made five years previously combined with a change made recently) is a potential shortcoming of change analysis. Not recognizing the introduction of gradual change as compared with immediate change also is possible. This technique may be adequate to determine the root cause of a relatively simple condition. In general, however, it is not thorough enough to determine all the causes of more complex conditions.

STRUCTURING THE ANALYSIS

Change analysis involves six steps. These steps are presented in the worksheet in Table 10.1. The following questions will help you complete the six steps and identify information required on the worksheet:

1. WHAT?
   a. What is the condition?
   b. What occurred to create the condition?
   c. What occurred prior to the condition?

d. What occurred following the condition?
e. What activity was in progress when the condition occurred?
f. What activity was in progress when the condition was identified?
   i. Operational evolution in the workspace?
      – Surveillance test?
      – Power increase/decrease?
      – Starting/stopping equipment?
   ii. Operational evolution outside the workspace?
      – Valve line-up?
      – Fuel handling?
      – Removing equipment from service?
      – Returning equipment to service?
   iii. Maintenance activity?
      – Surveillance?
      – Corrective maintenance?
      – Modification installation?
      – Troubleshooting?
   iv. Training activity?
      – Classroom activities?
      – Hands-on activities?
      – On-the-job?
      – Job instruction training?
g. What equipment was involved in the condition?
   i. What equipment initiated the condition?
   ii. What equipment was affected by the condition?
   iii. What equipment mitigated the condition?
   iv. What is the equipment's function?
   v. How does it work?
   vi. How is it operated?
   vii. What failed first?
   viii. Did anything else fail due to the first problem?
   ix. What form of energy caused the equipment problem?
   x. What recurring activities are associated with the equipment?
   xi. What corrective maintenance has been performed on the equipment?
   xii. What modifications have been made to the equipment?
h. What system or controls (barriers) should have prevented the condition?
i. What barrier(s) mitigated the consequences of the condition?

2. WHEN?
   a. When did the condition occur?
   b. What was the facility's status at the time of occurrence?
   c. When was the condition identified?
   d. What was the facility's status at the time of identification?
   e. What effects did the time of day have on the condition? Did it affect:
      i. Information availability?
      ii. Personnel availability?

   iii. Ambient lighting?
   iv. Ambient temperature?
 f. Did the condition involve shift-work personnel? If so:
   i. What type of shift rotation was in use?
   ii. Where in the rotation were the personnel?
 g. For how many continuous hours had any involved personnel been working?

3. WHERE?
 a. Where did the condition occur?
 b. What were the physical conditions in the area?
 c. Where was the condition identified?
 d. Was location a factor in causing the condition?
   i. Human factor:
     – Lighting?
     – Noise?
     – Temperature?
     – Equipment labeling?
     – Exposure levels?
     – Personal protective equipment required in the area?
     – Accessibility?
     – Indication availability?
     – Other activities in the area?
     – What position is required to perform tasks in the area?
   ii. Equipment factor:
     – Humidity?
     – Temperature?
     – Cleanliness?

4. HOW?
 a. Was the condition an inappropriate action or was it caused by an inappropriate action?
   i. An omitted action?
   ii. An extraneous action?
   iii. An action performed out of sequence?
   iv. An action performed to a too small of a degree? To a too large of a degree?
 b. Was the use of a procedure a factor in the condition?
 c. Was there an applicable procedure?
 d. Was the correct procedure used?
 e. Was the procedure followed?
   i. Followed in sequence?
   ii. Followed “blindly” (without thought?)
 f. Was the procedure:
   i. Legible?
   ii. Misleading?
   iii. Confusing?

    iv. An approved, current revision?
    v. Adequate to do the task?
    vi. In compliance with other applicable codes and regulations?
  g. Did the procedure:
    i. Have sufficient detail?
    ii. Have sufficient warnings and precautions?
    iii. Adequately identify techniques and components?
    iv. Have steps in the proper sequence?
    v. Cover all involved systems?
    vi. Require adequate work review?

5. WHO?
  a. Which personnel:
    i. Were involved with the condition?
    ii. Observed the condition?
    iii. Identified the condition?
    iv. Reported the condition?
    v. Corrected the condition?
    vi. Mitigated the condition
    vii. Missed the condition?
  b. What were:
    i. The qualifications of these personnel?
    ii. The experience levels of these personnel?
    iii. The work groups of these personnel?
    iv. The attitudes of these personnel?
    v. Their activities at the time of involvement with the condition?
  c. Did the personnel involved:
    i. Have adequate instruction?
    ii. Have adequate supervision?
    iii. Have adequate training?
    iv. Have adequate knowledge?
    v. Communicate effectively?
    vi. Perform correct actions?
    vii. Worsen the condition?
    viii. Mitigate the condition?

SUMMARY

Change analysis asks the following question: Is there anything that is different or has changed from the normal process or procedures that were in use? To find change, many questions must be answered. Although the change may seem minuscule to the investigator, remember that everyone is reticent to change. Many people find any change disturbing, even the smallest degree of change. So it is imperative to at least determine if change was the culprit in the accident under investigation.

REFERENCES

Bullock, M. G. *Change control and analysis* (SSDC-21). System Safety Development Center. Idaho Falls: EG&G Idaho, Inc., August 1981.

United States Department of Energy, Office of Nuclear Energy. Root Cause Analysis Guidance Document. Washington, D.C.: U.S. Department of Energy, February 1992.

11 Barrier Analysis

INTRODUCTION

The first step in most traditional accident prevention programs has been to look at information to determine where the industry or company appears to be having problems. Collecting and analyzing data is a first step in finding out what is going on, and it makes good sense to begin at this point. Because it prepares for the main course of analysis, this procedure is called a Preliminary Data Analysis (PDA). Two techniques for performing this analysis are presented in this chapter; they are the Physical Barrier Analysis (PBA) and the Human Barrier Analysis (HBA) (see Figure 11.1). These techniques result in two distinct considerations:

1. The thought process or way of thinking about the data to get usable information from it
2. The mechanics of documenting the results of that thinking in the form of organizers, tally sheets, etc.

The first concern, thinking about accidents, is a critical step in the analysis of accidents at any operation, whether it is large or small, with many or with few accidents. The second concern, staying organized while working with large amounts of data, results in the use of several recording aids to keep track of the products of various phases of the analysis. In an operation where there are a large number of accidents, these tools, forms, and aids are a necessary part of the analysis process. In an operation that has few accidents, where the amount of data to be analyzed is not so overwhelming, these organizing aids are probably not necessary.

You can use the PBA and HBA tools as appropriate, but the use of physical and human barriers conceptually will provide you with a different perspective on accident causes and their solutions. Using the different perspectives of PBA and HBA helps to

1. Identify the types of accidents as a function of how they can be prevented.
2. Begin to understand what may be causing the absence of or deficiencies in these preventive measures.

From this perspective it can be said that industrial accidents fall into one of two types:

1. Accidents that could have been prevented by some type of physical barrier
2. Accidents that could have been prevented by the individuals involved

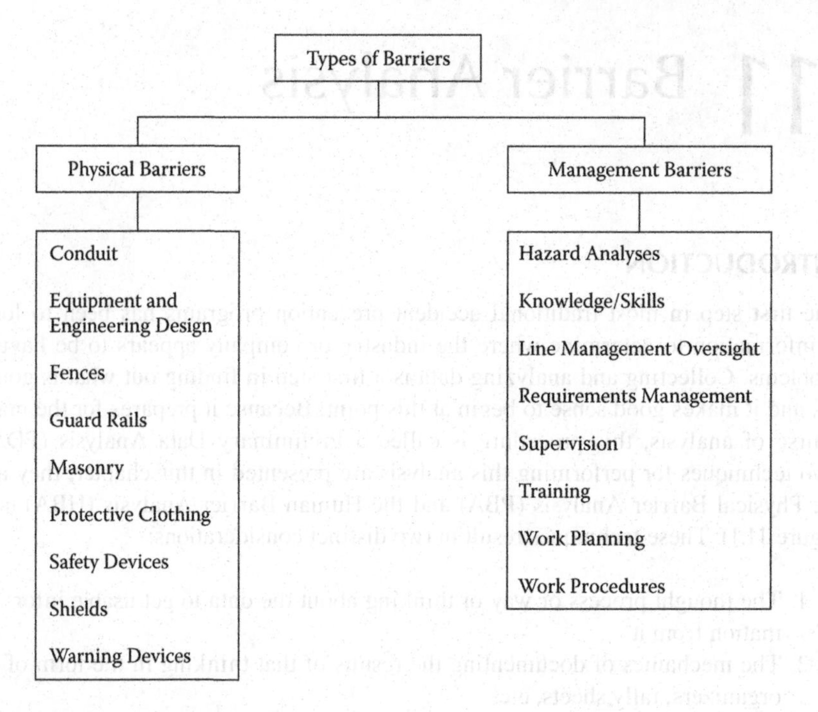

FIGURE 11.1 Types of barriers to protect workers from hazards. (*Source:* Courtesy of the U.S. Department of Energy.)

Industrial operations differ with respect to numbers of accidents, specific contributors to accidents, and the relative magnitude of the human-versus-physical condition problem. Each accident, at any given operation, is a manifestation of either the presence of a human barrier or the absence of a physical barrier. Any accident that can be prevented will be prevented by one or a combination of the following actions:

- Introduce a physical barrier:
  - On the energy source
  - Between the energy source and the person
  - On the person
  - That separates energy from people (time or space)
- Remove a human barrier that causes a lack of or deficiency in a person:
  - Information
  - Tools or equipment
  - Knowledge
  - Capacity
- Incentive

The PDA allows us to determine which barriers may be provided, modified, or removed to prevent accidents. A clearer picture of the distinct difference between physical and human preventive measures is provided by the following swimming pool analogy:

I have a two-year-old child. My neighbor owns a swimming pool.

- Is this a recipe for an accident?
- How can my neighbor and I prevent it?

Most likely, our list will include some of the following:

1. Teach the child to swim.
2. Put a fence around the pool.
3. Put a leash on the child.
4. Punish the child for going near the pool.
5. Supervise the child closely.
6. Move.
7. Fill in the pool.
8. Cover the pool.
9. Drain the pool.
10. Make the salt concentration of the water so high the child would float.
11. Put a life preserver on the child.

As can be seen from this swimming pool analogy, solutions on the list fall into two general types: either make a physical change (as suggested in 2, 3, 5, 6, 7, 8, 9, 10, and 11) or making a change in the child (as suggested in 1 and 4). Accidents at an industrial operation can be prevented in the same two general ways: change the physical environment or change the person. Physical environments can be changed by bolting, cutting, welding, etc. People can be changed by altering factors that shape their behavior, their knowledge, skill, understanding, etc.

Most safety professionals concur that a well-engineered industrial operation is the foundation for a safe operation. This means that the appropriate physical barriers should be in place and functional. PBA takes a quick look at the potential for improved engineering at an industrial operation. It is reasonable to assume that some operations make better use of barriers than others to prevent accidents. A PBA's primary purpose is to determine the extent to which these barriers, preventive measures, lack of utilization can be attributed to the cause of an accident.

Barriers prevent the release of unwanted energy, reduce its forcefulness, or prevent energy from contacting the individual. Unwanted energy is whatever figuratively or literally "draws blood." If the barrier is adequate, individuals should not be injured and thus no accident will occur. There are four classes of barriers that can be used singly or in combination to prevent workers from becoming injured by unwanted energy.

1. *Barriers on the energy source:*
   The primary purpose is to prevent the release of unwanted energy.
2. *Barriers between the person and the energy source:*
   The primary purpose is to prevent unwanted energy from contacting the person.
3. *Barriers on the person:*
   The primary purpose is to reduce the forcefulness of the unwanted energy.

4. *Barriers that separate the person from the unwanted energy by time or space:* The primary purpose is to prevent the unwanted energy from contacting the person.

Generally, barriers in 1 provide more positive protection than those in 2, and 2 more than 3, etc.

Some of the barriers are not clear-cut. Barriers on the person could very easily prevent the unwanted energy from contacting the person. Sometimes they serve to reduce the potential injury. For example, goggles will keep objects out of a person's eyes. But if the object has any force behind it, it could break the goggles. However, the amount of damage to the eye should be less as a result of the amount of impact the goggles have absorbed.

Barriers on the person can also be distinguished from barriers between the person and the energy source, if the former is considered as being extensions of the person. Gloves are extensions of the hands and skin; goggles are extensions of the eyes. This is in contrast to a guard or fence, neither of which is an extension of the person, but they do protect more parts of the body than personal protective devices. Any personal protective equipment is a barrier on the person. A more difficult discrimination exists between barriers that separate the person from the energy source, and barriers that separate the person from the energy by time or space. The latter is, in fact, a special case of the former. When we think of barriers between the person and energy source, we think of barriers that are constructed of some substance engineered for the purpose of separating the person and the energy source.

Time and distance are always available and become a barrier, often based on someone's judgment. Because time and space are easy to make into barriers, there needs to be clarification of time or space barriers. A time or space barrier should be based on a written policy or regulation, or must signal that the time or space barrier is in effect. This criterion limits the person's judgment. The signal or policy tells the person what the safest waiting period or distance is.

PHYSICAL BARRIER ANALYSIS

Physical Barrier Analysis (PBA) is based on four assumptions:

1. There are four types of physical barriers. When a physical intervention can prevent an accident, that intervention will always be one, or a combination, of the four types of physical barriers.
2. Accidents not caused by the lack of a physical barrier cannot be prevented by introducing such a barrier.
3. Being able to identify accidents that a physical barrier could have prevented and those that a physical barrier could not have prevented allows the analyst to fit remedies to the causes of accidents more accurately.
4. Accidents related to physical barriers are almost always the easiest type to prevent.

These four assumptions place physical barriers in proper perspective. While stressing their importance, their limitations are also defined. In order to perform a physical barrier analysis, the following will be needed:

1. Accident information
2. Groups of physical barrier analysis statements
3. A physical barrier analysis matrix
4. Instructions
5. Familiarity with the practices, procedures, methods, and strategies of the industrial operation being analyzed

Using the company's accident investigation forms, the first thing to do in analyzing the accident is to read the narrative and visualize what happened. Often, the report is sketchy, but most of the time you will be able to draw on your experience to read between the lines and get the picture. Once you have visualized the situation and what occurred, then ask: "What could have prevented this accident?" To help determine the appropriate preventive measures, make use of the list of physical barrier analysis statements. The following statements refer to each of the four types of physical barriers. For each type, there are four possibilities. Select the one that best represents a valid opinion of the accident.

1. A physical barrier on the energy source was:
   (Prevent Release)
   A. Not possible and practical.
   B. Practical and possible, but not provided.
   C. Provided, but not used.
   D. Used, but failed.
2. A physical barrier between the energy source and the person was:
   (Prevent Contact)
   A. Not possible and practical.
   B. Practical and possible, but not provided.
   C. Provided, but not used.
   D. Used, but failed.
3. A physical barrier on the person was:
   (Prevent Forcefulness)
   A. Not possible and practical.
   B. Practical and possible, but not provided.
   C. Provided, but not used.
   D. Used, but failed.
4. A barrier of time or space between the energy source and the person was:
   (Prevent Contact)
   A. Not possible and practical.
   B. Practical and possible, but not provided.
   C. Provided, but not used.
   D. Used, but failed.

Explanation of PBA

If in Statement 1 it is thought that a physical barrier on the energy source was not possible and practical, then select (A). Once (A) has been selected, it is not necessary to check (B), (C), and 4. Simply go to Statement 2. In looking at Statement 2, decide if a physical barrier between the energy source and the person was not possible and practical, then again choose (A) and proceed to Statement 3.

In Statement 3, (A) is probably not the proper option because safety glasses possibly would have prevented the accident. Thus, proceed through the remaining statements (B), (C), and (D) to determine the proper selection. Because it is believed that a physical barrier on the person was possible and practical, it must be decided if it was not provided, provided but not used, or used but did not protect the person. Making that choice depends on personal experience and knowledge of the operation.

Therefore, if personal knowledge and experience lead to the belief that the operation normally does not provide safety glasses, then select (B) and go to Statement 4. If the belief is that the industrial operation routinely provides safety glasses but workers usually do not use them, then select (C) and proceed to Statement 4. If the belief is that the industrial operation is a stickler for safety glasses, then it might be suspected that glasses were worn but failed, and then (D) should be selected and then proceed to Statement 4. If in looking at Statement 4 the belief is that a time or space barrier was not possible or practical, then again select (A).

The choices have now been made: 1 was (A), 2 was (A), 3 for purposes of the example was (C), and 4 was (A). Record the answers on the PBA matrix (see Table 11.1).

The PBA matrix is simply the PBA statements in a different format. Normally, many accidents will be analyzed. When this occurs, the accidents are numbered so that choices for each accident will equal four. If the example was accident number 1, then (1) would be placed in the appropriate cells corresponding to the selections that were made. In this example, a (1) would be placed in cells 1A, 2A, 3C, and 4A. This now completes the Physical Barrier Analysis of one accident. A listing of the steps for a PBA that is normally followed is:

1. Gather the accidents you wish to analyze.
2. Number them sequentially.
3. Starting with number 1, read the abstract; then the accident should be visualized.
4. Ask yourself the first group of four questions on the PBA question list. Choose one of the four that best represents the opinion.
5. Ask yourself each of the remaining four groups of questions. Make one selection from each group.
6. Transfer your selections to the PBA matrix.

HUMAN BARRIER ANALYSIS

When discussing physical barriers, the indication is that the absences of a physical barrier increase the likelihood of accidents. Therefore, putting the appropriate physical barriers in place should reduce or prevent those kinds of accidents.

TABLE 11.1
Physical Barrier Analysis (PBA) Matrix

	1	2	3	4
	Barrier on the Energy Source	**Barrier between Energy Source and the Person**	**Barrier on the Person**	**Time/Space Barrier Separating the Energy & the Person**
A Was Not Possible and Practical				
B Was Possible and Practical but Not Provided				
C Was Provided but Not Used				
D Was Provided and Used, but Failed				

Human barriers work the opposite way. The presence of human barriers increases the likelihood of accidents and prevents an individual from making the appropriate response. Human barriers increase the likelihood of an individual doing something that should not be done (e.g., walking into the path of a moving vehicle). Human barriers decrease the likelihood of an individual doing something that should be done (e.g., locking out and tagging disconnecting devices before working on electrical equipment).

When physical barriers are used, the physical environment in which the person works is changed. At times, there are some solutions that do not involve changing the physical environment. Those solutions involve doing something to the worker. When dealing with human barriers, try to change factors that directly impact the person, or try to change the way the person responds.

In doing the HBA, there are three assumptions to make. First, there are logical, understandable reasons why people perform in ways that lead to an accident. Second, those reasons usually take the form of barriers that cause them to do things they should not do. Third, the reasons that cause them not to do things they should not do. Identifying and eliminating these barriers will increase the likelihood of performance that reduces the possibility of accidents and reduces the likelihood of the performance that causes accidents.

These assumptions are fairly straightforward. There may be some question about the first one. Often, the performance of others is judged as illogical, incomprehensible, and sometimes stupid. When the analyst begins to look at performance from someone else's point of view and looks at it in more depth, it is found to be logical and understandable. The HBA helps to discover these reasons. To perform a human barrier analysis, the following are needed:

1. A completed PBA matrix
2. Four HRA tally sheets lettered A, B, C, and D
3. A working knowledge of human barrier definitions
4. Instructions
5. Familiarity with the practices, procedures, methods, and strategies of the operation you are analyzing

There are five major barriers to human performance. Individuals do not perform appropriately because of a deficiency in or lack of

1. Information
2. Proper tools
3. Knowledge
4. Capacity
5. Incentives

To perform an HBA, look at accidents that could have been prevented by the individual involved. You may wish to know if the presence of one of these barriers caused the accident. As in PBA, our experience with the operation and our knowledge of the industry and the operating procedures provide a basis for answering the HBA questions in Table 11.2.

Information

When people are not informed of what is expected of them, when they are not told how well they are performing, or when they are not given any guidance or clear direction on how to accomplish a task, an information barrier is present. The lack of consistent, practical operating procedures or the lack of feedback creates situations where people respond based on their own judgments of how things should be done. Often, the behavior is incompatible with other aspects of the operation. To determine if there was an information barrier, the question is asked: "Was the person given direction and guidance on how to do the job?" Direction and guidance are used as key words in the question because they cover both expected performance and feedback for performance.

Here is an example of an information barrier. An employee was welding and cutting zinc and galvanized tubing for main fan installation. The employee worked 2 days at this, and then became sick, dizzy, and sleepy from the fumes. Clearly, this individual lacked safety information.

TABLE 11.2
Human Barrier Analysis (HBA) Questions

- What barriers existed between the second, third, etc. condition/situation and the second, third, etc. problems?
- If there were barriers, did they perform their functions? Why?
- Did the presence of any barriers mitigate or increase the occurrence severity? Why?
- Were any barriers not functioning as designed? Why?
- Was the barrier design adequate? Why?
- Were there any barriers in the condition/situation source(s)? Did they fail? Why?
- Were there any barriers on the affected component(s)? Did they fail? Why?
- Were the barriers adequately maintained?
- Were the barriers inspected prior to expected use?
- Why were any unwanted energies present?
- Is the affected system/component designed to withstand the condition/situation without the barriers? Why?
- What design changes could have prevented the unwanted flow of energy? Why?
- What operating changes could have prevented the unwanted flow of energy? Why?
- What maintenance changes could have prevented the unwanted flow of energy? Why?
- Could the unwanted energy have been deflected or evaded? Why?
- To what other controls are the barriers subjected? Why?
- Was this event foreseen by the designers, operators, maintainers, or anyone?
- Is it possible to have foreseen the occurrence? Why?
- Is it practical to have taken further steps to reduce the risk of the occurrence?
- Can this reasoning be extended to other similar systems/components?
- Were adequate human factors considered in the design of the equipment?
- What additional human factors could be added? Should be added?
- Is the system/component user friendly?
- Is the system/component adequately labeled for ease of operation?
- Is there sufficient technical information for operating the component properly? How do you know?
- Is there sufficient technical information for maintaining the component properly? How do you know?
- Did the environment mitigate or increase the severity of the occurrence? Why?
- What changes were made to the system/component immediately after the occurrence?
- What changes are planned to be made? What might be made?
- Have these changes been properly, adequately analyzed for effect?
- What related changes to operations and maintenance have to be made now?
- Are expected changes cost effective? Why? How do you know?
- What would you have done differently to prevent the occurrence, disregarding all economic considerations (as regards operation, maintenance, and design)?
- What would you have done differently to prevent the occurrence, considering all economic concerns (as regards operation, maintenance, and design)?

Guidance and direction can come from a variety of sources. Written plans, policies, and procedures can be direct and guide behavior. Verbal direction and guidance from supervisors and co-workers provide other information. Whatever the source, however, direction and guidance that are not clear, consistent, or accurate can be an even greater barrier to human performance than no information at all.

Tools or Equipment

When the appropriate tools or equipment to do a job are not available or are improperly designed, a barrier to human performance is created. Where improper design is the cause for problems, the general talk is about human factors. Tools are not physical barriers because they generally do not function to protect the person from the energy source. Tools and equipment are extensions of the person that are instrumental in getting the required performance. To determine if tools or equipment are a barrier, the question to ask is: "Were the appropriate tools or equipment available, and were they properly designed for the job?"

An example of tools as a barrier is a piece of rebar was used as a pry bar because the correct bar was not provided.

Knowledge

Sometimes people do unsafe things because they do not know any better. When they do not know how to perform safely, a knowledge barrier to human performance exists. Knowing how to do things safely is acquired principally through education, training, and experience. Deficient, limited, or nonexistent education, training, or experience may indicate that a knowledge barrier exists. The question to ask to see if there is a knowledge barrier is: "Did the person know how to do the job as it is supposed to be done?"

An example of a knowledge barrier is a newly hired, inexperienced worker with very little training is put to work driving a forklift. If the forklift is not driven properly because the worker does not know how, then a knowledge barrier to human performance exists. Knowledge is not the same as information. It implies much more. It implies that the person can use information, and can put education, training, and experience together to act appropriately on information. Knowledge is, therefore, internal to the person, something that the person brings into the work environment. Externals, such as information and incentives, operate on the person's use of knowledge.

Capacity

Capacity refers to physical ability, concentration, spreading too thin, and habit intrusion. Capacity is also internal to the person, and is both mental and physical. When a task exceeds the capacity of the individual performing it or when something impairs the person's capacity, a barrier to performance exists.

The impairment of capacity can take many forms. If the job requires an extended high level of concentration, a person with a short attention span will be at a disadvantage. The task exceeds the person's capacity. A job may also become so routine that it is done by habit. This habit may be so strong that changes in the environment that require a different response are not recognized. In this case, habit has impaired the mental capacity of the individual to attend to the environment. More obvious examples include the person who is asked to lift more than could possibly be handled, or the person who comes to work under the influence of drugs or alcohol.

These impairments of physical and mental capacities have the potential of turning hazardous situations into a disaster for the individual and others. To determine if there is a capacity barrier, the question is asked: "Did the task exceed the capacity of the individual performing it?"

Incentive

One of the most powerful means of strengthening, maintaining, and changing a person's responses is the use of incentives. Our responses result in rewards and punishment, which can create barriers to human performance. When unsafe performance has been rewarded or safe performance punished, it can be expected that inappropriate performance will occur in the future. If cutting corners on safety in order to increase short-term production is rewarded, the frequency of cutting corners is going to increase. To determine if incentives have created a barrier to human performance, the question to ask is: "Has the person been rewarded for doing the job incorrectly or punished for doing it correctly?"

An example of an incentive barrier is a worker doing piecework gets a bonus for the number of parts finished over quota. The worker knows that her machine is due for scheduled maintenance but she continues to operate it longer than recommended in order to receive more bonus money. The supervisor knows what is going on but decides not to intervene. The worker is rewarded for unsafe performance resulting in an incentive barrier to proper performance. The worker is reinforced by the bonus money because misuse of equipment allows more production. Note also a second thing: The supervisor ignores the unsafe behavior and therefore does not provide direction and guidance. There are times when there may be a fine line between determining whether the barrier is misuse of incentive or lack of direction or guidance. However, the knowledge of the investigator regarding the operation of the industry should help make those determinations.

There is a tendency, when discussing incentives in industry, to think in terms of the industry's or the company's incentive program. While the incentive program is important, it is a small part of the concept being presented. The major concerns are those little things that occur or fail to occur after the person responds. While the opportunity for bonuses, promotions, jackets, and the like are important, events such as praise from one's supervisor or co-workers, or treatment from one's work crew, are powerful determinants of behavior. These events, which occur in normal, everyday interactions at the worksite, should be studied to see if an incentive barrier exists.

Example: An individual wears safety glasses while the rest of the work crew does not. The crew may poke fun at the person until he or she stops wearing the glasses.

Being part of a group and getting support from the group are powerful incentives in determining performance.

Summary

When analyzing accidents for human barriers, investigators must realize that the five barriers discussed often operate together. It is an advantage to present them independently for understanding. However, it is also important to make five judgments as to which barrier may be the primary contributor to an accident. Each type of barrier not only requires a different type of solution, but the implementation costs of the solutions are quite different. The barriers to human performance are

1. Information (Direction and Guidance)—Was the person given direction and guidance in regard to performing the job?
2. Tools and Equipment (Appropriate Tools or Equipment)—Were the appropriate tools or equipment available and properly designed?
3. Knowledge (Know-How)—Did the person know how to do the job as it was supposed to be done?
4. Capacity (Physical Condition/Impairment)—Did the task exceed the capacity?
5. Incentive (Reward or Punishment)—Has the person been rewarded for doing the job incorrectly or punished for doing it right?

Performing the Human Barrier Analysis

A human barrier prevents workers from performing the way they should. When human behavior becomes an issue, a degree of complexity transpires. The barrier may include some of the ones found in Table 11.3.

TABLE 11.3
Potential Human Barriers

Perceptions	Drugs/alcohol
Limitations—physical, skill, intelligence	State of mind
Language	Peer pressures
Ability to read	Stressors
Disabilities	Understanding expectations
Failure to be trained	Social pressure
Policies/procedures	Familial pressure
Poor management	Level of expectations
Supervisors	Fail to provide proper tools
Disease	Faulty/flawed information
Effects of illness	What's in it for me?
Medications	

The HBA proceeds from data collected on the PBA matrix (see Table 11.2). The data of concern are those found in rows

A. Was not possible and practical.
B. Was possible and practical but not provided.
C. Was provided but not used.
D. Was provided and used, but failed.

Transfer accidents from the PBA matrix to the respective HBA tally sheet by doing the following:

1. Accidents in A—record any accident number on the HBA tally sheet that appears four times in the A category.
2. Accidents in B, C, and D—record all accidents on the respective HBA tally sheet that appears in these categories. (If they appear more than once, record them more than once.)

The PBA matrix categorized the accidents into two groups. In one group, a physical barrier would have prevented the accident (rows B and D). In the second group (rows A and C), human behavior played a significant role. Accidents A and C are generally caused by individuals who did something they should not have or did not do something they should have. All industries have accidents of this type. The HBA is our tool for further analysis of the accidents in A and D, where human behavior played a role. The HBA helps us answer the following question: If a person could have done something to prevent the accident, why did he or she not?

After recording the accidents from rows A and C on the HBA Tally Sheets (see Tables 11.4 through 11.7), the data are ready to do an HBA. Go back to the accident abstract and for each accident, ask the questions to determine which human barrier was operating.

When analyzing each accident, it puts one in a forced choice situation. But only select one of the five possible human barriers as the cause of the accident. Using one's experience in the industry, and the knowledge of the industrial operations and procedures, decide which of the human barriers you think probably led to the performance and contributed to the accident. Then place a checkmark in the column under the appropriate barrier for that accident. For each accident on the HBA Tally Sheet, only one checkmark can be in one of the five columns.

After you have recorded the accidents from rows B and D on the HBA Tally Sheets, complete those forms by filling in the blanks with the barrier that was not provided and the barrier that failed, respectively.

WRAP-UP OF PRELIMINARY DATA ANALYSIS

The elements of the PDA have been reviewed. By now it should be realized that this system is not doing the thinking for us, but, instead, provides us with the tools to help us think in a clearer, more consistent manner. Its design intends to steady our

TABLE 11.4
HBA Tally Sheet A

Which human barrier will have to be eliminated or modified to prevent this accident?

GROUP A

Accident Number	Information	Tools and Equipment	Knowledge	Capacity	Incentives
TOTALS					

TABLE 11.5
HBA Tally Sheet B

GROUP B

Accident No.	Barrier that Was Not Provided

Total No. ___________

TABLE 11.6
HBA Tally Sheet C

GROUP C

	Human barrier that helps explain why it was not used					
Accident Number	Physical Barrier that Was Not Used	Information	Tools and Equipment	Knowledge	Capacity	Incentives
TOTALS						

TABLE 11.7
HBA Tally Sheet D

GROUP D

Accident No.	Barrier that Failed

Total No. __________

course and at the same time encourages flexibility in pursuing areas of potential use. The supervisory role of the organizer and director of the project should also be coming into focus. Consistent, progressive, efficient programs do not fall apart by themselves. A poorly managed program is normally required to ensure such things.

There are additional areas to think about that will give us a chance to solidify our roles and responsibilities to make the system work. The purpose of the PBA is to familiarize us with accidents in an effort to establish some broad perspective of what may be causing them. The PBA and HBA are only for fast screening. As in all fast screening mechanisms, we must be willing to trade off a degree of accuracy for expediency. If we are 10 percent off in the placement of accidents into groups, this is certainly understandable and acceptable considering the lack of information on many of the accidents.

When looking at major groupings or clusters, it really will not make or break us if we miss an accident that probably should have gone in Group A instead of Group C. The result may have made Group A contain 75 accidents instead of 76—that is still considered acceptable.

If we are inclined to lean one way or the other, we should be given the benefit of the doubt for all groups except Group A. Because Group A will normally be the largest, it is easy to begin to throw everything into Group A without just consideration of the other categories.

Take what is given about the accident, read into it what is needed to make sense out of it, trust your instincts, and when finished, the margin of error will be well within the acceptable limits. Primarily, you must be consistent. Lack of consistency really shows up when one puts Accident #1 into Group A and Accident #73 into Group D when they are identical accidents. Lack of consistency can jeopardize any potential usefulness of the PBA and HBA.

There is no clear-cut number of accidents one should set out to analyze. Generally, 100 accidents is broad enough coverage to exhibit the breakdown of accidents into their representative groups. If one begins with ten accidents, for example, it is difficult to get any meaning from the fact that three accidents are in Group 2, two in another, etc. There just is not enough difference between Group 3 and Group 2 to make viable assumptions based on that difference.

If you have an operation with few accidents, say ten to twenty, that you are interested in, then you would probably handle the analysis a little differently. Because the number of accidents is small, you can go to the operation and get additional information about each accident, and then factor them into their respective groups as per the PBA and HBA.

However, it is important to remember that HBA and PBA are more than a set of tools. They are different in perspective, a different way of looking at accidents. Even if the industrial operation only had one accident, you would still want to look at it in terms of PBA, and HBA, although you would not go through the paperwork of PBA and HBA.

There is no hard-and-fast rule on how far back to go. Is there a tendency to look at accidents and incidents dating back 1 year, 2 years, or 3 years? If you were dealing with a large operation that ranks above the national average for incident rate, then you may not have to go beyond 1 or 2 months to get 100 accidents.

Has there been a sudden increase in accidents? If so, in order to find out which group accounts for that increase, you will have to analyze the accidents before the increase to find out what types were happening previously, analyze those happening after the increase, and then compare the two results to see how they differ. You must study as many accidents as resources permit over the time period that you think will yield the most useful insight. When putting the puzzle together, the pieces can be put together to make something out of them, but first find them and develop a workable system. PBA yields the first piece.

Before any program of this nature is taken to an industrial setting, something or someone must initiate it. Sometimes people learning PBA and HBA become overwhelmed at the time it takes to learn the techniques. From that, they assume a lot of time and human resources are required. When someone has learned to use HBA and PBA, and is comfortable with it, it is very easy for one person to analyze 100 accidents in a couple of hours.

INTERPRETATION OF PBA AND HBA

If our wish is to use PBA and HBA data to direct us in carrying out a systematic investigation at the industrial site and resources are limited, the fast screen provided by PBA and HBA will allow us to direct resources better at the site. PBA and HBA will lead us to the things we want to look at and the questions we want to ask when we arrive at the operation.

For example, from analysis of accidents there may appear to be a good proportion of accidents that resulted from physical barriers not being provided. It is very likely at such an operation that supervisors are not providing those barriers or taking action to ensure that barriers are provided. Remember that the purpose of PBA and HBA is to create a statement to guide your planning of on-site activities. If we suspect a supervisory problem, then we can start by concentrating on that assumption rather than using resources elsewhere. Then come back and analyze other potential causes later.

Using past experience in industry and knowledge of the particular industrial operations and procedures, ask: "What could be going on?" Maybe management's support of safety is not getting down to the supervisors. Maybe the supervisors know management's safety posture but are being rewarded for cutting corners to increase production. Could it be that safety procedures simply are not clear? Maybe the worker is being rewarded for cutting corners, or discouraged or punished when work is done safely?

While we do not know exactly why the incentive and communications systems are inadequate, we may suspect they are potential causes for many accidents. This offers guidance in our planning. For example, it is unlikely we would look at worker training if an inadequate reward system is suspected. We may also have a few accidents in our HBA that indicated a knowledge and capacity barrier; then we may choose to not use those accidents to determine potential causes. We more likely will decide to use the limited resources the best way possible. In this case, HBA indicates that the most mileage can be gained by looking at information and incentives.

We can now better define what to look at when the operation is visited. The information comes, again, from our experience and knowledge. Continue the inquiry into the existence of barriers and how they have impacted the accidents that are being analyzed. Although all the answers may not be available to us, barrier analysis can certainly provide much data on which to proceed during the investigation.

REFERENCES

United States Department of Labor, Mine Safety and Health Administration. N/NM Accident and Problem Identification Course (MSHA IG 68). Washington, D.C.: U.S. Department of Labor, 1988.

United States Department of Energy, Office of Nuclear Energy. Root Cause Analysis Guidance Document. Washington, D.C.: U.S. Department of Energy, February 1992.

12 Job Safety/Hazard Analysis

INTRODUCTION

A job safety/hazard analysis (JSA/JHA) is a procedure that integrates accepted safety and health principles and practices into a specific task or job procedure. In a JSA/JHA, each basic step of the job is to identify potential hazards and to recommend the safest way to do the job. Jobs that should have JSA/JHA conducted on them and receive attention first are

- Jobs with the highest injury or illness rates
- Jobs with the potential to cause severe or disabling injuries or illnesses, even if there is no history of previous accidents
- Jobs in which one simple human error could lead to a severe accident or injury
- Jobs that are new to the operation or have undergone changes in processes and procedures
- Jobs complex enough to require written instructions

Much of the information within this chapter comes from the United States Department of Labor's Mine Safety and Health Administration material entitled *The Job Safety Analysis Process: A Practical Approach*. The precept behind using JSA/JHA is that fatalities, accidents, and injuries can be reduced by working together and sharing safety knowledge. An accident prevention method that has proven effective in industry is the Job Safety/Hazard Analysis program.

JSA/JHA is a basic approach to developing improved accident prevention procedures by documenting the first-hand experience of workers and supervisors and, at the same time, it tends to instill acceptance through worker participation. JSA/JHA can be a central element in a safety program; and the most effective safety programs are those that involve employees. Each worker, supervisor, and manager should be prepared to assist in the recognition, evaluation, and control of hazards. Worker participation is important to efficiency, safety, and increased productivity. Through the process of JSA/JHA, these benefits are fully realized.

This process can begin by

- Involving employees
- Reviewing accident history
- Conducting a preliminary job review
- Listing, ranking, and setting priorities for hazardous jobs
- Outlining the steps or tasks

Job Safety Analysis, commonly known as JSA and also called Job Hazard Analysis (JHA), is a process used to determine the hazards of, and safe procedures for, each step of a job. A specific job, or work assignment, can be separated into a series of relatively simple steps. The hazards associated with each step can be identified and solutions can be developed to control each hazard.

PERFORMING A JSA/JHA

To perform a JSA/JHA, the following question should be asked:

- What can go wrong?
- What are the consequences?
- How could it happen?
- What are the contributing factors?
- How likely is it that the hazard will occur?

The answers to these questions will help in hazard identification. Some of the other accident specific questions are

- Can any body part get caught in or between objects?
- Do tools, machine, or equipment present any hazards?
- Can the worker make harmful contact with moving objects?
- Can the worker slip, trip, or fall?
- Can the worker suffer strain from lifting, pushing, or pulling?
- Is excessive noise or vibration a problem?
- Is the worker exposed to extreme heat or cold?
- Is there danger from falling objects?
- Is lighting a problem?
- Can weather conditions affect safety?
- Is harmful radiation a possibility?
- Can contact be made with hot, toxic, or caustic substances?
- Are there dusts, fumes, mists, or vapors in the air?

FOUR BASIC STEPS OF A JSA/JHA

JSA/JHA involves four basic steps:

1. Select a job to be analyzed.
2. Separate the job into its basic steps.
3. Identify the hazards associated with each step.
4. Control each hazard.

Looking at these four steps in detail will help explain the process and value of this type of analysis.

SELECTING A JOB TO ANALYZE

The first step in JSA/JHA is to select a job to analyze. The sequence in which jobs are analyzed should be established when starting a JSA/JHA program. Potential jobs for analysis should have sequential steps and a work goal when these steps are performed.

To use the JSA/JHA program effectively, a method must be established to select and prioritize the jobs to be analyzed. The jobs must be ranked in the order of greatest accident potential. Jobs with the highest risks should be analyzed first. Workers and supervisors may or may not be involved with the ranking process, but if asked to rank or prioritize jobs to be analyzed, the following criteria should be used:

Accident Frequency

A job that has repeatedly produced accidents is a candidate for an immediate JSA/JHA. The greater the number of accidents associated with the job, the greater its priority should be. These jobs should be analyzed as soon as possible.

Accident Severity

Every job that has produced an injury that resulted in lost time, restricted work activity, or required medical treatment, should be analyzed. These injuries prove that the preventive action taken prior to their occurrence was not successful.

Judgment and Experience

Many jobs qualify for immediate JSA/JHA because of the potential hazards involved. For example, the lifting of heavy equipment frequently poses hazards. Because jobs rely on experience and judgment, these jobs present a real ongoing hazard to workers.

New Jobs, Nonroutine Jobs, or Job Changes

Prime candidates for JSA/JHA result from jobs that are not done often and some have never been done at a particular worksite. The hazards of the job might not be fully known. By applying the JSA/JHA process to these jobs, the likelihood of an accident occurring is greatly reduced.

Routine Jobs

In routine or repetitive jobs with inherent hazards, the employee is exposed repeatedly to hazards. For example, exposures to high levels of noise over a period of time will affect the hearing of a worker.

Accident statistics, the ability to recognize hazards, and good common sense coupled with the guidelines that have been discussed will help you prioritize which JSA/JHA should be developed first. The goal is that all jobs with sequential steps will ultimately be analyzed. New jobs will be analyzed as they are introduced into the workplace.

THE JOB SAFETY/HAZARD ANALYSIS WORKSHEET

After a job has been selected and the JSA/JHA has been initiated, a worksheet is prepared listing the basic job steps, the corresponding hazards, and the safe procedures for each step. The basic form generally has three columns. In the left column, the Sequence of Basic Job Steps is listed in the order in which they occur. The middle column describes all Potential Accidents or Hazards. The right column lists the Recommended Safe Job Procedures that should be followed to guard against these hazards in order to prevent potential accidents (see Figure 12.1).

SEPARATING THE JOB INTO ITS BASIC STEPS

In the JSA process, it is easiest to deal with each column of the form separately. Logically, the job should be broken down into its basic steps first. Each step or activity should briefly describe what is done. Each activity should be listed on the form in the order it is accomplished.

Avoid the common errors of making the breakdown so detailed that an unnecessarily large number of steps result, or making it so general that basic steps are omitted. If a large number of steps result from the analysis (i.e., more than fifteen), consider breaking down that job into more than one JSA/JHA.

At this point in the analysis, you need to do two things: first, observe the job actually being performed (if possible by more than one person); and second, involve at least one employee who does the job regularly in the analysis. The employee(s) selected to assist in the analysis should be briefed on the purpose of JSA/JHA and the mechanics of how a JSA/JHA is developed. There are several ways to get the JSA/JHA process started with an employee. Have him or her review a list of basic job steps that have developed from observations, or have the employee explain what the job steps are. The important thing is that the JSA/JHA creator and the employee or supervisor mutually agree that the JSA/JHA accurately reflects the steps involved in the job.

SEQUENCE OF BASIC JOB STEPS

The first step might be to do a walk-around inspection. When outlining the job steps, it will be tempting to get very detailed and list how to do the job rather than the basic job steps. An example of this would be to list each component of a machine that the worker will check (visual examination of tires, light lens glass, guards, hoses, etc.) as a job step. If there is a desire to include this information in a JSA/JHA, it is best included in the third column ("Recommended Safe Job Procedures" in Table 12.1). Another alternative is to add a fourth column to the form for this purpose.

The list of job steps in the Sequence of Basic Job Steps column of the JSA/JHA form (Table 12.1) will continue to be broken down into manageable steps. Again, the general rule of thumb is to try to keep the JSA/JHA between ten and fifteen basic steps. If more than fifteen steps exist, the JSA/JHA creator may be trying to include too much information in a single JSA/JHA.

TABLE 12.1
Job Safety/Hazard Analysis (JSA/JHA) Worksheet

Job Safety Analysis Worksheet

Title of Job/Operation ______________ Date ________ No. ________

Position/Title(s) of Person(s) Who Does Job ________ Name of Employee Observed ________

______________________________ Analysis Made By ________

Department ______________________________ ________

Section ______________________________ Analysis Approved By ________

Sequence of Basic Job Steps	Potential Accidents or Hazards	Recommended Safe Job Procedures

1. Struck-By (SB)
2. Struck-Against (SA)
3. Contacted-By (CB)
4. Contact-With (CW)
5. Caught-On (CO)
6. Caught-In (CI)
7. Caught-Between (CBT)
8. Fall-Same-Level (FS)
9. Fall-To-Below (FB)
10. Overexertion (OE)
11. Exposure (E)

Source: Courtesy of Mine Safety and Health Administration.

It is important that the JSA/JHA accurately describe the work. There may be some disagreement about basic job steps. Some may think that checking fluid levels is part of the walk-around inspection—and on some machines this may be true. However, on other machines, checking fluid levels presents an entirely new set of hazards that may be overlooked if grouped with the general walk-around inspection. Such hazards include working from an elevated position and coming into contact with hot fluids or surfaces. Keep in mind that a JSA/JHA that is too broad is also a poor JSA/JHA.

IDENTIFYING THE HAZARDS ASSOCIATED WITH EACH JOB STEP

After all basic steps of the operation of a piece of equipment or job procedure have been listed, each job step should be examined to identify hazards associated with it. The purpose is to identify and list the hazards that are possible in each step of the job. Some hazards are more likely to occur than others, and some are more likely to produce serious injuries than others. Consider all reasonable possibilities when identifying hazards.

To make this task manageable work with the basic types of accidents, the question to ask is: "Can any of these accident types or hazards inflict injury on a worker?" There are eleven basic types of accidents:

1. Struck-against
2. Struck-by
3. Contact-with
4. Contacted-by
5. Caught-in
6. Caught-on
7. Caught-between
8. Fall-same-level
9. Fall-to-below
10. Overexertion
11. Exposure

Look at each of these basic accident types in more detail. Analyze each job step (first column of the JSA/JHA form in Table 12.1) separately; look for only one kind of hazard or accident at a time.

STRUCK-AGAINST TYPE OF ACCIDENTS

Look at these first four basic accident types (struck-against, struck-by, contact-with, and contacted-by) in more detail, keeping the job step walk-around inspection in mind. Can the worker strike against anything while doing the job step? Think of the worker moving and contacting something forcefully and unexpectedly (an object capable of causing injury). Can he or she forcefully contact anything that will cause injury? This forceful contact may be with machinery, timber or bolts, protruding objects, or sharp, jagged edges. Identify not only what the worker can strike against, but also how the contact can come about. This does not mean that every object around the worker must be listed.

Struck-By Type of Accidents

Can the worker be struck by anything while doing the job step? The phrase "struck by" means that something moves and strikes the worker abruptly with force. Study the work environment for what is moving in the vicinity of the worker, what is about to move, or what will move as a result of what the worker does. Is unexpected movement possible from normally stationary objects? Examples are ladders, tools, containers, supplies, etc.

Contact-With and Contact-By Types of Accidents

Can the worker be contacted by anything while doing the job step? The contact-by accident is one in which the worker could be contacted by some object or agent. This object or agent is capable of injuring by nonforceful contact. Examples of items capable of causing injury are chemicals, hot solutions, fire, electrical flashes, and steam.

Can the worker come in contact with some agent that will injure by nonforceful contact? Any type of work that involves materials or equipment that may be harmful by nonforceful contact is a source of contact-with accidents. There are two kinds of work situations that account for most of the contact-with accidents: one situation is working on or near electrically charged equipment, and the other is working with chemicals or handling chemical containers.

The subtle difference between contact-with and contact-by injuries is that in the former, the agent moves to the victim, while in the latter, the victim moves to the agent.

Caught-In, Caught-On, and Caught-Between Types of Accidents

The next three accident types involve "caught" accidents. Can the person be caught in, caught on, or caught between objects?

- A caught-in accident is one in which the person, or some part of his or her body, is caught in an enclosure or opening of some kind.
- Can the worker be caught on anything while doing the job step? Most caught-on accidents involve the worker's clothing being caught on some projection of a moving object. This moving object pulls the worker into an injury contact. Or, the worker may be caught on a stationary protruding object, causing a fall.
- Can the worker be caught between any objects while doing the job step? Caught-between accidents involve having a part of the body caught between something moving and something stationary, or between two moving objects. Always look for pinch points.

Fall-to-Same-Level and Fall-to-Below Types of Accidents

Slip, trip, and fall accident types are one of the most common accident types occurring in the workplace. Can the worker fall while doing a job step? Falls are such frequent accidents that we need to look thoroughly for slip, trip, and fall hazards. Consider whether the worker can fall from something above ground level, or whether the worker can fall to the same level. Two hazards account for most fall-to-same-level accidents: slipping hazards and tripping hazards. The fall-to-below accidents occur in situations where employees work above ground or above floor level, and the results are usually more severe.

Overexertion and Exposure Types of Accidents

The next two accident types are overexertion and exposure. Can the worker be injured by overexertion? That is, can he or she be injured while lifting, pulling, or pushing? Can awkward body positioning while doing a job step cause a sprain or strain? Can the repetitive nature of a task cause injury to the body? An example of this is excessive flexing of the wrist, which can cause carpal tunnel syndrome (which is abnormal pressure on the tendons and nerves in the wrist).

Finally, can exposure to the work environment cause injury to the worker? Environmental conditions such as noise, extreme temperatures, poor air, toxic gases and chemicals, or harmful fumes from work operations should also be listed as hazards.

CONSIDER HUMAN PROBLEMS IN THE JSA/JHA PROCESS

While working with the middle column, "Potential Accidents or Hazards," we also need to consider the following points related to human problems:

- What effects could there be if equipment is used incorrectly?
- Can the worker take shortcuts to avoid difficult, lengthy, or uncomfortable procedures?

Normally, the job steps for a JSA/JHA are listed in logical sequence. Some workers, however, may wish to change the sequence for one reason or another. For example, one operator may choose to check fluid levels before he or she does a general walk-around. This type of flexibility is good for worker morale and productivity. However, on the other hand, there are times when the sequence of the job steps or deviations from the job steps are critical to safe performance of the job. An example of this is that the walk-around inspection must be made and safety deficiencies corrected before the machine is put into service for the day. It would not be safe or proper to do the walk-around inspection after the machine has been put into service.

ELIMINATING OR CONTROLLING THE HAZARDS

There are ways to eliminate the hazards by choosing a different process, modifying an existing process, substituting with a less hazardous substance, or improving the environment. In developing solutions, the following steps can be taken:

- Find a new way to do the job.
- Change the physical conditions that create the hazard.
- Change the procedure to eliminate the hazard.
- Reduce the frequency of performing that specific job. Reduce exposure.
- Contain the hazard with enclosure, barrier, or guards.
- Use protective devices such as personal protective equipment.

CHANGE JOB PROCEDURES

If the sequence of job steps or the deviations from established job steps are critical to the safe performance of a job, this should be noted in the JSA/JHA.

The next part of the JSA process is to develop the "Recommended Safe Job Procedure" to eliminate or reduce potential accidents or hazards that have been identified. (This is the third column in Table 12.1.) The following four points should also be considered for each hazard identified for the job step:

1. Can a less hazardous way to do the job be found?
2. Can an engineering revision take place to make the job or work area safer?
3. Is there a better way to do the job? This requires determining the work goal and then analyzing various ways to reach the goal to see which way is safest.
4. Are there work-saving tools and equipment available that can make the job safer?

What to consider with this group of questions is how the equipment and work area can be changed, or provided with additional tools or equipment, to make the job safer. An example of an engineering revision to a machine could be to modify both the ladders and platform on it with some type of anti-slip surface. An example of adding a work-saving tool might be to purchase special hand trucks for use in a shop or an area where fifty-five-gallon drums are now moved by hand.

If a new, less hazardous way to do the job cannot be found, physical conditions should be studied carefully.

CHANGE THE FREQUENCY OF PERFORMING THE JOB

Can the physical conditions that created the hazard be changed? Physical conditions may be tools, materials, and equipment that may not be right for the job. These conditions can be corrected by either engineering revisions or administrative revisions, or a combination of both.

As an example of changing physical conditions, consider the following scenario: A company had several injuries while putting lime into a new hopper in a water treatment plant. Upon investigation it was found that the hopper opening is 5 feet off the floor and the lime is packaged in 100-pound bags. An administrative revision may be to purchase 50-pound bags of lime; and an engineering revision would be to build a platform and steps so that the bags would only have to be lifted waist-high. Obviously, a combination of both of these would be the best solution.

If hazards cannot be engineered out of the job, can the job procedure be changed? Be careful here because changes in job procedures to help eliminate the hazards must be carefully studied. If the job changes are too difficult, long, or uncomfortable, then the employee will take risks or shortcuts to avoid these procedures. Caution must be exercised when changing job procedures to avoid creating additional hazards.

An example of changing job procedures might be to service equipment at the beginning of the shift rather than at the end of the shift. This will allow the engine and other components of the machine to cool down and thus will eliminate contact with or contact by hazards.

Can the necessity of doing the job, or the frequency of performing the job, be reduced? Often, maintenance jobs requiring frequent service or repair of equipment are hazardous. To reduce the necessity of such a repetitive job, ask what can be done to eliminate the cause or condition that makes excessive repair or service necessary. For example, a guard keeps vibrating loose on a piece of equipment, thereby requiring reinstallation. Different types of bolts and nuts, or some other type of fasteners, may eliminate the problem.

PERSONAL PROTECTIVE EQUIPMENT

Finally, can personal protective equipment be used? The use of personal protective equipment should always be the last consideration in reducing the hazards of a job. The usefulness of personal protective equipment depends entirely on the worker's willingness to use it faithfully. It is always better to control the hazards of a job by administrative or engineering revisions. Personal protective equipment should only be considered a temporary solution to protecting a worker from a hazard, or as supplemental protection to other solutions.

During the JSA/JHA process, safety problems are going to surface. Some of these problems will be easily solved with suggestions made to upper management. Administrative revisions are the easiest to make because there is little if any capital outlay. New, better, or additional personal protective equipment normally requires minimal expenditures and can be instituted promptly. Work-saving tools and other equipment may require large expenditures and might be phased in over time as tools or equipment are replaced. Engineering revisions may take time to design and install. Changes in physical conditions may have to be engineered into the next upgrade or redesign.

SUMMARY

The steps involved in a JSA/JHA process have been outlined in the previous pages. It should be especially clear that the main point of doing a JSA/JHA is to prevent accidents by anticipating and eliminating hazards. JSA/JHA is a procedure for determining the sequence of basic job steps, identifying potential accidents or hazards, and developing recommended safe job procedures. Table 12.2 provides an example JSA/JHA.

The JSA/JHA is an accident prevention technique used in many successful safety programs. The JSA/JHA process is not difficult if it is undertaken with a

TABLE 12.2
JSA/JHA on Cleaning a Small Chemical Reactor by Experienced and Trained Personnel

Sequence of Basic Job Steps	Potential Accidents or Hazards	Recommended Safe Job Procedure
1. Check with lab manager for proper cleaning solvents and PPE.	1. Putting an incompatible chemical into the reactor may cause a violent reaction. Residual contents of reactor may have harmful effects.	1. The lab manager knows the last material in the reactor and will determine proper cleaning solvent along with appropriate PPE for that solvent. Lab manager will also provide proper batch operating temperatures.
2. Bring drum of material to reactor along with empty drum for waste. (For these purposes, the solvent will be acetone.)	2. Moving drums manually can cause back or muscle strain. Drums can tip easily when being moved manually.	2. Use drum truck. Use proper technique when loading and unloading drums. Wear leather gloves when handling drums or drum truck to avoid hand abrasion from rough surfaces.
3. Check ventilation system for draw.	3a. If ventilation system is not working, any fumes from drums or reactor will stay in area. 3b. If the drive motor belts break, the motor will continue to run but no ventilation will be provided. 3c. A visual signal is always a good indicator that the system is operational.	3a. Check the system running lights to ensure that they are all green. 3b. Check the alarm system to guarantee that the alarm will sound if air movement stops. 3c. Hold a tissue or wipe in front of the local exhaust port to visually ascertain that there is draw.
4. Visually check the reactor system. Make sure that all valves feeding the reactor are closed, and that digital and manual temperature indicators agree.	4. Loading material on top of unknown material can cause a violent reaction. A discrepancy in temperature indicators indicates that one is out of calibration. Running in this mode can cause the batch to over- or underheat.	4. Make certain that reactor is empty by looking through glass port on top of vessel. Call an instrument person to verify the calibration of the temperature sensors.
5. Inert the reactor. (Pull vacuum on reactor. When 28" reached, turn off vacuum and pressure with nitrogen to 15 psig. Repeat 3 times.)	5. Adding a solvent to a reactor containing oxygen creates two-thirds of the fire triangle. Friction from the stirring shaft, static, and heat from the reactor seal can ignite the mixture and result in fire or explosion.	5. Reducing the oxygen content in the vessel to below 2% will not allow combustion to take place.

continued

TABLE 12.2 (continued)
JSA/JHA on Cleaning a Small Chemical Reactor by Experienced and Trained Personnel

Sequence of Basic Job Steps	Potential Accidents or Hazards	Recommended Safe Job Procedure
6. Ground and bond acetone drums near reactor.	6. Static electricity can ignite acetone in drums.	6. Grounding or drums will give zero static charge to drum. Bonding the drum to the reactor will ensure that the drum and the reactor are at the same static charge. (Reactors are permanently grounded.) Solvents are not allowed to freefall. All loadings are from the bottom of the vessel or through dip tubes in the reactor.
7. Charge material into reactor to approximately 25–50% full by eye.	7a. Drums may be pressurized, causing material to be sprayed when cap is loosened. 7b. Vapors from mixture can get into vacuum pump and cause explosion.	7a. Wear full-face shield and butyl gloves, in addition to normal PPE. Bring local ventilation hose next to drum bung. Open bung slowly to vent any pressure. 7b. Use residual vacuum when loading materials.
8. Turn on cooling water to condenser.	8. Without sufficient cooling in condenser, vapors will reach scrubber blower motor and cause a fire.	8. Open cooling water valves to condenser. Watch spinner in line to make certain there is water flow.
9. Open vent on reactor.	9. Opening a reactor vent valve while under vacuum will cause any material in vent line to be pulled into reactor, possibly creating a violent reaction.	9. Increase reactor pressure to 5 psig with nitrogen. Shut off nitrogen flow and open reactor vent valve slowly.
10. Heat batch until acetone refluxes (56°C).	10. Putting too much heat into the batch can cause vapors to pass through the condenser, reach the scrubber blower motor, and cause a fire.	10. Use 70°C hot water in reactor jacket to heat batch, using maximum water flow. Watch vapor condensation level in condenser and adjust hot water temperature to make level about half to two-thirds up the condenser.

TABLE 12.2 (continued)
JSA/JHA on Cleaning a Small Chemical Reactor by Experienced and Trained Personnel

Sequence of Basic Job Steps	Potential Accidents or Hazards	Recommended Safe Job Procedure
11. After 1 hour at reflux, inert the system and cool the batch to 20–25°C.	11a. If oxygen gets into system, a fire can occur, sparked by heat from the mechanical seal or the agitator. 11b. Glass will break if temperature extreme across system is too great.	11a. Close the reactor vent valve and apply 3 psig continuous nitrogen pressure. 11b. Use city water to cool batch. Do not use refrigerated water or refrigeration.
12. Ground and bond empty drum in front of reactor.	12. Static electricity can ignite acetone in drums.	12. Grounding of drums will give zero static charge to drum. Bonding the drum to the reactor will ensure that the drum and the reactor are at the same static charge.
13. Drain material from reactor into drum.	13a. Material can spray out of drum. 13b. Vapors can be present. 13c. When reactor is drained, reactor pressure will be relieved into drum, causing spraying of material. 13d. Material being removed can be contaminated with previously loaded material, causing health risk.	13a. Material is drained out slowly. Protective shielding is placed around bung to protect operator. 13b. Local ventilation is brought to drum bung to remove any vapors. 13c. Once the draining has started, stop the nitrogen pressure. The loss in reactor volume will reduce the pressure to almost atmospheric by the time the reactor is almost empty. 13d. Check with lab supervisor for proper PPE.
14. Close drum and reactor.	14a. If drum is too hot when closed, drum will implode, causing possible leakage. If drum is too cold, it will warm up and expand, causing possible leakage. 14b. Material can drain from reactor to floor, causing exposure issues.	14a. Using local ventilation, let drum equilibrate with room temperature, usually about 1 hour, before capping. 14b. Close bottom-out valve on reactor.

common-sense approach on a step-by-step basis. The JSA/JHA should be reviewed often and updated with input from both supervisors and workers who do the job every day. The implementation of the JSA/JHA process will mean continuous safety improvements at your workplace with the ultimate goal of zero accidents. JSA/JHA takes a little extra effort, but the results are positive and helpful for everybody.

There are many advantages in using the JSA/JHA. It provides training to new employees on safety rules and specific instructions on them and how the rules are to be applied to their work. This training is provided before the new employees perform the job task(s). The JSA/JHA also instructs new employees in safe work procedures.

With JSA/JHA, experienced employees can maintain safety awareness behavior and receive clear instructions for job changes or new jobs. Benefits also include updating current safety procedures and instructions for infrequently performed jobs.

It is important to involve workers in the JSA/JHA process. Workers are familiar with the jobs and can combine their experience to develop the JSA. This results in a more thorough analysis of the job. A complete JSA/JHA program is a continuing effort to analyze one hazardous job after another until all jobs with sequential steps have a written JSA/JHA. Once established, the standard procedures should be followed by all employees.

REFERENCES

Reese, C. D. and J. V. Eidson. *Handbook of OSHA construction safety & health.* Boca Raton, FL: CRC Press/Lewis Publishers, 1999.

United States Department of Labor, Occupational Safety and Health Administration, Office of Training and Education. *OSHA Voluntary Compliance Outreach Program: Instructors Reference Manual.* Des Plaines, IL: U.S. Department of Labor, 1993.

United States Department of Labor, Occupational Safety and Health Administration. Job Hazard Analysis, (OSHA 3071). Washington, D.C.: U.S. Department of Labor, 1992.

United States Department of Labor, Mine Safety and Health Administration. Accident Prevention (Safety Manual No. 4). Beckley, WV: U.S. Department of Labor, revised 1990.

United States Department of Labor, Mine Safety and Health Administration. Job Safety Analysis: A Practical Approach (Instruction Guide No. 83). Beckley, WV: U.S. Department of Labor, 1990.

United States Department of Labor, Mine Safety and Health Administration. Job Safety Analysis (Safety Manual No. 5). Beckley, WV: U.S. Department of Labor, revised 1990.

United States Department of Labor, National Mine Health and Safety Academy. Accident Prevention Techniques: Job Safety Analysis. Beckley, WV: U.S. Department of Labor, 1984.

United States Department of Labor, Mine Safety and Health Administration. Safety Observation (MSHA IG 84). Beckley, WV: U.S. Department of Labor, revised 1991.

13 Safe Operating Procedures (SOPs)

INTRODUCTION

A Safe Operating Procedure (SOP) or Standard Operating Procedure should include safety and health as a part of the standard operating practices delineated within it. The SOP is a set of written instructions that documents a routine or repetitive activity followed by an organization. The development and use of SOPs are an integral part of a successful quality system as it provides workers with information to perform a job properly and safely while facilitating consistency in the quality and integrity of a product or end result.

Workers may not automatically understand a task just because they have experience or training. Thus, many jobs, tasks, and operations are best supported by an SOP. The SOP walks the worker through the steps of how to do a task or procedure in a safe manner and calls attention to the potential hazards at each step. The development and use of SOPs minimizes variation and promotes quality through consistent implementation of a process or procedure within the company, even if there are temporary or permanent personnel changes. It minimizes opportunities for miscommunication and can address safety and health concerns. SOPs are frequently used as checklists by inspectors when auditing procedures. Ultimately, the benefits of a valid SOP are reduced work effort, safe performance, along with improved comparability, credibility, and legal defensibility.

You might ask why an SOP is needed if the worker has already been trained to do the job or task. As you may remember from the previous chapter, a job safety analysis usually keys in on those particular jobs that pose the greatest risk of injury or death. These are high-risk types of work activities and definitely merit the development and use of an SOP. There are times when an SOP, or step-by-step checklist, is useful, including the following:

1. New worker is performing a job or task for the first time.
2. Experienced worker is performing a job or task for the first time.
3. Worker has not performed job in a long time.
4. Experienced worker is performing a job that he or she has not done recently.
5. Mistakes could cause damage to equipment or property.
6. Job is done on an intermittent or infrequent basis.
7. New piece of equipment or different model of equipment is obtained.
8. Supervisors need to understand safe operation to be able to evaluate performance.

9. Procedure or action within an organization is repetitive and is carried out in the same way each time.
10. Procedure is critically important, no matter how seldom performed, and must be carried out exactly according to detailed, step-wise instructions.
11. Need to standardize the way a procedure is carried out for ensuring quality control or system compatibility.

When airline pilots fly, the most critical parts of the job are takeoffs and landings. Because these are such crucial aspects of flying, a checklist for proceeding in a safe manner is used to mitigate the potential for mistakes. It is vital to provide help when a chance for error can result in grave consequences. Similarly, laminated SOPs should be placed on equipment, machines, and vehicles for individuals who need a refresher prior to operation because they have not used the equipment or have performed a task on an infrequent basis.

Few people or workers want to admit that they do not know how to perform a job or task. They will not ask questions, let alone ask for help in doing an assigned task. This is the time when a laminated SOP or checklist could be placed at the worksite or attached to a piece of equipment. This can prove to be a very effective accident prevention technique. It can safely walk a worker through the correct sequence of necessary steps and thus avoid the exposure to hazards that can put the worker at risk of injury, illness, or death.

These SOPs could be used when helicopters are employed for lifting, industrial forklifts are utilized, materials are moved manually, etc. They should list the sequential steps required in order to perform the job or task safely, the potential hazards involved, and the personal protective equipment needed. Each step in the SOP should provide all the information required to accomplish the task in a safe manner.

If there is no annual training, then the use of SOPs may instill confidence in the workers that they have a way of refreshing their memories on how to do their tasks. When changing procedures, the SOP should be updated to ensure that the changed procedure is performed safely. A checklist is one form of SOP. The checklist is very effective when you are trying to make sure that every step is being accomplished.

SOPs are only useful when they are up-to-date and readily accessible at the actual job or task site. Because we have the ability to store SOPs in the computer, revisions can be made simply and are therefore easily changed if you receive good suggestions from supervisors or workers. Figure 13.1 shows a picture of a forklift and Table 13.1 shows an example SOP for a forklift. Using the format from this example, develop your own SOPs for procedures, jobs, tasks, or equipment.

An SOP is only one accident prevention technique or component of any safety and health initiative. There are specific jobs or tasks that lend themselves well to this approach. Make sure that SOPs are used when they benefit the type of work being performed the most, and not as a cure-all for all accidents and injuries. Use them as one of the many tools for accident prevention.

COMPONENTS OF AN SOP

SOPs are intended to provide clear instructions for safely conducting activities involved in each procedure covered. They help ensure consistent work activity using

FIGURE 13.1 Forklift.

the appropriate manufacturer's guidelines and instructions, and they provide other pertinent safety information from other resources. They also allow individuals with specific safety knowledge and expertise to address at least the following elements:

1. Steps for each operating phase:
   a. Initial start-up
   b. Normal operations
   c. Temporary operations
   d. Emergency shutdown, including the conditions under which emergency shutdown is required, and the assignment of shutdown responsibility to qualified operators or workers to ensure that emergency shutdown is executed in a safe and timely manner
   e. Emergency operations
   f. Normal shutdown
   g. Start-up following a turnaround, or after an emergency shutdown
2. Operating limits:
   a. Consequences of deviation
   b. Steps required for correcting or avoiding deviations
3. Safety and health considerations:
   a. Properties of, and hazards presented by, the chemicals and materials used in the process and hazards involved in the task
   b. Precautions necessary to prevent exposure, including engineering controls, administrative controls, and personal protective equipment
   c. Control measures to take if physical contact or airborne exposure occurs
   d. Quality control for raw materials, control of hazardous chemical inventory levels, and any other special or unique hazards
4. Safety systems and their functions

SOPs should be readily accessible to employees who work with or maintain a process or operation. They should be reviewed as often as necessary to ensure that they reflect current operating practice, including changes that result from changes in the process or procedures due to new understanding or requirements of technology, equipment, and facilities.

The employer should develop and implement safe work practices to provide for the control of hazards during operations such as equipment/machine operation, lockout/tagout, confined space entry, opening process equipment or piping, and control over an entrance into a facility by maintenance, contractor, laboratory, or other support personnel. These safe work practices must apply to both company employees and contract employees.

GUIDELINES FOR WRITING AN SOP

SOPs are often poorly written because little thought is given or effort made to get them right. At times they are mandated as a quick fix for a perceived problem. An organized and thoughtful approach will yield SOPs that are truly usable by the workers.

SOPs should be written in a concise, step-by-step, and easy-to-read format. The information presented should be unambiguous and not overly complicated. The active voice and present verb tense should be used. The term "you" should not be used, but implied. The document should not be wordy, redundant, or overly lengthy. Keep it simple and short. Information should be conveyed clearly and explicitly to remove any doubt as to what is required. Also, use a flowchart to illustrate the process being described. Here are some guidelines for writing an SOP:

1. Decide what SOPs must be written based upon a review of organizational functions.
2. Check to see if there is an existing SOP that can be revised or updated.
3. Gather information on the procedure from reference sources and knowledgeable employees. When possible, contact other agencies performing similar functions to see if they have an SOP, and request a copy to use as a guide or source of ideas.
4. Select a suitable format for the SOP to be written. (See example in Table 13.1.)
5. Assemble blank forms and any other documents to refer to in the SOP.
6. Write a draft of the SOP. Include copies of any blank forms or specific references referred to in the SOP.
7. Review, or have a fellow employee review, the draft SOP for technical adequacy.
8. Request that the draft SOP be reviewed for administrative adequacy by the supervisor or the officer in charge.
9. Incorporate any changes indicated by the reviews into a final draft.
10. Date, sign, assign a file number, and distribute the new or revised SOP. The final copy should be signed both by the official responsible for preparing the SOP and by the official's supervisor or the official in charge. Copies should be provided to supervisors and officers in charge of the immediate organization; they should be posted in an SOP file for ready reference.

TABLE 13.1
Safe Operating Procedure: Powered Industrial Forklift

What to Do	How to Do It	Key Points
Perform A Pre-Start-Up Inspection	Walk around the vehicle, checking overall condition of:	
	1. Tires	1. Check tires to ensure adequate tread, no cuts/missing chunks, all tire bolts are present and are tight.
	2. Fluid leaks	2. Check hydraulic hose fittings for evidence of fluid leak. Look beneath vehicle for fluid on the floor.
	3. Overhead guard	3. Check for missing bolts, bent frame.
	4. Lifting forks and load backrest	4. Check the lifting forks and backrest for damage. Check that the lifting fork's width adjustment lock pins are in good condition and are working smoothly.
	5. P/M sticker	5. If P/M expiration date is not valid, do not operate vehicle.
Perform Operating Controls Inspection	1. Sit in driver's seat and operate controls.	1. Check to ensure parking brake is engaged and gear shift is in park or neutral.
	2. Adjust seat for effective operation and comfort.	2. Seat should be adjusted to allow foot brake pedal to be depressed without reaching with foot.
	3. Fasten seat belt.	3. Seat belt should fit snugly across hips.
	4. Turn on power to vehicle.	4. Check 360 degrees around vehicle to ensure no one is standing near vehicle.
	5. Turn on headlights.	5. Headlights must be bright and in position to ensure being seen by other vehicle operators or pedestrians.
	6. Depress brake foot pedal.	6. The pedal must be firm and brake lights must function and be bright.
	7. Depress horn button.	7. Horn should function easily and be loud.
	8. Elevate lifting forks and tilt by pulling back on control levers.	8. Hydraulic controls shall operate smoothly.
	9. Lower lifting forks to 2"–4" inches above floor.	9. Keep lifting forks 2"–4" above surface when in motion.
	10. Report any safety check failure to supervisor immediately for repair.	10. Do not operate if any safety check fails. Ensure the vehicle is not operated until repaired.

continued

TABLE 13.1 (continued)
Safe Operating Procedure: Powered Industrial Forklift

What to Do	How to Do It	Key Points
Operating Procedure: Traveling to Destination	1. Depress foot brake pedal.	1. Check 360 degrees around vehicle to ensure no one is standing near vehicle. Keep lifting forks 2"–4" above surface when in motion.
	2. Release parking brake.	2. —
	3. Select direction of travel.	3. Engage gear drive. Check travel direction to ensure path is clear of pedestrians or other vehicles.
	4. Remove foot from brake pedal.	4. Remove foot slowly.
	5. Depress accelerator pedal.	5. Depress accelerator pedal slowly to avoid quick, jerky start. Keep lifting forks 2"–4" above surface when in motion.
	6. Obey safety rules and regulations.	6. Travel at speeds that allow vehicle to be under control at all times under any condition. • Travel single-file, keeping to the right. • Pedestrians have right of way. • Emergency vehicles have right of way at all times. • Use lights and horn when necessary. • Allow at least 15 feet or three vehicle lengths between you and powered industrial vehicle (PIV) in front. • Do not pass other vehicles traveling in the same direction at intersections or blind spots, or narrow passages. • Ensure there is adequate overhead clearance. • Do not travel over objects. • Avoid sudden stops, except in emergencies. • Stop at all stop signs, blind corners, or when entering intersecting aisle and look for pedestrians and vehicle traffic.
Operating Procedure: Material Pick-Up	1. Approach material slowly.	1. Reduce speed to avoid sudden stops.
	2. Stop vehicle, depress foot brake pedal.	2. Apply slow, steady pressure until vehicle stops.
	3. Shift to park or neutral.	3. Never select direction while in motion.
	4. Engage parking brake.	4. If on an incline, block wheels.
	5. Unfasten seat belt and dismount; set forks for maximum load width.	5. Know the vehicle's capacities and load weights.

TABLE 13.1 (continued)
Safe Operating Procedure: Powered Industrial Forklift

What to Do	How to Do It	Key Points
	6. Remount and fasten seat belt.	6. Seat should be adjusted to allow foot brake pedal to be depressed without over extending.
	7. Depress foot brake pedal.	7. Check 360 degrees around vehicle to ensure no one is standing near vehicle.
	8. Release parking brake.	8. —
	9. Select direction of travel.	9. Engage gear drive. Check travel direction to ensure path is clear.
	10. Remove foot from brake pedal.	10. Remove foot slowly.
	11. Depress accelerator pedal.	11. Depress accelerator pedal slowly to avoid quick, jerky start.
	12. Approach the load with lifting forks level.	12. Lifting forks should be parallel with walking surface.
	13. Penetrate forks to back of pallet.	13. The pallet should be set against the backrest.
	14. Depress foot brake pedal, bringing vehicle to a stop.	14. Apply slow, steady pressure until vehicle stops.
	15. Raise the lifting forks until pallet is 2"–4" above walking surface.	15. Never raise forks while in motion.
	16. Tilt the lifting forks back slightly.	16. Tilt the forks backward slightly to prevent the load from falling forward.
	17. Select direction of travel.	17. Engage gear drive. Check travel direction to ensure path is clear of pedestrians or other vehicles.
	18. Remove foot from brake pedal.	18. Remove foot slowly.
	19. Depress accelerator pedal.	19. Depress accelerator pedal slowly to avoid quick, jerky start.
	20. Travel carefully to destination, obeying rules and regulations.	20. Travel at speeds that allow vehicle to be under control at all times under any condition. • Travel single-file, keeping to the right. • Pedestrians have right of way. • Emergency vehicles have right of way at all times. • Use lights and horn when necessary. • Allow at least 15 feet or three vehicle lengths between you and PIV in front. • Do not pass other vehicles traveling in the same direction at intersections or blind spots, or narrow passages.

continued

TABLE 13.1 (continued)
Safe Operating Procedure: Powered Industrial Forklift

What to Do	How to Do It	Key Points
		• Ensure there is adequate overhead clearance. • Do not travel over objects. • Avoid sudden stops, except in emergencies. • Stop at all stop signs, blind corners, or when entering intersecting aisle and look for pedestrians and vehicle traffic. • Always drive with load facing uphill. • Drive backward when view is obstructed by large loads.
Operating Procedure: Material Drop-Off	1. Depress foot brake pedal bringing vehicle to a stop.	1. Apply slow, steady pressure until vehicle stops.
	2. Shift to park or neutral.	2. Never select reverse while in motion.
	3. Tilt forks forward until parallel with walking surface.	3. Never lower forks while in motion.
	4. Lower lifting forks until pallet bottom is resting on surface and forks no longer support load.	4. Lower load slowly to prevent sudden drop.
	5. Engage reverse drive.	5. Check 360 degrees around vehicle to ensure no one is standing near vehicle.
	6. Remove foot from brake pedal.	6. Engage gear drive. Check travel direction to ensure path is clear of pedestrians or other vehicles.
	7. Depress accelerator pedal.	7. Depress accelerator pedal slowly to avoid quick, jerky start.
	8. Travel enough distance until lifting forks can clear pallet.	8. Travel in reverse until there is enough distance between the end of lifting forks and pallet.
	9. Depress foot brake pedal, bringing vehicle to a stop.	9. Apply slow steady pressure until vehicle stops.
	10. Engage forward drive.	10. Check 360 degrees around vehicle to ensure no one is standing near vehicle. Never select directional change while in motion.
	11. Remove foot from brake pedal.	11. Check travel direction to ensure path is clear of pedestrians or other vehicles.
	12. Depress accelerator pedal.	12. Depress accelerator pedal slowly to avoid quick, jerky start.

TABLE 13.1 (continued)
Safe Operating Procedure: Powered Industrial Forklift

What to Do	How to Do It	Key Points
	13. Travel carefully to destination, obeying rules and regulations.	13. Travel at speeds that allow vehicle to be under control at all times under any condition. • Travel single-file, keeping to the right. • Pedestrians have right of way. • Emergency vehicles have right of way at all times. • Use lights and horn when necessary. • Allow at least 15 feet or three vehicle lengths between you and PIV in front. • Do not pass other vehicles traveling in the same direction at intersections or blind spots, or narrow passages. • Ensure there is adequate overhead clearance. • Do not travel over objects. • Avoid sudden stops, except in emergencies. • Stop at all stop signs, blind corners, or when entering intersecting aisle and look for pedestrians and vehicle traffic.
Operation Procedure: Shutdown	1. Depress foot brake pedal, bringing vehicle to a stop.	1. Apply slow, steady pressure until vehicle stops.
	2. Shift to park or neutral.	2. Never select directional changes while in motion.
	3. Engage parking brake.	3. If on an incline, block wheels.
	4. Lower lifting forks slowly until resting on walking surface.	4. Never lower forks while in motion.
	5. Turn off power.	5. Remove key.
	6. Release seat belt.	6. —
	7. Dismount.	7. Wheels must be blocked if parked on an incline.
	8. Perform walk-around inspection, noting damage or operational problems.	8. Report all operation problems to supervisor for repair.
	9. Remove all trash.	

The most common problems and errors found in SOPs are summarized in the following list:

1. *Enumerating responsibilities for carrying out a procedure rather than stating who does the procedure and how.* Regulations are the place for delineating responsibilities, not SOPs.
2. *Failure to clearly state who carries out which step in the procedure.* The "Who" is as important as the "What?"
3. *Inclusion of steps or procedures performed by persons outside the organization.* This information has no place in an SOP because it involves actions that are beyond the direct control of the organization. Include only those steps that are carried out by the employees in the immediate organization; all else is irrelevant.
4. *Vagueness and imprecision.* What if the reader of the SOP cannot figure out exactly who (job description) is required to carry out a step in the procedure, and furthermore cannot determine precisely how it is to be carried out? Obviously, then, the SOP has failed in its primary objective, communication. This is why a prime function of the reviewer is to check to see if the writer has conveyed the message clearly and unequivocally.

HOW SOPS WORK

Safe or standard operating procedures describe tasks to be performed, data to be recorded, operating conditions to be maintained, samples to be collected, and safety and health precautions to be taken. The procedures must be technically accurate, understandable to employees, and revised periodically to ensure that they reflect current operations. Operating procedures should be reviewed by engineering staff and operating personnel to ensure that they are accurate and provide practical instructions on how to actually carry out job duties safely.

Operating procedures include specific instructions or details on what steps should be taken or followed in carrying out the stated procedures. These instructions for each operating procedure should include the applicable safety precautions and should contain appropriate information on safety implications. For example, operating procedures should address operating parameters and contain operating instructions about pressure limits, temperature ranges, flow rates, what to do when an upset condition occurs, what alarms and instruments are pertinent if an upset condition occurs, and other subjects (e.g., operating instructions to drive a powered vehicle properly during start-up or shutdown involve unique processes). In some cases, different parameters from those of normal operation are required. These operating instructions must clearly indicate the distinctions between normal and deviant operations, such as instructions and parameters on the appropriate allowances for driving while fully loaded.

Operating procedures and instructions are important for training operating personnel. Operating procedures are the standard operating procedures (SOPs) for operations. Operators and operating staff, in general, need to have a full understanding of operating procedures. If workers are not fluent in English, then procedures and instructions should be prepared in a second language understood by those workers. In addition, operating procedures should be changed when there is a change in the process as a result of management decision or procedural changes. The consequences of operating procedure changes must be fully evaluated and communicated to personnel. For example, mechanical changes to a process made by the maintenance department (e.g., changing a valve from steel to brass or other subtle changes) should be evaluated to determine if operating procedures and practices should also be changed.

All management changes and actions must be coordinated and integrated with current operating procedures, and operating personnel must be oriented to the changes in procedures before the changes are made. When the process is shut down in order to make the changes, the operating procedures must be updated before start-up of the process.

Training in how to handle upset conditions must be accomplished as well as what operating personnel are to do in emergencies (such as when a pump seal fails or a pipeline ruptures). Communication must also be maintained between operating personnel and workers performing work within the operating or production area, especially with nonroutine tasks. Hazards of the task must be conveyed to both operating personnel (e.g., supervisors), in accordance with established procedures, and to those workers performing the actual tasks.

SUMMARY

An SOP (Standard or Safe Operating Procedure) is a document that describes the regularly recurring operation relevant to quality and safe performance. The purpose of the SOP is to carry out the operations correctly and always in the same manner. An SOP should be available at the place where the work is performed.

SOPs are organizational tools that provide a foundation for training new employees, for refreshing the memories of management and experienced employees, and for ensuring that important procedures are carried out in a standard, specified manner. The principal function of an SOP is to provide detailed, step-by-step guidance to employees who are required to carry out a certain procedure. In this instance, the SOP serves not only as a training aid, but also as a means of helping to ensure that the procedure is carried out in a standard, approved manner.

Another important function of the SOP is to keep management informed about the way functions are performed in areas under their supervision. A complete file of well-written, up-to-date SOPs is an indication of good management and provides management with instant access to information on functional details of the organization for which they are responsible. This is of enormous benefit during inspections and management reviews, to say nothing of providing timely answers to unanticipated questions from superiors.

An SOP is a compulsory instruction. If deviations from this instruction are allowed, the conditions for these should be documented, including who can give permission for this and what exactly the complete procedure will be. The original should be kept in a secure place, while working copies should be authenticated with stamps or signatures of authorized persons.

The overridingly important feature of a good SOP is that it communicates what is to be done in a clear, concise, and step-wise manner. The most important person to whom it must communicate is typically the new employee who may have little or no experience with the procedure in question. Therefore, it is imperative that the writer of an SOP figuratively place him- or herself in the position of a new, inexperienced employee in order to appreciate what must be communicated and how to communicate it.

The content of an SOP should be comprehensive in terms of how to get the procedure accomplished, but should not encompass matters not directly relevant because to digress does not directly address the issue of how to get the procedure accomplished and exactly who is to do it.

Perhaps the best advice concerning the content of an SOP is this: Ask yourself the questions Who? What? Where? When? How? If the SOP answers all these questions, it is complete. If not, revise it until it does in as clear and logical an order as possible.

REFERENCES

Reese, C. D. and J. V. Eidson. *Handbook of OSHA construction safety & health.* Boca Raton, FL: CRC Press/Lewis Publishers, 1999.

United States Department of Labor, Occupational Safety and Health Administration, Office of Training and Education. OSHA Voluntary Compliance Outreach Program: Instructors Reference Manual. Des Plaines, IL: U.S. Department of Labor, 1993.

United States Department of Labor, Mine Safety and Health Administration, Accident Prevention (Safety Manual No. 4). Beckley, WV: U.S. Department of Labor, revised 1990.

United States Department of Labor, Mine Safety and Health Administration, Job Safety Analysis: A Practical Approach (Instruction Guide No. 83). Beckley, WV: U.S. Department of Labor, 1990.

United States Department of Labor, Mine Safety and Health Administration, Job Safety Analysis (Safety Manual No. 5). Beckley, WV: U.S. Department of Labor, revised 1990.

United States Department of Labor, Mine Safety and Health Administration, Safety Observation (MSHA IG 84). Beckley, WV: U.S. Department of Labor, revised 1991.

14 Job Safety Observation

INTRODUCTION

Much of the material on job safety observation (JSO) in this chapter was taken from the United States Department of Labor, Mine Safety and Health Administration's publication entitled "Safety Observation (MSHA IG 84)." Job safety observations are one of the accident prevention techniques that can be used to assess safe work performance.

There are many categories of accident causes and many terms used to describe these causes. To precisely determine the causes for each category, the terms "person causes" and "environmental causes" are often used. The "actual" and "potential" causes of accidents are generally accepted as the key factors in a successful loss prevention effort. Actual causes—direct and indirect—can only be considered after an accident has occurred. They can be found by asking the question: What caused the accident? Potential causes may be avoided before an accident actually occurs by asking the following question: What unsafe conditions (environmental causes) and/or unsafe procedures (person causes) could cause an accident? Working with actual causes is similar to firefighting, with after-the-fact analysis and hindsight. The process of understanding, determining, and correcting potential causes is comparable to fire prevention or foresight.

All categories of accident causes must be considered and used in any effective loss prevention program. Safety observations and inspections are necessary phases in the overall safety effort. A job safety observation is the process of watching a person (worker) perform a specific job to detect unsafe behavior or recognize safe job performance. Making a safety inspection is the process of visually examining the work area and work equipment to detect unsafe conditions (environmental causes). Detecting and eliminating potential causes of accidents may best be accomplished when supervisors understand safety observations, and when safety inspections become separate phases of the loss prevention work. This chapter deals primarily with making job safety observations.

The safety observation phase is initiated when a written set of procedures is prepared by management and safety health personnel. The procedures should include a prepared job safety analysis ready for use, step-by-step safe job operating procedures, and the training of all supervisors in observation procedures. Objectives must be established for each step of the JSO program. The establishment of definite goals at all levels of management will give direction to the safety and health effort. Management should outline the purpose and types of job safety observations, how to select a job or task for planned safety observations, preparing for a planned safety observation, a checklist of activities to observe, unsafe procedures, the employees' role in the observation process, the action to take after the observation, and procedure for dealing with unsafe behavior and performance.

PURPOSE OF JOB SAFETY OBSERVATION

The basic idea of a JSO is simple: It is a special effort to see how employees do their jobs. Planned safety observation involves more effort than an occasional or incidental observation of job procedures. This chapter describes various ways of determining unsafe practices and violations of safety rules. This accident prevention method emphasizes the importance of a proper supervisor-employee relationship. Becoming more interested in the employee through observations will lead to greater cooperation in the safety and health program.

TYPES OF JOB SAFETY OBSERVATIONS

There are many different types of JSOs. The two that are reviewed in this chapter are incidental observation and planned observation.

Incidental Safety Observations

Detecting and correcting any unsafe behavior is an important duty of any supervisor. Supervisors must give this type of observation as much attention as possible by pausing occasionally to discuss the actions of employees and observe employees, and by observing both for safe and unsafe behaviors. This observation is often done by a casual glance. These actions are incidental safety observations. This is a very informal observation approach.

Planned Safety Observations

In a planned observation, the supervisor selects the employee and the job to observe. The supervisor also decides the most suitable time. Some supervisors may want to make assignments for planned job safety observations. The basic tool for making a planned observation is job safety/hazard analysis (JSA/JHA). (If a JSA/JHA is not used, the supervisor must be completely familiar with the job steps, job hazards, and safe job procedures.) The supervisor should observe the employee doing a complete job cycle, paying attention to safe or unsafe procedures and conditions. A Planned Job Safety Observation Form should be used (see Table 14.1). All safe practices noted should result in a sincere compliment to the employee involved while any unsafe actions call for appropriate corrective measures. In either instance, the supervisor should make an observation record. A planned safety observation is a valuable loss prevention tool.

SELECTING A JOB OR TASK FOR A PLANNED JOB SAFETY OBSERVATION

To conduct a planned safety observation, a method must be established to select jobs to observe. The same selection method used in job safety analysis can be used in the job observation selection. These selections are divided into four areas listed below.

TABLE 14.1
Planned Job Safety Observation Form

JOB SAFETY OBSERVATION

Observation completed by: ________________ **Title:** ________________ **Date:** ________

1. Job being observed: ________________
2. Worker observed (Name): ________________
3. Experience of worker at job or task ______yrs. ______mos.
4. Is worker dressed appropriately for the job? ______yes ______ no

Comments: ________________

5. Is worker wearing all required personal protective equipment? ______yes ______no

Comments: ________________

Steps in performing the job or task should be marked (S) Satisfactory, (R) Reobserve, or (U) Unsatisfactory.

Steps (Describe)	**Hazard Involved**	**Worker Performance**
1.		
2.		
3.		
4.		
5.		

6. Did the worker perform the job according to the safety operating procedure?
 ______yes______no
 Comments: ________________
7. Did the worker follow the safety and health rules? ______yes ______no
 Comments: ________________
8. Did the worker perform the job or task safely? ______yes ______no
 Comments: ________________
9. Did the worker have a good safety attitude? ______yes ______no
 Comments: ________________
10. Does the worker need training?
11. Should disciplinary action be taken?
12. Should the worker be removed from the job or task?
13. Was the worker told about any deficiency?
14. Your final recommendations, if any.

Source: Reese, C. D. and J. V. Eidson. *Handbook of OSHA construction safety & health.* Boca Raton, FL: CRC/Lewis Publishers, 1999.

Selection Based on Jobs

Select jobs or tasks where accidents have occurred. Jobs may be selected where there have been repeated accidents or lost time due to accidents, or jobs where accidents have required medical treatment. Other selections may be jobs that are not done where the hazards may not be fully known. Some operations may have a system within their own safety program whereby violations are issued when a job is done improperly. These violations may serve as an indicator that a certain job(s) should be the subject of a planned safety observation.

Selection Based on Individual Needs

Supervisors may have reasons for observing certain employees. For example, repeat offenses or observed unsafe behavior are examples of an area where a job would be selected because of individual needs. The supervisor should select the repeat offender for regular planned safety observations. Such planned observations can show reasons why some employees have more accidents than others. Another area selected for observation includes chronically unsafe employees. These may be workers who take unnecessary chances, violate safety rules, and develop improper work methods. Workers suspected of being physically or mentally incapable of safe work may also fit into the area of individual needs.

Experienced Employees

Many accidents involve experienced employees, and planned safety observations can detect the reasons. Some workers who have been doing a job for many years will often develop short-cuts and effort-saving practices that are hazardous. Because accidents have not happened before, they will insist that the short-cuts are safe.

Inexperienced workers usually follow examples set by supervisors or experienced workers. When these people work in an unsafe manner, it has an influence on the inexperienced worker. All employees should be included in the safety observation program so that all unsafe practices can be found and corrected.

Inexperienced Employees

Detecting unsafe behavior and practices quickly in the inexperienced employee is important and will allow the supervisor to take corrective measures quickly. This will prevent accidents and will deter employees from developing unsafe work habits. It is much easier and more effective to correct an inexperienced worker when he or she is observed performing unsafely.

Inexperienced workers require frequent planned safety observations to establish exactly what they know and do not know about doing their work safely. Further instruction and guidance are usually the required procedures after observations show the need for additional safety training.

PREPARING FOR A PLANNED SAFETY OBSERVATION

A planned safety observation gives supervisors a positive means of determining the effectiveness of safety instructions. It also serves as a learning tool for the supervisor. The supervisor learns more about each job, each worker, and areas requiring closer supervision. The supervisor will learn to be more perceptive in all areas of responsibility that will result in the best distribution of supervisory time. The supervisor will learn to devote time and effort where needed. The primary need to accomplish an ongoing observation system and continually improving safety and health program starts with a written procedure for the planned job safety observation. Tailor the program to the needs of the operation. Include the following:

1. The reasons for the plan (must include orientation of all employees about the purpose of the program)
2. The objectives of the program
3. Training requirements for all levels of supervision
4. Safety and Health Department's responsibilities
5. Preparation procedures for making a job safety observation
6. How to make a planned safety observation
7. Recording the observation outcomes
8. Holding a post-observation conference
9. Follow-up procedures (including enforcement policy)

Because supervising workers means observing, the supervisor is the most qualified to make job safety observations. He or she knows the workers and the training they have received, and knows the jobs and how these jobs should be done. With proper preparation, the supervisor can conduct a thorough observation. Other management personnel may take part in the planned safety observation but the supervisor must ultimately be responsible for observation.

The supervisor should select the person to observe usually because of some specific reason such as inexperience, a reputation for taking chances, because of a history of accidents, or routine observation to check compliance with procedures. After identifying the worker to observe, the first question to answer is, "What job should this person be doing?"

Some jobs involve only a few simple, routine tasks. Others involve doing many tasks, some more often than others. It is important to keep records of observations. See Table 14.1 for an example of a Job Safety Observation Form. The results will tell each supervisor just what has been or should be done about a particular worker. A Job Safety Observation Form contains the employee's name, department, job title, and a list of the tasks performed when doing that job. If the employee has completed the task safely, place an "S" in the appropriate column to indicate that the task was performed satisfactorily. An "R" indicates that, for some reason(s), the employee should be observed again at a later time. Use the comments space to record any observed unsafe practices, or any other information about the task(s) or the worker. To choose the task to observe, consider whether (1) it involves some new procedures because of a recent JSA revision, (2) there has been a change in equipment or machinery, (3) it poses an exceptional hazard, or (4) it is a job infrequently done but is complex. A "U" indicates unsatisfactory performance at a particular task and corrective action is required.

A detailed description of the direct, indirect, and basic causes of accidents can be found in Chapter 7. Remember, the unplanned or unwanted release of excessive amounts of energy or hazardous materials causes most accidents. With few

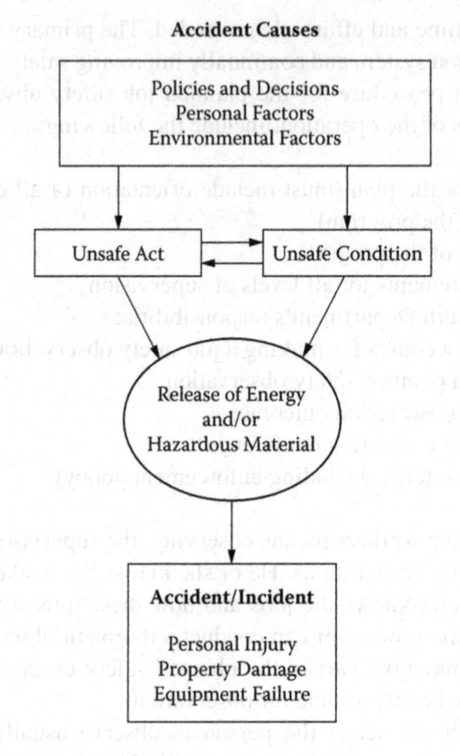

FIGURE 14.1 Accident causes (direct, indirect, and basic).

exceptions, these releases are caused by unsafe acts and unsafe conditions. An unsafe act or an unsafe condition may trigger the release of large amounts of energy or hazardous materials. This may cause the accident (see Figure 14.1). Particular attention should be paid to these potential causal factors.

We often think of unsafe acts and unsafe conditions as the basic causes of accidents. They are only symptoms of failure. The basic causes are poor management policies and decisions, and personnel and environmental factors. Fortunately, most employers now realize that safety and health must be a necessary part of the total operating system. These employers take the responsibility to prepare a written safety policy and guidelines to instill safety awareness in their employees. Selection of best employees for the job, training, employee placement, and the purchase of safe equipment and supplies are important to a successful accident prevention program.

CHECKLIST OF ACTIVITIES TO OBSERVE

A checklist or form should be used for recording observations. This form should record what the worker has been observed doing, when the observation was made, and what action or actions were taken or planned as a result. The checklist should include key points of the particular job (see Table 14.2).

If a JSA/JHA is available, it should be used as the basic information and guideline for the observation. The supervisor should watch each step of the job and be alert to

TABLE 14.2
An Example of a Completed Job Safety Observation

JOB SAFETY OBSERVATION

Observation completed by: James McCoy **Title:** Shift Foreman **Date:** March 3, 1998

1. Job being observed: Replacement of rotary actuator
2. Worker observed: Mike Murphy
3. Experience of worker at job was 1 month.
4. The worker was dressed appropriately for the job. He wore comfortable fitting maintenance uniform.
5. Worker needed to wear additional PPE: Safety glasses, steel-toed shoes, coveralls, rubber work gloves.

Steps taken to Accomplish Task:	Hazard Involved:	Worker Performance:
A. Lockout Machine	Un-isolated energy sources could be lethal.	S
B. Removed Rotary Table	Hydraulic lines must be removed, double-check pressure gages. Table is heavy and could swing or drop.	S
C. Secure Rotary to Table	Table was to be placed securely on steel sawhorses.	R
D. Change Rotary Actuator	Minimal risk of injury.	S
E. Replace Table	Table is heavy and could swing or drop. Body parts could be pinched or crushed.	S

Note: Needed to discuss with the mechanic how to secure rotator to mitigate the chances of a mishap. Also, a conclusion was to enlist the aid of another mechanic to assist in guiding the table into place.

Steps Taken to Accomplish Task:	Hazard Involved:	Worker Performance:
F. Re-energize Machine.	Improper unlocking sequence of the equipment could lead to property damage and/or personal injury.	S

6. The worker performed the job according to an unwritten safe operating procedure with a few mistakes. When in doubt, he asked for advice. He was conscientious about the methods he used to accomplish the task at hand.
7. Safety and health rules were followed in most instances.
8. Not only was the worker conscious about his own safety, but he watched out for those around him.
9. It is my opinion that the worker needs no further training in safe work habits, but should be reobserved in two weeks. He completed the task twice with mistakes.
10. Final Recommendation: Step-by-step safe operating procedures for repairing items involving the Rotary Tables would enable ALL maintenance personnel to perform their tasks more efficiently and consistently. This mechanic spent quite a bit of time in the planning of the task at hand. I'm afraid that he is not within the norm. Standard procedures would have enabled him to get right down to the task at hand with minimal delay.

compliance with the safe job procedures. The supervisor should recognize the need to revise the job safety (hazard) analysis if necessary.

When a JSA/JHA is not available, the supervisor should make notes beforehand in preparation for the planned observation. A list of possible unsafe actions for the job would provide a guide to use. The supervisor must keep an open mind for possibilities that are not on the list. Most checklists do not cover every conceivable unsafe action. Use a general checklist with information similar to Table 14.1.

A checklist should also include specific safety behaviors to look for. Some general questions to consider are

- How does the worker handle tools and equipment?
- Does the worker show concern for doing a good job?

A checklist can also detect hazards caused by inappropriate clothing and inadequate personal protective equipment. Whether clothing should be considered hazardous may depend on the type of work done, the work area, the weather, or company regulations. A checklist may detect hazards such as

1. Loose or ragged clothing that could be caught in moving machinery
2. Rings, bracelets, necklaces, or wristwatches that could be hazardous when climbing, using certain tools, working around heat or chemicals, or doing electrical work
3. Long hair that could catch in or on moving or rotating machinery, or that could become harmful in other ways

Personal protective equipment can also be checked during observations. Items can include

- Proper head protection, eye protection, and foot protection
- When needed, proper gloves, respirators, and suitable ear protection
- The use of protective clothing
- Respirators and other items required for specific areas

UNSAFE PROCEDURES

Any specific unsafe behavior on the job will correspond to basic types of unsafe procedures. These unsafe procedures can be included in an observation checklist:

1. Operating or using equipment without authority
2. Failure to secure against unexpected movement
3. Failure to utilize lockout/tagout or blocking
4. Operating or working at an unsafe speed
5. Using unsafe tools and equipment
6. Using tools and equipment unsafely
7. Failure to warn or signal as required
8. Assuming an unsafe position or unsafe posture
9. Removing or making safety devices inoperable
10. Repairing, servicing, or riding equipment in a hazardous manner
11. Failure to wear required personal protective equipment
12. Wearing unsafe personal attire
13. Violation of known safety rules and safe job procedures
14. Indulging in practical jokes, fighting, sleeping, creating distractions, etc.

THE OBSERVATION

The employee observed should be aware of the observation from time to time. Inform all employees, as part of their orientation, that planned job safety observations are a phase of the safety and health program. All employees must fully understand the reason for incidental and planned observations.

With this in mind, there is another consideration in preparing for the planned safety observation. Should you inform the employee in advance about the observation? Under certain conditions, the employee should have advance notice.

If you want to find what an employee knows and does not know about job safety procedures, tell the employee in advance. If any unsafe practices are revealed under this condition, the supervisor can conclude that the employee does not know safety procedures. If the supervisor tells the employee in advance, he or she will learn what is known or not known about job safety procedures.

If the employee is new, or one who has relatively little experience, tell him or her in advance. This is also true for an employee who may have experience but has never been checked using a planned safety observation.

Do not tell the employee in advance about the observation if the objective is to learn how the employee normally does the job. When the supervisor knows that an employee understands how to do a job safely, the supervisor must then find out how the employee works when no one is observing. To determine this, the employee should not be told in advance about the job safety observation. Most employees work safely under the eyes of the supervisor, especially when they are told of the observation. If the employee does not show unsafe practices, the supervisor can assume that the employee usually works safely. If the same person is later observed doing some part of the job unsafely, the supervisor must believe that work is performed unsafely at other times. The supervisor may know that the employee can perform the job safely from past observation. From observing employees without their knowledge, the supervisor may learn that they are not putting into practice their job knowledge.

The supervisor should never try to observe an employee from a hidden position. The supervisor may learn by the observation, but respect and human relations may be lost. After informing the employee of the observation, the supervisor must stand clear of the employee's work area. The employee must be given plenty of room to work. The supervisor must not create a hazard by standing too close or distracting the employee.

The supervisor should avoid work interruptions unless absolutely necessary. It is better to save all minor corrections until the observation is complete. However, if the employee does something in an unsafe manner, stop him or her immediately. The supervisor should explain what has been done wrong and why it is unsafe. Call the employee's attention to the JSA/JHA when training was originally conducted within the framework of a JSA/JHA program. Check to see if the employee fully understands the explanation. This can often be done by having the employee tell and show the safe way to do the job.

AFTER THE OBSERVATION

A brief post-observation discussion should be held with the employee. One of two things should be done. When the job was done as required, the employee should be complimented. Recognition, when deserved and when sincerely given, will reinforce safe behavior. This acts as a positive motivation for the worker. Any unsafe behavior seen during the observation should be discussed. If the employee performed in an unsafe manner, the supervisor should discuss what occurred and get the employee's reasons for doing the job as it was done.

The concern now is not with what has been done, but with changing the employee's behavior so the same thing is not done in the future. The supervisor must take required corrective action and should give the reasons behind the measures. The post-observation conference should always end with the feeling that things are right between the supervisor and the employee.

DEALING WITH UNSAFE BEHAVIORS OR POOR PERFORMANCE

The "dealing with unsafe performance" diagram in Figure 14.2 may be helpful in determining why unsafe practices are continuing. This diagram gives a step-by-step procedure to follow in correcting unsafe performance. The supervisor should record the specific unsafe performance that has been observed and determine the type of accident that could result from this action. Because this action and the possible results have already been discussed with the employee after or during the first observation, the supervisor must decide what has caused this action to continue.

Let's trace the steps of the diagram in Figure 14.2. The first step is to identify the performance problem. In this area, the performance problem may be someone's actual performance and his or her desired performance.

Is the problem worth solving? Not every performance problem is worth solving. The problem must be evaluated realistically. What impact will the performance problem have on doing the job safely? If the problem has or can have no adverse impact, then ignore it.

In determining the cause, you need to ask or try to find out the cause of the performance problem. Could this person do it if he or she really had to? If the answer to this question is "No," then the following question should be asked: "Does the employee have the potential to perform this job safely?" Both physical and mental capabilities must be assessed in this area. If the answer to this question is "No," then the job may have to be changed to fit the employee's ability or if possible the employee can be placed in a job that fits his or her capabilities. If the employee has the potential to do the job, then the following question should be asked: Has the person ever performed the job? If the answer to the question is "No," then classroom or on-the-job training will be necessary. If the person has performed the job, refresher training may be all that is necessary.

If the employee could do the job if he or she really had to, then a non-training solution will help identify and solve the performance problem. One way to solve the non-training problem is to provide feedback. Feedback should come from

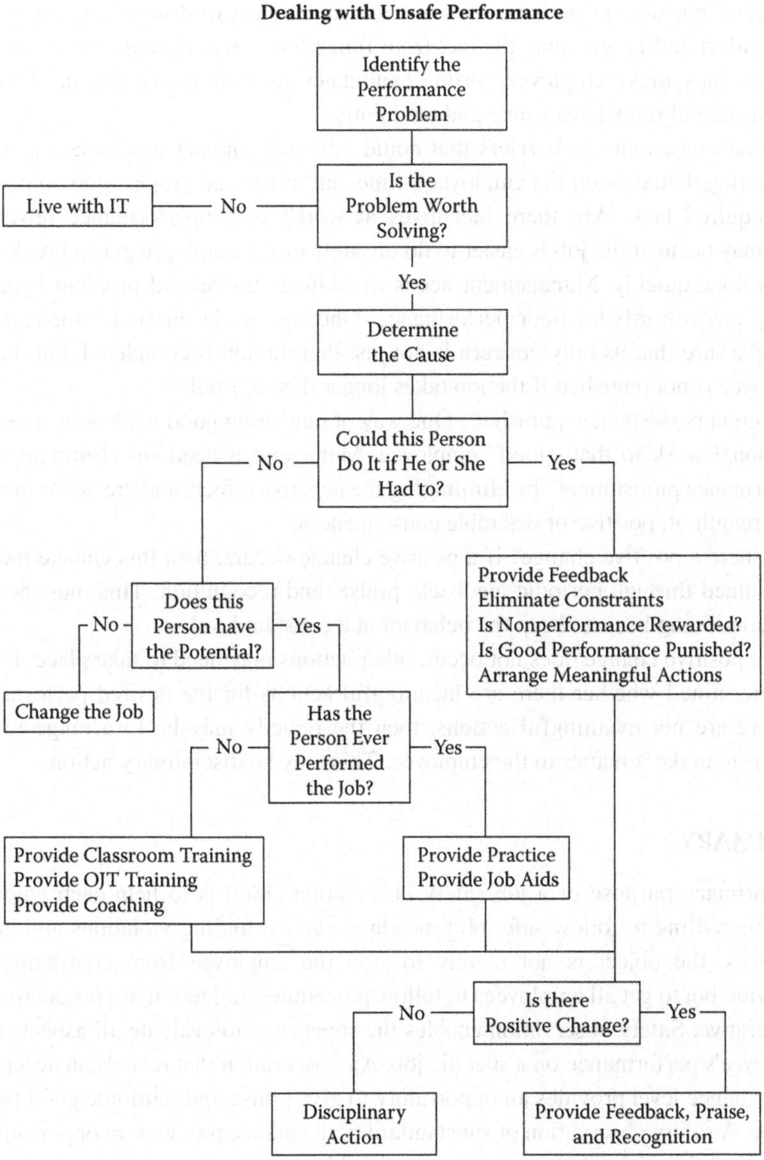

Dealing with Unsafe Performance

FIGURE 14.2 Dealing with unsafe performance diagram. (*Source:* Courtesy of the Mine Safety and Health Administration.)

management being aware of the problem, observing the employee, or simply making the employee aware that he or she is expected to do the job correctly.

Another way to solve the problem is to eliminate constraints or barriers. Tools, equipment, and time might fall into this category because these constraints can take many forms. Ergonomic considerations in job design are an example of a potential constraint. Employees forced to stoop, bend, twist, turn, stretch, lean, reach, or assume unnatural postures to perform routine tasks are subject to discomfort and fatigue. Discomfort and fatigue may distract from immediate hazards, and chronic physical stresses may make employees prime candidates for cumulative trauma disorders such as carpal tunnel syndrome and tendonitis.

Other constraints or barriers that could adversely impact performance include conflicting demands on the employee's time, authority, and proper tools to perform the required task. Are there incentives at work? Is nonperformance rewarded? This may occur if the job is easier to do unsafely or the employee gets a break if the job is done quickly. Management needs to address this reward problem by eliminating any rewards for poor performance. One way to do this is for management to make sure that its only concern is not just that the job is completed, but that the employee is not punished if the job takes longer than normal.

Is good performance punished? One way of punishing good work is by assigning additional work to that "good" employee. Management needs to eliminate "good performance punishment" by eliminating the negative effects and create, or increase the strength of, positive or desirable consequences.

Is there a positive change? If a positive change occurs, then this change must be maintained through periodic feedback, praise, and recognition. This must be done on a continuing basis to keep the behavior at a constant level.

If a positive change does not occur, other actions may need to take place. It must be determined whether there are meaningful actions for the desired performance. If there are not meaningful actions, then the remedy may be to arrange to take actions to make it matter to the employee. This may be disciplinary action.

SUMMARY

The primary purpose of a job safety observation (JSO) is to help each employee become willing to follow safe job procedures. When finding violations and unsafe practices, the object is not merely to stop the employee from continuing this behavior, but to get all employees to follow procedures and to remain productive and cooperative. Safety observation enables the supervisor to evaluate all aspects of an employee's performance on a specific job. An observation that reveals an acceptable performance level provides an opportunity to give praise and reinforce good performance. A safety observation of substandard performance provides an opportunity to take corrective action before unnecessary problems or losses occur.

All companies must establish and enforce rules of conduct and rules of safety for everyone involved in their operation. Because supervisors are responsible for following safety and health rules, a JSO is a valuable method for accomplishing an effective job safety observation program that can enhance all aspects of the accident/incident program, such as job safety analysis, safety inspections, and safety orientation and safety and health training programs.

REFERENCES

Kavianian, H. R. and C. A. Wentz, Jr. *Occupational and environmental safety engineering and management.* New York: Van Nostrand Reinhold, 1990.

Mager, Robert F. *Analyzing performance problems.* Belmont, CA: Fearson Publishers, Inc., 1970.

Reese, C. D. and J. V. Eidson. *Handbook of OSHA construction safety & health.* Boca Raton, FL: CRC Press/Lewis Publishers, 1999.

United States Department of Labor, Mine Safety and Health Administration. Accident Investigation (Safety Manual No. 10). Beckley, WV: U.S. Department of Labor, reprinted 1989.

United States Department of Labor, Mine Safety and Health Administration. Accident Prevention (Safety Manual No. 4). Beckley, WV: U.S. Department of Labor, revised 1990.

United States Department of Labor, Mine Safety and Health Administration. Job Safety Analysis: A Practical Approach (Instruction Guide No. 83). Beckley, WV: U.S. Department of Labor, 1990.

United States Department of Labor, Mine Safety and Health Administration. Job Safety Analysis (Safety Manual No. 5). Beckley, WV: U.S. Department of Labor, revised 1990.

United States Department of Labor, National Mine Health and Safety Academy. Accident Prevention Techniques: Job Observation. Beckley, WV: U.S. Department of Labor, 1984.

United States Department of Labor, Mine Safety and Health Administration, Safety Observation (MSHA IG 84). Beckley, WV: U.S. Department of Labor, revised 1991.

15 Safety and Health Audits

INTRODUCTION

Workplace audits are inspections that are conducted to evaluate certain aspects of the work environment regarding occupational safety and health. The use of safety and health audits has been shown to have a positive effect on a company's loss control initiative. In fact, companies that perform safety and health audits have fewer accidents/incidents than companies that do not perform such audits.

Safety and health audits (inspections), which are often conducted in workplaces, serve a number of evaluative purposes. Audits or inspections can be performed to

1. Check compliance with company rules and regulations.
2. Check compliance with OSHA rules
3. Determine the safety and health condition of the workplace
4. Determine the safe condition of equipment and machinery
5. Evaluate supervisor's safety and health performance
6. Evaluate worker's safety and health performance
7. Evaluate progress regarding safety and health issues and problems
8. Determine the effectiveness of new processes or procedural changes

THE NEED FOR AN AUDIT

First, determine what needs to be audited. You might want to audit specific occupations (e.g., machinist), tasks (e.g., welding), topic (e.g., electrical), team (e.g., rescue), operator (e.g., crane operator), part of the worksite (e.g., loading/unloading), compliance with an OSHA regulation (e.g., Hazard Communication Standard), or the complete worksite. You may want to perform an audit if any of the previous items or activities have unique identifiable hazards, new tasks involved, increased risk potential, changes in job procedures, areas with unique operations, or areas where comparisons can be made regarding safety and health factors.

In the process of performing audits, you may discover hazards that are in a new process, hazards once the process has been instituted, a need to modify or change processes or procedures, or situational hazards that may not exist at all times. These audits may verify that job procedures are being followed, and identify work practices that are both positive and negative. They may also detect exposure factors—both chemical and physical—and determine monitoring and maintenance methods and needs.

Sometimes, audits are driven by the frequency of injury; the potential for injury; the severity of injuries; new or altered equipment, processes, and operations; and excessive waste or damaged equipment. These audits may be continuous, ongoing, planned, periodic, intermittent, or dependent upon specific needs. Audits may often determine employee comprehension of procedures and rules, and the effectiveness of workers' training. Audits may be used to assess the work climate or perceptions held by workers and others and evaluate the effectiveness of a supervisor regarding his or her commitment to safety and health.

At many active workplaces, daily site inspections are performed by the supervisor in order to detect hazardous conditions, equipment, materials, or unsafe work practices. At other times, periodic site inspections are conducted by the site safety and health officer. The frequency of inspections is established in the workplace safety and health program. The supervisor, in conjunction with the safety and health officer, determines the required frequency of these inspections, based on the level and complexity of the anticipated activities and on the hazards associated with these activities. In a review of worksite conditions/activities, site hazards, and protecting site workers, the inspections should include an evaluation of the effectiveness of the company's safety and health program. The safety and health officer should revise the company's safety and health program as necessary to ensure the program's continued effectiveness.

Prior to the start of each shift or new activity, a workplace and equipment inspection should take place. This should be done by the workers, crews, supervisor, and other qualified employees. At a minimum, they should check the equipment and materials that they will be using during the operation or shift for damage or defects that could present a safety hazard. In addition, they should check the work area for new or changing site conditions or activities that could also present a safety hazard.

All employees should immediately report any identified hazards to their supervisors. All identified hazardous conditions should be eliminated or controlled immediately. When this is not possible,

1. Interim control measures should be implemented immediately to protect workers.
2. Warning signs should be posted at the location of the hazard.
3. All affected employees should be informed of the location of the hazard and the required interim controls.
4. Permanent control measures should be implemented as soon as possible.

When a supervisor is not sure how to correct an identified hazard, or is not sure if a specific condition presents a hazard, he or she should seek technical assistance from a competent person, a site safety and health officer or other supervisors or managers.

Safety and health audits should be an integral part of your safety and health effort. Anyone conducting a safety and health audit must know the workplace, the procedures or processes being audited, the previous accident history, and the company's policies and operations. This person should also be trained in hazard recognition and interventions regarding safety and health.

WHEN TO AUDIT

The supervisor or project inspector must perform daily inspections of active worksites to detect hazardous conditions resulting from equipment or materials or unsafe work practices. The supervisor, inspector, or site safety and health officer must perform periodic inspections of the workplace at a frequency established in the worksite's specific safety and health program. The supervisor, in conjunction with the site safety and health officer, should determine the required frequency of these inspections based on the level and complexity of anticipated work activities and on the hazards associated with these activities. In addition to a review of worksite conditions and activities, these inspections must include an evaluation of the effectiveness of the worksite safety and health program in addressing site hazards and in protecting site workers. The safety and health officer may need to revise the safety and health program as necessary to ensure the program's continued effectiveness. Work crew supervisors, foremen, and employees need to inspect their workplace prior to the start of each work shift or new activity. At a minimum,

1. Supervisors and employees should check the equipment and materials that they will use during the operation or work shift for damage or defects that could present a safety hazard.
2. Supervisors and employees should check the work area for new or changing site conditions or activities that could present a safety hazard.
3. Employees should immediately report identified hazards to their supervisors.

WHAT TO AUDIT

The complexity of the worksite and the myriad of areas, equipment, tasks, materials, and requirements can make the content of most audits overwhelming. As evidenced in Table 15.1, the audit topics that could be targeted on a worksite are expansive.

Safety and health audits/inspections can be done on or for the entire plant (e.g., manufacturing), a department (e.g., quality control), a specific worker unit (e.g., boiler repair), a job or task (e.g., diving), a certain work environment (e.g., confined spaces), a specific piece of equipment (e.g., forklift), a worker performing a task (e.g., power press operator), or prevention of an event (e.g., fire). It is important to ensure that inspections and audits be tailored to meet the needs of your company regarding auditing for safety and health concerns.

TYPES OF AUDIT INSTRUMENTS

The topics of audits and the form you use will depend on the type of work being performed. There are four primary audit formats: checklist, evaluation, narrative, or compliance. A short example of a checklist is shown below. This checklist requires a simple yes or no response (see Table 15.2).

_____Yes _____No Chemical goggles are worn when working with chemicals.
_____Yes _____No A face shield is used when secondary eye protection is needed.

The evaluation format requires the person completing the audit to complete the form as follows: 1 (excellent), 2 (very good), 3 (good), 4 (fair), and 5 (very poor) (see Table 15.3). An example of this type of format is:

TABLE 15.1
Audit Topics

Acids	Fire extinguishers	Personal services and first aid
Aisles	Fire protection	Power sources
Alarms	Flammables	Power tools
Atmosphere	Forklifts	Radiation
Automobiles	Fumes	Railroad cars
Barrels	Gas cylinders	Respirators
Barriers	Gas engines	Safety devices
Boilers	Gases	Signs
Buildings	Generators	Scaffolds
Cabinets	Hand tools	Shafts
Catwalks	Hard hats	Shapers
Caustics	Hazardous chemical processes	Shelves
Chemicals	Heavy equipment	Solvents
Compressed gas cylinders	Hoists	Stairways
Containers	Horns and signals	Steam systems
Controls	Hoses	Storage facilities
Conveyors	Housekeeping	Tanks
Cranes	Jacks	Transportation equipment
Confined spaces	Ladders	Trucks
Docks	Lifting	Ventilation
Doors	Lighting	Walkways
Dusts	Loads	Walls and floor openings
Electrical equipment	Lockout/tagout	Warning devices
Elevators	Machines	Welding and cutting
Emergency procedures	Materials	Work permit
Environmental factors	Mists	Working surfaces
Explosives	Noise	Unsafe conditions
Extinguishers	Piping	Unsafe acts
Fall protection	Platforms	X-rays
Fibers	Personal protective equipment	

1 2 3 4 5 Eye protection is worn by all workers (all workers or some).
1 2 3 4 5 When handling chemicals, goggles are worn.
1 2 3 4 5 When grinding, face shields are worn over safety glasses.

A narrative format requires a written response (see Table 15.4). An example of this type of format is:

1. Protective eyewear is worn in all work locations.
   Explain: ______________________________
2. Chemical goggles are worn.
   Explain: ______________________________
3. Face shields are provided.
   Explain: ______________________________

TABLE 15.2
Checklist Audit for General Workplace

WORKPLACE SURVEY

Use this checklist to survey the safety and health conditions at your operation.

Housekeeping

Is work area clean and orderly?	☐ Yes	☐ No	☐ N/A
Is work free of tripping hazards?	☐ Yes	☐ No	☐ N/A
Are aisles and egress routes free of clutter and clearly identified?	☐ Yes	☐ No	☐ N/A
Are waste receptacles identified?	☐ Yes	☐ No	☐ N/A

Comments: ________________________________

Environment

Is the lighting adequate?	☐ Yes	☐ No	☐ N/A
Is there excessive noise?	☐ Yes	☐ No	☐ N/A
Is there adequate ventilation?	☐ Yes	☐ No	☐ N/A
Is this a climate-controlled area?	☐ Yes	☐ No	☐ N/A
Is the temperature comfortable?	☐ Yes	☐ No	☐ N/A

Comments: ________________________________

Material Handling

Are cranes and hoists available?	☐ Yes	☐ No	☐ N/A
Are cranes and hoist inspections and Periodic/Monthly (P/M) current?	☐ Yes	☐ No	☐ N/A
Are cranes and hoist load limits labeled and legible?	☐ Yes	☐ No	☐ N/A
Are slings in good condition, no evidence of fraying?	☐ Yes	☐ No	☐ N/A
Do slings have load limit labels?	☐ Yes	☐ No	☐ N/A
Are cranes/hoist/slings in use? If so, are they being used properly?	☐ Yes	☐ No	☐ N/A
Are forklifts in use?	☐ Yes	☐ No	☐ N/A
Are forklift operator inspections P/M current?	☐ Yes	☐ No	☐ N/A
Are operators trained and industrial vehicles license current?	☐ Yes	☐ No	☐ N/A

Comments: ________________________________

Equipment and Tools

Are hand tools in good operating condition?	☐ Yes	☐ No	☐ N/A
Have electrical hand tools, requiring fault testing, been inspected?	☐ Yes	☐ No	☐ N/A
Are machines adequately guarded to protect employees from hazards?	☐ Yes	☐ No	☐ N/A
Are flexible cords and cables used for machine power cables in good condition and protected from damage?	☐ Yes	☐ No	☐ N/A
Is there a LO/TO procedure for de-energizing the equipment?	☐ Yes	☐ No	☐ N/A
Are employees trained in equipment operating procedures?	☐ Yes	☐ No	☐ N/A

Comments: ________________________________

PPE/Emergency Safety Equipment

Are the employees wearing PPE?	☐ Yes	☐ No	☐ N/A
Are safety showers and eyewash stations present?	☐ Yes	☐ No	☐ N/A
Is the safety shower and eyewash stations inspection current?	☐ Yes	☐ No	☐ N/A
Are employees trained to use safety shower and eyewash stations?	☐ Yes	☐ No	☐ N/A

Comments: ________________________________

continued

TABLE 15.2 (continued)
Checklist Audit for General Workplace

WORKPLACE SURVEY

Electrical

Is the area free of exposed wires?	☐ Yes	☐ No	☐ N/A
Are electrical panel and switch boxes accessible?	☐ Yes	☐ No	☐ N/A
Are electrical power source covers secured to prevent unauthorized entry?	☐ Yes	☐ No	☐ N/A
Are overhead switch box emergency shut-off poles available?	☐ Yes	☐ No	☐ N/A
Is there a minimum of three feet clearance in front of electrical cabinets?	☐ Yes	☐ No	☐ N/A

Comments: ______

Fire Protection

Are fire extinguishers available?	☐ Yes	☐ No	☐ N/A
Are fire extinguisher locations identified?	☐ Yes	☐ No	☐ N/A
Are the fire extinguishers the correct type for controlling hazards in the area?	☐ Yes	☐ No	☐ N/A
Are inspections of fire extinguishers current?	☐ Yes	☐ No	☐ N/A
Are employees trained to use fire extinguishers?	☐ Yes	☐ No	☐ N/A

Comments: ______

Compressed Gas Cylinders

Are cylinders secured?	☐ Yes	☐ No	☐ N/A
Are cylinder valve caps in place and secure?	☐ Yes	☐ No	☐ N/A
Are oxygen and acetylene welding tanks equipped with shut-off valves?	☐ Yes	☐ No	☐ N/A

Comments: ______

Material Handling/Storage

Are all chemicals properly labeled and stored?	☐ Yes	☐ No	☐ N/A
Are all flammable liquids properly stored?	☐ Yes	☐ No	☐ N/A
Are employees trained in hazardous material handling?	☐ Yes	☐ No	☐ N/A
Is appropriate PPE readily available?	☐ Yes	☐ No	☐ N/A

Comments: ______

A compliance checklist is used to determine the degree of compliance that exists relevant to a specific regulation. It will provide an indication that you have touched all facets of the regulation and allows you to evaluate your level of compliance (see Table 15.5).

DEVELOP AND EVALUATE AUDIT SCORES

Select the appropriate type of instrument; this is tied to its intended use. Instruments are developed or revised if

- Change has occurred.
- Accident/incidents have occurred.
- A serious new hazard has evolved.
- Previous audits show a need for new audit approaches or instruments.

TABLE 15.3
Evaluative Audit for a Construction Site

EVALUATION: SAFETY AUDIT CHECKLIST OF A CONSTRUCTION SITE

JOB: ______________________ **LOCATION:** ______________

COMPLETED BY: ______________________ **DATE:** ______________

In order to evaluate the degree or amount of compliance with safety and health rules and regulations. Circle the numerical rating value that best represents what is observed:

1 = excellent
2 = good
3 = fair or average
4 = poor or deficient
N/A = not applicable

1. Safety and health protection poster (OSHA Poster) displayed on the job.	1	2	3	4	N/A
2. Emergency telephone numbers posted.	1	2	3	4	N/A
3. First aid kit and supplies on job site; checked weekly.	1	2	3	4	N/A
4. Supply tags/locks available to use on unsafe tools or equipment and prevent use.	1	2	3	4	N/A
5. Hard hats worn by all employees at all times.	1	2	3	4	N/A
6. Emergency action plan communicated to all employees.	1	2	3	4	N/A
7. Eye and face protection available and used when necessary.	1	2	3	4	N/A
8. All hand tools are in a state of good repair.	1	2	3	4	N/A
9. Site lighted (if necessary) to minimum illumination requirements.	1	2	3	4	N/A
10. Grounding program implemented and current protectors or GFI on all circuits.	1	2	3	4	N/A
11. Temporary lighting (if needed) properly installed and protected.	1	2	3	4	N/A
12. Electrical services and all electrical equipment properly grounded and fused.	1	2	3	4	N/A
13. Tools with abrasive wheels properly installed and guarded.	1	2	3	4	N/A
14. Safety guards properly installed on all power equipment and tools.	1	2	3	4	N/A
15. All fuels (gasoline LPG, acetylene) stored and handled according to standards.	1	2	3	4	N/A
16. Transportation, movement, storage, and use of compressed gas cylinders according to standard.	1	2	3	4	N/A
17. Employees instructed in safe welding and cutting practices.	1	2	3	4	N/A
18. Fire equipment readily available; fire watch for cutting, welding, burning.	1	2	3	4	N/A
19. Rigging equipment inspected before use; workers instructed in proper use.	1	2	3	4	N/A
20. All housekeeping materials stored properly.	1	2	3	4	N/A
21. Scrap and debris removed from work area daily, or as required.	1	2	3	4	N/A
22. Aisles, passageways, and roadways free and clear of materials, equipment.	1	2	3	4	N/A
23. Ladders have no broken or missing rungs or side rails.	1	2	3	4	N/A
24. Ladder use; 1:4 pitch; secured at top and bottom; side rails extended 36" above the landing platform.	1	2	3	4	N/A
25. Scaffold platforms planked full width except for necessary entrance opening.	1	2	3	4	N/A
26. All scaffold sides and ends provided with guardrails, midrails, and toeboards.	1	2	3	4	N/A

continued

TABLE 15.3 (continued)
Evaluative Audit for a Construction Site

EVALUATION: SAFETY AUDIT CHECKLIST OF A CONSTRUCTION SITE					
27. Scaffold footing secure and prevented from movement or shifting.	1	2	3	4	N/A
28. Erection, movement, or dismantling of scaffolding under supervision of competent person.	1	2	3	4	N/A
29. Scaffolds properly tied to buildings or other appropriate structure.	1	2	3	4	N/A
30. Rebar and other pointed objects under scaffolds, guarded or bent down.	1	2	3	4	N/A
31. Proper ladder or other access available to scaffold platforms.	1	2	3	4	N/A
32. All stairways equipped with handrails and guardrails.	1	2	3	4	N/A
33. Adequate supply of potable water, cups, and containers for used cups.	1	2	3	4	N/A
34. Perimeter guarding protection for areas where 6-foot falls possible.	1	2	3	4	N/A
35. Toilet facilities available and in compliance with standards.	1	2	3	4	N/A
36. Adequate ventilation provided where required.	1	2	3	4	N/A
37. All temporary heating devices properly installed and inspected. No LPG stored inside.	1	2	3	4	N/A
38. All flammable liquids stored and transported in approved containers with proper labeling.	1	2	3	4	N/A
39. All equipment is equipped with operating lights and audible warning.	1	2	3	4	N/A
40. All equipment and trucks equipped with fire extinguishers that are accessible from operator position.	1	2	3	4	N/A
41. Operating rules for material hoist, cranes, forklifts, and other material handling machinery are posted in operator's station.	1	2	3	4	N/A
42. Wire rope in cranes and hoists is inspected daily and before use.	1	2	3	4	N/A
43. Operator's station on all machinery is provided with overhead protection.	1	2	3	4	N/A
44. Protection provided where persons are required to work or pass under scaffolding in use.	1	2	3	4	N/A
45. There is adequate security protection provided at the site during nonworking hours.	1	2	3	4	N/A
46. Employees are adequately trained on hazard communications program requirements.	1	2	3	4	N/A
47. Containers of chemicals and hazardous materials properly labeled.	1	2	3	4	N/A
48. Inventory of chemicals and hazardous materials and MSDSs are available at site.	1	2	3	4	N/A
49. Limited access zone established when necessary.	1	2	3	4	N/A
50. Masonry walls adequately braced when needed.	1	2	3	4	N/A
51. Employees and equipment are the necessary distance from overhead power lines.	1	2	3	4	N/A
52. Employees in excavations more than 5 feet deep protected by shoring, sloping, and other means.	1	2	3	4	N/A
53. Work areas examined for potential confined spaces.	1	2	3	4	N/A

The following items require immediate correction/attention:

____

____

____

____

TABLE 15.4
Narrative Audit for Chemical Spill

CHEMICAL SPILL AUDIT

Using the Appropriate Material Safety Data Sheet (MSDS)

1. What is the chemical that was spilled? ____________________
2. How much chemical was spilled? ____________________
3. What emergency procedures are being followed? ____________________
____
4. What hazard is involved with the chemical?
   Physical? ____________________
   Health? ____________________
5. What personal protective equipment is being used or required? ____________________
____
6. Are there unique handling problems? ____________________
____
7. How is the chemical being disposed of? ____________________
____
8. Does transporting of the waste present any problems? ____________________
____
9. Are there any lingering issues after cleanup? ____________________
____

It is important to put weight on items in an audit. The weighted instrument allows for targeting of potential problems as well as tracking those areas where the serious accidents/incidents are most likely to transpire. Numerical scores are much more powerful than hearsay types of data, results, and outcomes. An individualized weighting system tailors audits to the company's specific needs. Inspection records and scores provide a record that affords the company the opportunity to assess progress or the lack of progress of the company's safety and health effort.

QUALIFICATIONS OF AUDITORS

Those individuals who are selected to perform audits or inspections should be qualified by having previous work experience in that area. They should also be trained and educated on the company's audit approach and the use of its audit instruments. The auditor must understand the technical aspects of operations and be familiar with all job procedures, manufacturing processes, and company policies. The individual(s) who perform the inspection should have good problem-solving skills and possess investigative abilities. Auditors must work independently, be objective in their assessments, and possess skills to communicate effectively.

SUMMARY

To prevent accidents on your jobsite, you definitely should develop audits that meet your company's individual needs and include components for your company's

TABLE 15.5
Regulatory Compliance Checklist

Lockout/Tagout Safety Checklist

This safety compliance checklist assists employers and supervisors in determining that procedures and equipment are available and personnel are trained in the control of hazardous energy sources. This checklist only addresses the minimum required standards. Where appropriate, it may be supplemented with local site- or shop-unique requirements. Relevant references are noted after each question.

	OK	Action Needed
Are all authorized employees, whose job requires them to perform service or maintenance on machines, systems, or equipment, trained on lockout/tagout procedures? [29 CFR 1910.147 (c)(7)(i)]	____	____
Is lockout/tagout training sufficient to ensure recognition of applicable hazardous energy sources? [29 CFR 1910.147 (c)(7)(i)]	____	____
Do authorized employees know the adequate methods and means of isolating hazardous energy sources? [29 CFR 1910.147 (c)(7)(i)(A)]	____	____
Are affected employees instructed by their supervisor on the purpose and use of energy control procedures? [29 CFR 1910.147 (c)(7)(i)(B)]	____	____
Are all employees whose duties require them to be in an area where energy control procedures are used, instructed on their purpose, the prohibitions of lockout/tagout, and about a change in equipment that presents a new hazard? [29 CFR 1910.147 (c)(7)(iii)(A)]	____	____
If random lockout/tagout inspections reveal problems, is retraining accomplished? [29 CFR 1910.147 (c)(7)(iii)(B)]	____	____
If a supervisor has reason to suspect there are inadequacies in the employee's knowledge of lockout/tagout procedures, is retraining accomplished? [29 CFR 1910.147 (c)(7)(iii)(B)]	____	____
Is lockout/tagout retraining sufficient to provide employee proficiency and introduce new or revised procedures? [29 CFR 1910.147 (c)(7)(iii)(B)]	____	____
As a minimum, is lockout/tagout training recorded with the employee's name, class attendance date, and his or her work area? [29 CFR 1910.147 (c)(7)(iii)(C)]	____	____
Lockout/Tagout Procedures		
Are lockout/tagout devices capable of withstanding the environment to which they are exposed? [29 CFR 1910.147 (c)(5)(ii)(A)(1)]	____	____
Are lockout/tagout devices easily recognizable and clearly visible? [29 CFR 1910.147 (c)(5)(ii)]	____	____
Do locks have substantial strength to prevent removal without applying excessive force such as bolt cutters? [29 CFR 1910.147 (c)(5)(ii)(C)(1)]	____	____
Are lockout/tagout devices standard in either shape, color, or format? [29 CFR 1910.147 (c)(5)(ii)(B)]	____	____
Are tags, tag attachment, and lock attachment mechanisms designed so that the probability of accidental removal is minimized? [29 CFR 1910.147 (c)(5)(ii)(C)(2)]	____	____
Are tag attachments self-locking and attachable by hand? [29 CFR 1910.147 (c)(5)(ii)(C)(2)]	____	____
Have facilities identified their requirements for tags, locks, and attachment hardware, and do they have an adequate supply on hand? [29 CFR 1910.147 (c)(5)(i)]	____	____

TABLE 15.5 (continued)
Regulatory Compliance Checklist

Lockout/Tagout Safety Checklist

Has an initial survey been completed to identify all primary and secondary equipment energy sources? [29 CFR 1910.147 (d)(1)] ____ ____

Are drawings, prints, and actual inspections used to assist in identifying all sources of equipment energy? [29 CFR 1910.147 (d)(1)] ____ ____

If an energy-isolating device is not capable of being locked out, can it be demonstrated that a tagout affords adequate protection? [29 CFR 1910.147 (c)(3)(ii)(B)] ____ ____

If using a tagout, can an additional means of protection be provided, such as blocking a control switch? [29 CFR 1910.147 (c)(3)(ii)(B)] ____ ____

Are all energy-isolating devices adequately labeled or marked to indicate their function? [29 CFR 1910.147 (d)(1)] ____ ____

Is all new or replacement equipment able to accept a lock device? [29 CFR 1910.147 (c) (2)(iii)] ____ ____

Prior to lockout/tagout implementation, are all affected employees notified of the work to be performed? [29 CFR 1910.147(c)(9)] ____ ____

If equipment complexity warrants, is a special lockout/tagout plan developed? [29 CFR 1910.147 (c)(4)(i)] ____ ____

Do affected employees review the plan of lockout/tagout sequences of complex operations? [29 CFR 1910.147 (c)(7) (i)(A)] ____ ____

Is there a written listing of all energy-isolating devices on shop equipment? [29 CFR 1910.147 (c)(4)] ____ ____

During lockout/tagout, are all operating controls turned off by an authorized employee? [29 CFR 1910.147(d)(1)] ____ ____

Is an approved lock or tag used to isolate each hazardous energy source? 29 CFR 1910.147 (d) (4)(i)] ____ ____

Are tagouts located in such a position that they will be immediately obvious to anyone attempting to operate an energy-isolating device? [29 CFR 1910.147 (c)(7)(ii)(C)] ____ ____

Is the equipment or system examined to detect and relieve any stored hazardous energy? [29 CFR 1910.147 (d)(5)(ii)] ____ ____

Is the equipment or system tested to determine if operation of the energy-isolation device is working? [29 CFR 1910.147 (d)(6)] ____ ____

Before energy is restored, is a visual inspection and personnel count of the work area conducted by an authorized employee? [29 CFR 1910.147 (e)(1)] ____ ____

Is each lockout/tagout device removed by the authorized employee who applied it? [29 CFR 1910.147 (e)(2)(ii)] ____ ____

Does the supervisor maintain a record of placement and removal of lockouts/tagouts? [29 CFR 1910.147 (c)(4)(ii)(B)] ____ ____

Special Lockout/Tagout Considerations

If a group lockout/tagout system exists, does the procedure provide the same protection that a single employee would receive? [29 CFR 1910.147 (f)(3)(i)] ____ ____

Is responsibility for a number of personnel working under the protection of a particular lockout/tagout vested with an authorized employee or supervisor? [29 CFR 1910.147 (f)(3)(ii)(A)] ____ ____

continued

TABLE 15.5 (continued)
Regulatory Compliance Checklist

Lockout/Tagout Safety Checklist

Are specific procedures established for lockout/tagout utilization during shift change? [29 CFR 1910.147 (E)(4)]	___	___
If outside contractors are working on-site, do our personnel ensure compliance with lockout/tagout procedures? [29 CFR 1910.147 (f)(2)]	___	___
Periodic Inspections		
Are periodic inspections of lockout/tagout procedures conducted at least annually? [29 CFR 1910.147 (c)(6)(i)]	___	___
Are inspections conducted by a supervisor or authorized employee, other than the person using the lockout/tagout procedure? [29 CFR 1910.147 (c)(6)(i)(A)]	___	___
Does the inspection include a review, between the inspector and the authorized/affected employees, of their responsibilities under the lockout/tagout program? [29 CFR 1910.147 (c)(6)(i)(D)]	___	___
Are inspections certified and recorded? [29 CFR 1910.147 (c)(6)(ii)]	___	___
As a minimum, do the inspections include the date of inspection, the employee and inspector names, and the equipment on which the lockout/tagout procedures are being used? [29 CFR 1910.147 (c)(6)(ii)]	___	___

particular type of work. Packaged audits that have been developed by others will not be, in all likelihood, very useful for you since each company is unique and has its own hazards and procedures.

REFERENCES

Reese, C. D. and J. V. Eidson. *Handbook of OSHA construction safety & health*. Boca Raton, FL: CRC Press/Lewis Publishers, 1999.

United States Department of Labor, Occupational Safety and Health Administration, Office of Training and Education. OSHA Voluntary Compliance Outreach Program: Instructors Reference Manual. Des Plaines, IL: U.S. Department of Labor, 1993.

16 Fleet Safety Program

INTRODUCTION

Fleet safety is often viewed as operator safety, which is definitely a key component of a company's attempt to protect its large dollar investment in vehicles and mobile equipment. It goes without saying that many of the accidents that occur are a direct result of driver error. But driver error is not the fault of the individual. It is the fault of management's failure to institute a fleet safety program that provides organization, direction, and accountability for the fleet of vehicles that the company owns.

Commitment to a fleet safety program communicates the value that the company places on their property and employees. The care given to both vehicles/equipment and employees conveys the company's true view of the value of accident prevention.

A fleet safety program should consist of the following:

- A written fleet safety program
- A vehicle/equipment maintenance procedure
- Recordkeeping process
- Operator selection process
- Operator training requirements
- Operator performance requirements

WRITTEN FLEET SAFETY PROGRAM

The company should clearly state its policy regarding fleet safety and delineate what is expected to transpire as a result of its program. The overall intent of this program must be stated clearly. The program should incorporate the many facets of any good accident prevention effort. The fleet safety program should provide the framework for safety management of the company's vehicles/equipment and employees. The company needs to communicate program goals to drivers and supervisory personnel.

There must be a designated person with responsibility for safety and compliance with regulations. This designated person must assume responsibility to comply with existing regulations and implement and enforce the company rules and policies. This person must oversee the qualifying of operators/drivers and the care and safety maintenance of the company's fleet.

It is management's responsibility to recruit and screen new drivers, monitor driver qualifications and safety infractions, and provide training to upgrade drivers' skills and knowledge.

Management should provide a formal mechanism for investigating and reviewing accidents and monitoring maintenance and equipment safety. Management should also implement safe driving incentives and offer recognition to drivers who meet the required standard of performance. The company must constantly monitor the effectiveness of its fleet safety program.

VEHICLE/EQUIPMENT MAINTENANCE

The cost of a fleet of vehicles is a staggering investment and a major cash outlay for companies. To reap the full benefits of such an investment, start with a thorough purchasing process. You want quality and dependability for your money. This will entail some research on your part to ensure that you are getting the most for your money. Once you have your fleet in place, then you will want to get the most mileage out of your purchase. This can only be accomplished by having a preventive maintenance program in place that includes regularly scheduled maintenance, follow-up to operator complaints, and daily preshift inspections of vehicles and equipment (see Table 17.1). You will need a recordkeeping system for your maintenance program that includes the following:

- Operator's inspection record: A checklist of things to be checked daily by operators and any corrections needed to ensure the safety of the vehicle. (This should go to the maintenance shop.)
- Schedule maintenance record: The maintenance shop record of routine or periodic service for each vehicle.
- Service record: To show all findings and results of the inspections, routine service, and repairs made, along with the date of each such maintenance procedure.
- Vehicle history record: A complete history of the vehicle including, but not limited to, any accidents in which it was involved, any catastrophic failure or repairs (i.e., engine change), and when tires were replaced.

REGULATIONS AND MOTOR VEHICLES

The Occupational Safety and Health Administration (OSHA) has jurisdiction over off-highway loading and unloading, such as warehouses, plants, grain-handling facilities, retail locations, marine terminals, wharves, piers, and shipyards. The DOT (U.S. Department of Transportation) has jurisdiction over interstate highway driving, commercial driving licensing (CDL), the hours of service, and roadworthiness of the vehicles. The EPA (U.S. Environmental Protection Agency) has jurisdiction over the natural environment and pollution prevention programs. OSHA has jurisdiction unless preempted by another federal agency such as the DOT or EPA, but can only be preempted in a specified activity or task. OSHA has the ultimate responsibility for the safety and health of all employees.

Other Agency Overview

When another federal agency has regulated a working condition, OSHA is preempted by Code of Federal Regulation-29CFR1910.4(b)1 from enforcing its regulations. For example,

- The DOT regulates driving over public highways, the health and safety of drivers involving their use of drugs and alcohol, hours of service, and use of seat belts. The DOT also regulates the road-worthiness of trucks and trailers and has specific requirements for the safe operation of trucks.
- The DOT has jurisdiction over interstate commerce while OSHA has jurisdiction over intrastate commerce except when handling hazardous materials. DOT has issued regulations regarding the shipping, packaging, and handling of these materials. However, if a truck driver becomes an emergency responder in the event of a spill or other disaster, then OSHA has jurisdiction.
- The Federal Aviation Administration (FAA) regulates flight crews and some other aspects of the safety of ground crews. If there is a clause that covers a working condition in an operational plan negotiated between the carrier and the FAA, the FAA has jurisdiction over that working condition. Otherwise, OSHA covers most of the working conditions of ground crews and baggage handlers.
- Due to the DOT brake regulation, OSHA does not cite for failure to chock trailer wheels if a vehicle is otherwise adequately secured. DOT regulations preempt enforcement and the DOT has jurisdiction. However, if the vehicle is an intrastate truck, OSHA has jurisdiction. Only another federal agency may preempt OSHA's jurisdiction.

Federal Highway Administration (FHWA)

The Federal Highway Administration (FHWA) coordinates highway transportation programs in cooperation with states and other partners to enhance the country's safety, economic vitality, quality of life, and the environment. Major program areas include the Federal-Aid Highway Program, which provides federal financial assistance to the states to construct and improve the National Highway System, urban and rural roads, and bridges; and the Federal Lands Highway Program, which provides access to and within national forests, national parks, Indian reservations, and other public lands. The FHWA also manages a comprehensive research, development, and technology program.

Federal Motor Carrier Safety Administration (FMSCA)

The primary mission of the Federal Motor Carrier Safety Administration is to prevent commercial motor vehicle-related fatalities and injuries. On its website, it has links to its regulations and Regulatory Guidance, a series of frequently asked questions that offer detailed answers to specific situations.

DRIVERS

Drivers in the trucking industry experienced the most fatalities of all occupations, accounting for 12 percent of all worker-related deaths according to www.OSHA.gov, 2010. About two-thirds of fatally injured drivers were involved in highway crashes. Drivers also had more nonfatal injuries than workers in any other occupation. Half of the nonfatal injuries were serious sprains and strains; this may be attributed to the fact that many drivers must unload the goods they transport.

Driver Illnesses and Injuries

Roughly 475,000 large trucks with a gross vehicle weight rating of more than 10,000 pounds are involved in crashes that result in approximately 5,360 fatalities and 142,000 injuries each year (www.OSHA.gov, 2010). Of the fatalities, about 74 percent were occupants of other vehicles (usually passenger cars), 3 percent were pedestrians, and 23 percent were occupants of large trucks. The unsafe actions of automobile drivers are a contributing factor in about 70 percent of fatal crashes involving trucks. More public awareness of how to share the road safely with large trucks is needed. Safe speeds save lives. Exceeding the speed limit was a factor in 22 percent of the fatal crashes. Greater speed enforcement is needed.

Common Driver Injuries

Common trucker injuries include

- Strains and sprains (50%) (see Figure 16.1)
- Bruises
- Fractures
- Cuts and lacerations
- Soreness and pain
- Multiple traumatic injuries

Events or Exposures Leading to Trucker Injury

Common events or exposures leading to trucker injury include

- Overexertion
- Contact with object
- Being struck by an object
- Falling (on the same level) (see Figure 16.2)
- Transportation accidents

OPERATOR RECRUITMENT AND SELECTION

Fleet safety may be viewed as vehicle safety or mechanical safety, but this type of safety depends on both maintenance and operators. An operator's job must include pre-operation inspections and, upon completing his or her use of the vehicle, the reporting of any defects. This should be normal operating procedure for any preventive maintenance program. Nevertheless, what you can expect depends on the quality of your operators. Operators are the center of your fleet safety program.

Thus, in a fleet safety program, it is important to select the best operators for the job. The operator is vital to the prevention of accidents, incidents, vehicle damage,

FIGURE 16.1 Drivers face sprains and strains from loading and unloading.

and injuries. Careful selection of the operator is paramount to an effective fleet safety program. The selection process should involve access to an operator's past employment history, driving record (including accidents), accommodations, or awards, as well as previous operator's experience, if any, on your type of equipment.

As a condition of employment and based on the criteria in a written job description, all potential operators should be able to pass a physical and mental examination and an alcohol/drug test. You will also need a well-written job description, and you may want your legal counsel and others to review it prior to its use.

To improve fleet safety, adequately qualified drivers must be recruited and their performance monitored. The great majority of preventable accidents can be shown to be directly related to the performance of the driver. It is therefore extremely productive to any fleet safety program to have careful new driver selection and adequate monitoring procedures for existing drivers.

An established formal procedure for interviewing, testing, and screening applicants should be in place. A defined standard of skill and knowledge should be met by successful applicants. Appropriate methods should be in place to check out previous employment histories and references of all potential operators. You must access and check the prior driving records of the applicants. Each applicant should undergo a physical examination that includes testing for drugs and alcohol.

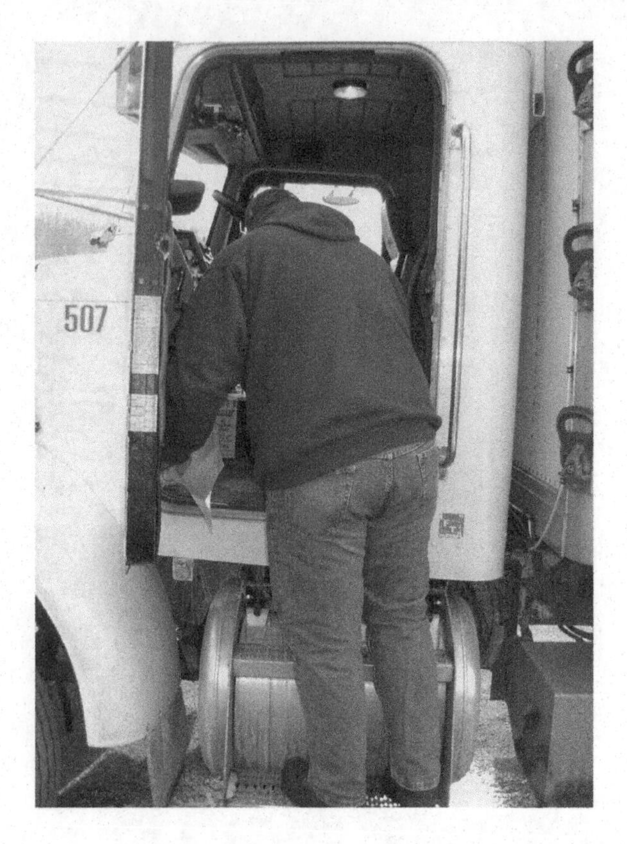

FIGURE 16.2 Drivers can slip and fall while mounting and dismounting vehicles.

Once an operator is hired, there should be a formal program for monitoring the operator's performance, a periodic review of the driving record, and a periodic review of the driver's health should be conducted. Operators should be monitored occasionally for drug and alcohol abuse. (*Note:* A company should have an established policy on drug and alcohol abuse.) A means should be in place for identifying deficiencies in an operator's skills and knowledge, and a procedure should be in place for remedial training. It is well worth the effort to establish a procedure for terminating unqualified operators.

RECORDS TO MAINTAIN

There are three types of records that must be maintained. They are vehicle, operator, and accident records. Vehicle records were discussed previously. Driver records should be kept, which include hiring, training, and job performance. Hiring records should include

- Application form completed for screening
- Previous employment record

- Character references
- Results of work reference checks
- Record of violations and accidents
- Results of interviews
- Results of tests given
- Physical examination report
- Information on previous driver training

Records should be kept on the training the company has provided to the operator by the company. These records should include such information as

- Subjects or curriculum covered
- Date of training
- Length of training
- Instructor's evaluation [test or hands-on (actual operation) assessment]
- Instructor's name
- Location of training
- Any special training aids used (virtual reality or mock trainer)
- Any remedial training
- Any refresher training
- Any prior training from a previous employment or school

Job performance records are important to maintain. These records are important for determining pay increases, promotions, and bonuses. They are also an important tool if you have to dismiss a driver or operator. There records should lend credence to your reason for any dismissal. These performance records should include the following:

- Supervisor's reports or other types of appraisals
- Any commendations
- Accident experience
- Overall driving record
- Any traffic violations
- Any supervised driving (supervisor rides with operator)
- Status in company's safe-driving award program
- Performance in maintenance functions
- Disciplinary actions taken
- Property damage
- Record of training received
- Complaints registered
- Vehicles driven or assigned
- Losses of cargo, money, etc.

The final record is the accident record, which should be an integral part of the company's accident investigation procedures. These records are generated when an operator has or reports an accident. It is usually the person responsible for safety or the supervisor who conducts an accident investigation to determine causes and preventions. This type of record allows the company to track its accidents/incidents and take action to mitigate recurrence or further damage. It is good practice to have all accident reports reviewed by a team composed of management and employees. These reports should include data on the accident, type, time, location, operator, direct cause, indirect cause, and basic cause as well as recommendations for intervention and prevention.

OPERATOR TRAINING

All operators should undergo training related to company and government policies and procedures. This training should include recordkeeping, accident/incident reporting, driving requirements, and defensive driving. After classroom training, each operator should be required to take a supervised driving test, or hands-on supervised operational drive, to determine his or her competence (Figure 16.3). This should be done before the operator is released for work-related driving assignments. Even after the operator is released from training status, he or she may have a supervisor accompany him or her on work assignments.

FIGURE 16.3 Qualified, trained, and safe drivers are the key to fleet safety.

Operator training should consist of the following items:

- Federal or state regulations
- Company's rules and policies
- Right-of-way
- Start-up/back-up
- Negotiating curves
- Passing
- Turning left and right
- Crossing intersections
- Using and changing lanes
- Parking
- Negotiating downgrades
- Driving in adverse conditions
- Emergency equipment and procedures
- Pedestrian interaction
- Passenger management
- Driver inspection reports and procedures
- Brake performance
- Tire inflation
- Tire wear and deterioration
- Wheel retention and deterioration
- Steering system performance
- Full trailer coupling
- Fifth wheel hitches and adjustable axles
- Vehicle lighting and visibility
- Payload characteristics
- Securing cargos

Operators should be observed and evaluated on a periodic basis, retrained if necessary, and supervised more closely. If you are sure that your vehicles are in proper and safe operating condition, your operators then become the key to your fleet safety program. Good, conscientious operators can prevent accidents from occurring; they are the focus of your fleet safety program.

COMPANY OPERATOR'S MANUAL

The company's operator manual is a key communication link between the company and its operators. It conveniently brings together information about the company and its policies and procedures. It is indispensable for training new operators and is a handy reference for existing operators. The manual should be progressively developed and continually updated.

The manual should describe the fleet safety program. It should set out the carrier's standards for safe driving. Procedures for reviewing and classifying accidents, established by the company, should be included. It should also explain the company's disciplinary procedures.

FIGURE 16.4 Another layer of safety is pre-operational inspection by drivers.

PRE-OPERATION INSPECTION

Prior to placing a vehicle in service, an operator should conduct a pre-operational inspection (Figure 16.4; see also Table 17.1). The operator should evaluate at least the following items:

- Apply parking brake
- Fuel tank and cap
- Side marker lights
- Reflectors
- Tires and wheels (lugs)
- Mirrors
- Steering wheel (excess play)
- Apply trailer brakes
- Turn on all lights, including four-way flashers
- Fire extinguisher and warning devices
- Headlights
- Clearance lights
- Identification lights
- Stoplights
- Turn signals and four-way flashers
- Rear end protection (bumper)
- Cargo tie-downs/doors
- Safety chains
- Hoses and couplers

- Electrical connectors
- Couplings (fifth wheel, chains, lock devices)
- Start engine
- Oil pressure (light or gage)
- Air pressure or vacuum (gage)
- Low air or vacuum warning device
- Instrument panel (lights or buzzers)
- Horn
- Backup alarm
- Windshield wipers and washer
- Heater and defroster

DRIVING TASKS

Operators must take great care not to make driving errors. This comes with patience, training, and understanding the role and responsibilities of a driver of a vehicle on the road. Some of the common mistakes that operators make include

- Following too closely
- Inattention or drowsiness
- Misjudging speed of oncoming traffic
- Misjudging speed and closeness of vehicles
- Wandering over lane dividers
- Failure to anticipate lane mergers
- Failure to signal
- Misjudging time for vehicle to clear intersection
- Failure to obey traffic control device
- Poor mirror adjustment
- Failure to use mirrors properly
- Failure to scan space to the sides
- Failure to use turn signals
- Turning from wrong lane
- Failure to give right-of-way to passing traffic
- Assuming other drivers will see and avoid their operating vehicle
- Adverse conditions
- Unable to judge safe speed for road conditions
- Aggressive braking on slippery road
- Failure to anticipate objects on road/bad road surface
- Failure to use headlights and running lights
- Sudden stopping in travel lane
- Parking in travel lane without use of emergency equipment
- Excessive speed for curves
- Inattentive to pedestrian traffic crossing
- Failure to anticipate a pedestrian error
- Running onto a curb
- Overuse of brakes on downgrade

- Overuse of trailer-only brakes
- Failure to respond to symptoms of drowsiness or illness during driving
- Failure to be alert to vehicles symptoms, which could result in hazard or damage
- Failure to start up slowly
- Failure to check all around the vehicle for clearance
- Failure to back immediately after checking clearance

Potential hazardous driving actions should be covered in detail during initial training and any time that refresher training, remedial training, or planned observation of driving skills is undertaken. Everyone tends to forget as time goes on, and all of us can use a reminder.

SAFE DRIVING RECOGNITION

To encourage safe driving and improve operator awareness of safety, safe driving recognition or incentive programs should be an integral part of a formal fleet safety program. Such programs identify superior driving performance and present the selected operator as examples to be emulated by the rest of the fleet. Such programs can be generated internally within the company. These types of programs are put in place for various reasons, such as to

- Encourage safe driving performance
- Heighten operator safety awareness
- Foster operator professionalism
- Focus on monitoring individual driver's performance and skills
- Help monitor fleet performance and effectiveness of the fleet safety program

PLANNING SCHEDULES, LOADS, AND ROUTES

Assisting the operator with pre-travel planning avoids overburdening the operator with unusual driving conditions caused by tight schedules, unusual cargoes, and unfamiliar or hazardous routes. Plans to minimize the need for excessive on-duty schedules and tight schedules and allowances made for adverse weather will help operators and reduce hazards that operators face. The company needs to match cargoes with vehicles during dispatching. The company should have procedures for handling the problem of overloading, improperly loaded, or poorly secured cargoes. Routes should be planned and drivers coached to avoid especially hazardous locations.

PREVENTING ACCIDENTS

A preventable accident is one that occurs because the operator fails to act in a reasonably expected manner to prevent it. In judging whether the operator's actions were reasonable, one seeks to determine whether the operator drove defensively and demonstrated an acceptable level of skill and knowledge. The judgment of what is reasonable can be based on a company-adopted definition, thus establishing a goal for its safety management program.

The concept of a preventable accident is a fleet safety management tool that achieves the following goals:

1. It helps establish a safe driving standard for the driver.
2. It provides a criterion for evaluating individual drivers.
3. It provides an objective for accident investigations and evaluations.
4. It provides a means for evaluating the safety performance of individual drivers and the fleet as a whole.
5. It provides a means for monitoring the effectiveness of fleet safety programs.
6. It assists in dealing with driver safety infractions.
7. It assists in the implementation of safe driving recognition programs.

Fleet safety driving performance depends on management's commitment to the implementation of a formal fleet safety program. An effective safety program will interact with most aspects of fleet operations and challenge the skills and knowledge of its supervisors and drivers.

REFERENCES

National Safety Council. *Motor fleet safety manual (3rd ed.)*. Itasca, IL, 1986.

Petersen, D. *Techniques of safety management (2nd ed.)*. New York: McGraw-Hill Book Company, 1978.

Reese, C. D. and J. V. Eidson. *Handbook of OSHA construction safety & health*. Boca Raton, FL: CRC Press/Lewis Publishers, 1999.

17 Preventive Maintenance Programs

INTRODUCTION

A Preventive Maintenance Program (PMP) depends heavily on an inspection form or checklist to ensure that a vehicle or equipment inspection procedure has been fully accomplished and its completion documented. It has long been noted that a PMP has benefits that extend beyond caring for equipment. Of course equipment is expensive and, if cared for properly and regularly, will last a lot longer, cost less to operate, operate more efficiently, and have fewer catastrophic failures.

Remember that properly maintained equipment is also safer and there is a decreased risk of accidents occurring. The degree of pride for having safe operating equipment will transfer to the workers in the form of better morale and respect for the equipment. Well-maintained equipment sends a strong message regarding safe operation of equipment.

If you allow operators to use equipment, machinery, or vehicles that are unsafe or in poor operating condition, you send a negative message that says: "I don't value my equipment, machinery, or vehicles and I don't value my workforce either." A structured PMP will definitely foster a much more positive approach regarding property and the workforce. Reasons for a PMP are to

1. Improve operating efficiency of equipment, machinery, or vehicles.
2. Improve attitudes toward safety by maintaining good/safe operating equipment.
3. Foster involvement of not only maintenance personnel, but also supervisors and operators, which forces everyone to have a degree of ownership.
4. Decrease risk for incidents or mishaps.

The first aspect of a PMP is to have a schedule for regular maintenance of all your equipment. Second, it motivates supervisors to make certain that all operators are conducting daily inspections. Third, it assures the company that all defects are reported immediately. Finally, it documents that repairs are made prior to operating vehicles or equipment. If this is impossible, the equipment should be tagged and removed from service.

COMPONENTS OF A PMP

You will need the following in order to have an effectively functional PMP:

1. A maintenance department that carries out a regular and preventive maintenance schedule
2. Supervisors and operators who are accountable and responsible
3. A pre-shift checklist for each type of equipment, machinery, or vehicle (an example vehicle checklist can be found in Table 17.1)
4. An effective response system when defects or hazards are discovered
5. A commitment by management that your PMP is important and will be achieved

PMPs are another example of an accident prevention technique. Of course, this technique would apply only if your company has equipment, machinery, or vehicles. A PMP is also an integral part of a fleet safety program.

PREVENTIVE MAINTENANCE

Preventive maintenance's ultimate purpose is to prevent accidents caused by vehicle deficiencies, machinery defects, or equipment hazards. Worn, failed, or incorrectly adjusted components can cause or contribute to accidents. Preventive maintenance and periodic inspection procedures help prevent failures from occurring while the vehicle/machine/equipment is being operated. Such procedures also reduce reliance on the operator, who may have limited skill and knowledge for detecting these deficiencies.

As an example of how a PMP works, the remainder of this chapter uses as an illustration a truck and its operator as the subject to demonstrate the PMP approach to maintenance.

MANAGEMENT'S ROLE

Management experiences large numbers of repairs to the vehicles in its fleet. This should be viewed as an indication of inadequate maintenance and inspection procedures, and an indication of potential unsafe vehicles, which could facilitate or cause accidents. The scheduling of periodic inspections and maintenance activities should be part of management's fleet safety program. Management should maintain a record on each vehicle regarding repairs, inspections, and maintenance.

When hazards or defects are found, there must be guidelines or rules for placing vehicles out of service until problems are corrected. Operators must be sufficiently trained and knowledgeable to detect maintenance and repair needs, and to report them to the appropriate authority.

Is the emphasis of your preventive maintenance and inspection program based on the importance of recognizing the vehicle components that directly affect vehicle control, such as braking systems, steering systems, couplers, tires and wheels, and suspension systems? Each of these could affect the vehicle's safe operation. You

TABLE 17.1
Pre-Shift Vehicle Checklist

PRE-SHIFT EQUIPMENT CHECKLIST

Check any of the following defects prior to operating your equipment and vehicle and report those defects to your supervisor or maintenance department.

Type of Equipment: ______________________ Identification Number: ____________

Date: ______________________

1. Walk-Around:
   - ___Tires
   - ___Broken lights
   - ___Oil leaks
   - ___Hydraulic leaks
   - ___Mirrors
   - ___Tracks
   - ___Damaged hose
   - ___Bad connections fittings
   - ___Cracks in windshields or other glass
   - ___Damaged support structures
   - ___Damage to body structures
   - ___Fluid levels
   - ___Oil
   - ___Hydraulics
   - ___Brakes
2. Operation:
   - ___Engine starts
   - ___Air pressure or vacuum gauge
   - ___Oil pressure
   - ___Brakes
   - ___Parking brakes
   - ___Horn
   - ___Front lights
   - ___Back lights
   - ___Directional lights
   - ___Warning lights
   - ___Back-up alarm
   - ___Noises or malfunctioning
   - ___Engine
   - ___Clutch
   - ___Transmission
   - ___Axles
   - ___Fuel level
   - ___Instrument panel
   - ___Windshield wipers or washers
   - ___Heater or defroster
   - ___Mirrors

continued

TABLE 17.1
Pre-Shift Vehicle Checklist

PRE-SHIFT EQUIPMENT CHECKLIST

3. Operation:
   ___Seat belts
   ___Steering wheel play or alignment
4. While Underway:
   ___Engine knocks, misses, overheats
   ___Brakes operate properly
   ___Steering loose, shimmy hard, etc.
   ___Transmission noisy, hard shifting, jumps out of gear, etc.
   ___Speedometer
   ___Speed control
5. Emergency Equipment:
   ___First aid kit
   ___Fire extinguisher
   ___Flags, flares, warning devices
   ___Reflectors
   ___Tire chains, if needed
6. Cargo-Related Equipment:
   ___Tiedowns
   ___Cargo nets
   ___Tarps
   ___Spare parts
7. Other Items:
   ___Hand tools
   ___Spare parts
   ____
   ____

No Defects Noted: Operator Signature: ______________ Date: __________ Time: ________

Describe Any Defects Noted: ____

____

____

Defects Corrected: (Initials) Defect Correction Unnecessary: (Initials) Defects Corrected By: (Signature) Date:

____

Defects Corrected: Operators Signature: Date:

____Yes ____No

should also have a program that can determine when vehicle component wear is at a point where replacement is necessary.

The PMP must be enforced by management. It should also be evaluated continuously to gauge it effectiveness in preventing accidents. Your fleet of vehicles should be capable of passing the minimum industry inspection standards.

THE PREVENTIVE MAINTENANCE PROGRAM (PMP)

A preventive maintenance and inspection program should recognize wear of consumable components that must be periodically replaced or serviced. It should take into account indicators of deterioration that can be monitored at the operator inspection level. The driver should be trained in troubleshooting. Special attention must be paid to the condition of components that cannot be easily observed by the driver. Maintenance supervisors and mechanics should inspect those components where problems can occur but are not easily discernible.

OPERATORS' INSPECTIONS

To ensure that vehicles are in safe operating condition while driven, the operator should check the whole vehicle carefully—both pre-travel and post-trip. These pre-travel and post-trip inspection reports are an important part of a PMP. If something seems wrong with the vehicle, stop and check it out. Do not continue with the trip until you are satisfied it is safe to do so.

The driver should be the one held ultimately responsible to make sure that the vehicle being driven is in a safe operating condition. Appropriate inspection procedures and reports assist in ensuring this. The operator is also in a position to detect vehicle deficiencies and refer them to maintenance for repairs. The driver should not operate a faulty vehicle. Federal and state laws require that the operator should not drive a vehicle unless fully satisfied that it is in a safe operating condition (see Figure 17.1).

FIGURE 17.1 Operators need to conduct pre-shift inspections and constantly look for potential problems such as this broken exhaust.

Brake Safety

The braking system is one of several key safety-related items. Catastrophic brake failure, such as sudden air loss, may lead to loss of control and the driver's inability to recover. Progressive brake deterioration, such as brake shoe wear without corresponding adjustment, can be even more troublesome because it may appear during normal driving and precipitate an accident during emergency braking applications.

The driver should test the brakes for stopping performance before going on a highway. The operator needs to make sure brakes are properly adjusted. The driver should be trained to determine if the air system is operating satisfactorily and be able to check to make sure that low air warning devices are functioning. It is good practice for the driver to test vehicle brakes during a trip. Before entering severe downgrades, the driver should stop and check brake adjustment.

Tire Safety

Tires are one of several key safety-related vehicle components. Improper tire pressure, either too little or too much, can lead to deterioration and eventual catastrophic failure. A tire that is worn or damaged may fail as a blowout and result in loss of control of the vehicle. The principal indicators of deterioration are tread wear, tread and sidewall damage, and air leakage.

During vehicle inspections, drivers must check tires to make sure that their condition is within company-established, out-of-service criteria. During an inspection or trip, the operator must monitor tires for road damage or deterioration such as tread or sidewall separation, cuts or gouges, flat spots or uneven wear, leaks (monitor tire inflation), and flat tires at duals. Operators should monitor tire inflation during extended trips. Operators should not operate on tires with inflation pressures other than those specified by the manufacturer.

Wheel Assembly Safety

Wheels are one of several key safety-related items. Incorrectly assembled, or damaged wheel components, can result in the collapse of a wheel assembly and consequent loss of control. When operators are inspecting the wheels, they should look for cracks in the wheels and rims, improperly seated lock rings, and rust around wheel nuts; check for wheel assembly tightness; check wheel nut tightness after recent tire change; and check for missing components.

Steering System Safety

The steering system can fail catastrophically or deteriorate progressively. As steering wheel play progressively increases, it becomes more difficult for the driver to steer. This should be viewed as an indicator of deteriorating steering system components, which may eventually lead to catastrophic failure. Steering wheel play can be detected by the driver during inspections. Operators should cite steering deficiencies on the vehicle inspection report.

Trailer Hitch Safety

Trailer separation can occur due to improper hitching, or inadequate or damaged equipment. Pintle hooks and ball hitches may uncouple if improperly latched. Hitch mounts can separate due to damage or lack of maintenance. The operator should check to see that hitch components are in good condition on the trailer and truck, and adjust the coupler if necessary. Each operator must ensure that the pintle hook or ball hitch is properly locked, and that safety chains are properly connected. All electric and air lines must also be properly connected to ensure safe usage.

Hitch and Axle Safety

Eliminate accidents caused by trailer separation, malfunctioning trailer brakes, inactive running lights, or trailer axle separation, by using proper coupling procedures of semi-trailers. This ensures that coupling equipment remains in good order, landing gear is not damaged, air lines and electric lines are hooked up, axle loads are balanced, and coupling is secure.

The operator must adjust the trailer height to minimize coupling impact. He or she should also examine the conditions of the kingpin and jaws, and check that the jaws are locked after coupling. Next, make sure that the landing gear is raised, and hook up the air and electric lines carefully. If the trailer axle is adjustable, make sure it is locked properly. Check also to see that the kingpin is not riding on top of the jaws.

If the tractor has an adjustable fifth wheel, the driver must ensure its adjustment is locked. No driver should pull a trailer with the slide stops. Before driving away, apply the trailer brakes and pull gently against them to check coupling.

Driving Light Safety

Trucks and tractor-trailer combinations, due to their length and lower maneuverability, may be struck by other vehicles because the other operator does not see the vehicle or misjudges its rate of movement. Such drivers can be assisted by making sure that the truck's lighting system and reflectors are adequate and clean. The truck driver should use extra care in crossing traffic lanes and making turns during adverse visibility conditions. Use extra care when pulling low-profile trailers, such as empty flatbed tractors, empty container chasses, construction equipment trailers, and pole trailers.

Load Safety

There is a real need to reduce the number of accidents caused by overloading, poor load distribution, and lack of clearance with fixed objects. Many accidents are caused by inadequate loading procedures or route planning. Heavy, high, and offset loads can precipitate rollovers during emergency steering maneuvers and when driving at excessive speeds. High trailers or outsized loads can also result in collisions when routes are not planned.

Drivers need to make sure that vehicle and axle weights are within legal limits. They should know the vehicle weight limits as well as tire inflation ratings; both must be compatible with the load and driving conditions. Suspension and coupling ratings for the vehicle shall be appropriate for the load.

When a trailer is being loaded with mixed cargo, heavier articles should be loaded on the bottom. Check to see that heavy articles are not offset to one side of the trailer. When driving with heavy or high loads, drivers should use reduced speeds. Road advisory signs for curve speed normally do not apply to heavily loaded commercial vehicles. Go slower. Drivers never know when they may have to make an emergency lane change. Also, be aware when trailer wheels are off track; they may collide with curbs, or track onto unimproved shoulders. Both can lead to the loss of control when the vehicle is heavily loaded. It is critical to know vehicle height and plan your route so that low bridges do not become a surprise. When picking up a sealed trailer, the driver should find out the load's characteristics.

Load Securing Safety

Loads that break loose on the road can create control difficulties for the driver and present a hazard for other drivers. Shifting cargo can cause loss of control and truck rollover. Drivers must check to make sure that the lading has been properly secured. Periodically check to see that bracing and tie-downs are still intact and that the load has not shifted. Some loads, cargo, or lading, such as liquids in tanks or portable tanks, have a tendency to shift; drivers should operate at reduced speeds during turns or braking to guard against loss of control. Pay particular attention to bracing and tie-downs when picking up unusual loads. Drivers should satisfy themselves that loading personnel have done their job properly.

MAINTENANCE

Maintenance crews should use established company and industry guidelines to determine whether components should be returned to service. They need to determine cause of damage or deterioration. Such analysis may help identify improper use of machinery, or faulty maintenance procedures, which should be corrected. All maintenance or repairs for each vehicle should be documented on a separate maintenance reporting form.

Never allow any vehicle that is not roadworthy to leave the maintenance facility. Such vehicles should be locked out and tagged out to prevent their use.

MANAGEMENT RESPONSIBILITY

Management's responsibility is to establish inspection and reporting procedures for drivers. It must ensure compliance with state, federal, and industry rules. Its job is to select drivers and provide them with adequate training for inspecting critical components and to determine whether their vehicle will be safe while on the road. Management must enforce both driver inspection and maintenance policies. It must see to it that maintenance personnel are responsive to driver-reported deficiencies, and it must establish company standards for placing vehicles out of service. It must

FIGURE 17.2 Preventive maintenance can help prevent costly repairs like this.

ensure that preventive maintenance procedures adequately detect and repair worn or defective components (see Figure 17.2).

Management must make sure drivers and maintenance personnel are trained and knowledgeable to make a determination during inspections as to whether or not a vehicle is safe or should be taken out of service. Neither set of employees (drivers or maintenance personnel) should perceive that management wants them to circumvent good vehicle safety practices.

SUMMARY

A preventive maintenance program can be developed for any industrial function that has a maintenance component. The same principles used in the truck operator example in this chapter can be used for other mobile equipment, machinery, or process maintenance (see Figure A.3). It matters little the type of maintenance. It is more important that the workers, operators, supervisors, and management realize that the care taken in prevention will pay dividends in cost savings, accident costs, less equipment damage, and better overall working conditions and employee attitudes. Care of your physical assets (physical plant, vehicles, machinery, etc.) is critical to the bottom line. The bonus is that you will reduce one of the factors that can contribute to your occupational injury rate.

REFERENCES

Reese, C. D. and J. V. Eidson. *Handbook of OSHA construction safety & health*. Boca Raton, FL: CRC Press/Lewis Publishers, 1999.

FIGURE 17.3 This chapter is applicable to all over-the-road vehicles like this delivery truck and not just long-haul trucks.

United States Department of Labor, Occupational Safety and Health Administration, Office of Training and Education. OSHA Voluntary Compliance Outreach Program: Instructors Reference Manual. Des Plaines, IL:, U.S. Department of Labor, 1993.

18 Special Emphasis Programs

INTRODUCTION

Special emphasis programs were briefly mentioned previously in Chapter 2 but should be reinforced as an effective accident prevention technique. Any time you institute a special program that targets a unique safety and health issue, an organized approach in prevention should be developed. The benefits of instituting a special program include the fact that the potential hazard remains on everybody's mind, management receives feedback, and workers receive reinforcement for the desired performance. A program can be developed in any area where the company feels there is a need. Some areas of focus could be ladder safety, back injuries, vehicle or equipment safety, power tool incidents, etc. For success, the program may contain goals to attain, rewards to receive, or even consequences for enforcement if the rules of the program are not followed. By setting up a program, the company is at least taking action to target accidents and prevent their occurrence.

An example of a special emphasis program is a ladder safety program. A ladder safety program will be utilized as an example to illustrate how this might be done.

LADDER SAFETY PROGRAM

MANAGEMENT'S COMMITMENT

The first step is to develop a policy that will be signed by top management regarding the commitment to the special emphasis program on ladder safety.

> This ________________ company is committed to preventing ladder-related accidents. It is expected that the company rules and policies for ladder safety will be followed by all employees. The company is committed to providing you with safe and appropriate ladders for your work activities. It is your responsibility to use ladders safely in accordance with the company's rules and policies and the ladder training that you have received.
>
> ________________________________
>
> Company President's Signature

HAZARD IDENTIFICATION

Prior to the use of a ladder, all employees are to inspect the ladder in accordance with the company ladder checklist (see Table 18.1).

TABLE 18.1
Company's Ladder Checklist

COMPANY'S LADDER CHECKLIST

Inspection:

_____1. Are the joints between the steps and side rails tight?
_____2. Is all hardware secure?
_____3. Are nonslip safety cleats securely attached to the side rails?
_____4. Do the locks, and other movable parts, move without binding or excessive play?
_____5. Are wooden ladders unpainted so as to reveal cracks and defects?
_____6. Are metal ladders free of corrosion?

Placing Ladders:

_____7. Are ladder bases set at a distance from the structure equal to about 1/4 of the ladder height?
_____8. Is the ladder resting on a firm level surface, not on boxes, barrels, or blocks?
_____9. Are the feet of the ladder braced or staked to prevent slipping?
____10. Are stepladders opened fully?
____11. Are there guards or barricades around any ladder used near foot or vehicle traffic?
____12. Is the area surrounding the top and bottom of the ladder clear?
____13. If the ladder is placed in front of a door or passage, is the door locked or the passage barricaded?
____14. Does the ladder extend 3 feet above the top level, and is it secured?

Workers:

____15. Are workers wearing clean shoes?
____16. Are the workers' hands free of grease, oil, and other slippery substances?
____17. Is a lift or hoist being used for heavy equipment or materials?
____18. Is the worker below the third rung from the top of the ladder while performing work?
____19. Is the worker avoiding overreaching, so as not to lose balance?
____20. If a stepladder is used, is the worker no higher than third step from the top?
____21. Is the worker afraid to work at heights?

Control and Prevention

As part of the company's effort to control and prevent ladder accidents, the posters in Tables 18.2, 18.3, and 18.4 should be posted around the work area as constant reminders of what should be occurring regarding the safe use of ladders.

If a damaged or unsafe ladder is identified, it should be removed from service and a tag should be affixed to it stating "Removed from Service. Do not Use." The supervisor must be informed of the unsafe ladder. The supervisor must ensure that the ladder is not used and that it is either disposed of or repaired to a safe condition. Any ladder accidents must be fully investigated by the supervisor and recommendations to prevent further occurrence made and implemented.

Weekly safety talks should be given to reinforce training, safety rules, safe practices, and unique ladder applications. Supervisors should give safety talks on ladders similar to the information that is found in Tables 18.5 and 18.6.

TABLE 18.2
Ladder Poster

LADDER POSTER

Choose a ladder that is or has
Unbroken rungs or steps
Safety feet
Meets ANSI standards
The right size to do the job
Functional spreaders that lock
Metal ladders marked indicating their ability to conduct electricity
Not been painted

TABLE 18.3
Ladder Setup Poster

LADDER SETUP

Place it on a level surface.
Use wide boards under it if you are on soft ground.
Place the feet parallel with the top support.
Anchor the top.
A straight ladder should extend 3 feet past the support point.
Tie or brace it at bottom or have someone hold it.
Keep the ladder the right distance from the wall or support (use the 4-to-1 rule).
Raise an extension ladder before extending it.
If you place a ladder in front of a door, make sure the door is locked or blocked.

TABLE 18.4
Ladder Do's and Don'ts Poster

LADDER DO'S AND DON'TS

DO:	DON'T:
Clean your shoes first.	Try to use a ladder if it is scaffolding that you really need.
Face the ladder while climbing up or down.	Carry objects while climbing; use a special belt/tool pouch or hoist materials up with a rope.
Hold the ladder with both hands while climbing.	Step on the top two steps of a stepladder or the top three ladder rungs of an extension ladder.
Hold the ladder with one hand while working.	Lean too far in either direction while working on a ladder.
Use a hanger or tool pouch for tools or a bucket.	Let your belt buckle go outside the rails.
Keep your weight centered between the rails.	Go near power lines or electricity with metal ladders.

TABLE 18.5
Safety Guidelines for Extension Ladders

BE LADDER WISE

Whether using a ladder at work or at home, these are some lessons that can be learned to prevent a serious injury from happening to you, a co-worker, or family member.

Safety Guidelines for Extension Ladders

INSPECT

Missing screws/rivets
Frayed ropes and damaged pulleys
Damaged rung locks
Damaged or missing support feet
Cracked, split, or damaged rungs, uprights, or braces

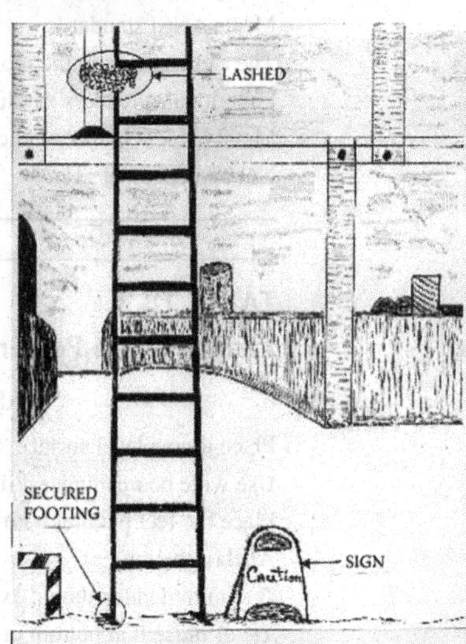

Do's

Inspect ladder; report damaged ladder immediately
Use the correct length ladder for job
Face ladder at all times
Use both hands when climbing
Climb and descend slowly
Use a tool pouch for tools
Use safety cones, signs, or barriers
Watch out for contact with electricity
Side rails should extend at least 3 feet above landing
Lash base or top of ladder or have co-worker hold the ladder
Use the 4-to-1 rule (count the number of rungs to the support point and divide by 4 – that's how many feet the ladder should be away from the wall)

Don'ts

Use damaged ladders
Stand on the top rung
Move ladder while you are on it
Lean too far to either side
Set up ladders on slick or slippery surface

Training

All employees, including management and supervisors, are to receive ladder safety training. The content of the training shall include

- Company rules and policies
- Accountability and responsibility
- Ladder hazard recognition
- Reporting and responding to ladder hazards
- Safe use of ladders
- Climbing safely

Our motto is "Climb Safely."

TABLE 18.6
Safety Guidelines for Stepladders

BE LADDER WISE

Safety Guidelines for Stepladders

INSPECT

Missing screws or rivets on steps
Damaged, loose, or broken hinges
Deformed, loose, or broken spreaders
Damaged or missing support feet
Side rails for cracks and damage

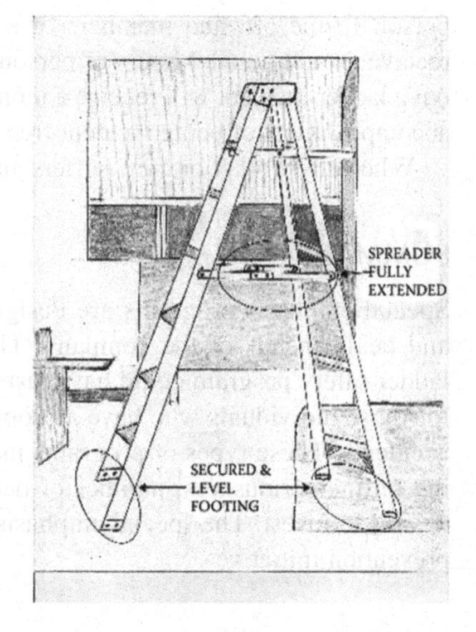

Do's

Inspect ladder
Report damaged ladder immediately
Use the correct length ladder for job
Face ladder at all times
Use both hands when climbing
Climb and descend slowly
Use a tool pouch for tools
Watch out for contact with electricity
Use safety cones, signs, or barriers when it makes sense
Set up on stable flat surface

Don'ts

Use damaged ladders
Stand on the top
Sit on the top
Move ladder while you are on it
Lean too far to either side
Throw tools from the ladder
Jump from the ladder
Set up ladders on slick or slippery surfaces

INCENTIVES

Anyone who experiences a ladder incident will receive retraining related to ladder safety. The company feels that workers and crews who do not experience a ladder incident over set periods of time should receive recognition for their safe performance.

All employees who go 1 month without an incident will receive a company ball cap. All work crews who go without an incident for a month will receive a company tee-shirt. All workers who go 3 months without an incident will receive a gift certificate for a lunch, and work crews who go 3 months without an incident will receive a pizza lunch.

Workers who go 5 months without an incident will receive a pocketknife, and work crews who go 5 months without an incident will receive a travel mug. All workers who go 6 months without a ladder incident will receive a gift certificate for dinner for two. All crews who go incident-free for 6 months will receive an additional certificate for another person for dinner.

At 9 months, workers who continue to not have a ladder incident will receive a $20 gift certificate from a local retail store. Likewise, all members of a work crew who have gone 9 months without an incident will receive an additional $10 gift certificate.

At the end of a year without a ladder incident, workers will receive 4 hours of personal time off, and members of a crew who are free of ladder incidents will receive an additional 4 hours of personal time off. All workers who go a year without a ladder incident will receive a monogrammed jacket and a dinner to honor their accomplishment of being incident-free.

When any worker or crew suffers an incident, the sequence will begin again.

SUMMARY

Special emphasis programs are designed to meet the unique occupational safety and health needs of the company. The previous example of a special emphasis ladder safety program could have just as easily been an accident repeater program for those individuals who have a propensity to suffer from recurring occupational accidents. These types of programs may be very simple or very complex, depending on the seriousness, number of occurrences, or amount of resources that one desires to invest. The special emphasis program definitely has its place in accident prevention initiatives.

REFERENCES

Reese, C. D. and J. V. Eidson. *Handbook of OSHA construction safety & health.* Boca Raton, FL: CRC Press/Lewis Publishers, 1999.

19 Using Safety and Health Consultants

INTRODUCTION

From time to time, each of us is faced with problems or issues that surpass our own training, experience, or expertise. An educated person is one who recognizes that he or she needs help. There are always plenty of individuals ready and willing to provide advice and help. However, one must make sure that the consultant is competent help that can truly and meaningfully provide the assistance that is desperately needed.

When the company realizes that it does not have the knowledge to do what is required, it is good to be confident enough in oneself to understand that limitations exist. Finding the appropriate individual to help the company may take as much effort as would have been used if the company had the expertise to solve its own problem, but cannot! The company will want to make sure that it is getting its money's worth.

NEED FOR A CONSULTANT

The consultant that is needed may be a specialist in a particular area of safety and health such as, for example, an ergonomist or an engineer who can help with redesigning issues. No matter the person you need, the company must proceed in an organized fashion in selecting that individual and finally obtaining a solution to the problem. Companies may use occupational safety and health consultants for a number of reasons, to include

- Identification of new safety or health problems that require technical or professional resources beyond what is available in the company
- A management initiative to redesign, streamline, and enhance current safety and health processes and programs
- A directive to outsource a non-core company function
- Regulatory-driven regulations requiring new or additional compliance measures
- Correct deficiencies in the safety and health program

A consultant will likely draw on a wide and diverse experience base in helping to address problems and issues. He or she is not influenced by politics and allegiance, and is therefore in a better position to make objective decisions. The professional consultant strives to provide cost-effective solutions because his or her repeat business is based on performance and professional reputation. Also, consultants are temporary employees, and the usual personnel issues are not applicable to them. Consultants work at the times of the day, weeks, or month when the company has a need and not at their own convenience. Most consultants have become qualified to perform these services for you through either education or experience.

In determining the need for a consultant, the company will want to consider whether using a consultant will be cost effective, faster, or more productive. Companies may also find the necessity for a consultant when they feel a need for outside advice, access to special instrumentation, an unbiased opinion or solution, or an assessment that supports the initial solution.

A consultant can address many of the company's safety and health issues. He or she can also act as an expert witness during legal action, but primarily the consultant is hired to solve a problem. Thus, the consultant must be able to identify and define the existing problem and then provide appropriate solutions for that problem. Consultants usually have a wide array of resources and professional contacts that they can access in order to assist in solving the problem.

To get names and recommendations of potential consultants, contact professional organizations, colleagues, and insurance companies that often employ loss control or safety and health personnel. Furthermore, do not overlook local colleges and universities; many times they also provide consultative services. In selecting a consultant, make sure to ask for

1. A complete résumé that provides the consultant's formal education, as well as his or her years of experience
2. A listing of previous clients and permission to contact them
3. The length of time the individual has been a consultant and his or her current status regarding existing business obligations
4. Verification of professional training
5. Documentation of qualifications, such as registered professional engineer, certified safety professional, or certified industrial hygienist
6. A listing of the consultant's memberships in professional associations
7. Any areas of specialization and ownership or access to equipment and certified testing laboratories

CONDUCTING THE INTERVIEW

It is important that proper interviews be conducted of the consultants who have made the final cut. Make sure that the following items are part of evaluating and ranking the potential consultants:

- Review credentials and qualifications to determine his or her experience and expertise. Health and safety experts should hold proper credentials.
- Does the consultant have experience with the company's type of problem or project?
- What procedures does the consultant do, and what do they subcontract?

- Review the scope of services to determine if the consultant has expertise for the company's requirements.
- Is the consultant's staff trained to properly conduct the scope of work?
- Has the consultant had other projects similar to the company's problem? Ask for references from previous clients that had similar needs. Has he or she had repeat business?
- Does the consultant have to come to speed, and is there a leaning curve?
- Ask for a list of subcontractors and check their performance.
- Does the consultant have professional, compensation, and liability insurance?
- How does the consultant tend to bill for project? Are charges *per diem, direct personnel cost (time and materials), cost plus fixed payment, fixed lump sum,* or *percent of cost?*
- Is the consultant a member of trade or professional organizations, and what is his or her status in these organizations?
- Carefully and thoroughly evaluate all references that the consultant provides.
- What is the consultant's present workload?
- Is the consultant and his or her staff trained in safety and health procedures in accordance with OSHA?
- Does the consultant have the equipment and testing instruments to carry out any needed procedures?
- Has the consultant met expectations, fulfilled the requirements, met deadlines, and provided a professional and accurate product or deliverables in the past?
- Are there any conflicts of interest?
- Does the consultant have the resources to deliver on the requirements in the scope of work?
- Is the consultant financially stable?
- Do the work practices and testing follow nationally recognized best practice standards?
- Is there a quality control facet, and are testing laboratories certified?
- Will legal representation be needed to safeguard the company's confidential data or trade secrets?

Evaluate more than one consultant so that some basis for comparison is available. This will help in selecting the consultant who is best suited for your particular situation. Pay close attention to the consultant's compensation procedures. The following financial information will be needed: Does he or she work hourly? What is the estimated total cost? What will be used as a retainer? How will the consultant bill for travel, shipping, report generation, computer time, and other services such as laboratory fees? The decision to employ the consultant should not be based purely on a cost comparison. It may be like comparing cooking apples and eating apples. They are both apples but only one has the qualities to meet the company's needs at that point in time, and expect to pay the price for the type of apple that meets your needs. It is appropriate to prequalify consultants prior to allowing them to submit a proposal, be interviewed, or make the cut for selection.

SCOPE OF WORK

The most important document in securing a consultant is the scope of work, which provides some guarantee of attaining the expected outcome when evaluating the consultant. The scope of work should include a description of the needs or project, the company's expectations, any unusual challenges for the consultant, a timeline for completion of the project, the provision of a detailed financial estimate for completing the project, and the expected outputs from the effort such as reports, data, results of testing, or other specified outcomes. The methods and action plan should be part of the response to the scope of work, as well as the resources, manpower, and specialized equipment needed to complete the requirements of the scope of work.

The consultant submitting a proposal for the scope of work might benefit from a visit to the site to consider the problem prior to providing a response to the scope of work. Be careful of proposals that appear overly optimistic or overreaching, quote extremely low bids or unrealistic timelines, contain conflicts of interest or hard-sell approaches, minimize or maximize potential technical or legal problems, or contain unsound approaches to solving the problem.

Although a verbal agreement may seem fine, it is recommended that a written agreement that spells out how much the service is going to cost or the maximum number of hours that the company will support be included in the signed contract. The company should provide a scope of work that sets out the steps that will be followed in order to solve the problem. It is appropriate to include, in this written document, the output expectations from the process. A minimum output would be a written report, but the company may not want to pay for this. If not, a verbal report would be the output in the agreement. All expected outputs should be part of this agreement. This may include drawings, step-by-step procedures, and follow-up. The company needs to obtain all the information required to solve your problem, based on the consultant's recommendations. The company may have to construct, implement, or redesign equipment, processes, or procedures. Unless some sort of a protective clause is contained within the agreement, there is no guarantee that the consultant's recommendations will fix the problem. If a guarantee is desired, expect to pay more because the consultant will be responsible for the implementation of his or her recommendations.

If the process of using a consultant is to work, it must be in full cooperation with the company; thus, the consultant can fail if he or she is not an integral part of the company's team. For this reason it is important to provide all the information needed. Clearly define the problem. Set objectives. Agree on realistic requirements. Get a clear and complete proposal from the consultant. Review the progress from time to time, and stay within the scope of the problem. The consultant can only be as effective as the company allows him or her to be. The company must do its part to ensure success. Make sure that a cost-benefit analysis is performed before pursuing a consultant. Agree on costs and fees prior to starting.

The consultant is not a friend. The company has hired him or her based on an evaluation that this is the best person to help solve the problem that the company cannot solve, does not have time to solve, or does not have the resources to solve. The consultant should define the problem, analyze the problem, and make recommendations for a solution. The consultant can be made a part of your team by providing full information and support. This will ensure that the greatest value is reaped from the investment in the consultant.

FINAL HIRING STEPS

To summarize, the primary steps for employing a consultant are

1. The company must know what it wants from the beginning.
2. A scope of work should be put in writing.
3. Use a bid process for selecting the consultant.
4. Delineate the timeline for completion of work.
5. Decide who the consultant will be reporting to.
6. Know the company's budget limitations (develop a formal budget).
7. Make sure that a written contract exists.
8. Have the consultant give his or her protocol for completing the work.
9. Discuss the evaluative approach being used by the consultant prior to the contract.
10. Have mechanisms in place to hold the consultant accountable for not completing the agreed-upon work.
11. Establish penalties for not meeting pre-set schedules.
12. Have a tracking system to evaluate the work's progress.
13. Be consistent in the expectations because each time the scope of work is changed or delayed inhibits progress, it will cost extra.
14. Follow good business practices and maintain a professional relationship.

Before selecting a consultant, the official in charge should do his homework. Make sure the consultant can service the company's geographic area or multiple site locations, depending on its needs. Hiring a consultant is not an easy process. Ensure that the job is not too large for one consultant and may require more than what a "friend down the street" can offer. Although the company may be familiar with the consultant from a social relationship, this does not mean that he or she can and will provide the service needed, unless the company is familiar with his or her work from previous efforts or work done for company. The company needs to consider what happens in case of illness, disability, or death of the consultant. A contingency plan should exist as to how the work will continue if some unfortunate event transpires. If special equipment is needed for sampling, for example, visit the consultant's office to ensure that he or she have the necessary resources to accomplish what he or she are committing to do.

In summary, check the background of the consultant; it should include the consultant's education, industrial experience, membership in professional organizations, and any certification. This is just a starting point. Procure and check the consultant's referrals. The more invested in a consultant, the more referrals and the more amount of homework that needs to be done. Obtain copies of previous contracts and reports, and review them for quality, professionalism, and depth of information. A responsible consultant will be happy to provide samples of his or her work.

Remember that company personnel will have to work with the consultant. The chemistry between the consultant and the company's team needs to be there. The attitude of the consultant may be as important as his or her abilities, as the consultant must work within your company's framework and with others and yourself. The consultant's inability to communicate effectively can doom the project.

SUMMARY

When a qualified and experienced safety and health consultant is selected, it is much more likely that

- Money and time will be saved.
- Projects will be done only once, because they will be done correctly the first time.
- Relationships with regulatory agencies will be beneficial and cooperative.
- Essential reporting to agencies will be completed in a timely manner.
- The client, the consultant, and the regulatory staff will be satisfied with the final outcome.
- Projects will be completed safely, and unexpected events will be minimized.

Consulting services are an important function in providing assistance relevant to safety and health issues. A consultant is always willing to provide safety and health services. However, depending on the urgency of the situation or the types of safety and health services needed, the cost may be quite high. Nevertheless, these costs are often justifiable because it is sometimes difficult to secure a consultant who has the expertise needed to assist with specific problems. Remember that the low bid for a consultant may not be in the company's best interest if he or she does not have the expertise to provide the solution needed. For a lower price, the company may get a less qualified consultant. Once a consultant is selected, a contractual agreement must be developed. This agreement should delineate the services to be provided, the outcomes or products to be attained, and the follow-up obligations required.

Most consultants are skilled safety and health professionals who bring to the table years of experience and unique expertise. The service that they offer is well worth the price. Using a planned approach will ensure that the company is getting the consultant that is needed for the job. However, be cautious! There are some individuals out there who are less scrupulous and will deceive the company and be happy to take its money, waste time, and leave the company with an inferior product or no product. As Linda F. Johnson advises, "As in all business decisions, do your homework."

REFERENCES

Johnson, L. F., Choosing a safety consultant. *Occupational Safety and Health,* July 1999.

Reese, C. D. and J. V. Eidson. *Handbook of OSHA construction safety & health.* Boca Raton, FL: CRC Press/Lewis Publishers, 1999.

20 Safety and Health Training

INTRODUCTION

Safety and health training is not the answer to accident prevention. Training cannot be the sum total answer to all accidents in the workplace. In fact, training is only applicable when a worker has not been trained previously, is new to a job or task, or safe job skills needed to be upgraded.

If a worker has the skills to do his or her job safely, then training will not address unsafe job performance. The problem is not the lack of skill to do the job safely, but rather the worker's unsafe behavior that he or she has elected to exhibit. Do not construe these statements to suggest that safety and health training does not have an important function as part of an accident prevention program. Without a safety and health training program, a vital element of workplace safety and health is missing.

The safety and health training needed should include not only workers, but also supervisors and management. Without training for managers and supervisors, it cannot be expected or assumed that they are cognizant of the safety and health practices of your company. Without this knowledge they will not know safe from unsafe, how to implement the loss control program, or even how to reinforce, recognize, or enforce safe work procedures unless proper training on occupational safety and health has transpired.

Safety and health training is critical to achieving accident prevention. Companies that do not provide new-hire training, supervisory training, and worker safety and health training have an appreciably greater number of injuries and illnesses than companies that carry out safety and health training. All types of programs may be offered that use many of the recognized accident prevention techniques, but without workers who are trained in their jobs and employ safe work practices, the efforts to reduce and prevent accidents and injuries will result in marginal success. If a worker has not been trained to do a job in a productive and safe manner, a very real problem exists. Do not assume that a worker knows how to do his or her job and will do it safely unless he or she has been trained to do so. Even with training, some may resist safety procedures and then you have deportment or behavioral problems—not a training issue.

It is always a good practice to train newly hired workers and experienced workers who have been transferred to a new job. It is also important that, any time a new procedure, new equipment, or extensive changes in job activities occur, workers receive training. Well-trained workers are more productive, more efficient, and safer.

Training for the sake of documentation is a waste of time and money. Training should be purposeful and goal or objective driven. An organized approach to on-the-job safety and health will yield the proper ammunition to determine your real training needs. These needs should be based on accidents/incidents, identified hazards, hazard/accident prevention initiatives, and input from your workforce. Tailor training to meet the company's needs and those of the workers.

Look for results from your training. Evaluate those results by looking at the reduced number of accidents/incidents, improved production, and good safety practices performed by your workforce. Evaluate the results using job safety observations and safety and health audits, as well as statistical information on the numbers of accidents and incidents.

Many OSHA regulations have specific requirements on training for fall protection, hazard communication, hazardous waste, asbestos and lead abatement, scaffolding, etc. It seems relatively safe to say that OSHA expects workers to have training on general safety and health provisions, hazard recognition, as well as task-specific training. Training workers regarding safety and health is one of the most effective accident prevention techniques.

WHEN TO TRAIN

There are appropriate times when safety and health training should be provided. They are when

- A worker lacks the safety skills
- A new employee is hired
- An employee is transferred to another job or task
- Changes have been made in normal operating procedures
- A worker has not performed a task for some period of time

TRAINING NEW HIRES

History has shown that individuals new to the workplace suffer more injuries and deaths than experienced workers. These usually occur within days or the first month on the job. It is imperative that new hires receive initial training regarding safety and health practices at the worksite. Some potential topics that should be covered during the training of a worker new to the workplace are

- Accident reporting procedures
- Basic hazard identification and reporting
- Chemical safety
- Company's basic philosophy on safety and health
- Company's safety and health rules
- Confined space entry
- Electrical safety
- Emergency response procedures (fire, spills, etc.)
- Eyewash and shower locations
- Fall protection
- Fire prevention and protection

- First aid/CPR
- Hand tool safety
- Hazard communications
- Housekeeping
- Injury reporting procedures
- Ladder safety
- Lockout/tagout procedures
- Machine guarding
- Machine safety
- Material handling
- Mobile equipment
- Medical facility location
- Personal responsibility for safety
- Rules regarding dress code, conduct, and expectations
- Unsafe acts/conditions reporting procedures
- Use of personal protective equipment (PPE)

TRAINING SUPERVISORS

The supervisor is the key person in the accomplishment of safety and health in the workplace. The supervisor will often be overlooked when it comes to safety and health training. Thus, the supervisor is frequently ill-equipped to be the lead person for the company's safety and health initiative. A list of suggested training topics for supervisors includes

- Accident causes and basic remedies
- Building attitudes favorable to safety
- Communicating safe work practices
- Cost of accidents and their effect on production
- Determining accident causes
- First aid training
- Giving job instruction
- Job instruction for safety
- Knowledge of federal and state laws
- Making the workplace safe
- Mechanical safeguarding
- Motivating safe work practices
- Number and kinds of accidents
- Organization and operation of a safety program
- Safe handling of materials
- Supervising safe performance on the job
- Supervisor's place in accident prevention
- The investigation and methods of reporting accidents to the company and government agencies

FIGURE 20.1 Supervisors often perform on-the-job training. (*Source:* Courtesy of the Occupational Safety and Health Administration.)

TRAINING EMPLOYEES

The training of all workers, regarding safety and health, has been demonstrated to reduce costs and increase the bottom line. Workers must be trained in safety and health, the same way as job skill training would be conducted in the form of on-the-job training (OJT) or job instruction training (JIT) (see Figure 20.1) Most employers and safety professionals realize that a skilled worker is a safe worker. The following are some recommendations that should be addressed to ensure that workers are trained or that training has been upgraded:

- Any new hazards or subjects of importance
- Basic skills training
- Explanation of policies and responsibilities
- Federal and state laws
- First aid training
- Importance of first aid treatment
- Methods of reporting accidents
- New safe operating procedures
- New safety rules and practices
- New skill training for new equipment, etc.
- Technical instruction and job descriptions
- Where to get first aid
- Where to get information and assistance

TABLE 20.1
Individual Worker's Training Form

INDIVIDUAL WORKER'S TRAINING

WORKER'S NAME: ______________________ SOC. SEC. # __________
CLOCK NUMBER: ______________________________________

Subject	Date:	Length of Training	Instructor	Worker's Signature
New Hire Orientation				
Hazard Communications				
Hazardous Waste				

***Keep this form in the employee's personnel file**

DOCUMENTING SAFETY AND HEALTH TRAINING

Each worker's safety and health training should be documented and placed in their permanent personnel file. The documentation should include the name of the worker, job title, social security number, clock number, date, topic or topics covered, length of training, and the trainer's name. This should be signed and dated by the trainer. This is the documentation of training that an OSHA inspector, legal counsel, or other interested parties will want see to verify the training that workers have received. Table 20.1 provides an example Individual Worker's Training Form, used to document training.

AFTER THE TRAINING

Communication is the key to occupational safety and health. The training message of accident prevention must constantly be reinforced. Use frequent reminders in the form of message boards, fliers, newsletters, paycheck inserts, posters, the spoken word, and face-to-face encounters. Make the message consistent with the policies and practices of the company in order for it to be believable. It is important to communicate safety and health goals and provide feedback on progress toward accomplishing those goals. Most workers want their information in short, easily digested units.

Face-to-face interactions personalize the communications, provide information immediately (no delays), and allow for two-way communications, which improves the accuracy of the message. Finally, these interactions provide performance and real-time feedback, and reinforce the company's safety and health stance.

SAFETY TALKS

Safety talks are especially important to supervisors in the workplace and on worksites because they afford each supervisor the opportunity to convey, in a timely manner, important information to workers. Safety talks may not be as effective as one-on-one communications, but they still surpass a memorandum or written message. In the 5 to 10 minutes prior to the workday, during a shift, at a break, or as needed, this technique helps communicate time-sensitive information to a department, crew, or work team.

In these short succinct meetings, supervisors convey changes in work practices; short training modules; facts related to an accident or injury; specific job instructions, policies, and procedures; rules and regulation changes; or other forms of information that the supervisor feels are important to every worker under his or her supervision.

Although safety talks are short, these types of talks should not become just a routine part of the workday. Thus, to be effective, they must cover current concerns or information, be relevant to the job, and have value to the workers. Plan safety talks carefully in order to effectively convey a specific message and a real accident prevention technique. Select topics applicable to the existing work environment; plan the presentation and focus on one issue at a time. Use materials to reinforce the presentation and clarify the expected outcomes. Some guidelines to follow:

1. Plan a safety talk training schedule in advance and post a notice.
2. Prepare supporting materials in advance.
3. Follow a procedure in the presentation: explain goals, try to answer questions, restate goals, and ask for action.
4. Make attendance mandatory.
5. Make each employee sign a log for each session.
6. Ask for feedback from employees on the topic at hand or other proposed topics.
7. Involve employees by reacting to suggestions or letting them make presentations when appropriate.
8. Reinforce the message throughout the workweek.

No matter how effective communications are with the workforce, there is still a need to ensure that workers have the competence to perform the basic skills of the tasks to which they have been assigned.

SAFETY TALKS AND MEETINGS

Safety talks, sometimes called toolbox talks, are an important training tool for the safety and health department and the supervisor. These safety talks can be used to cover a wide range of important safety topics in real time or immediately after mishaps or near-misses have occurred. Safety talks have the benefit of incorporating specific company issues and concerns.

Safety talks cannot be done in a helter-skelter fashion. They should be approached in an organized manner using a planned approach or they become nonfunctional. Safety talks must have "meat" to them. Thus, they should not become a gripe session or have the appearance of being a "seat-of-the-pants" presentation.

Safety talks, as with other training, should be documented. This documentation should include the date of the talk, presenter's name, topic, a list of those in attendance, and any materials used that should be attached to the Record of Safety Meeting Form (see Table 20.2).

TABLE 20.2
Record of Safety Meeting Form

RECORD OF SAFETY MEETING

SUBJECT: ________________

PRESENTER: ________________

DATE: ________________

LENGTH OF TIME: ________________

WORKER'S NAME	SIGNATURE	SOC. SEC #

NOTE: Staple any handouts or materials used during the toolbox meeting.

OSHA TRAINING REQUIREMENTS

Many standards promulgated by the Occupational Safety and Health Administration (OSHA) explicitly require the employer to train employees in the safety and health aspects of their jobs. Other OSHA standards make it the employer's responsibility to limit certain job assignments to employees who are "certified," "competent," or "qualified" (meaning that they have had special previous training, in or out of the workplace). The term "designated" personnel means selected or assigned by the employer or the employer's representative as being qualified to perform specific duties. These requirements reflect OSHA's belief that training is an essential part of every employer's safety and health program for protecting workers from injuries and illnesses. Many researchers conclude that those who are new on the job have a higher rate of accidents and injuries than more experienced workers. If ignorance of specific job hazards and of proper work practices is even partly to blame for a higher injury rate, then training will help to provide a solution.

As an example of the trend in OSHA's safety and health training requirements, the Process Safety Management of Highly Hazardous Chemicals standard (Title 29 Code of Federal Regulations, Part 1910.119) contains several training requirements. This standard was promulgated under the requirements of the Clean Air Act Amendments of 1990. The Process Safety Management Standard requires the employer to evaluate or verify that employees comprehend the training given to them. This means that the training to be given must have established goals and objectives regarding what is to be accomplished. Subsequent to the training, an evaluation would be conducted to verify that employees understand the subjects presented or acquired the desired skills or knowledge. If the established goals and objectives of the training program were not achieved as expected, the employer would then revise the training program to make it more effective, or conduct frequent refresher training or some combination of these. Requirements of the Process Safety Management Standard follow concepts embodied in the OSHA training guideline.

The length and complexity of OSHA standards may make it difficult to find all the references to training. So, to help employers, safety and health professionals, training directors, and others with a need to know, OSHA's training-related requirements are found in Appendix D.

It is usually a good idea for the employer to keep a record of all safety and health training. Records can provide evidence of the employer's good faith and compliance with OSHA standards. Documentation can also supply an answer to one of the first questions an accident investigator will ask: Was the injured employee trained to do the job?

Training in the proper performance of a job is time and money well spent, and the employer might regard it as an investment rather than an expense. An effective program of safety and health training for workers can result in fewer injuries and illnesses, better morale, and lower insurance premiums, among other benefits.

OSHA TRAINING GUIDELINES

The Occupational Safety and Health Act of 1970 does not specifically address the responsibility of employers to provide health and safety information and instruction to employees, although Section 5(a)(2) does require that each employer "...shall comply with occupational safety and health standards promulgated under this Act." However, more than a hundred of the Act's current standards do contain training requirements.

Therefore, OSHA has developed voluntary training guidelines to assist employers in providing the safety and health information and instruction needed for their employees to work at minimal risk to themselves, to fellow employees, and to the public. The guidelines are designed to help employers

1. Determine whether a worksite problem can be solved by training
2. Determine what training, if any, is needed
3. Identify goals and objectives for the training
4. Design learning activities
5. Conduct training
6. Determine the effectiveness of the training
7. Revise the training program based on feedback from employees, supervisors, and others

OSHA guidelines provide employers with a model for designing, conducting, evaluating, and revising training programs. The training model can be used to develop training programs for a variety of occupational safety and health hazards identified in the workplace. Additionally, it can assist employers in their efforts to meet the training requirements in current or future occupational safety and health standards.

A training program designed in accordance with these guidelines can be used to supplement and enhance the employer's other education and training activities. The guidelines afford employers significant flexibility in the selection of content, training, and program design. OSHA encourages a personalized approach to the informational and instructional programs at individual worksites, thereby enabling employers to provide the training that is most needed and applicable to local working conditions.

Assistance with training programs or the identification of resources for training is available through such organizations as OSHA full-service area offices, state agencies that have their own OSHA-approved occupational safety and health programs, OSHA-funded state on-site consultation programs for employers, local safety councils, the OSHA Office of Training and Education, and OSHA-funded New Directions grants.

LEGAL ASPECT OF TRAINING

The adequacy of employee training may also become an issue in contested cases where the affirmative defense of unpreventable employee misconduct is raised. Under case law, well-established in the Occupational Safety and Health Review Commission and the courts, an employer may successfully defend against an otherwise valid citation by demonstrating that all feasible steps were taken to avoid the occurrence of a hazard, and that actions of the employee involved in the violation were a departure from a uniformly and effectively enforced work rule, of which the employee had either actual or constructive knowledge.

In either type of case, the adequacy of training given to employees in connection with a specific hazard is a factual matter that can be decided only by considering all the facts and circumstances surrounding the alleged violation. The guidelines are not intended, and cannot be used, as evidence of the appropriate level of training in litigation involving either the training requirements of OSHA standards (refer to Appendix D) or affirmative defenses based on employer training programs.

OSHA TRAINING MODEL

The OSHA training model is designed to be one that even the owner of a business with very few employees can use without having to hire a professional trainer or purchase expensive training materials. Using this model, employers or supervisors can develop and administer safety and health training programs that address problems specific to their own business, fulfill the learning needs of their own employees, and strengthen the overall safety and health program of the workplace.

Determining if Training Is Needed

The first step in any training process is a basic one: to determine whether a problem can be solved by training. Whenever employees are not performing their jobs properly, it is often assumed that training will bring them up to standard procedure. However, it is possible that other actions (such as hazard abatement or the implementation of engineering controls) can enable employees to perform their jobs properly.

Ideally, safety and health training should be provided before problems or accidents occur. This training should cover both general safety and health rules, and specific work procedures; and training should be repeated if an accident or near-miss incident occurs.

Problems that can be addressed effectively by training include those that arise from lack of knowledge of a work process, unfamiliarity with equipment, or incorrect execution of a task. Training is less effective (but still can be used) for problems arising from an employee's lack of motivation or lack of attention to the job. Whatever its purpose, training is most effective when designed in relation to the goals of the employer's total safety and health program. An example of a training needs assessment instrument that could be used with supervisors or workers can be found in Table 20.3.

Identifying Training Needs

If the problem is one that can be solved, in whole or in part, by training, then the next step is to determine what training is needed. For this, it is necessary to identify what the employee is expected to do and in what ways, if any, the employee's performance is deficient. This information can be obtained by conducting a job analysis that pinpoints what an employee needs to know in order to perform a job.

When designing a new training program, or preparing to instruct an employee in an unfamiliar procedure or system, a job analysis can be developed by examining engineering data on new equipment, or the safety data sheets on unfamiliar substances. The content of the specific federal or state OSHA standard applicable to a business can also provide direction in developing training content. Another option is to conduct a Job Hazard Analysis (see Chapter 12). This is a procedure for studying and recording each step of a job, identifying existing or potential hazards, and determining the best way to perform the job in order to reduce or eliminate risks. Information obtained from a Job Hazard Analysis can be used as the content for training activity.

If an employer's training needs can be met by revising an existing training program rather than developing a new one, or if the employer already has some knowledge of the process or system to be used, appropriate training content can be developed through such means as

1. Using company accident and injury records to identify how accidents occur and what can be done to prevent them from recurring.
2. Requesting employees to provide, in writing and in their own words, descriptions of their jobs. These should include the tasks performed and the tools, materials, and equipment used.
3. Observing employees at the worksite as they perform tasks, asking about the work, and recording their answers.
4. Examining similar training programs offered by other companies in the same industry, or obtaining suggestions from such organizations as the National Safety Council, the Bureau of Labor Statistics, OSHA-approved state programs, OSHA full-service Area Offices, OSHA-funded state consultation programs, or the OSHA Office of Training and Education.

TABLE 20.3
Training Needs Assessment Instrument

SAFETY AND HEALTH TRAINING NEEDS ASSESSMENT

In trying to determine the safety and health training needs for supervisors and workers at XYZ Company, we are asking you to give us assistance by completing the following questionnaire. Please place on the back of this sheet under "Other Comments" any guidance or concerns which you have.

I. In order to provide you with an opportunity to have input on XYZ's safety and health training, select only 12 of the following topics that you deem most important for such a training program.

☐ Safety and Health Management	☐ Tracking Safety Performance
☐ Ergonomics	☐ Ladder Safety
☐ Welding and Cutting Safety	☐ Fall Protection
☐ Electrical Safety	☐ Crane Safety
☐ Lifting Safety	☐ Hazardous Chemicals
☐ Confined Spaces	☐ Mobile Equipment
☐ Hazard Communications	☐ Powered Industrial Trucks
☐ Mobile Work Platforms and Lifts	☐ Working around Water
☐ Haulage Equipment	☐ Material Handling
☐ Fire Prevention and Protection	☐ Industrial Hygiene
☐ Environmental Management	☐ Accident Investigation
☐ Promoting Safety and Health	☐ Walking and Working Surfaces
☐ Rigging Safety	☐ Personal Protective Equipment
☐ Safety and Health Training	☐ Equipment and Machine Guarding
☐ Scaffolding Safety	☐ Motivating Safety and Health
☐ Communicating Safety and Health	☐ Lockout/Tagout
☐ Hand and Power Tool Safety	☐ Accountability and Responsibility for Safety and Health

II. Are there any other specific topics that you think should be covered that were not part of the previous list?

____

____

____

III. List and rank the five most important topics to you (1 would be the most important and 5 the least important):

____

____

IV. How should the training be conducted to make it most effective?

____

____

V. How long should each training module last?

____

VI. OTHER COMMENTS: What other recommendations or guidance do you have for the training?

____

____

Employees themselves can provide valuable information on the training they need. Safety and health hazards can be identified through the employees' responses to such questions as whether anything about their jobs frightens them, have they had any near-miss incidents, do they feel they are taking risks, or do they believe that their jobs involve hazardous operations or substances.

Identifying Goals and Objectives

Once the need for training has been determined, it is equally important to determine what kind of training is needed. Employees should be made aware of all the steps involved in a task or procedure, but training should focus on those steps on which improved performance is needed. This avoids unnecessary training and tailors the training to meet the needs of the specific employees and processes.

Once the employees' training needs have been identified, employers can then prepare objectives for the training. Instructional objectives, if clearly stated, will tell employers what they want their employees to do, to do better, or to stop doing.

Learning objectives need not necessarily be written; but in order for the training to be as successful as possible, clear and measurable objectives should be thought out before the training begins. For an objective to be effective, it should identify as precisely as possible what the individuals will do to demonstrate that they have learned, or that the objectives have been reached. They should also describe the important conditions under which the individual will demonstrate competence and define what constitutes acceptable performance.

Using specific, action-oriented language, the instructional objectives should describe the preferred practice or skill and its observable behavior. For example, rather than using the statement, "The employee will understand how to use a respirator" as an instructional objective, it would be better to say, "The employee will be able to describe how a respirator works and when it should be used." Objectives are most effective when worded in sufficient detail that other qualified persons can recognize when the desired behavior is exhibited.

Developing Learning Activities

Once employers have stated precisely what the objectives for the training program are, then learning activities in the form of training methods and materials can be identified and described. Methods and materials for the learning activity can be as varied as the employer's imagination and available resources will allow. The learning activities should enable employees to demonstrate that they have acquired the desired skills and knowledge.

Start by establishing objectives. To ensure that employees transfer the skills or knowledge from the learning activity to the job, the learning situation should simulate the actual job as closely as possible. Thus, employers may want to arrange the objectives and activities in a sequence, which corresponds to the order in which the tasks are to be performed on the job, if a specific process is to be learned.

For example, if an employee must learn the start-up processes of using a machine, the sequence might be as follows:

1. Check that the power source is connected.
2. Ensure that safety devices are in place and operative.
3. Know when and how to throw the off switch; and so on.

Next, a few factors will help determine the training methods to be incorporated into employee training. One factor concerns the training resources available to the employer. For example, can a group training program using an outside trainer and a film be organized, or should the employer have individuals train employees on a one-to-one basis? Another factor concerns the kind of skills or knowledge to be learned. Is the learning oriented toward physical skills (such as the use of special tools) or toward mental processes and attitudes? Such factors will influence the type of learning activity designed by employers. For example, the training activity can be group oriented with lectures, role playing, and demonstrations; or it can be designed for the individual with self-paced instruction.

Finally, decide on materials; the employer may want to use materials such as charts, diagrams, manuals, slides, films, viewgraphs (overhead transparencies), videotapes, audiotapes, a simple chalkboard, or any combination of these, and other instructional aids. Whatever the method of instruction, learning activities should be developed in such a way that the employees come away with the ability to clearly demonstrate that they have acquired the desired skills or knowledge.

Conducting the Training

After organizing the learning activity, the employer is ready to begin conducting training. To the extent possible, training should be presented so that its organization and meaning are clear to employees. This will help reinforce the message and the long-term recall. Employers or supervisors should

1. Provide overviews of the material to be learned.
2. Relate, wherever possible, the new information or skills to the employees' goals, interests, or experience.
3. Reinforce what the employees have learned by summarizing both the program's objectives and key points of information covered. These steps will assist employers in presenting training in a clear, unambiguous manner.

In addition to organizing content, employers must also develop the structure and format of the training. Content developed for the program should be closely related to the nature of the workplace (or other training site) and the resources available for training. Having planned this, employers will be able to determine for themselves the frequency of training activities, the length of sessions, the instructional techniques, and the individual(s) best qualified to present the information (see Figure 20.2).

FIGURE 20.2 Competent and qualified instructors are needed. (*Source:* Courtesy of the Occupational Safety and Health Administration.)

To motivate employees to pay attention and learn form the training activities provided by the employer or supervisor, each employee must be convinced of the importance and relevance of the material. Some ways to develop motivation are

1. Explain the goals and objectives of instruction.
2. Relate the training to the interests, skills, and experiences of the employees.
3. Outline the main points to be presented during the training session(s).
4. Point out the benefits of training (e.g., the employee will be better informed, more skilled, and thus more valuable both on the job and in the labor market; or the employee will, if he or she applies the skills and knowledge learned, be able to work at reduced risk).

An effective training program allows employees to participate in the training process and practice their skills or knowledge. This will help ensure that they are learning the required knowledge or skills, and it permits correction if necessary. Employees can become involved in the training process by participating in discussions, asking questions, contributing their knowledge and expertise, learning through hands-on experiences, and through role-playing exercises.

Evaluating Program Effectiveness

To make sure that the training program is accomplishing its goals, an evaluation of the training can be valuable. Training should have, as one of its critical components, a method of measuring the effectiveness of the training. A plan for evaluating the training session(s), either written or thought out by the employer, should be developed when the course objectives and content are developed. It should not be delayed until the training has been completed. Evaluation will help employers and supervisors determine the amount of learning achieved and whether an employee's performance has improved on the job. Among the methods of evaluating training are

1. *Student opinion questionnaires.* Informal discussions with employees can help employers determine the relevance and appropriateness of the training program.
2. *Supervisors' observations.* Supervisors are in good positions to observe an employee's performance both before and after the training and note improvements or changes.
3. *Workplace improvements.* Success of a training program may be changes throughout the workplace that result in reduced injury or accident rates.

However it is conducted, an evaluation of training can give employers the information necessary to decide whether or not employees achieved the desired results, and whether the training session should be offered again at some future date.

Improving the Program

If, after evaluation, it is clear that the training did not give the employees the level of knowledge and skill that was expected, then it may be necessary to revise the training program or provide periodic retraining. At this point, asking questions of employees and of those who conducted the training may be of some help. Among the questions that could be asked are

1. Were parts of the content already known, and therefore unnecessary?
2. What material was confusing or distracting?
3. Was anything missing from the program?
4. What did the employees learn, and what did they fail to learn?

It may be necessary to repeat steps in the training process, that is, to return to the first steps and retrace one's way through the training process. As the program is evaluated, the employer should ask the following:

1. If a job analysis was conducted, was it accurate?
2. Was any critical feature of the job overlooked?
3. Were the important gaps in knowledge and skill included?
4. Was material already known by the employees intentionally omitted?
5. Were the instructional objectives presented clearly and concretely?
6. Did the objectives state the level of acceptable performance that was expected of employees?
7. Did the learning activity simulate the actual job?
8. Was the learning activity appropriate for the kinds of knowledge and skills required on the job?
9. When presenting the training, was the organization of the material and its meaning made clear?
10. Were the employees motivated to learn?
11. Were the employees allowed to participate actively in the training process?
12. Was the employer's evaluation of the program thorough?

A critical examination of the steps in the training process will help employers determine where course revision is necessary.

MATCHING TRAINING TO EMPLOYEES

While all employees are entitled to know as much as possible about the safety and health hazards to which they are exposed, and employers should attempt to provide all relevant information and instruction to all employees, the resources for such an effort frequently are not, or are not believed to be, available. Thus, employers are often faced with the problem of deciding who is in the greatest need of information and instruction.

One way to differentiate between employees who have priority needs for training and those who do not is to identify employee populations that are at higher levels of risk. The nature of the work will provide an indication that such groups should receive priority for information on occupational safety and health risks.

IDENTIFYING EMPLOYEES AT RISK

One method of identifying employee populations at high levels of occupational risk (and thus in greater need of safety and health training) is to pinpoint hazardous occupations. Even within industries that are hazardous in general, there are some employees who operate at greater risk than others. In other cases, the hazardousness of an occupation is influenced by the conditions under which it is performed, such as noise, heat, cold, safety, or health hazards in the surrounding area. In these situations, employees should be trained not only on how to perform their job safely, but also on how to operate within a hazardous environment.

A second method of identifying employee populations at high levels of risk is to examine the incidence of accidents and injuries, both within the company and within the industry. If employees in certain occupational categories are experiencing higher accident and injury rates than other employees, training may be one way to reduce that rate. In addition, accident investigation can identify not only specific employees who could benefit from training, but also company-wide training needs.

Research has identified the following variables as being related to a disproportionate share of injuries and illnesses at the worksite on the part of employees:

1. The age of the employee (younger employees have higher incidence rates).
2. The length of time on the job (new employees have higher incidence rates).
3. The size of the firm (in general terms, medium-size firms have higher incidence rates than smaller or larger firms).
4. The type of work performed (incidence and severity rates vary significantly by Standard Industrial Classification (SIC) Code).
5. The use of hazardous substances (by SIC Code).

These variables should be considered when identifying employee groups for training in occupational safety and health.

In summary, information is readily available to help employers identify which employees should receive safety and health information, education, and training, and who should receive it before others. Employers can request assistance in obtaining information by contacting such organizations as OSHA Area Offices, the Bureau of Labor Statistics, OSHA-approved state programs, state on-site consultation programs, the OSHA Office of Training and Education, or local safety councils.

TRAINING EMPLOYEES AT RISK

Determining the content of training for employee populations at higher levels of risk is similar to determining what any employee needs to know, but more emphasis is placed on the requirements of the job and the possibility of injury. One useful tool for determining training content from job requirements is the Job Hazard Analysis described earlier. This procedure examines each step of a job, identifies existing or potential hazards, and determines the best way to perform the job in order to reduce or eliminate the hazards. Its key elements are

1. Job description
2. Job location
3. Key steps (preferably in the order in which they are performed)
4. Tools, machines, and materials used
5. Actual and potential safety and health hazards associated with these key job steps
6. Safe and healthful practices, apparel, and equipment required for each job step

Material Safety Data Sheets (MSDSs) can also provide information for training employees in the safe use of materials. These data sheets, developed by chemical manufacturers and importers, are supplied with manufacturing or construction materials and describe the ingredients of a product, its hazards, protective equipment to be used, safe handling procedures, and emergency first-aid responses. The information contained in these sheets can help employers identify employees in need of training (i.e., workers handling substances described in the sheets) and train employees in safe use of the substances. MSDSs are generally available from suppliers, manufacturers of the substance, large employers that use the substance on a regular basis, or they may be developed by employers or trade associations. MSDSs are particularly useful for those employers that are developing training in safe chemical use as required by OSHA's Hazard Communication Standard.

SUMMARY

In an attempt to assist employers with their occupational health and safety training activities, OSHA has developed a set of training guidelines in the form of a model. This model is designed to help employers develop instructional programs as part of their total education and training effort. The model addresses the questions of who should be trained, on what topics, and for what purposes. It also helps employers determine how effective the program has been and enables them to identify employees who are in greatest need of education and training. The model is general enough to be used in any area of occupational safety and health training, and allows employers to determine for themselves the content and format of training. Use of this model in training activities is just one of many ways that employers can comply with the OSHA standards that relate to training and enhance the safety and health of their employees.

REFERENCES

Reese, C. D. and J. V. Eidson. *Handbook of OSHA construction safety & health.* Boca Raton, FL: CRC Press/Lewis Publishers, 1999.

United States Department of Labor. Training Requirements in OSHA Standards and Training Guidelines (OSHA 2254). Washington, D.C.: U.S. Department of Labor, 1998.

United States Department of Labor, Occupational Safety and Health Administration, Office of Training and Education. OSHA Voluntary Compliance Outreach Program: Instructors Reference Manual. Des Plaines, IL: U.S. Department of Labor, 1993.

21 Analyzing Accident Data

INTRODUCTION

The importance of monitoring and tracking the overall safety and health efforts of companies cannot be overstated. If companies do not keep records on all aspects of their accident prevention efforts, they will not have the information needed to obtain continuous funding for accident prevention or justification of its performance value. Those companies will not be able to explain where budget dollars were expended, nor will they have a way to assess the loss control progress within the company. The data for comparing companys' performance regarding trends and tendencies of occupational injuries and illnesses occurring in other companies or nationally come from the records being kept by individual companies.

Tracking should not just be a count of recordable incidents. Companies should track near-misses, equipment damage, and first-aid events.

Tracking means that company officials must pay close attention to workers' compensation cases as well as their associated costs. The longer an individual is off the job, the more the cost and the greater the likelihood that the worker will not return to work.

OSHA RECORDKEEPING

As part of the tracking process, a company must comply with the Occupational Safety and Health Administration's (OSHA's) recordkeeping requirements for occupational injuries and illnesses. Any occupational illness that has resulted in an abnormal condition or disorder caused by exposure to environmental factors, which may be acute or chronic due to inhalation, absorption, ingestion or direct contact with toxic substances or harmful agents, and any repetitive motion injury is to be classified as an illness. All illnesses are recordable, regardless of severity. Injuries are recordable when

1. An on-the-job death occurs (regardless of length of time between injury and death).
2. One or more lost workdays occur.
3. Restriction of work or motion transpires.
4. Loss of consciousness occurs.
5. The worker is transferred to another job.
6. The worker receives medical treatment beyond first aid.

Most employers with more than ten employees are required to complete the Occupational Injuries and Illnesses (OSHA 301) or a comparable form and retain it for 5 years as well as complete the Log and Summary of Occupational Injuries and Illnesses (OSHA 300A log) and post it yearly from February 1 to April 30. The OSHA 301, 300, and 300A logs are the tracking records that must be kept for compliance with OSHA regulations.

COMPANY RECORDS

Many companies conduct accident investigations and keep accident records and other data on the company's safety and health initiatives. If a company has a sufficient number of accidents/incidents and enough detail in their occupational injury/illness investigation data, the company can begin to examine trends or emerging issues relevant to their safety and health intervention/prevention efforts. The analysis of this data can be used to evaluate the effectiveness of safety and health at various workplaces, jobsites, or for groups of workers. The safety and health data can be used by a company to compare to that of other companies that perform similar work, employ a comparable workforce, or compete in the same kind and size of market on a state, regional, national, or international basis.

By analyzing their accidents/incidents, companies are in a better position to compare apples to apples rather than apples to oranges. Companies will be able to identify not only the types of injuries, types of accidents, and types of causes, but they will also be able to intervene and provide recommendations for preventing these accidents/incidents in the future. Companies will be able to say with confidence that "I do" or "I do not" have a safety and health problem. If a company finds that it has a problem, the analysis and data will be essential, especially if it is trying to elicit advice on how to address health and safety needs.

Gathering and analyzing accident/incident data is not the company's entire safety and health program, but a single element. Data provides feedback and evaluative information as companies proceed toward accomplishing their safety and health goals; thus, data contributes an important component in the analysis process.

IMPORTANT ANCILLARY DATA NEEDED FOR MORE COMPLETE ANALYSIS

Careful and complete analysis of the data collected following an incident is critical to the accurate determination of an accident's causal factors and an important part of preventing a reoccurrence. The results of comprehensive analyses provide the basis for corrective and preventive measures.

The analysis portion of the accident investigation is not a single, distinct part of the investigation. Instead, it is the central part of the process that includes collecting facts and determining causal factors. Well-chosen and carefully performed analysis is important because it provides results that can be used by a company to improve its safety and health performance.

When collecting data after an incident, the volume seems to be ever increasing as the incident is evaluated. The accuracy and relevance of the data is critical to an effective analysis process.

Some of the data item that should be acquired is

- Age
- Sex
- Race
- Date
- Time of the incident
- Day of the week
- The shift
- Occupation of victim
- Task being performed at time of incident
- Type of accident
- Nature of injury or illness: names the principal physical characteristic of a disabling condition, such as sprain/strain, cut/laceration, or carpal tunnel syndrome
- Part of body affected: directly linked to the nature of injury or illness cited, such as back, finger, or eye
- Source of injury or illness: the object, substance, exposure, or bodily motion that directly produced or inflicted the disabling condition cited (examples include lifting a heavy box; exposure to a toxic substance, fire, or flame; and bodily motion of an injured or ill worker)

Health and Safety Statistics Data

The Bureau of Labor Statistics' (BLS) Occupation Injury and Illness Classification Manual provides a much more detailed breakout of the type data that is most useful in analysis of occupational incidents. The four major areas of data that the BLS believes are important to collect during review or investigation of occupational safety and health incidents are Nature, Part of Body Affected, Source, and Event or Exposure. The following provide more detailed information with regard to the content of these data sources:

- The *nature* of an injury or illness identifies the principal characteristic(s) of work-related injury or illness. The divisions for the nature of the injury or illness are as follows:
  - 0 Traumatic Injuries and Disorders
  - 1 Systemic Disease or Disorders
  - 2 Infectious and Parasitic Diseases
  - 3 Neoplasms, Tumors, and Cancer
  - 4 Symptoms, Signs, and Ill-Defined Conditions
  - 5 Other Conditions or Disorders
  - 8 Multiple Diseases, Condition, or Disorders
  - 9999 Nonclassifiable

- The *part of body affected* identifies the part of the body affected by the previously identified nature of injury or illness. The divisions for part of body for the injury or illness are as follows:
  - 0 Head
  - 1 Neck, Including Throat
  - 2 Trunk
  - 3 Upper Extremities
  - 4 Lower Extremities
  - 5 Body Systems
  - 8 Multiple Body Parts
  - 9 Other Body Parts
  - 9999 Nonclassifiable

- The *source* of an injury or illness identifies the object, substance, bodily motion, or exposure that directly produced or inflicted the previously identified injury or illness. The divisions for source for injury or illness are as follows:
  - 0 Chemical and Chemical Products
  - 1 Containers
  - 2 Furniture and Fixtures
  - 3 Machinery
  - 4 Parts and Materials
  - 5 Persons, Plants, Animals, and Minerals
  - 6 Structures and Surfaces
  - 7 Tools, Instruments, and Equipment
  - 8 Vehicles
  - 9 Other Sources
  - 9999 Nonclassifiable

- The *event or exposure* describes the manner in which the injury or illness was produced or inflicted by the source of injury or illness. The divisions for event or exposure for injury and illness are as follows:
  - 0 Contact with Objects and Equipment
  - 1 Falls
  - 2 Bodily Reaction and Exertion
  - 3 Exposure to Harmful Substances or Environments
  - 4 Transportation Accidents
  - 5 Fires and Explosions
  - 6 Assaults and Violent Acts
  - 9 Other Events or Exposures
  - 9999 Nonclassifiable

Each of theses is further broken down in greater detail in BLS's Occupation Injury and Illness Classification Manual. The BLS also has a manual that details occupations.

The principal challenge in collecting data is distinguishing between accurate and erroneous information in order to focus on areas that will lead to identifying the accident's causal factors. These causal factors are key to accident/incident prevention. This can be accomplished by

- Understanding the activity that was being performed at the time of the accident
- Personally conducting a walkthrough of the accident scene
- Challenging "facts" that are inconsistent with other evidence (e.g., physical)
- Corroborating facts through interviews
- Testing or inspecting pertinent components to determine failure modes and physical evidence
- Reviewing policies, procedures, and work records to determine the level of compliance or implementation

Careful and complete analysis of the data collected following an accident is critical to the accurate determination of an accident's causal factors. The results of comprehensive analyses provide the basis for corrective and preventive measures.

STATISTICAL ANALYSIS FOR COMPARISONS

Companies and federal officials frequently utilize the following statistical pieces of information designed to allow the company to compare its safety and health performance with others: the incident rate, illness rate, lost workday cases rate or severity rate, and restricted workday case rate. These rates, respectively, answer the questions of "How often or frequently are accidents occurring?" and "How bad are the injuries/illnesses that are occurring?" The number of times that occupational injuries/illnesses happen is the determinant for the incident rate, while the number of days away from work (lost-time workdays) or restricted workdays are the prime indicator of the severity rate. Both of these rates provide unique information regarding your safety and health effort.

To find the injury/illness rate, count the number of distinct events, which resulted in injuries/illnesses. To compare your incident rate to other companies, you must normalize your data. This is accomplished by using a constant of 200,000 work-hours, which was established by the Bureau of Labor Statistics. The 200,000 work-hours are the number of hours that 100 full-time workers would work during 50 weeks at 40 hours per week. Thus, a company's incident rate or illness rate can be calculated in the following manner:

$$\text{Incident rate} = \frac{\text{Number of your OSHA recordable injuries/illnesses} \times 200{,}000 \text{ (work hour constant)}}{\text{Total number hours that your employees worked during the year}}$$

The incident rate could be a rate calculated for recordable (combined) injuries and illnesses, recordable injuries, recordable illnesses, all injuries with lost workdays, all illnesses with lost workdays or restricted workdays, injuries requiring only medical treatment, or first-aid injuries. These calculated rates would not normally be calculated on a national basis, but could be used to compare your progress on a yearly basis, or between jobsites or projects.

The severity rate, which is often called the "lost-time workday rate," is used to determine how serious the injuries and illnesses are. The same formula can be used to calculate the restricted workday case rate. A company may have a low incident rate or few injuries and illnesses but, if the injuries and illnesses that are occurring result in many days away from work or restricted workdays, the lost-time workdays or restricted workday cases can be as costly as, or more costly than, having a large number of no lost workdays or restricted workday injuries or illnesses, which have only medical costs associated with them. Lost-time workday cases can definitely have a greater impact on your workers' compensation costs and premiums.

Calculation of the severity rate is similar to the incident rate except that the total number of lost-time workdays or restricted workdays is used in place of the number of OSHA recordable injuries/illnesses. The severity rate for a company can be calculated in the following manner:

$$\text{Severity rate} = \frac{\text{Number of your lost-time workdays} \times 200{,}000 \text{ (work hour constant)}}{\text{Total number of hours that your employees worked during the year}}$$

The incident and severity rates are both expressed as a rate per 100 full-time workers. This provides a standard comparison value for a company whether it has 20 or 1,000 workers. Thus, both the 20-employee company and 1,000-employee companies can compare their safety and health performance to each other.

Sometimes the temptation exists to focus on only the lost-time workday cases, but how can you identify which injury or illness is going to result in a lost-time workday? At times, the difference between a medical treatment injury and a lost-time workday injury may only be a matter of inches or chance. Thus, it is more logical to address your total injury problem.

WORKERS' COMPENSATION

Each employer is expected to provide protection for his or her workers who become ill or are injured by something within the workplace. The premiums that the employer pays are to provide medical treatment and supplemental income when a worker is unable to return to work immediately. This supplemental income is usually 66.6 percent of the worker's wages and is not taxed. Once a worker files a claim for workers' compensation, the employee is not usually allowed to sue his or her employer.

Thus, workers' compensation premiums paid by the employer act like a protection or insurance policy against liability when a worker suffers injury or illness due to a hazard within the workplace.

The employer's premium is based on the type of industry as well as the injury and illness history of the employer's workplace. The more injuries and illnesses that employers experience, the higher the premium paid by the employer. Also, the more days away from work experienced by the workers, the more the premium increases.

Thus, as a cost factor, the reduction in occupational injuries and illnesses is important to cost containment. Any employer that is responsible understands the relationship between the bottom-line profit and the number of accidents.

Employers must set up a tracking system that monitors each workers' compensation claim filed by their workers. They also need to work diligently to return workers back to the workforce as soon as medically feasible. As part of the healing process, employers must ensure that the workers are receiving their medication and going for medical treatment. The longer a worker is away from the workplace, the more unlikely it is that he or she will not return to the workforce. It has been documented that workers who are off work for 6 months have a probability of a 50 percent of returning to work. For those off a year, the probability is 25 percent; and after two years or longer, the probability of returning to work is 0 percent.

Each workers' compensation claim is either directly or indirectly a drain on the profit margin of the company. This is the primary business reason to track workers' compensation claims. Many employers who have a less than stellar experience modification rate (EMR) may have difficulty in procuring work, or bidding on work, when perspective buyers assess their safety and health performance.

The prevention of occupational injuries and illnesses will definitely reduce the cost of workers' compensation premiums. It does not matter whether the company pays into the state workers' compensation system or is self-insured. What motivates most employers to reduce workplace injuries and illnesses is usually dollars. Some studies show that other benefits result from a decrease in the number of injuries and illnesses. For example, employers can expect to increase attendance, morale, and productivity. These are just a few of the side effects of reduced workers' compensation costs.

COST OF ACCIDENTS

The direct (insured) cost of accidents is by far the easiest to track. These direct costs are for medical care, repairing or replacing damaged equipment, and workers' compensation premiums. There is no possible way to not see the cost of an ambulance, hospital bill, or repair bill when it comes. Likewise, employers know the dollar amounts being put out for workers' compensation.

Data from the National Safety Council (NSC) for 2008 indicates that the costs of work-related injuries and deaths were $183.0 billion. Wage and productivity loss accounted for $88.4 billion, medical cost for $38.3 billion, and employer cost equaled $12.7 billion. The average cost of a workplace death was put at $1,310,000 and a disabling injury cost at $48,000. A look at other injury costs provided by the National Safety Council indicates that a reasonable, serious, non-disabling injury would have an average cost of $22,674 (2006–2007) (NSC Injury Fact, 2010).

Many safety and health experts estimate that the indirect (uninsured) costs of accidents, and the costs associated with them, are equal to five to ten times the direct cost of the accidents. These indirect costs are caused by many of the following:

- Time lost from work by the injured
- Loss in earning power
- Economic loss to injured worker's family

- Lost time by fellow workers
- Loss of efficiency due to break-up of crew
- Lost time of supervisor
- Cost of breaking in a new worker
- Damage to tools and equipment
- The time-damaged equipment is out of service
- Spoiled work
- Loss of production
- Spoilage from fire, water, chemical, explosives, etc.
- Failure to fill orders
- Overhead cost (while work was disrupted)
- Miscellaneous: there are at least a hundred other cost items that appear one or more times with every accident

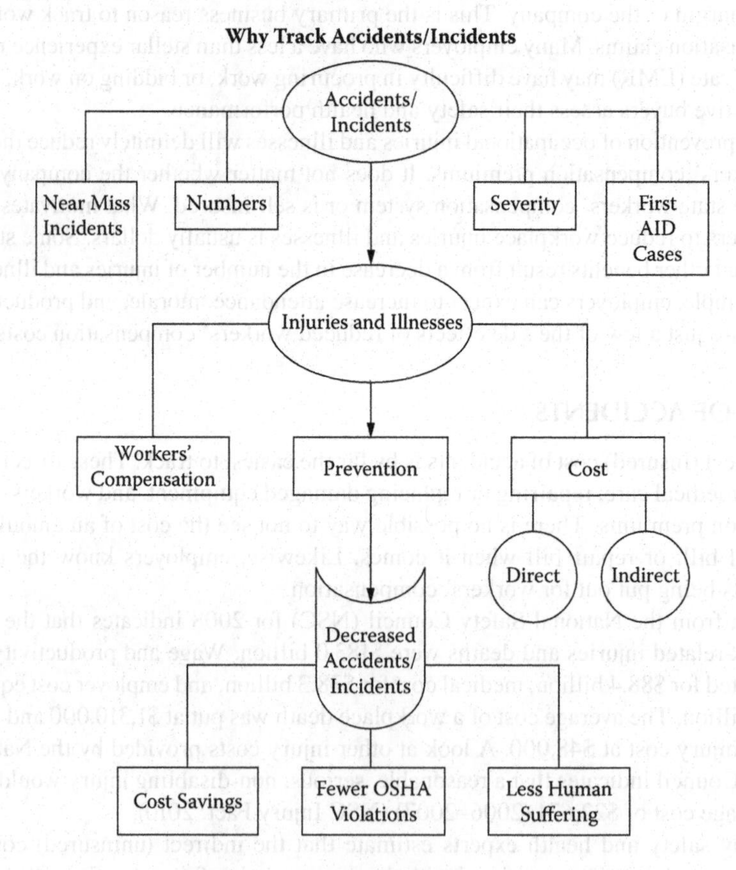

FIGURE 21.1 Tracking occupational accident/incidents.

SUMMARY

Until you track your accidents, their related costs, and the resulting injuries and illnesses, you have little baseline information upon which to lobby for better accident prevention, which can result in a multitude of benefits (see Figure 21.1). Some of these benefits are

- Better OSHA compliance
- Economic improvement
- Humanitarian
- Less legal liability
- Better employer/employee relations
- More time for production

The benefits of tracking a company's workplace safety and health accidents/incidents provides the data to assess the effectiveness of any accident prevention initiative and allows the company to determine the cost benefits of the safety and health dollars that are being spent by the company. Thus, recordkeeping and tracking makes good business sense.

The analysis of industry-related hazards and the accident/incidents they cause are an important step in the overall process of reducing occupation-related injuries, illnesses, and deaths. Only after a systematic look at the hazards and accidents can you hope to integrate the accident prevention techniques and tools that can have an impact on a company's safety and health initiative.

REFERENCES

Reese, C. D. and J. V. Eidson. *Handbook of OSHA construction safety & health.* Boca Raton, FL: CRC Press/Lewis Publishers, 1999.

United States Department of Labor, Occupational Safety and Health Administration, Office of Training and Education. OSHA Voluntary Compliance Outreach Program: Instructors Reference Manual. Des Plaines, IL: U.S. Department of Labor, 1993.

22 Prevention and OSHA Regulations

INTRODUCTION

A worker should be able to go to work each day and expect to return home uninjured and in good health. There is no logical reason why a worker should be part of workplace carnage. Workers do not have to become one of the yearly workplace statistics.

Workers who know the occupational safety and health rules and safe work procedures, and follow them, are less likely to become one of the 4,340 (2009) occupational trauma deaths, one of 90,000 estimated occupational illness deaths, or even one of the 3.7 million (2008) nonfatal occupational injuries and illnesses.

The essence of workplace safety and health should not rest upon Occupational Safety and Health Administration (OSHA) regulations, as they should not be the driving forces behind workplace safety and health. OSHA has limited resources for inspection and a limited number of inspectors. Enforcement is usually based on serious complaints, catastrophic events, and workplace deaths. Employers with a good safety and health program and a good safety record have a better opportunity of procuring contracts and orders because of their workplace safety and health record, and reap the benefits of low insurance premiums for workers' compensation and liability. Usually, safety and health are linked to the bottom line, which is seldom perceived as humanitarian.

This chapter provides answers to many of the questions that are frequently asked regarding OSHA compliance, workplace safety and health, and how workers and employers are to mesh with the intent of a having a safe and healthy workplace.

FEDERAL LAWS

Congress establishes federal laws (legislation or acts) and the President of the United States signs them into law. These laws often require that regulations (standards) be developed by federal agencies that are responsible for the intent of the law.

OSHAct

The Occupational Safety and Health Act (OSHAct) of 1970 is just such a law establishing a federal agency; it is also called the Williams-Steiger Act. It was signed by President Richard Nixon on December 29, 1970, and became effective on April 29, 1971. (The OSHAct was not amended until November 5, 1990, by Public Law 101-552.) The OSHAct assigned the responsibility of implementing and enforcing the law to a newly created agency, the Occupational Safety and Health Administration (OSHA), located within the United States Department of Labor (DOL).

Content of the OSHAct

Prior to the OSHAct, there were some state laws, a few pieces of federal regulation, and a small number of voluntary programs by employers. Most of the state programs were limited in scope and the federal laws only partially covered workers.

Another important reason for the OSHAct was the increasing number of injuries and illnesses within the workplace. Thus, the OSHAct was passed with the express purpose of insuring that every working man and woman in the nation would be provided safe and healthful work conditions while preserving this national human resource: the American worker. The OSHAct is divided into thirty-four sections, with each having a specific purpose. The full text of the OSHAct is approximately thirty-one pages and a copy can be obtained from your local OSHA office.

REGULATION PROCESS

OSHA was mandated to develop, implement, and enforce regulations relevant to workplace safety and health and the protection of workers. Time constraints prevented the newly formed OSHA from developing brand-new regulations. Therefore, OSHA adopted previously existing regulations from other government regulations, consensus standards, proprietary standards, professional groups' standards, and accepted industry standards. This is the reason that, today, the hazardous chemical exposure levels, with a few exceptions, are the same as the existing Threshold Limit Values (TLVs) published by the American Conference of Governmental Industrial Hygienist in 1968. Once these TLVs were adopted, it became very difficult to revise them. Although research and knowledge in the past 40 years has fostered newer and safer TLVs, they have not been adopted by OSHA.

As stated previously, the original OSHA standards and regulations have come from three main sources: consensus standards, proprietary standards, and federal laws that existed when the OSHAct became law. Consensus standards are industry-wide standards developed by organizations. They are discussed and substantially agreed upon through industry consensus. OSHA has incorporated into its standards the standards of two primary groups: the American National Standards Institute (ANSI) and the National Fire Protection Association (NFPA). Proprietary standards are prepared by professional experts from specific industries, professional societies, and associations. The proprietary standards are determined by a straight membership vote, not by consensus.

Some of the preexisting federal laws that are enforced by OSHA include the Federal Supply Contracts Act (Walsh-Healy), the Federal Service Contracts Act (McNamara-O'Hara), the Contract Work Hours and Safety Standard Act (Construction Safety Act), and the National Foundation on the Arts and Humanities Act. Standards issued under these Acts are now enforced in all industries where they apply.

Standards are sometimes referred to as being either "horizontal" or "vertical" in their application. Most standards are "horizontal" or "general." This means they apply to any employer in any industry. Fire protection, working surfaces, and first aid standards are examples of "horizontal" standards. Some standards are only relevant to a particular industry and are called "vertical" or "particular" standards.

Through the newspapers and conversations, it certainly sounds as if OSHA is producing new standards each day that will impact the workplace. This simply is not true. The regulatory process is very slow. When OSHA needs to develop a new regulation or even revise an existing one, it becomes a lengthy and arduous process, often taking more than a decade to come into being. The steps for developing a new regulation are as follows:

1. The Agency (OSHA) opens a Regulatory Development Docket for a new or revised regulation.
2. This indicates that OSHA believes a need for a regulation exists.
3. An Advanced Notice of Proposed Rulemaking (ANPRM) is published in the *Federal Register* and written comments are requested to be submitted within thirty to sixty days.
4. The comments are analyzed.
5. A Notice of Proposed Rulemaking (NPRM) is published in the *Federal Register* with a copy of the proposed regulation.
6. Another public comment period transpires usually for thirty to sixty days.
7. If no additional major issues are raised by the comments, the process continues to Step 10.
8. If someone raises some serious issues, the process goes back to Step 4 for review and possible revision of the NPRM.
9. Once the concerns have been addressed, it continues forward to Steps 5 and 6 again.
10. If no major issues are raised, a Final Rule (FR) will be published in the *Federal Register*, along with the date when the regulation will be effective (usually 30 to 120 days).
11. There can still be a Petition of Reconsideration of the Final Rule. There are times when an individual or industry may take legal action to bar the regulation's promulgation.
12. If the agency does not follow the correct procedures or acts arbitrarily or capriciously, the court may void the regulation and the whole process must be repeated.

If you desire to comment on a regulation during the development process, you should feel free to do so; your comments are important. You should comment on the areas where you agree or disagree. This is your opportunity to speak up. If no one comments, it is assumed that nobody cares one way or the other. You must be specific. Give examples, be precise, give alternatives, and provide any data or specific information that can back up your opinion. Federal agencies always welcome good data that substantiates your case. Cost-benefit data are always important in the regulatory process, and any valid cost data that you are able to provide may be very beneficial. But make sure that your comments are based on what is published in the *Federal Register* and not based upon hearsay information. Remember that the agency proposing the regulation may be working under specific constraints. Make sure you understand these constraints. Due to restrictions, the agency may not have the power to do what you think ought to be done.

FEDERAL REGISTER

The *Federal Register* is the official publication of the U.S. Government. If you are involved in regulatory compliance, you should obtain a subscription to the *Federal Register.* The reasons for obtaining this publication are clear. It is official, comprehensive, and not a summary done by someone else. It is published daily and provides immediate accurate information. The *Federal Register* provides early notices of forthcoming regulations, informs you of comment periods, and gives the preamble and responses to questions raised about a final regulation. It provides notices of meetings, gives information on obtaining guidance documents, supplies guidance on findings, gives information on cross references, and gives the yearly regulatory development agenda. It is the "Bible" for regulatory development, recognizable by brown paper and newsprint quality printing (see Figure 22.1).

PURPOSE OF OSHA

To ensure, as much as possible, a healthy and safe workplace and conditions for workers in the United States, the Occupational Safety and Health Administration was created by the OSHAct to

1. Encourage employers and employees to reduce workplace hazards by implementing new and improving existing safety and health programs.
2. Provide for research in occupational safety and health to develop innovative ways of dealing with occupational safety and health problems.
3. Establish separate but dependent responsibilities and rights for employers and employees for the achievement of better safety and health conditions.
4. Maintain a reporting and recordkeeping system to monitor job-related injuries and illnesses.
5. Establish training programs to increase the number and competence of occupational safety and health personnel.
6. Develop mandatory job safety and health standards and enforce them effectively.
7. Provide for the development, analysis, evaluation, and approval of state occupational safety and health programs.
8. Provide technical and compliance assistance, training and education, and cooperative programs and partnerships to help employers reduce worker accidents and injuries.

CODE OF FEDERAL REGULATIONS

Probably one of the most common complaints from people who use the U.S. Code of Federal Regulations (CFR) is: How do you wade through hundreds of pages of

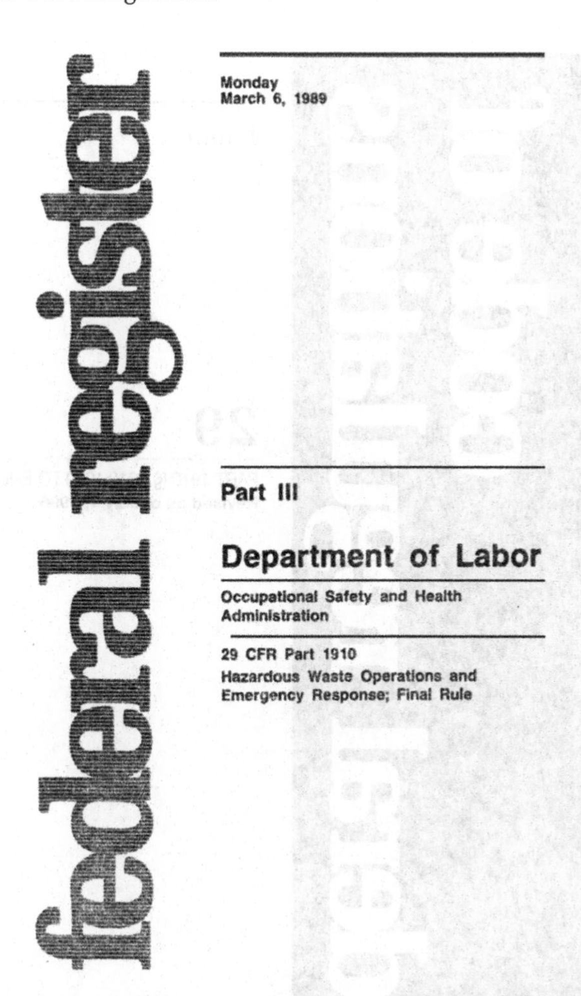
federal register

Monday
March 6, 1989

Part III

Department of Labor

Occupational Safety and Health
Administration

29 CFR Part 1910
Hazardous Waste Operations and
Emergency Response; Final Rule

FIGURE 22.1 Sample cover for the *Federal Register.* (*Source:* Courtesy of U. S. Department of Labor.)

standards and make sense out of them? From time to time you may have experienced this frustration and been tempted to throw the standards in the "round file."

The Code of Federal Regulations (CFR) is a codification of the general and permanent rules published in the *Federal Register* by departments and agencies of the executive branch of the federal government. The code is divided into fifty titles that represent broad areas that are subject to federal regulations. Each title is divided into chapters that usually bear the name of the issuing agency. Each chapter is further subdivided into parts, covering specific regulatory areas. Based on this breakdown, the Occupational Safety and Health Administration is designated Title 29 (Labor), Chapter XVII (Occupational Safety and Health Administration), and Part 1910 for the General Industry Sector.

Each volume of the CFR is revised at least once each calendar year and is issued on a quarterly basis. OSHA issues regulations at the beginning of the fourth

FIGURE 22.2 Sample of the cover of the U.S. Code of Federal Regulations. (*Source:* Courtesy of the U. S. Government.)

quarter, or July 1 of each year (the approximate revision date is printed on the cover of each volume).

The CFR is kept up-to-date by individual revisions issued in the *Federal Register* (see Figure 22.2). These two publications must be used together to determine the latest version of any given rule.

To determine whether there have been any amendments since the revision date of the U.S. Code volume in which you are interested, the following two lists must be consulted: (1) the "Cumulative List of CFR Sections Affected," issued monthly; and

(2) the "Cumulative List of Parts Affected," appearing daily in the *Federal Register*. These two lists refer you to the *Federal Register* page where you can find the latest amendment of any given rule. The pages of the *Federal Register* are numbered sequentially from January 1 to January 1 of the next year.

As stated previously, Title 29, Chapter XVII has been set aside for the Occupational Safety and Health Administration. Chapter XVII is broken down into parts. Part 1910 contains the General Industry Standards. The General Industry Standards are further broken down into subparts, sections, and paragraphs.

CFR NUMBERING SYSTEM

To use the CFR, you need an understanding of the hierarchy of the paragraph numbering system. The numbering system is a mixture of letters and numbers. Prior to 1979, italicized lowercase letters and lowercase roman numerals were used. A change was made after 1979.

CFR Numbering Hierarchy

<1979	1980
(a)	(a)
(1)	(1)
(i)	(i)
Italicized *(a)*	*(A)*
Italicized *(1)*	*(1)*
Italicized *(i)*	*(i)*

When trying to make use of the regulations, having knowledge of the regulatory numbering system will help avoid a lot of the headaches. This should make the regulation easier to comprehend and more user friendly. The following illustrates and explains the numbering system using an example from the 29 CFR 1910.110:

29 CFR 1910.110 (b)(13)(ii)(b)(7)(iii)

Portable containers shall not be taken into buildings except as provided in paragraph (b)(6)(i) of this section.

Title	Code of Fed. Reg.	Part	Subpart	Section	Paragraph
29	CFR	1910	H	.110	

As can be seen from this example, the first number ("29") stands for the title. Next in sequence is "CFR," which of course stands for the Code of Federal Regulations, followed by "1910," which is the part and "H" the subpart as all subparts are alphabetically lettered in 1910. Finally, there is a period that is followed by an Arabic number. This will always be the section number. In this case, Section .110 is the storage and handling of liquefied petroleum gases standard. If the number had been .146, the section would pertain to permit-required confined spaces.

29 CFR 1910.110 (b)(13)(ii)(b)(7)(iii)

Portable containers shall not be taken into buildings except as provided in paragraph (b)(6)(i) of this section.

Title	Code of Fed. Reg.	Part	Subpart	Section	Paragraph
29	CFR	1910	H	.110	(b)

This means that the next breakdown of paragraphs will be sequenced using lowercase letters in parentheses (a), (b), (c), etc. If you had three major paragraphs of information under a section, they would be lettered .110(a), .110(b), and 110(c).

29 CFR 1910.110 (b)(13)(ii)(b)(7)(iii)

Portable containers shall not be taken into buildings except as provided in paragraph (b)(6)(i) of this section.

Title	Code of Fed. Reg.	Part	Subpart	Section	Paragraph and Subparagraph
29	CFR	1910	H	.110	(b)(13)

The next level of sequencing involves the use of Arabic numbers. As illustrated, if there were three paragraphs of information between subheadings (a) and (b), they would be numbered (a)(1), (a)(2), and (a)(3).

29 CFR 1910.110 (b)(13)(ii)(b)(7)(iii)

Portable containers shall not be taken into buildings except as provided in paragraph (b)(6)(i) of this section.

Title	Code of Fed. Reg.	Part	Subpart	Section	Paragraph and Subparagraph
29	CFR	1910	H	.110	(b)(13)(ii)

The next level uses the lowercase Roman numerals. An example would be between paragraphs (2) and (3). If there were five paragraphs of information pertaining to paragraph (2), they would be numbered (2)(i), (2)(ii), (2)(iii), (2)(iv), and (2)(v).

Since 29 CFR 1910.110 was promulgated prior to 1979, all subparagraph numbering beyond this is italicized letters and numbers [e.g., (*b*)(*7*)(*iii*)].

If after 1979 there are subparagraphs to the lowercase Roman numerals, then a capital or uppercase letter is used, such as (A), (B), … (F). Any other subparagraph falling under an uppercase letter is numbered using brackets. For example, (*1*), (*5*), …, (*23*), and any subparagraph to the bracketed numbers would be denoted by an italicized Roman numeral as follows: (*i*), (*iv*), …, (*ix*).

OSHA STANDARDS COVERED

OSHA standards cover Construction and the General Industry, which include Manufacturing, Transportation and Public Utilities, Wholesale and Retail Trades,

Finance, Insurance and Services, as well as other industrial sectors (e.g., Longshoring). Some of the specific areas covered by regulations are

Lockout/Tagout	Electrical Safety	Housekeeping
Training Requirements	Noise Exposure	Fire Prevention
Hazard Communication	Confined Spaces	Personal Protection
Ventilation	Equipment Requirements	Sanitation
Medical and First Aid	Fall Protection	Working with Hazardous Chemical
Emergency Planning	Hazardous Substances	Recordkeeping
Guarding	Use of Hand Tools	Machine Safety
Ladders and Scaffolds	Equipment Safety	Radiation
Explosives/Blasting	Hazardous Chemicals	Lead

COPIES OF THE OSHA STANDARDS

The standards for Occupational Safety and Health are found in Title 29 of the Code of Federal Regulations (CFR). The Standards for Specific Industries are found in Title 29 of the Code of Federal Regulations Part:

- General Industry—29 CFR PART 1910
- Shipyard Employment—29 CFR PART 1915
- Marine Terminals—29 CFR PART 1917
- Longshoring—29 CFR PART 1918
- Gear Certification—29 CFR PART 1919
- Construction—29 CFR PART 1926
- Agriculture—29 CFR PART 1928
- Federal Agencies—29 CFR 1960

RELIEF (VARIANCE) FROM AN OSHA STANDARD

Variances can be obtained for OSHA standards for the following reasons:

1. The employer may not be able to comply with the standard by its effective date.
2. An employer may not be able to obtain the materials, equipment, or professional or technical assistance.
3. The employer has in place processes or methods that provide protection to workers that is "at least as effective as" the standard's requirements.

A "temporary variance" meeting the criteria above may be issued until compliance is achieved or for one year, whichever is shorter. It can also be extended or renewed for 6 months (twice). Employers can obtain a "permanent variance" if the employer can document with a preponderance of evidence that existing or proposed methods, conditions, processes, procedures, or practices provide workers with protections equivalent to or better than the OSHA standard. Employers are required to post a copy of the variance in a visible area in the workplace as well as make workers aware of request for a variance.

OSHACT PROTECTS

Usually all employers and their employees are considered under the OSHAct. However, this statement is not entirely true because the following are not covered under the OSHAct:

1. Self-employed persons
2. Farms at which only immediate family members are employees
3. Workplaces already protected by other federal agencies under federal statues, such as the United States Department of Energy and the Mine Safety and Health Administration
4. Federal employees and state and local employees

ROLE OF THE NATIONAL INSTITUTE FOR OCCUPATIONAL SAFETY AND HEALTH (NIOSH)

NIOSH is one of the Centers for Disease Control under the Department of Health and Human Services with headquarters in Atlanta, Georgia; it is not part of OSHA. Its functions include

1. Recommend new safety and health standards to OSHA
2. Conduct research on various safety and health problems
3. Conduct Health Hazard Evaluations (HHEs) of the workplace when called upon (see Table 22.1)
4. Publish an annual listing of all known toxic substances and recommended exposure limits
5. Conduct training that will provide qualified personnel under the OSHAct

ROLE OF THE OCCUPATIONAL SAFETY AND HEALTH REVIEW COMMISSION (OSHRC)

The Occupational Safety and Health Review Commission (OSHRC) was established, when the OSHAct was passed, to conduct hearings when OSHA citations and penalties are contested by employers or by their employees.

EMPLOYERS ARE RESPONSIBLE FOR WORKERS' SAFETY AND HEALTH

The employer is held accountable and responsible under the OSHAct. The "General Duty Clause," Section 5(a)(1) of the OSHAct, states that employers are obligated to provide a workplace that is free of recognized hazards that are likely

TABLE 22.1
Health Hazard Evaluation Form

U.S. DEPARTMENT OF HEALTH AND HUMAN SERVICES — Form Approved
U.S. Public Health Service — OMB No. 0920-0260
Centers for Disease Control and Prevention — Expires January 31, 2012
National Institute for Occupational Safety and Health

Request for Health Hazard Evaluation (HHE)

(also available at: www.cdc.gov/niosh/hheform.html)

Establishment Where Possible Hazard Exists

1. Company Name: ____________
2. Address: ____________

   City: ____________ State: ____________ Zip Code: ____________
3. What product or service is provided at this workplace? ____________
4. Specify the particular work area, such as building or department, where the possible hazard exists:

   ____
5. How many employees are exposed? ____________
6. Duration of exposure (hrs/day)? ____________
7. What are the occupations of the exposed employees; what is the process/task?

   Occupations: ____________

   Process/task: ____________
8. To your knowledge, has NIOSH, OSHA, MSHA, or any other government agency previously evaluated this workplace? ☐ YES ☐ NO
9. Is a similar request currently being filed with, or is the problem under investigation by, any other local, state, or federal agency? ☐ YES ☐ NO
10. If either question 8 or 9 is answered yes, give the name and location of each agency. ____________

    ____

    ____
11. Which company official is responsible for employee health and safety?

    Name: ____________ Title: ____________ Phone: ____________
12. How did you learn about the NIOSH HHE program? ☐ Company representative ☐ Co-worker ☐ Union ☐ Other employee representative ☐ NIOSH Website ☐ CDC 800 Number (CDC-INFO) ☐ News media (TV, radio, newspaper, magazine) ☐ Other (please list) ____________

Description of the Possible Hazard or Problem

13. Please list all substances, agents, or work conditions that you believe may contribute to the possible health hazard. (Include chemical names, trade names, manufacturer, or other identifying information, as appropriate.) ____________

    ____
14. In what physical form(s) do(es) the substance exist? ☐ Dust ☐ Gas ☐ Liquid ☐ Mist ☐ Other
15. How are the affected employees exposed? (route of exposure) ☐ Breathing ☐ Skin contact ☐ Swallowing ☐ Other (please list) ____________

[Send completed form to address listed on the reverse side]

continued

TABLE 22.1 (continued)
Health Hazard Evaluation Form

This form is provided in assist in requesting a health hazard evaluation from the U.S. Department of Health and Human Services. Public reporting burden for this collection of information is estimated to average 12 minutes per response, including the time for reviewing instructions, searching existing data sources, gathering and maintaining the data needed, and completing and reviewing the collection of information. An agency may not conduct or sponsor, and a person is not required to respond to, a collection of information unless it displays a currently valid OMB control number. Send comments regarding this burden estimate or any other aspect of this collection of information, including suggestions for reducing this burden, to DHHS Reports Clearance Officer; Paperwork Reduction Project (0920-0102); Rm 531, Hubert H. Humphrey Building, 200 Independence Ave., SW, Washington, D.C. 20201. (See Statement of Authority on reverse.)

16. What health problem(s) do employees have as a result of these exposures? (Please circle the one of most concern.) ______________________

17. Use the space below to supply any additional relevant information. ______________________

____

____

____

Submitting the HHE Request

18. Requester's Signature: ______________________ Date: __________

19. Type or print name: ______________________

20. Address: ______________________
    City: ______________ State: __________ Zip Code: __________

21. a) Business phone: b) Home phone: __________
    c) Best time of day to call: d) E-mail: __________

22. Check and complete only one of the following three boxes:

☐ I am a **current employee** of the employer, and an **authorized representative of two or more*** **other current employees** in the workplace where the exposures are found. Two additional employee signature are required for a valid request.*	**Please provide additional signatures.** Signature: __________ Phone: __________ E-mail: __________
* Additional signatures are not necessary if you are 1 of 3 or fewer employees n the affected workplace.	Signature: __________ Phone: __________ E-mail: __________

☐ I am an authorized representative, or an officer of the **union** or other organization representing the employees for collective bargaining purposes	**Name and address of this organization:**

☐ I am an employer representative.	Title: __________

TABLE 22.1 (continued)
Health Hazard Evaluation Form

23. Please indicate your desire:
    - ☐ I do not want my name revealed to the employer.
    - ☐ My name may be revealed to the employer.

SEND COMPLETED FORM TO:
National Institute for Occupational Safety and Health
Hazard Evaluation and Technical Assistance Branch
4676 Columbia Parkway, Mail Stop R-9
Cincinnati, OH 45226-1988
Phone:(513)841-4382 Fax:(513)841-4488

STATEMENT OF AUTHORITY: Sections 20 (a) (3-6) of the Occupational Safety and Health Act (29 USC 669 (a) (6-9)), and section 501 (a) (11) of the Federal Mine Safety and Health Act (30 USC 951 (a) (11)). The identity of the requester will not be revealed if he or she so indicates on the application form in accordance with the provisions of 42 CFR Part 85.7. The voluntary cooperation of the respondent requester is required to initiate the Health Hazard Evaluation.

Source: Courtesy of the National Institute for Occupational Safety and Health.

to cause death or serious physical harm to employees. Employers are responsible for the following:

1. Abide by and comply with the OSHA standards.
2. Maintain records of all occupational injuries and illnesses.
3. Maintain records of workers' exposure to toxic materials and harmful physical agents.
4. Make workers aware of their rights under the OSHAct.
5. Provide, at a convenient location and at no cost, medical examinations to workers when the OSHA standards require them.
6. Report to the nearest OSHA office within eight hours all occupational fatalities or a catastrophe where three or more employees are hospitalized.
7. Abate cited violations of OSHA standards within the prescribed time period.
8. Provide training on hazardous materials and make MSDSs available to workers upon request.
9. Ensure workers are adequately trained under the regulations.
10. Post information required by OSHA, such as citations, hazard warnings, and injury/illness records.

WORKERS' RIGHTS

Workers have many rights under the OSHAct. These rights include the following:

1. The right to review copies of appropriate standards, rules, regulations, and requirements that the employer should have available at the workplace
2. The right to request information from the employer on safety and health hazards in the workplace, precautions that may be taken, and procedures to be followed if an employee is involved in an accident or is exposed to toxic substances
3. The right to access relevant worker exposure and medical records
4. The right to be provided personal protective equipment
5. The right to file a complaint with OSHA regarding unsafe or unhealthy workplace conditions and request an inspection
6. The right not to be identified to the employer as the source of the complaint
7. The right to not be discharged or discriminated against in any manner for exercising rights under the OSHAct related to safety and health
8. The right to have an authorized employee representative accompany the OSHA inspector and point out hazards
9. The right to observe the monitoring and measuring of hazardous materials and see the results of the sampling, as specified under the OSHAct and as required by OSHA standards
10. The right to review the occupational injury and illness records (OSHA No. 300) at a reasonable time and in a reasonable manner
11. The right to have safety and health standards established and enforced by law
12. The right to submit to NIOSH a request for a Health Hazard Evaluation (HHE) of the workplace
13. The right to be advised of OSHA actions regarding a complaint and request an informal review of any decision not to inspect or issue a citation
14. The right to participate in standard development
15. The right of a worker to talk with the OSHA inspector related to hazards and violations during the inspection
16. The right of the worker filing a complaint to receive a copy of any citations and the time for abatement
17. The right to be notified by the employer if the employer applies for a variance from an OSHA standard and testify at a variance hearing and appeal the final decision
18. The right to be notified if the employer intends to contest a citation, abatement period, or penalty
19. The right to file a Notice of Contest with OSHA if the time period granted to the company for correcting the violation is unreasonable, if it is contested within 15 working days of the employer's notice
20. The right to participate at any hearing before the OSHA Review Commission or at any informal meeting with OSHA when the employer or a worker has contested the abatement date
21. The right to appeal the Review Commission's decisions in the U.S. Court of Appeals
22. The right to obtain a copy of the OSHA file on the facility or workplace

WORKERS' RESPONSIBILITIES UNDER THE LAW

Workers have certain responsibilities that they should adhere to, but the ultimate responsibility for safety and health and the assurance that workers are performing work in a safe manner rests with the employer. Workers should do the following:

1. Comply with OSHA regulations and standards.
2. Do not remove, displace, or interfere with the use of any safeguards.
3. Comply with the employer's safety and health rules and regulations.
4. Report any hazardous condition to the supervisor or employer.
5. Report any job-related injuries and illness to the supervisor or employer.
6. Cooperate with the OSHA Inspector during inspections when requested to do so.

THE RIGHT NOT TO BE DISCRIMINATED AGAINST

Workers have the right to expect safety and health on the job without fear of punishment. This is spelled out in Section 11(c) of the OSHAct. The law states that employers shall not punish or discriminate against workers for exercising rights such as

1. Complaining to an employer, union, or OSHA (or other government agency) about job safety and health
2. Filing a safety and health grievance
3. Participating in an OSHA inspection, conferences, hearing, or OSHA-related safety and health activity

THE RIGHT TO KNOW

The "right to know" means that the employer must establish a written, comprehensive hazard communication program that includes provisions for container labeling, materials safety data sheets, and an employee training program. The program must include

1. A list of the hazardous chemicals in the workplace
2. The means the employer uses to inform employees of the hazards of non-routine tasks
3. The way the employer will inform other employers of the hazards to which their employees may be exposed

Workers have the right to information regarding the hazards to which they are, or will be, exposed. They have the right to review plans such as the Hazard Communication Program. They have a right to see a copy of a Material Safety Data Sheet (MSDS) during their shift and receive a copy of an MSDS when requested. Also, information on hazards that may be brought to the workplace by another employer should be available to workers. Other forms of information, such as exposure records, medical records, etc., are to be made available to workers upon request.

ENVIRONMENTAL MONITORING RESULTS

Workers have the right to receive the results of any OSHA test for vapors, noise, dusts, fumes, or radiation. This includes observation of any measurement of hazardous materials in the workplace.

PERSONAL PROTECTIVE CLOTHING

Workers must be provided, at no cost, with the proper and well-maintained personal protective clothing, when appropriate for the job.

OSHA INSPECTIONS

OSHA can routinely initiate an unannounced inspection of a business. Inspections may occur due to routine inspections or through complaints. These occur during normal working hours.

Workers have the right to request an inspection. The request should be in writing (either by letter or by using the OSHA Complaint Form to identify the employer and the alleged violations). Send the letter or form to the area director or state OSHA director. If workers receive no response, they should contact the OSHA regional administrator. It is beneficial to call the OSHA office to verify its normal operating procedures. If workers allege an imminent danger, they should call the nearest OSHA office.

These inspections will include checking company records, reviewing compliance with hazard communication standards, evaluating fire protection and personal protective equipment, and reviewing the company's health and safety plan. This inspection will include conditions, structures, equipment, machinery, materials, chemicals, procedures, and processes. OSHA's priorities for scheduling an inspection are as follows:

1. Situations involving imminent danger
2. Catastrophes or fatal accidents
3. Complaint by workers or their representatives
4. Regular inspections targeted at high-hazard industries
5. Follow-up inspections

OSHA can give an employer advance notice of a pending inspection at certain times:

1. In case of an imminent danger
2. When it would be effective to conduct an inspection after normal working hours
3. When it is necessary to ensure the presence of the employer, specific employer representative, or employee representatives
4. When the area director determines that an advance notice would enhance the probability of a more thorough and effective inspection

No inspection will occur during a strike, work stoppage, or picketing unless the area director approves such action and, usually, this would be due to extenuating circumstances (e.g., an occupational death inside the facility). The steps of an OSHA inspection include the following:

1. The inspector becomes familiar with the operation, including previous citations, accident history, and business demographics. The inspector gains entry. OSHA is forbidden to make a warrantless inspection without the employer's consent. Thus, the inspector may have to obtain a search warrant if reasonable grounds for an inspection exist.
2. The inspector will hold an opening conference with the employer or a representative of the company. It is required that a representative of the company be with the inspector during the walk-around and a representative of the workers be given the opportunity to accompany the inspector.
3. The inspection tour may take from hours to days, depending on the size of the operation. The inspector will usually cover every area within the operation while assuring compliance with OSHA regulations.
4. A closing conference will be conducted for the employer to review what the inspector has found. The inspector will request an abatement time for the violations from the employer.
5. The area director will issue the written citations with proposed penalties along with the abatement dates to the employer. This document is called the "Notification of Proposed Penalty."

OSHA RECEIVES A COMPLAINT

OSHA gathers information and decides whether to send a compliance office (inspector) or inform the employee about a decision not to inspect. The time period for response is based on the seriousness of the complaint. The usual times are

1. Within 24 hours if the complaint alleges an imminent danger
2. Within 3 days if the complaint is serious
3. Within 20 days for all other complaints

CITATIONS

If violations of OSHA standards are detected, then the citations will include the following information:

1. The violation
2. The workplace affected by the violation
3. Denote specific control measures
4. The abatement period or the time to correct the hazard

Copies of the citation should be posted near the location of the violation for at least three days or until the violation is abated, whichever is longer.

TYPES OF VIOLATIONS

Violations are categorized in the following manner:

De Minimis	No Penalty
Nonserious	$1,000–$7,000/violation
Serious	$1,500–$7,000/violation
Willful, No Death	Up to $70,000 and $7,000/day for each day it remains
Willful, Repeat Violations	Same as Willful, No Death
Willful, Death Results	Up to $70,000 and 6 months in jail
Willful, Death Result, Second Violation	$70,000 and 1 year in jail
Failure to Correct a Cited Violation	$7,000/day
Failure to Post Official Documents	$l,000/poster
Falsification of Documents	$70,000 and 6 months in jail

CHALLENGING CITATIONS, PENALTIES, AND OTHER ENFORCEMENT MEASURES

Upon receipt of penalty notification, the employer has fifteen days to submit a Notice of Contest to OSHA. This Notice must be given to the workers' authorized representative; or if no workers' representative exists, then the Notice of Contest must be posted in a prominent location in the workplace. An employer that has filed a Notice of Contest may withdraw it prior to the hearing date by

1. Showing that the alleged violation has been abated or will be abated
2. Informing the affected employees, or their designated representative, of the withdrawal of the contest
3. Paying the fine that had been assessed for the violation

Employers can request an informal hearing with the area director to discuss these issues, and the area director can enter into a settlement agreement if the situation merits it. But, if a settlement cannot be reached, then the employer must notify the area director in writing with a Notice of Contest of the citation, penalties, or abatement period within 15 days of receipt of the citation.

WORKERS GET THE RESULTS OF AN INSPECTION

The workers or their representative can request that the inspector conduct a closing conference for labor, and all citations should be posted by the employer for the workers' information.

Workers can contest the length of the time period for abatement of a citation and the employer's petition for modification of abatement in order to extend the time for correcting the hazard (workers must do this within 10 working days of posting). Workers cannot contest the following:

1. The employer's citations
2. Employer's amendments to citations

3. Penalties for the employer's citations
4. The lack of penalties

Two items can be challenged by workers.:

1. The time element in the citation for abatement of the hazard
2. An employer's Petition for Modification of Abatement (PMA); workers have 10 days to contest the PMA

DETERMINING PENALTIES

Penalties are usually based on four criteria:

1. The seriousness or gravity of the alleged violations
2. The size of the business
3. The employer's good faith in genuinely and effectively trying to comply with the OSHAct before the inspection and effort to abate and comply with the law during and after the inspection
4. The employer's history of previous violations

STATE PROGRAMS

Some states elect to enforce occupational safety and health themselves. They must develop a program that OSHA will review and approve. Approximately twenty-three states have such programs (see Appendix E for a list and addresses for state plans). If a state has a state plan (program) that is approved, the following conditions must exist:

1. The state must create an agency to carry out the plan.
2. The state's plan must include safety and health standards and regulations. The enforcement of these standards must be at least as effective as the federal plan.
3. The state plan must include provisions for the right of entry and inspection of the workplace, including a prohibition on advance notice of inspections.
4. The state's plan must also cover state and local government employees.

If a state has specific standards or regulations, they must be at least as stringent as the federal standards and regulations. Some states have standards and regulations that go beyond the requirements of the existing federal standards and regulations.

WORKERS' TRAINING

Many standards promulgated by OSHA specifically require the employer to train employees in the safety and health aspects of their jobs. Other OSHA standards make it the employer's responsibility to limit certain job assignments to employees who are "certified," "competent," or "qualified" (meaning that they have had special previous training, in or out of the workplace). OSHA's regulations imply that an employer has assured that a worker has been trained prior to being designated as the individual to perform that task).

Some OSHA standards have specific requirements for training. A listing can be found in Appendix D.

To completely address this issue, one would have to go directly to the regulation that applies to the specific type of activity. The regulation may mandate hazard training, task training, and length of training as well as specifics to be covered by the training.

It is always a good idea for the employer to keep records of training. These may be used by a compliance inspector during an inspection, after an accident resulting in injury or illness, as a proof or good intention to comply by an employer, or when a worker goes to a new job.

OCCUPATIONAL INJURIES AND ILLNESSES

Any occupational illness that has resulted in an abnormal condition or disorder caused by exposure to environmental factors may be acute or chronic. Inhalation, absorption, ingestion, direct contact with toxic substances or harmful agents, and any repetitive motion injury are classified as an illness. All illnesses are recordable, regardless of severity. Injuries are recordable when

1. An on-the-job death occurs (regardless of the length of time between the injury and death).
2. One or more lost workdays occur.
3. Restriction of work or motion transpires.
4. Loss of consciousness occurs.
5. The worker is transferred to another job.
6. The worker receives medical treatment beyond first aid (see Figure 22.3).

Employers with more than ten employees are required to complete and maintain occupational injury and illness records. The OSHA 301 "Injury and Illness Incident Report," or equivalent, must be completed within 7 days of the occurrence of an injury at the worksite and the OSHA 301 must be retained for 5 years. Also, the OSHA 300 "Log of Work-Related Injuries and Illnesses" is to be completed within 7 days when a recordable injury or illness occurs, and retained for 5 years. The OSHA 300A "Summary of Work-Related Injuries and Illnesses" must be posted yearly from February 1 to April 30. OSHA forms can now be maintained on the computer until they are needed.

MEDICAL AND EXPOSURE RECORDS

Medical examinations are required by some OSHA regulations for workers before they can perform certain types of work. This work includes, at present, the following:

- Asbestos abatement
- Lead abatement
- Hazardous waste remediation
- Physicians may require medical exams prior to wearing a respirator

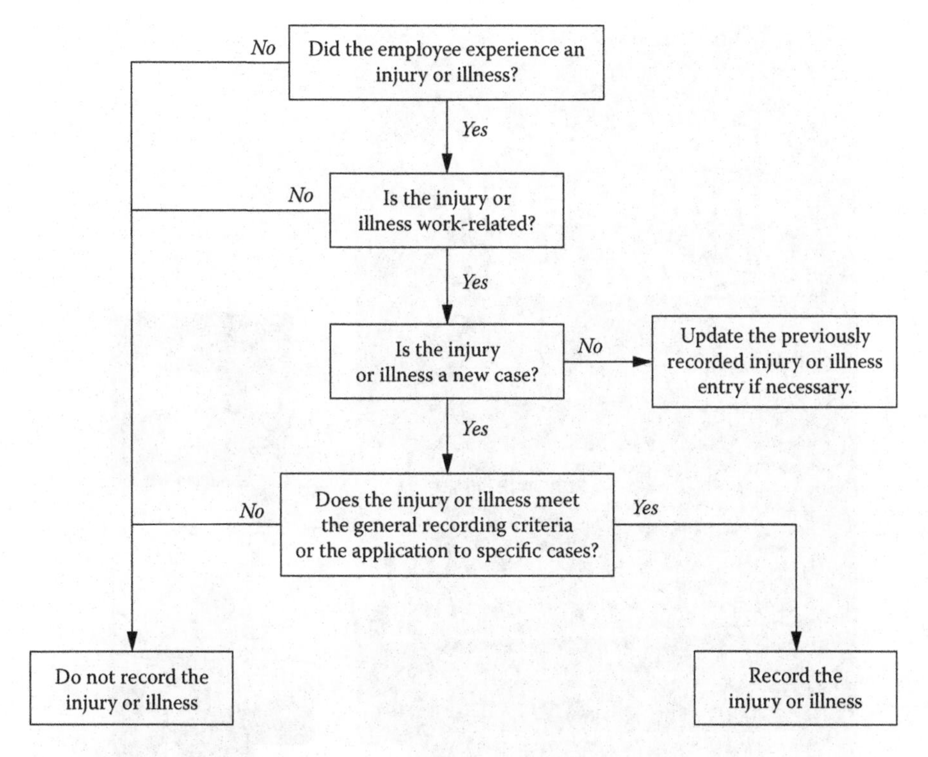

FIGURE 22.3 Diagram for determining OSHA Recordable Illness and Injury. (Courtesy of the Occupational Safety and Health Administration.)

Exposure records (monitoring records) are to be maintained by the employer for 30 years. Medical records are to be maintained by the employer for the length of employment plus 30 years.

Workers must make a written request to obtain a copy of their medical records or make them available to their representative or physician. Workers' medical records are considered confidential and require a request in writing from the worker to the physician in order for the records to be released.

POSTING

Employers are required to post in a prominent location the following:

1. Job Safety and Health Protection workplace poster (OSHA 3165) or state equivalent (see Figure 22.4)
2. Copies of any OSHA Citations of Violations of the OSHA standards are to be at or near the location of the violation for at least three days or until the violation is abated, whichever is longer
3. A copy of summaries of petitions for variances from any standard and this includes recordkeeping procedures
4. The summary portion of the "Log and Summary of Occupational Injuries and Illnesses" (OSHA Form 300A) annually from February 1 to March 1

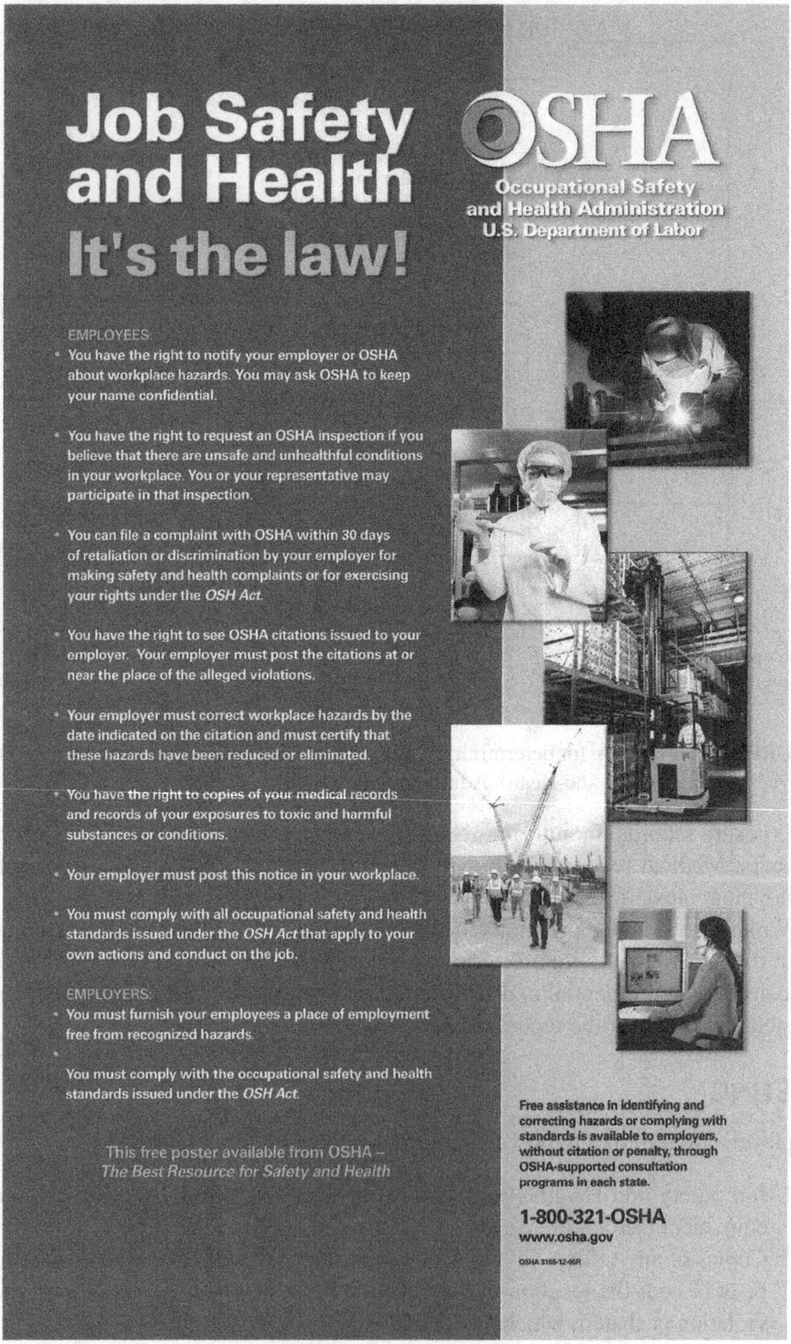

Job Safety and Health

It's the law!

OSHA
Occupational Safety and Health Administration
U.S. Department of Labor

EMPLOYEES:

- You have the right to notify your employer or OSHA about workplace hazards. You may ask OSHA to keep your name confidential.
- You have the right to request an OSHA inspection if you believe that there are unsafe and unhealthful conditions in your workplace. You or your representative may participate in that inspection.
- You can file a complaint with OSHA within 30 days of retaliation or discrimination by your employer for making safety and health complaints or for exercising your rights under the *OSH Act*.
- You have the right to see OSHA citations issued to your employer. Your employer must post the citations at or near the place of the alleged violations.
- Your employer must correct workplace hazards by the date indicated on the citation and must certify that these hazards have been reduced or eliminated.
- You have the right to copies of your medical records and records of your exposures to toxic and harmful substances or conditions.
- Your employer must post this notice in your workplace.
- You must comply with all occupational safety and health standards issued under the *OSH Act* that apply to your own actions and conduct on the job.

EMPLOYERS:

- You must furnish your employees a place of employment free from recognized hazards.
- You must comply with the occupational safety and health standards issued under the *OSH Act*.

This free poster available from OSHA –
The Best Resource for Safety and Health

Free assistance in identifying and correcting hazards or complying with standards is available to employers, without citation or penalty, through OSHA-supported consultation programs in each state.

1-800-321-OSHA
www.osha.gov

OSHA 3165-12-06R

FIGURE 22.4 Job Safety and Health Poster. (Courtesy of the Occupational Safety and Health Administration.)

WHAT TO DO WHEN OSHA COMES KNOCKING

When an inspector from OSHA or a corporate or insurance company's safety and health professional shows up at a project or jobsite, there is nothing to worry about if a safety and health program has been implemented and its mandates are being enforced.

To start, the following items should be in place:

1. A job safety and health protection poster (OSHA 3165) is posted on a bulletin board that is visible to all workers.
2. Summaries are available of any petitions for variances.
3. Copies exist of any new or unabated citations.
4. A summary of the OSHA 300A Summary that has been posted during the month of February through April.
5. The following should be available for the workers' and inspectors' examination:
    - Any exposure records for hazardous materials
    - The results of medical surveillance
    - All NIOSH research records for exposure to potentially harmful substances
6. Verification that workers have been told the following:
    - If exposures have exceeded the levels set by the standard and if corrective measures are being taken
    - If there are hazardous chemicals in their work area
7. Training records are available at the time of inspection.

Inspection Process

The inspection process should be handled in a professional manner, and mutual respect between the inspector and the employer or representative should be developed in a short period of time. It is appropriate to

1. Check the compliance officer's credentials and secure security clearance, if required.
2. Discuss company's safety and health program and its implementation.
3. Delineate activities and initiatives taken to improve safety and health on the job, as well as worker protection.
4. Ask for recommendations and advice that will improve what is being done.
5. Discuss any consultation programs or voluntary participation programs and pursue any inspection exemptions.
6. Ask the purpose, scope, and applicable standards for the inspection and obtain a copy of the employee complaint, if that is what triggered the inspection.
7. Make sure the employer's representative who accompanies the inspector is knowledgeable.
8. Include, if possible, an employees' representative.

9. The employer's representative must be familiar with the project and should try to choose an appropriate route for the inspection. However, the inspector's route cannot be dictated; he or she can choose the route for the inspection, if desired.
10. Make sure all observations, conversations, photographs, readings, and records examined are duplicated. Take good notes and ask appropriate questions.
11. Have records available for the inspector such as the OSHA 300 Log, OSHA 301, exposure records, and training records.
12. Pay close attention to unsafe or unhealthy conditions that are observed. Discuss how to correct them with the inspector and take corrective actions immediately, if possible.
13. Never at any time interfere with employee interviews with the inspector.

Mitigating the Damage

There is no turning back; the inspection will occur. It is imperative that the inspection's outcome results in as little damage as possible. This can be accomplished in many ways during the inspection process. Some actions may seem redundant, but they need to be reinforced:

1. Ask for an OSHA consultation service or pursue an exemption if the inspector cannot tell you how to abate or correct a violation.
2. Know the jobsite and be familiar with all the processes and equipment.
3. Try to select the inspector's route, if possible. Save the known or suspect problem areas for last.
4. Take good notes and document the inspection process completely. Photograph anything that the inspector does.
5. Many benefits are gained from good recordkeeping.
6. Correct apparent violations immediately, if possible.
7. Maintain updated copies of any required written programs.

Closing Conference

During the closing conference, when the culmination of the inspection process occurs, adhere to the following items to maintain the overall continuity of the process:

1. Listen actively and carefully to the discussion of unsafe or unhealthy conditions and apparent violations.
2. Ask questions for clarification so as to avoid confrontation. Confrontation will accomplish nothing.
3. Make sure the inspector discusses the appeal rights, informal conference procedure rights, and procedures for contesting a citation.
4. Produce documentation to support the company's compliance efforts or special emphasis programs.
5. Provide information that will guide the inspector in setting the times for abatement of citations.

After the Inspector Leaves

Citations and notices will arrive by certified mail and should be posted at or near the area where the violation occurred for at least 3 days or until abated, whichever is longer. Any notice of contest or objection must be received in writing by the OSHA area director within 15 days of receipt of any citations. The area director will forward the notice of contest to the Occupational Safety and Health Review Commission (OSHRC). It is also a good idea to request an informal meeting with the area director during the 15-day period.

The notice of contest will be assigned to an administrative law judge by the OSHRC. Once the judge rules on the contest notice, further review by the OSHRC may be requested. If necessary, the OSHRC ruling can be appealed to the U.S. Court of Appeals.

Remember that all citations or violations must be corrected or abated by the prescribed date unless the citation or abatement date is formally contested. If the response to a citation or violation cannot be abated in the time allotted, due to factors that are beyond reasonable control, a petition to modify the time for abatement must be filed with the area director to extend the date.

Make the Inspection a Positive Experience

A proactive safety and health preparation can make for a quality safety program. This is a safeguard for property, equipment, profits, and liability, as well as for workers. Working with workers to correct deficiencies can foster better safety attitudes. A safer workplace is also a more productive workplace. This also safeguards a very important asset: “the worker.” Remember that OSHA has a great deal of expertise within its ranks. Use OSHA as a resource to improve your safety and health program.

Prior to the knock by OSHA, it is necessary to implement a safety and health program. This includes, as stated earlier, the following:

1. Formal written program or safety manual
2. Standard operating procedures (SOPs) that incorporate OSHA standards
3. Worker and supervisor training
4. Standard recordkeeping procedures
5. Workplace inspections, audits, job observations, and job safety analyses
6. Safety and health committees, if possible
7. Accident or incident investigation procedures
8. Hazard recognition or reporting procedures
9. First aid and medical facility availability
10. Employee medical surveillance or examinations
11. Consultation services available

With these as a prerequisite, an OSHA inspection quickly becomes a positive learning experience from which many benefits will be reaped. It can produce higher morale, better production, a safer workplace, and a better bottom line because many negatives will have been avoided by good pre-activity and planning.

SUMMARY

Although it is ultimately the responsibility of the employer to maintain workplace safety and health, adherence to OSHA occupational safety and health rules is the foundation upon which a good safety and health program can be built. OSHA holds all responsible for the well-being of those who work there, including the employer, managers, supervisors, the worker and his or her fellow workers such that all must abide by the workplace safety and health rules and the OSHA standards.

Although many employers would balk at the idea that OSHA is an asset to them, without OSHA many employers would not have a foundation upon which to enforce their safety and health program. After all, the employer can say, when questioned about a safety and health regulation or rule, that it is OSHA's fault and not his or her (employer's) fault. "OSHA made me do it!" Without OSHA's threats of enforcement, employers would have no leverage upon which to enforce many safety and health policies.

Together, and through cooperation, all parties can ensure a safe and healthy workplace by following good safety and health practices for that specific workplace. A safe and healthy home away from home is, and should be, the ultimate goal.

REFERENCES

Blosser, F. *Primer on occupational safety and health.* Washington, D.C.: The Bureau of National Affairs, Inc., 1992.

Reese, C. D. and J. V. Eidson. *Handbook of OSHA construction safety & health.* Boca Raton, FL: CRC Press/Lewis Publishers, 1999.

United States Department of Labor, Occupational Safety and Health Administration. OSHA 10 and 30 Hour Construction Safety and Health Outreach Training Manual. Washington, D.C.: U.S. Department of Labor, 1991.

United States Department of Labor, Occupational Safety and Heath Administration, General Industry. *Code of Federal Regulations* (*Title* 29, *Part 1910*). Washington, D.C.: Government Printing Office, 1998.

United States Department of Labor, Occupational Safety and Heath Administration, Construction. *Code of Federal Regulations* (*Title 29, Part 1926*). Washington, D.C.: Government Printing Office, 1998.

United States Department of Labor, Occupational Safety and Health Administration, Office of Training and Education. OSHA Voluntary Compliance Outreach Program: Instructors Reference Manual. Des Plaines, IL: U.S. Department of Labor, 1993.

23 Health Hazard Prevention

INTRODUCTION

Occupational illnesses have always been viewed as being under-counted. They have been estimated at somewhere between 50,000 and 90,000 cases annually. Occupational illnesses have some unique features that can be attributed to their under-reporting, including the following:

- They are more difficult to recognize or diagnose.
- There is often a latency period between exposure and the occurrence of symptoms.
- There is not always a clear cause and effect relationship.
- Root or basic causes are not clearly apparent.
- It is difficult to determine a sequence of events.

In the arena of prevention, the objectives for Healthy People 2020 from the U.S. Department of Health and Human Services (DNHS) are as follows:

- Reduce the rate of injury and illness cases involving days away from work due to overexertion or repetitive motion.
- Reduce pneumoconiosis.
- Reduce the proportion of workers with elevated blood lead concentration from occupational exposure.
- Reduce occupation skin diseases or disorders among full-time workers.
- Reduce new cases of work-related noise-induced hearing loss.
- Increase the proportion of employees who have access to workplace programs that prevent or reduce employee stress.

Results from the 2010 Healthy People objectives indicated that skin disorder more than met the goals. The goals for overexertion, repetitive motion, and workplace stress met approximately 50 percent of the goal. The results for pneumoconiosis and blood lead level were less favorable. Data for noise-induced hearing loss is not yet available.

Health hazards that cause occupational illnesses present a more complex issue than safety hazards. Health-related hazards must be identified (recognized), evaluated, and controlled in order to prevent occupational illnesses, which come from exposure to them. Health-related hazards come in a variety of forms, such as chemical, physical, ergonomic, or biological:

1. *Chemical hazards* arise from excessive airborne concentrations of mists, vapors, gases, or solids that are in the form of dusts or fumes. In addition to

the hazard of inhalation, many of these materials may act as skin irritants or may be toxic by absorption through the skin. Chemicals can also be ingested although this is not usually the principal route of entry into the body.

2. *Physical hazards* include excessive levels of nonionizing and ionizing radiations, noise, vibration, and extremes of temperature and pressure.
3. *Ergonomic hazards* include improperly designed tools or work areas. Improper lifting or reaching, poor visual conditions, or repeated motions in an awkward position can result in accidents or illnesses in the occupational environment. Designing the tools and the job to be done to fit the worker should be of prime importance. Intelligent application of engineering and biomechanical principles is required to eliminate hazards of this kind.
4. *Biological hazards* include insects, molds, fungi, viruses, and bacterial contaminants (sanitation and housekeeping items such as potable water, removal of industrial waste and sewage, food handling, and personal cleanliness can contribute to the effects from biological hazards). Biological and chemical hazards can overlap in some cases.

These health-related hazards can often be difficult and elusive to identify. A common example of this is a contaminant in a building that has caused symptoms of illness. Even the evaluation process may not be able to detect a contaminant that has dissipated before a sample can be collected. This leaves nothing to control and possibly no answer to what caused the illness.

Health hazards also present a problem in that some can cause acute effects, while others can lead to chronic disease at some time much later. This chapter cannot cover all the potential causes and risks of occupational illness, but it does review some of the most common health-related risks in the workplace.

ASBESTOS

Asbestos was a widely used, mineral-based material that is resistant to heat and corrosive chemicals. Typically, asbestos appears as a whitish, fibrous material that may release fibers that range in texture from coarse to silky; however, the airborne fibers that can cause health damage may be too small to see with the naked eye.

An estimated 1.3 million employees in the construction and general industry face significant asbestos exposure on the job. Heaviest exposures occur in the construction industry, particularly when removing asbestos during renovation or demolition. In the past, employees were also likely to be exposed during the manufacture of asbestos products (such as textiles, friction products, insulation, and other building materials) and during automotive brake and clutch repair work.

Exposure to asbestos can cause asbestosis (scarring of the lungs resulting in loss of lung function that often progresses to disability and to death), mesothelioma (cancer affecting the membranes lining the lungs and abdomen), lung cancer, and cancers of the esophagus, stomach, colon, and rectum.

The U.S. Occupational Safety and Health Administration (OSHA) has issued revised regulations covering asbestos exposure in general industry and construction. Both standards set a maximum exposure limit and include provisions for engineering

ASBESTOS WASTE DISPOSAL SITE
BREATHING ASBESTOS DUST MAY CAUSE LUNG DISEASE AND CANCER

FIGURE 23.1 Asbestos warning sign.

controls, respirators, protective clothing, exposure monitoring, hygiene facilities and practices, warning signs, labeling, recordkeeping, and medical exams. Nonasbestiform tremolite, anthophyllite, and actinolite were excluded from coverage under the asbestos standard.

Here are some of the highlights of the revised rules in 29 CFR 1910.1001, promulgated October 11, 1994:

1. *Permissible exposure limit (PEL):* In both general industry and construction, workplace exposure must be limited to 0.1 fibers per cubic centimeter of air (0.1 f/c^3), averaged over an 8-hour work shift. The excursion or short-term limit is one fiber per cubic centimeter of air (1 f/c^3) averaged over a sampling period of 30 minutes.
2. *Methods of compliance:* In both general industry and construction, employers must control exposures using engineering controls, to the extent feasible. Where engineering controls are not feasible to meet the exposure limit, they must be used to reduce employee exposures to the lowest levels attainable and must be supplemented by the use of respiratory protection.
3. *Respirators:* In general industry and construction, the level of exposure determines what type of respirator is required; the standards specify the respirator to be used.
4. *Regulated areas:* In general industry and construction, regulated areas must be established where the 8-hour Time Weighted Average (TWA) or 30-minute excursion values for airborne asbestos exceed the prescribed permissible exposure limits. Only authorized persons wearing appropriate respirators can enter a regulated area. In regulated areas, eating, smoking, drinking, chewing tobacco or gum, and applying cosmetics are prohibited. Warning signs must be displayed at each regulated area and must be posted at all approaches to regulated areas. (See Figure 23.1.)
5. *Labels:* Caution labels must be placed on all raw materials, mixtures, scrap, waste, debris, and other products containing asbestos fibers.
6. *Recordkeeping:* The employer must keep an accurate record of all measurements taken to monitor employee exposure to asbestos. This record is to include the date of measurement, operation involving exposure, sampling and analytical methods used, and evidence of their accuracy; number, duration, and results of samples taken; type of respiratory protective devices worn; name, social security number, and the results of all employee exposure measurements. This record must be kept for 30 years.
7. *Protective clothing:* For any employee exposed to airborne concentrations of asbestos that exceed the PEL, the employer must provide and require

the use of protective clothing such as coveralls or similar full-body clothing (usually Tyvek), head coverings, gloves, and foot covering. Wherever the possibility of eye irritation exists, face shields, vented goggles, or other appropriate protective equipment must be provided and worn.

8. *Construction:* In construction (29 CFR 1626.1101), there are special regulated-area requirements for asbestos removal, renovation, and demolition operations. These provisions include a negative pressure area, decontamination procedures for workers, and a "competent person" with the authority to identify and control asbestos hazards. The standard includes an exemption from the negative pressure enclosure requirements for certain small-scale, short-duration operations, provided special work practices, prescribed in the standard, are followed, including the following:
9. *Hygiene facilities and practices:* Clean change rooms must be furnished by employers for employees who work in areas where exposure is above the TWA and/or excursion limit. Two lockers or storage facilities must be furnished and separated to prevent contamination of the employee's street clothes from protective work clothing and equipment. Showers must be furnished so that employees may shower at the end of the work shift. Employees must enter and exit the regulated area through the decontamination area. The equipment room must be supplied with impermeable, labeled bags and containers for the containment and disposal of contaminated protective clothing and equipment.
10. *Lunchroom facilities:* Employers must provide a positive pressure, filtered air supply and a readily accessible lunchroom for employees. Employees must wash their hands and face prior to eating, drinking, or smoking. The employer must ensure that employees do not enter lunchroom facilities with protective work clothing or equipment unless surface fibers have been removed from the clothing or equipment. Employees may not smoke in work areas where they are occupationally exposed to asbestos.
11. *Medical exams:* In general industry, exposed employees must have a preplacement physical examination before being assigned to an occupation exposed to airborne concentrations of asbestos at or above the action level or the excursion level. The physical examination must include chest x-ray, medical and work history, and pulmonary function tests. Subsequent exams must be given annually and upon termination of employment, although chest x-rays are required annually only for older workers whose first asbestos exposure occurred more than 10 years ago. In construction, examinations must be made available annually for workers exposed above the action level, or excursion limit, for 30 or more days per year, or who are required to wear negative pressure respirators; chest x-rays are at the discretion of the physician.

All workers must receive asbestos abatement training usually equivalent to 32 hours and yearly refresher training.

Although asbestos is currently not being manufactured or installed in the United States, there are many facilities and locations where it still exists. Damaged or friable asbestos should be removed, while asbestos that is in good condition can be managed in places using enclosures or sealants. The hazard involved with asbestos should not be taken lightly, as the long-term effects can be disastrous.

BACK INJURIES

Most of the safety and health community considers back injuries a one-time event. It seems more realistic to consider such injuries as caused by a series of microtraumas (strains and sprains) over time that eventually results in what is commonly called a back injury. This concept moves back injuries into the realm of cumulative trauma disorders. Many companies in the past hired younger workers who may be less susceptible to injury to do continuous or repetitive lifting task rather than use older or more experienced workers who might be more likely to be injured. These young workers had not had the micro-trauma injuries that set them up for a back injury.

At some time, eight out of ten adults will suffer back pain. In the work world, a sore back is probably normal for most workers. In most industries where material handling occurs, back injuries are due to overexertion (see Figure 23.2).

The spine (back) is like a stack of kids' blocks (vertebrae) supported by string, wire, and sheets of plastic (tendons, ligaments, tissues, and muscles). Like any material, these tendons, ligaments, and muscles are pulled, stretched, and twisted as they support the back. It is no wonder that, when they are not maintained or rested, and/or receive a lot of strain or pressure, they are sprained or strained and cause pain. This is why many back injury cases require little or no medical treatment other than to avoid lifting and allow the torn body tissue time to heal.

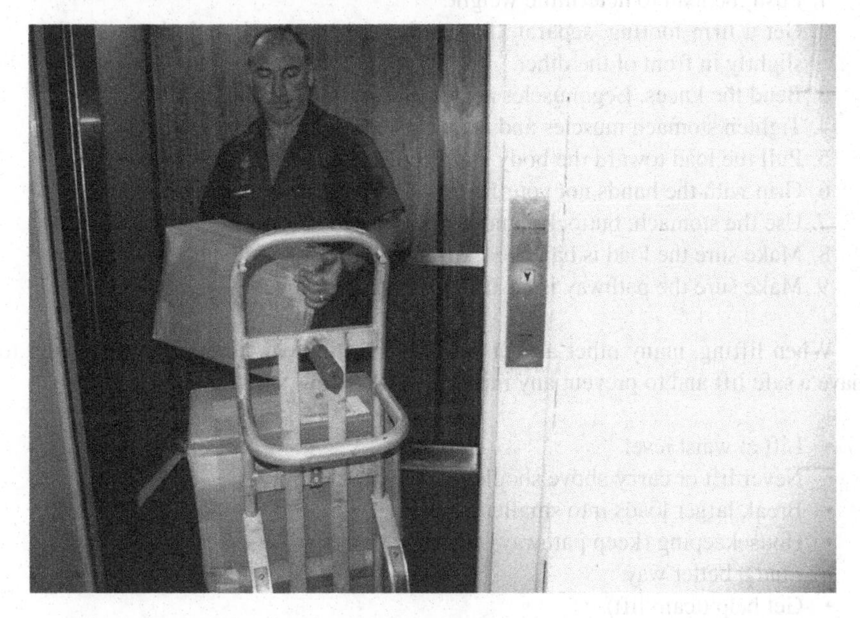

FIGURE 23.2 Safe lifting techniques can help prevent back injuries.

The lower back supports 70 percent to 80 percent of the body's weight above the hips. If the weight of an object to be lifted or the worker is overweight with a potbelly or out of shape, a great deal of stress is being placed on the back, which is often more than the back can support. Much can be done to prevent an injury to the back; some of these are

- Good posture is important (do not slouch).
- Regular exercise is important (normal work may not be enough).
- Warm up and stretch before you lift.
- Use good lifting techniques.
- Don't overdo it (get help).

Poor posture can cause back pain and can be corrected by good posture. Over a long period of time, poor posture can cause permanent damage to the back. Strengthening and flexibility exercises will help prevent new and recurring injuries. Many companies have begun pre-work exercising and warm-up exercise programs that have been successful in reducing back injuries as well as their severity. There are times when exercise is not appropriate, such as when there is pain in the leg below the knee; weakness, numbness, or pins and needles in feet or toes; back pain after a recent accident; bladder problems after recent back pain; or when a feeling of less wellness in conjunction with back pain exists. A physician should be contacted immediately when any of these symptoms exist.

Warm up before lifting, like athletes always do before strenuous exertion; then use good lifting techniques (see Figure 23.3), such as

1. Push the load to determine weight.
2. Get a firm footing, separate feet, point toes outward, and place one foot slightly in front of the other.
3. Bend the knees. Leg muscles are stronger than back muscles.
4. Tighten stomach muscles and keep back straight/upright.
5. Pull the load toward the body and keep it close to the body.
6. Grip with the hands not your fingers. Wear tight-fitting gloves.
7. Use the stomach, buttocks, and leg muscles to lift and to set the load down.
8. Make sure the load is balanced; lift it straight up and do not twist the body.
9. Make sure the pathway is clear in the path of travel.

When lifting, many other aspects of proper lifting can be employed in order to have a safe lift and to prevent any injuries while lifting, including

- Lift at waist level.
- Never lift or carry above shoulder level.
- Break larger loads into smaller ones.
- Housekeeping (keep pathways clear).
- Find a better way.
- Get help (team lift).
- Push, do not pull a load.

A
As you approach the load determine its weight, size and shape. Consider your physical ability.

B
Stand close to the object with feet 8 to 12 inches apart for good balance.

C
Bend the knees and get a firm grasp on the object.

D
Using both leg and back muscles lift the load straight up. Keep the object close to the body.

E
Do not twist or turn until the object is in carrying position.

F
Rotate body by turning your feet and make sure path of travel is clear.

G
To set the object down, use leg and back muscles and lower the object by bending the knees.

FIGURE 23.3 Steps in making a proper lift. (*Source:* Courtesy of the U.S. Bureau of Mines.)

- Solve heavy load lifts and repetitive lifts using lifting devices.
- Use mechanical help.

In summary, workers must be careful, especially if they are overweight; they should eat moderately and diet sensibly, exercise to strengthen muscles and improve flexibility, plan each lift before starting, lift naturally, bend the knees, and let the legs—not the back—power the lift up smoothly (do not twist while lifting, and don't overdo it). Get help with loads that are too heavy or awkward for one person to move comfortably. If attention is not paid to proper lifting and handling, back injuries are sure to follow, as the back is the most common body part injured and the most frequent type of occupational injury.

BLOODBORNE PATHOGENS

OSHA has promulgated a regulation to minimize serious health risks faced by workers exposed to blood and other potentially infectious materials. Among the risks are human immunodeficiency virus (HIV), hepatitis B, and hepatitis C.

Emphasis on engineering controls is based on ways to better protect workers from contaminated needles or other sharp objects. Many safety medical devices are already available and effective in controlling these hazards, and wider use of such devices would reduce thousands of injuries each year. OSHA issued a final regulation (29 CFR 1910.1030) on occupational exposure to bloodborne pathogens in 1991, to protect nearly six million workers in healthcare and related occupations at risk of exposure to bloodborne diseases. Occupational exposure to blood and other potentially infectious materials that may contain bloodborne pathogens, or microorganisms that can cause bloodborne diseases, is of concern to those workers who are at risk for exposure.

An annual review of the exposure control plan by employers must ensure that their plans reflect consideration and use of commercially available safer medical devices. An emphasis should be placed on the use of effective engineering controls, to include safer medical devices, work practices, administrative controls, and personal protective equipment. Employers should rely on relevant evidence, in addition to Food and Drug Administration (FDA) approval, to ensure the effectiveness of devices designed to prevent exposure to bloodborne pathogens.

Multi-employer worksites should receive special attention, such as employment agencies, personnel services, home health services, independent contractors, and dentist and physicians in independent practice. The purpose of the regulation is to limit occupational exposure to blood and other potentially infectious materials, as any exposure can result in the transmission of bloodborne pathogens and can lead to disease or death. It covers all employees who can be "reasonably anticipated," as the result of performing their job duties, to face contact with blood and other potentially infectious materials or body fluids. OSHA has not attempted to list all occupations where exposures could occur. "Good Samaritan" acts, such as assisting a co-worker with a nosebleed, would not be considered occupational exposure.

Infectious materials include semen, vaginal secretions, cerebrospinal fluid, synovial fluid, pleural fluid, pericardial fluid, peritoneal fluid, amniotic fluid, saliva in dental procedures, any body fluid visibly contaminated with blood, and all body fluids in situations where it is difficult or impossible to differentiate between body fluids. They also include any unfixed tissue or organ, other than intact skin, from a human (living or dead); any human immunodeficiency virus- (HIV-) containing cell or tissue cultures; organ culture; and HIV or hepatitis- (HBV-) containing culture media or other solutions as well as blood, organs, or other tissues from experimental animals infected with HIV or HBV.

This regulation requires employers to develop an exposure control plan to identify, in writing, tasks and procedures as well as job classifications where occupational exposure to blood occurs without regard to personal protective clothing and equipment. It must also set forth the schedule for implementing other provisions of the standard and specifying the procedure for evaluating circumstances surrounding exposure incidents. The plan must be accessible to employees and available to OSHA. Employers must review and update it at least annually—more often if necessary to accommodate workplace changes.

The regulation mandates universal precautions (e.g., treating body fluids/materials as if infectious) and emphasizes engineering and work practice controls. It stresses handwashing and requires employers to provide facilities, and ensure that employees use them, following exposure to blood. It sets out procedures to minimize needlesticks, minimize the splashing and spraying of blood, ensure appropriate packaging of specimens and regulated wastes, and decontaminate equipment or label it as contaminated before shipping to servicing facilities.

Employers must provide, at no cost, and require employees to use appropriate personal protective equipment, such as gloves, gowns, masks, eye/face protection, mouthpieces, and resuscitation bags. They must clean, repair, and replace personal protective equipment when necessary. Gloves are not necessarily required for routine phlebotomies in volunteer blood donation centers, but must be made available to employees who want them.

The regulation requires a written schedule for cleaning and for identifying the method of decontamination to be used, in addition to cleaning, following contact with blood or other potentially infectious materials. It specifies methods for disposal of contaminated sharp objects and provides standards for containers for these items and other regulated waste. Further, the standard includes provisions for handling contaminated laundry to minimize exposures.

The HIV and HBV research laboratories and production facilities must follow standard microbiological practices and specify additional practices intended to minimize exposures accessible to employees working with concentrated viruses. These practices also reduce the risk of accidental exposure for other employees at the facility. In turn, these facilities must include required containment equipment and an autoclave for decontamination of regulated waste, and they must be constructed to limit risks and enable easy clean-up. Additional training and experience are required of workers in these facilities.

OSHA also requires vaccinations (i.e., hepatitis B) to be made available to all employees who have occupational exposure to blood. Vaccinations must be given within 10 working days of assignment, at no cost, at a reasonable time and place, under the supervision of a licensed physician or healthcare professional, and according to the latest recommendations of the U.S. Public Health Service (USPHS). Prescreening may not be required as a condition of receiving the vaccine. Employees must sign a declination form if they choose not to be vaccinated, but may later opt to receive the vaccine at no cost to the themselves. Should booster doses later be recommended by the USPHS, employees must be offered them.

Specific procedures for post-exposure and follow-up must be made available to all employees who have had an exposure incident, plus any laboratory tests must be conducted by an accredited laboratory at no cost to the employee. Follow-ups must include a confidential medical evaluation documenting the circumstances of exposure, identifying and testing the source individual if feasible, testing the exposed employee's blood if he or she consents, post-exposure prophylaxis, counseling, and evaluation of the reported illnesses. Healthcare professionals must be provided specific information to facilitate the evaluation, and their written opinion on the

FIGURE 23.4 Biological hazard symbol.

need for hepatitis B vaccination must follow the exposure. Information, such as the employee's ability to receive the hepatitis B vaccine, must be supplied to the employer. All diagnoses must remain confidential.

Warning labels are required, including the orange or orange-red biohazard symbol. They must be affixed to containers of regulated waste, refrigerators and freezers, and other containers that are used to store or transport blood or other potentially infectious materials (see Figure 23.4). Red bags, or containers, may be used instead of labeling. When a facility uses universal precautions in its handling of all specimens, labeling is not required within the facility. Likewise, when all laundry is handled with universal precautions, the laundry need not be labeled. Blood that has been tested and found free of HIV or HBV and released for clinical use, and regulated waste that has been decontaminated, need not be labeled. Signs must be used to identify restricted areas in HIV and HBV research laboratories and production facilities.

There is a mandate requiring training initially upon assignment and annually (employees who have received appropriate training within the past year need only receive additional training in items not previously covered). Training must include making accessible a copy of the regulatory text of the standard and an explanation of its contents, general discussion on bloodborne diseases and their transmission, the exposure control plan, engineering and work practice controls, personal protective equipment, hepatitis B vaccine, response to emergencies involving blood, how to handle exposure incidents, the post-exposure evaluation and follow-up program, and explanation of signs/labels/color-coding. There must be opportunity for questions and answers, and the trainer must be knowledgeable in the subject matter. Laboratory and production facility workers must receive additional specialized initial training.

The maintenance of medical records, to be kept for each employee who has had occupational exposure for the duration of employment plus 30 years, must be confidential and must include names and social security numbers; hepatitis B vaccination status (including dates); results of any examinations, medical testing, and follow-up procedures; a copy of the healthcare professional's written opinion; and a copy of information provided to the healthcare professional. Training records must be maintained for 3 years and must include dates, contents of the training program or a summary, trainer's name and qualifications, and names and job titles of all persons attending the sessions. Medical records must be made available to the employee, anyone with written consent of the employee, and OSHA and NIOSH (they are not available to the employer). Disposal of records must be in accord with OSHA's standard covering access to records.

As can be seen, the whole intent of this bloodborne pathogen regulation is based on prevention of diseases that can result from exposure in the workplace to these organisms.

CARCINOGENS

Cancer exposure risk for the general public is set at 1 in 1,000,000 by the U. S. Environmental Protection Agency while the Occupational Health and Safety Administration exposure standards are usually less protective of workers at 1 cancer case in 1,000. This is a "fence line" mentality that seems to suggest that workers can withstand high levels of exposure.

Carcinogens are substances or agents that have the potential to cause cancer. Whether these chemicals or agents have been shown to only cause cancer in animals makes little difference to workers. Workers should consider these as causing cancer on a precautionary basis as all is not known regarding their long-term effects on humans. Since most scientists say that there is no known safe level of a carcinogen exposure, then no exposure should be the goal of workplace health and safety. Do not let the label "suspect" carcinogen or agent put the mind at ease. This chemical or agent can cause cancer. The Occupational Safety and Health Administration has identified thirteen chemicals as carcinogens. They are as follows:

- 4-Nitrobiphenyl, Chemical Abstracts Service Register Number (CAS No.) 92933
- alpha-Naphthylamine, CAS No. 134327
- Methyl chloromethyl ether, CAS No. 107302
- 3,3′-Dichlorobenzidine (and its salts), CAS No. 91941
- bis-Chloromethyl ether, CAS No. 542881
- beta-Naphthylamine, CAS No. 91598
- Benzidine, CAS No. 92875
- 4-Aminodiphenyl, CAS No. 92671
- Ethyleneimine, CAS No. 151564
- beta-Propiolactone, CAS No. 57578
- 2-Acetylaminofluorene, CAS No. 53963
- 4-Dimethylaminoazobenzene, CAS No. 60117
- N-Nitrosodimethylamine, CAS No. 62759

There are many other chemicals that probably should be identified as carcinogens, but have not stood the scrutiny of the regulatory process to make them such. This is probably, in many cases, due to special interest by manufacturers and other groups.

The OSHA regulation 29 CFR 1910.1003 pertains to solid or liquid mixtures containing less than 0.1 percent by weight or volume of 4-nitrobiphenyl, methyl chloromethyl ether, bis-chloromethyl ether, beta-naphthylamine, benzidine, or 4-aminodiphenyl, and solid or liquid mixtures containing less than 1.0 percent by weight or volume of alpha-naphthylamine, 3,3′-dichlorobenzidine (and its salts), ethyleneimine, beta-propiolactone, 2-acetylaminofluorene, 4-dimethylaminoazobenzene, or N-nitrosodimethylamine.

Any equipment, material, or other item taken into or removed from a regulated area shall be done so in a manner that does not cause contamination in nonregulated areas or the external environment. Decontamination procedures shall be established and implemented to remove carcinogens addressed by this section of the CFR from the surfaces of materials, equipment, and the decontamination facility.

Dry sweeping and dry mopping are prohibited for 4-nitrobiphenyl, alpha-naphthylamine, 3,3′-dichlorobenzidine (and its salts), beta-naphthylamine, benzidine, 4-aminodiphenyl, 2-acetylaminofluorene, 4-dimethylaminoazobenzene, and N-nitrosodimethylamine.

Entrances to regulated areas shall be posted with signs bearing the following legend:

CANCER-SUSPECT AGENT AUTHORIZED PERSONNEL ONLY

Entrances to regulated areas containing operations covered by OSHA's regulations are to be posted with signs bearing the following legend:

CANCER-SUSPECT AGENT EXPOSED IN THIS AREA

The OSHA requirements for areas containing a carcinogen require that regulated areas shall be established by an employer where a carcinogen is manufactured, processed, used, repackaged, released, handled, or stored. All such areas shall be controlled in accordance with the requirements for the following category, or categories, describing the operation involved:

1. *Isolated systems:* Employees working with a carcinogen addressed by this section of the CFR within an isolated system such as a "glove box" shall wash their hands and arms upon completion of the assigned task and before engaging in other activities not associated with the isolated system.
2. *Closed system operation:* Within regulated areas, where the carcinogens are stored in sealed containers, or contained in a closed system, including piping systems, with any sample ports or openings closed while the carcinogens are contained within, access shall be restricted to authorized employees only.
3. *Open-vessel system operations:* Open-vessel system operations are prohibited.

Employees exposed to 4-nitrobiphenyl; alpha-naphthylamine; 3,3′-dichlorobenzidine (and its salts); beta-naphthylamine; benzidine; 4-aminodiphenyl; 2-acetylaminofluorene; 4-dimethylaminoazobenzene; and N-nitrosodimethylamine shall be required to wash hands, forearms, face, and neck upon each exit from the regulated areas, close to the point of exit, and before engaging in other activities.

Transfer from a closed system, charging or discharging point operations, or otherwise opening a closed system is prohibited. In operations involving "laboratory-type hoods" or in locations where the carcinogens are contained in an otherwise "closed system," but is transferred, charged, or discharged into other normally closed containers, the provisions that apply are

1. Access shall be restricted to authorized employees only.
2. Each operation shall be provided with continuous local exhaust ventilation so that air movement is always from ordinary work areas to the operation. Exhaust air shall not be discharged to regulated areas, non-regulated areas, or the external environment unless decontaminated. Clean makeup air shall be introduced in sufficient volume to maintain the correct operation of the local exhaust system.
3. Employees shall be provided with, and required to wear, clean, full body protective clothing (smocks, coveralls, or long-sleeved shirt and pants), shoe covers, and gloves prior to entering the regulated area.
4. Employees engaged in handling operations involving the carcinogens addressed by this section of the CFR must be provided with, and required to wear and use, a half-face filter-type respirator with filters for dusts, mists, and fumes, or air-purifying canisters or cartridges. A respirator affording higher levels of protection than this respirator may be substituted.
5. Prior to each exit from a regulated area, employees shall be required to remove and leave protective clothing and equipment at the point of exit and at the last exit of the day, to place used clothing and equipment in impervious containers at the point of exit for purposes of decontamination or disposal. The contents of such impervious containers shall be identified.
6. Drinking fountains are prohibited in the regulated area.
7. Employees shall be required to wash hands, forearms, face, and neck on each exit from the regulated area, close to the point of exit, and before engaging in other activities. Employees exposed to 4-nitrobiphenyl, alpha-naphthylamine, 3,3′-dichlorobenzidine (and its salts), beta-naphthylamine, benzidine, 4-aminodiphenyl, 2-acetylaminofluorene, 4-dimethylamino-azobenzene, and N-nitrosodimethylamine shall be required to shower after the last exit of the day.

For maintenance and decontamination activities that occur during the clean-up of leaks or spills, during maintenance or repair operations on contaminated systems or equipment, or during any operation involving work in an area where direct contact with a carcinogen addressed by this section of the CFR could result, each authorized employee entering that area shall

1. Be provided with and required to wear clean, impervious garments, including gloves, boots, and continuous-air supplied hood
2. Be decontaminated before removing the protective garments and hood
3. Be required to shower upon removing the protective garments and hood

The requirements for general regulated areas are

1. The employer must implement a respiratory protection program in accordance with 29 CFR 1910.134.

2. In an emergency, immediate measures including, but not limited to, the following requirements of paragraphs (d)(2)(i) through (v) shall be implemented.
3. The potentially affected area shall be evacuated as soon as the emergency has been determined.
4. Hazardous conditions created by the emergency shall be eliminated and the potentially affected area shall be decontaminated prior to the resumption of normal operations.
5. Special medical surveillance by a physician shall be instituted within 24 hours for employees present in the potentially affected area at the time of the emergency. A report of the medical surveillance, and any treatment, shall be included in the incident report.
6. Where an employee has a known contact with a carcinogen addressed by this section of the CFR, such employee shall be required to shower as soon as possible, unless contraindicated by physical injuries.
7. An incident report on the emergency shall be completed.
8. Emergency deluge showers and eyewash fountains, supplied with running potable water, shall be located near, within sight of, and on the same level with locations where a direct exposure to ethyleneimine or β-propiolactone would be most likely as a result of equipment failure or improper work practice.

Storage or consumption of food; storage or use of containers of beverages; storage or application of cosmetics; smoking; storage of smoking materials, tobacco products, or other products for chewing, and the chewing of such products are prohibited in regulated areas. Where employees are required to wash, washing facilities, or shower facilities shall be provided in accordance with 29 CFR 1910.141. Where employees wear protective clothing and equipment, clean changing rooms shall be provided for the number of such employees required to change clothes, in accordance 29 CFR 1910.141. Where toilets are in regulated areas, such toilets shall be in a separate room.

Except for outdoor systems, regulated areas shall be maintained under negative pressure with respect to nonregulated areas. Local exhaust ventilation may be used to satisfy this requirement. Clean makeup air in equal volume shall replace the air removed.

Appropriate signs and instructions shall be posted at the entrance to, and exit from, regulated areas, informing employees of the procedures that must be followed in entering and leaving a regulated area.

Containers of a carcinogen must be accessible only to, and handled only by, authorized employees, or by other trained employees. Many containers may have their contents identification limited to a generic or proprietary name or other proprietary identification of the carcinogen and percent, and identification that includes the full chemical name and Chemical Abstracts Service Registry. Containers shall have the warning words "CANCER-SUSPECT AGENT" displayed immediately under or adjacent to the contents identification. Containers whose contents are carcinogens addressed by this section of the CFR with corrosive or irritating properties shall have label statements warning of such hazards, noting, if appropriate, particularly sensitive or affected portions of the body. Lettering on signs and instructions is required and should be a minimum letter height of 2 inches. Labels on containers are required and must not be less than one-half the size of the largest lettering on the package and not less than 8-point type in any instance, and in some cases such required lettering must be more than 1 inch in height. No statement shall appear on or near any required sign, label, or instruction that contradicts, or detracts from, the effect of any required warning, information, or instruction.

Each employee, prior to being authorized to enter a regulated area, must receive training and an indoctrination program including, but not necessarily limited to, the nature of the carcinogenic hazards of a carcinogen addressed by this CFR section. Training and indoctrination must include local and systemic toxicity; the specific nature of the operation involving a carcinogen that could result in exposure; and the purpose for and application of the medical surveillance program, including, as appropriate, methods of self-examination. Every employee must learn the purpose for and application of decontamination practices, the purpose for and significance of emergency practices and procedures, and the employee's own role in emergency procedures. Specific information must also be taught to aid the employee in recognizing and evaluating conditions and situations that may result in the release of a carcinogen addressed by this CFR section. It should also be required that the employee know the purpose for and application of specific first aid procedures and practices. A review of this CFR section covering the employee's first training and indoctrination program should be conducted annually thereafter.

Specific emergency procedures shall be prescribed and posted. Employees must familiarize themselves with their terms and rehearse their application. All materials relating to the program must be provided upon request to authorized representatives of the Assistant Secretary and the Director of OSHA.

Any changes in such operations, or other information, must be reported in writing within 15 calendar days of such change and include

1. A brief description and in-plant location of the area(s) regulated and the address of each regulated area
2. The name(s) and other identifying information as to the presence of a carcinogen addressed by this section in each regulated area
3. The number of employees in each regulated area during normal operations, including maintenance activities
4. The manner in which carcinogens addressed by this section are present in each regulated area; for example, whether it is manufactured, processed, used, repackaged, released, stored, or otherwise handled

Incidents that result in the release of a carcinogen into any area where employees may be potentially exposed must be reported. A report of the occurrence of the incident and the facts obtainable at that time, including a report on any medical treatment of affected employees, shall be made within 24 hours to the nearest OSHA

Area Director. A written report shall be filed with the nearest OSHA Area Director within 15 calendar days thereafter and shall include

1. A specification of the amount of material released, the amount of time involved, and an explanation of the procedure used in determining this figure
2. A description of the area involved, and the extent of known and possible employee exposure and the known and possible areas of contamination
3. A report covering of any medical treatment of affected employees and any medical surveillance program implemented
4. An analysis of the circumstances of the incident and measures taken or to be taken, with specific completion dates, to avoid further similar releases

At no cost to the employee, a program of medical surveillance must be established and implemented to cover both employees considered for assignment to enter regulated areas, and for authorized employees. Before an employee is assigned to enter a regulated area, a physical examination by a physician must be provided. The examination must include the personal history of the employee, family and occupational background, and genetic and environmental factors. Authorized employees shall be provided with periodic physical examinations, no less often than annually, following the pre-assignment examination.

In all physical examinations, the examining physician should consider whether there exist conditions of increased risk. These conditions may include reduced immunological competence, additional risk of those undergoing treatment with steroids or cytotoxic agents, pregnancy, and cigarette smoking.

Employers must maintain complete and accurate records of all such medical examinations. Records shall be maintained for the duration of the employee's employment. Upon termination of the employee's employment, including retirement or death, or in the event that the employer ceases business without a successor, records, or notarized true copies thereof, shall be forwarded by registered mail to the Director of NIOSH. Records shall be provided upon request to the employee, his or her designated representatives, and the Assistant Secretary of OSHA. Any physician who conducts a medical examination is to furnish to the employer a statement of the employee's suitability for employment in the specific area with an exposure potential.

The specific nature of the previous requirements is an indicator of the danger presented by exposure to, or working with, carcinogens that are regulated by the OSHA. There are other carcinogens, which OSHA regulates, that are not part of the original thirteen. They include the following:

- Vinyl chloride (1910.1017)
- Inorganic arsenic (1910.1018)
- Cadmium (1910.1027 and 1926.1127)
- Benzene (1910.1028)
- Coke oven emissions (1910.1029)
- 1,2-Dibromo-3-chloropropane (1910.1044)
- Acrylonitrile (1910.1045)
- Ethylene oxide (1910.1047)

- Formaldehyde (1910.1048)
- Methylenedianiline (1910.1050)
- 1,3-Butadiene (1910.1051)
- Methylene chloride (1910.1052)

Recently, OSHA has reduced the PEL for methylene chloride from 400 ppm to 25 ppm. This is a huge reduction in the PEL, equal to a fifteen-times reduction that a worker can face as a potential exposure. This reduction indicates the potential of methylene chloride to cause cancer in workers and should raise the flag that chemicals that are believed to cause cancer are not to be taken lightly. Information and research are continuously evolving and providing new insight into the dangers of these chemicals and agents.

COLD STRESS

Temperature is measured in degrees Fahrenheit (°F) or Celsius (°C). Most people feel comfortable when the air temperature ranges from 66°F to 79°F and the relative humidity is about 45 percent. Under these circumstances, heat production inside the body equals the heat loss from the body, and the internal body temperature is kept around 98.6°F. For constant body temperature, even under changing environmental conditions, rates of heat gain and heat loss should balance. Every living organism produces heat. In cold weather the only source of heat gain is the body's own internal heat production, which increases with physical activity. Hot drinks and food are also a source of heat.

The body loses heat to its surroundings in several different ways. Heat loss is greatest if the body is in direct contact with cold water. The body can lose twenty-five to thirty times more heat when in contact with cold, wet objects than under dry conditions or with dry clothing. The higher the temperature differences between the body surface and cold objects, the faster the heat loss. Heat is also lost from the skin by contact with cold air. The rate of loss depends on the air speed and the temperature difference between the skin and the surrounding air. At a given air temperature, heat loss increases with air speed. Sweat production and its evaporation from the skin also cause heat loss. This is important when performing hard work.

Small amounts of heat are lost when cold food and drink are consumed. As well, heat is lost during breathing by inhaling cold air, and through evaporation of water from the lungs.

The body maintains heat balance by reducing the amount of blood circulating through the skin and outer body parts. This minimizes cooling of the blood by shrinking the diameter of blood vessels. At extremely low temperatures, loss of blood flow to the extremities may cause an excessive drop in tissue temperature, resulting in damage such as frostbite. Shivering, which increases the body's heat production, provides a temporary tolerance for cold but cannot be maintained for long periods of time.

Overexposure to cold causes discomfort and a variety of health problems. Cold stress impairs performance of both manual and complex mental tasks. Sensitivity and dexterity of fingers lessen in cold. At lower temperatures still, cold affects deeper muscles, resulting in reduced muscular strength and stiffened joints. Mental alertness is reduced due to cold-related discomfort. For all these reasons, accidents are more likely to occur in very cold working conditions.

The main cold injuries are frostnip, frostbite, immersion foot, and trenchfoot, which occur in localized areas of the body. *Frostnip* is the mildest form of cold injury. It occurs when ear lobes, noses, cheeks, fingers, or toes are exposed to cold. The skin of the affected area turns white. Frostnip can be prevented by warm clothing and is treated by simple rewarming.

Frostbite is a common injury caused by exposure to extreme cold or contact with extremely cold objects. It occurs when the tissue temperature falls below the freezing point. Blood vessels may be severely and irreparably damaged, and blood circulation may stop in the affected tissue. In mild cases, the symptoms include inflammation of the skin in patches, accompanied by slight pain. In severe cases, there could be tissue damage without pain, or there could be burning or prickling sensations resulting in blisters. Frostbitten skin is highly susceptible to infection, and therefore gangrene is possible. In cases of suspected frostbite, the body should be slowly warmed to normal temperature. Frostbitten limbs should be warmed first in cold water at 50°F to 60°F. The water temperature should increase by 10°F every five minutes to a maximum of 104°F. Complete recovery takes several days. Residual effects such as cold feet, numbness, abnormal skin color, and pain in joints may continue for several days, especially in winter.

Immersion foot occurs in individuals whose feet have been wet, but not freezing cold, for days or weeks. The primary injury is to nerve and muscle tissue. Symptoms are numbness, swelling, or even superficial gangrene. Trenchfoot is "wet cold disease" resulting from exposure to moisture at or near the freezing point for one to several days. Symptoms are similar to immersion foot (swelling and tissue damage).

Hypothermia can occur in moderately cold environments; the body's core temperature does not usually fall more than 2°F to 3°F below the normal 98.6°F because of the body's ability to adapt. However, in intense cold without adequate clothing, the body is unable to compensate for the heat loss, and the body's core temperature starts to fall. The sensation of cold, followed by pain, in exposed parts of the body is the first sign of cold stress. The most dangerous situation occurs when the body is immersed in cold water. As the cold worsens or the exposure time increases, the feeling of cold and pain starts to diminish because of increasing numbness (loss of sensation). If no pain can be felt, serious injury can occur without the victim noticing. Next, muscular weakness and drowsiness are experienced. This condition is called *hypothermia* and usually occurs when body temperature falls below 92°F. Additional symptoms of hypothermia include interruption of shivering, diminished consciousness, and dilated pupils. When body temperature reaches 80°F, coma (profound unconsciousness) sets in. Heart activity stops around 60°F and the brain stops functioning around 63°F. The hypothermia victim should be immediately warmed, either by being moved to a warm room or by the use of blankets. Rewarming in water at 104°F to 108°F has been recommended in cases where hypothermia occurs after the body was immersed in cold water.

Susceptibility to cold injury varies from person to person. In general, people in good physical health are less susceptible to cold injury. Conditions that worsen the risk of cold injury include the following:

- Sex
- Old age
- Diseases of the blood circulation system
- Previous cold injury
- Raynaud's disease
- Fatigue
- Consumption of alcohol or nicotine
- Use of certain drugs or medication
- Injuries resulting in blood loss or altered blood flow

Although people easily adapt to hot environments, they do not acclimatize well to cold. However, frequently exposed body parts can develop some degree of tolerance to cold. Blood flow in the hands, for example, is maintained in conditions that would cause extreme discomfort and loss of dexterity in unacclimatized persons. This is noticeable among fishermen who are able to work with bare hands in extremely cold weather.

Workplace conditions that influence cold stress are the severity of cold stress, which depends on the air temperature, wind speed, and intensity of physical activity. Air temperature is measured by an ordinary thermometer in degrees Fahrenheit or Celsius. Different types of commercially available anemometers are used to measure wind speed or air movement. These are calibrated in miles per hour (mph). The following is a suggested guide for estimating wind speed, if accurate information is not available:

- Light flag moves (5 mph)
- Light flag fully extended (10 mph)
- Raises newspaper sheet (15 mph)
- Causes blowing and drifting snow (20 mph)

The production of body heat by physical activity (metabolic rate) is difficult to measure. At any temperature, one feels colder as the wind speed increases. The combined effect of cold air and wind speed is expressed as wind-chill temperature in degrees Fahrenheit. This is essentially the air temperature that would produce the same cooling effect on exposed human flesh as the given combination of air temperature and wind speed (see Table 23.1).

In the United States there are no OSHA exposure limits for cold working environments. It is often recommended that work warm-up schedules be developed. In most normal cold conditions, a warm-up break every 2 hours is recommended, but, as temperatures and wind increase, more warm-up breaks are needed.

For continuous work in temperatures below the freezing point, heated warming shelters such as tents, cabins, and rest rooms should be made available. The pace of work should not be so heavy as to cause excessive sweating. If such work is

TABLE 23.1
Windchill Factor: Cooling Power of Wind on Exposed Flesh Expressed as an Equivalent Temperature (under calm conditions)

Estimated Wind Speed (mph)	Actual Thermometer Reading (°F)											
	50	40	30	20	10	0	−10	−20	−30	−40	−50	−60
	Equivalent Temperature Reading (°F)											
Calm	50	40	30	20	10	0	−10	−20	−30	−40	−50	−60
5	48	37	27	16	6	−5	−15	−26	−36	−47	−57	−68
10	40	28	16	4	−9	−24	−33	−46	−58	−70	−83	−95
15	36	22	9	−5	−18	−32	−45	−58	−72	−85	−99	−112
20	32	18	4	−10	−25	−39	−53	−67	−82	−96	−110	−124
25	30	16	0	−15	−29	−44	−59	−74	−88	−104	−118	−133
30	28	13	−2	−18	−33	−48	−63	−79	−94	−109	−125	−140
35	27	11	−4	−20	−35	−51	−67	−82	−98	−113	−129	−145
40	26	10	−6	−21	−37	−53	−69	−85	−100	−116	−132	−148
Wind speeds greater than 40 mph have little added effect.	Little danger (for properly clothed person). Maximum danger of false sense of security.				Increasing danger. Danger from freezing of exposed flesh.			Great danger.				

Trenchfoot and immersion foot may occur at any point on this chart.

Source: NAVMED Bulletin 5052-29.

necessary, proper rest periods in a warm area should be allowed for changing into dry clothes. New employees should be given enough time to become accustomed to cold and protective clothing before assuming a full workload.

The risk of cold injury can be minimized by proper equipment design, protective clothing, and safe work practices. For work performed below the freezing point, metal handles and bars should be covered by thermal insulating material. Also, machines and tools should be designed so that they can be operated without having to remove gloves.

Protective clothing is needed for work at or below 40°F. Clothing should be selected to suit the degree of cold, level of activity, and job design. Clothing should be worn in multiple layers, which provides better protection than a single thick garment. The layer of air between clothing provides better insulation than the clothing itself. In extremely cold conditions, where face protection is used, eye protection must be separated from respiratory channels (nose and mouth) to prevent exhaled moisture from fogging and frosting eye shields.

For work in wet conditions, the outer layer of clothing should be waterproof. If the work area cannot be shielded against wind, an easily removable windbreaker garment should be used. Under extremely cold conditions, heated protective clothing should be made available. Clothing should be kept clean, as dirt fills air cells in fibers of clothing and destroys its insulating ability. Clothing must be dry. Moisture should be kept off clothes by removing snow prior to entering heated shelters. While the worker is resting in a heated area, perspiration should be allowed to escape by opening the neck, waist, sleeves, and ankle fasteners.

If fine manual dexterity is not required, gloves should be used below 40°F for light work and below 20°F for moderate work. For work below 0°F, mittens should be used. Felt-lined, rubber-bottomed, leather-topped boots with removable felt insoles are best suited for heavy work in the cold. Almost 50 percent of body heat is lost through the head. A wool-knit cap or a liner under a hard hat reduces excessive heat loss.

The contact of bare skin with cold surfaces (especially metallic) below 20°F should be avoided, as should skin contact when handling evaporative liquids (e.g., gasoline, alcohol, and cleaning fluids) below 38°F. Sitting or standing still for prolonged periods should be avoided. Alcohol should not be consumed in cold environments, as it causes rapid loss of body heat and thus increases the risk of hypothermia. For warming purposes, hot nonalcoholic beverages or soups are suggested. Coffee drinking should be limited because it increases urine production and blood circulation. Both effects result in an increased loss of body heat. Balanced meals and adequate liquid intake are essential to the production of body heat and the prevention of dehydration.

Because work in extreme cold presents a potential health hazard, workers and supervisors involved with work in cold environments should be well informed. Each must know about the symptoms of cold strain and cold injury, measures of environmental conditions, proper clothing habits, safe work practices, physical fitness requirements for work in the cold, and emergency procedures in case of cold injury.

The feeling of cold is a combined effect of air temperature and wind speed. The effects of cold increase as wind speed increases. Uncomfortably cold working conditions can lead to reduced work efficiency and higher accident rates. Cooling of body parts may result in cold injuries such as frostbite, frostnip, and trenchfoot. Extremities can become frostbitten, even while the rest of the body remains sufficiently warm, because the blood supply is relatively poor in these body parts. The most severe cold injury is hypothermia, which occurs as a result of excessive loss of body heat and the consequent lowering of the inner core temperature. If immediate medical attention is not provided, hypothermia can be fatal.

As stated before, hypothermia and other cold injuries can be prevented by wearing sufficient winter clothing, including head protection, gloves or mittens, and insulated boots. It is important that all clothing remain dry. Loose, layered clothing provides better protection from cold because of the added insulating properties of air trapped within the layers. Periodic exercise helps keep the extremities warm. However, once limbs are frostbitten, exercise can result in increased damage. The preferred treatment is slow warming of affected parts. Workers should be educated to identify signs of cold injuries and instructed in methods of providing emergency care to affected co-workers.

ERGONOMICS

Ergonomics is the science of fitting the job to the worker, not fitting the worker to the job. When there is a mismatch between the physical requirements of the job and the physical capacity of the worker, work-related musculoskeletal disorders (WMSDs) can result. Workers who must repeat the same motion throughout their workday; who must do their work in an awkward position, use a great deal of force to perform their jobs, or repeatedly lift heavy objects; or who face a combination of these risk factors are most likely to develop WMSDs.

In 1996, U.S. workers experienced more than 647,000 lost workdays due to WMSDs. WMSDs now account for 34 percent of all lost workday injuries and illnesses. These injuries cost businesses $15 to $20 billion in workers' compensation costs each year. Indirect costs may run as high as $45 to $60 billion.

Workers who experience WMSDs may be unable to perform their jobs or even simple household tasks. WMSDs represent real workplace problems faced by real people. The scientific basis for the relationship between work and development of WMSDs and for addressing ergonomic problems in the workplace is well established.

Real solutions have been demonstrated in workplaces of all sizes across a broad range of industries. Many employers have developed effective ergonomic programs and common-sense solutions to address WMSDs in their workplaces. Often, WMSDs can be prevented by simple and inexpensive changes in the workplace. Adjusting the height of working surfaces, varying tasks for workers, and encouraging short rest breaks can reduce risks. Reducing the size of items that workers must lift or providing lifting equipment also may aid workers. Specially designed equipment, such as curved knives for poultry processors, may help.

Although OSHA has been developing a regulation that calls for employers to establish ergonomic programs to prevent WMSDs, it has not been promulgated. The agency has pledged to focus on jobs where injuries are high and solutions well demonstrated. In consultation with stakeholders, OSHA has identified significant problems for workers involved in production operations in manufacturing and manual handling throughout general industry. Job-related musculoskeletal disorders also occur in other jobs. At a minimum, employers that have workers experiencing injuries need to address the problem.

One size does not fit all. That is why OSHA would like to see a tailored program approach. That is also why no one will ever be able to say that X number of repetitions or lifting X pounds will result in injury, or conversely that Y number of repetitions or Y pounds will definitely *not* result in injury for anyone, any time, anywhere. However, many employers have proven that establishing a systematic program to address such issues as repetition, excessive force, awkward postures, and heavy lifting results in fewer injuries to workers.

OSHA has identified the following critical elements: management leadership and employee participation, hazard identification and information, job hazard analysis and control, employee training, medical management, and program evaluation as effective components of an ergonomic program. The keys to success are simple: reduce repeated motions, forceful hand exertions, and prolonged bending or working above shoulder height (see Figure 23.5). Reduce vibration. Rely on equipment (not backs) for heavy or repetitive lifting. Provide "micro" breaks to allow muscles to recover.

FIGURE 23.5 Worker performing task above shoulder level.

HAZARDOUS CHEMICALS

Hazardous and toxic substances can be defined as those chemicals present in the workplace that are capable of causing harm (see Figure 23.6). In this definition, the term "chemicals" includes dusts, mixtures, and common materials such as paints, fuels, and solvents. OSHA currently regulates exposure to approximately 400 substances. The OSHA Chemical Sampling Information file contains a listing for approximately 1,500 substances. The EPA's Toxic Substance Chemical Act: Chemical Substances Inventory lists information on more than 62,000 chemicals or chemical substances. Some libraries maintain files of Material Safety Data Sheets (MSDSs) for more than 100,000 substances. It is not possible at this time to address or regulate the hazards associated with each of these chemicals that can potentially be found in the workplace.

Because there is no evaluation instrument that can identify the chemical or the amount of chemical contaminant present, it is not possible to make a real-time assessment of a worker's exposure to potentially hazardous chemicals. If this is not enough of a problem, the Threshold Limit Values (TLVs) provided by the American Conference of Governmental Industrial Hygienist (ACGIH) in 1968 is the basis of OSHA's Permissible Exposure Limits (PELs). In the early 2000s, workers are being provided protection with chemical exposure standards that are more than 40 years old. The ACGIH regularly updates and changes its TLVs based upon new

FIGURE 23.6 Hazardous chemicals can be found on most worksites.

scientific information and research, but OSHA must do a cost-benefit analysis for each chemical in order to upgrade the PEL.

The U.S. Environmental Protection Agency allows for one death or one cancer case per million people exposed to a hazardous chemical. Certainly the public needs these kinds of protections. Using the existing OSHA PELs risk factor is only as protective as one death due to exposure in 1,000 workers. This indicates that there exists a "fence line mentality" that suggests that workers can tolerate higher exposures than the general public is expected to endure. As one illustration of this, the exposure to sulfur dioxide for the public is set by the EPA at 0.14 ppm average over 24 hours, while the OSHA PEL is 5 ppm average over 8 hours. Certainly, there is a wide margin between what the public can be subjected to and what a worker is supposed to be able to tolerate. The question is this: Is there a difference between humans in the public arena and those in the work arena? Maybe workers are assumed to be more immune to the effects of chemicals when they are in the workplace than when they are at home because of workplace regulations and precautions.

A more significant issue is that regarding mixtures. The information does not exist to show the risk of illnesses, long-term illnesses, or the toxicity of combining these hazardous chemicals. At present, it is assumed that the most dangerous chemical of the mixture has the most potential to cause serious health-related problems, then the next most hazardous, and so on. However, little consideration is given regarding the increase in toxicity, long-term health problems, or present hazards. Because most chemicals used in industry are mixtures formulated by manufacturers, it makes it even more critical to have access to the MSDSs and take a conservative approach to the potential for exposure. This means that any signs or symptoms of exposure should be addressed immediately; worker complaints should be addressed with sincerity and true concern; and employers should take precautions beyond those called for by MSDSs if questions prevail.

Actually, the amount of information that exists on dose/response for chemicals and chemical mixtures is limited. This is especially true for long-range or latency effects. If a chemical kills or makes a person sick within minutes or hours, the dose/response is easily understood. However, if chemical exposure over a long period of time results in an individual's death or illness, then the dose needed to do this is, at best, a guess. It most certainly does not take into account other chemicals the worker was exposed to during his or her work life and whether they exacerbated the effects or played no role in the individual's death or illness. This is why it is critical for individual workers to keep their exposure to chemicals as low as possible. Even then, there are no guarantees that they may not come down with an occupational disease related to chemical exposure.

Many employers and workers as well as physicians are not quick or trained to identify the symptoms of occupational exposure to chemicals. In one case, two men painted for 8 hours with a paint containing 2-nitropropane in an enclosed environment. At the end of their shift, one of the workers felt ill and stopped at the emergency center at the hospital. After examination, he was told to go home and rest and would probably be better in the morning. Later that evening, he returned to the hospital and died of liver failure from 2-nitropropane exposure. The other worker suffered irreparable liver damage but survived. No one had asked the right questions regarding workers' occupational exposure. The symptoms were probably similar to a common cold or flu, which is often the case unless some detective work is done. Often, those with chemical poisoning go home and off-gas or excrete the contaminant during the 16 hours when they have no exposure. They feel better the next day and return to work and are re-exposed. Thus, the worker does not truly recognize this as a poisoning process. Being aware of the chemicals used, reviewing the MSDSs, and following the precautions recommended that are critical to the safe use of hazardous chemicals are important.

With this point made, it becomes critical that employers know the dangers of the chemicals being used present to their workforce. Employers need to obtain and review the MSDSs for all chemicals in use on their worksite and take the proper precautions recommended by those MSDSs. Also, it behooves workers to get copies of and review the MSDSs for chemicals that they use.

HAZARDOUS WASTE

The term "hazardous waste" indicates, by its definition, that it is no longer a useful product. It also places it in the class of hazardous chemicals. In the past, these types of chemicals were dumped down drains, scattered on the soil, or buried. Because of

FIGURE 23.7 Workers must be fully protected when dealing with hazardous waste.

the pollution and potential health effects these procedures cause, these procedures have become unacceptable.

Today, hazardous waste must be collected, placed in containers, and transported to a storage or treatment facility for proper disposal. Individuals who must remove and work with hazardous waste must be trained in accordance with the Hazardous Waste Operation and Emergency Response (29 CFR 1910.120) regulation (see Figure 23.7). This regulation addresses the requirements and procedures for the remediation of a hazardous waste site that ensures both protection of the public and the workforce. Approximately 140 chemicals have been identified as the most common chemicals found on hazardous waste sites by the Agency for Toxic Substances and Disease Registry (ATSDR).

A special U.S. Department of Transportation Hazardous Waste Manifest is needed to transport these hazardous chemicals across the nation's highways. Drivers of such vehicles must possess a Commercial Drivers License (CDL) and be trained in the safe transport of hazardous waste.

Once the hazardous waste arrives at a disposal facility, it may be stored, buried, or incinerated. How it will be disposed of depends on the degree of hazard presented by the waste. Hazardous materials such as polychlorinate biphenyls (PCBs) are usually incinerated. The PCBs must be destroyed to a level of six 9s or 99.9999 percent burned to ensure that little or no toxic residue survives. Other materials such as asbestos, which is a regulated waste, can be buried because it is very unlikely due to it stability to get into groundwater and spread.

The United States is a chemical-using country, and much has been done to minimize the amount of chemicals being used. However, it is not feasible to do away with all the chemicals that are needed in our industrial environment. Thus, it is necessary to ensure that proper procedures are in place to handle hazardous waste and prevent further contamination, illnesses of the public and workers due to exposure, and organize long-range protective programs to prevent events (e.g., spills, etc.) that could cause catastrophic health problems.

HEAT STRESS

Operations involving high air temperatures, radiant heat sources, high humidity, direct physical contact with hot objects, or strenuous physical activities have a high potential for inducing heat stress in employees engaged in such operations. Such places include iron and steel foundries, nonferrous foundries, brick-firing and ceramic plants, glass products facilities, rubber products factories, electrical utilities (particularly boiler rooms), bakeries, confectioneries, commercial kitchens, laundries, food canneries, chemical plants, mining sites, smelters, and steam tunnels. Outdoor operations conducted in hot weather, such as construction, refining, asbestos removal, and hazardous waste site activities, especially those that require workers to wear semipermeable or impermeable protective clothing, are also likely to cause heat stress among exposed workers.

Age, weight, degree of physical fitness, degree of acclimatization, metabolism, use of alcohol or drugs, and a variety of medical conditions, such as hypertension, all affect a person's sensitivity to heat. However, even the type of clothing worn must be considered. Prior heat injury predisposes an individual to additional injury. It is difficult to predict just who will be affected and when, because individual susceptibility varies. In addition, environmental factors include more than the ambient air temperature. Radiant heat, air movement, conduction, and relative humidity all affect an individual's response to heat.

There is no OSHA regulation for heat stress. The American Conference of Governmental Industrial Hygienists (1992) has stated that workers should not be permitted to work when their deep body temperature exceeds 100.4°F.

Heat is a measure of energy in terms of quantity. A calorie is the amount of heat required to raise 1 gram of water 1°C (based on a standard temperature of 61.7°C to 63.5°C). Conduction is the transfer of heat between materials that contact each other. Heat passes from the warmer material to the cooler material. For example, a worker's skin can transfer heat to a contacting surface if that surface is cooler, and vice versa. Convection is the transfer of heat in a moving fluid. Air flowing past the body can cool the body if the air temperature is cool. On the other hand, air that exceeds 95°F can increase the heat load on the body. Evaporative cooling takes place when sweat evaporates from the skin. High humidity reduces the rate of evaporation and thus reduces the effectiveness of the body's primary cooling mechanism. Radiation is the transfer of heat energy through space. A worker whose body temperature is greater than the temperature of the surrounding surfaces radiates heat to these surfaces. Hot surfaces and infrared light sources radiate heat that can increase the body's heat load.

The main complications that transpire when workers suffer from heat exposure are

1. *Heat stroke*, which occurs when the body's system of temperature regulation fails and body temperature rises to critical levels. This condition is caused by a combination of highly variable factors, and its occurrence

is difficult to predict. Heat stroke is a medical emergency. The primary signs and symptoms of heat stroke are confusion; irrational behavior; loss of consciousness; convulsions; a lack of sweating (usually); hot, dry skin; and an abnormally high body temperature (e.g., a rectal temperature of 105.8°F). If body temperature is too high, it causes death. Elevated metabolic temperatures caused by a combination of workload and environmental heat load, both of which contribute to heat stroke, are highly variable and difficult to predict.

If a worker shows signs of possible heat stroke, professional medical treatment should be obtained immediately. The worker should be placed in a shady area and the outer clothing should be removed. The worker's skin should be wetted and air movement around the worker should be increased to improve evaporative cooling until professional methods of cooling are initiated and the seriousness of the condition can be assessed. Fluids should be replaced as soon as possible. The medical outcome of an episode of heat stroke depends on the victim's physical fitness along with the timing and effectiveness of first aid treatment. Regardless of the worker's protests, no employee suspected of being ill from heat stroke should be sent home or left unattended unless a physician has specifically approved such an order.

2. The signs and symptoms of *heat exhaustion* are headache, nausea, vertigo, weakness, thirst, and giddiness. Fortunately, this condition responds readily to prompt treatment. Heat exhaustion should not be dismissed lightly for several reasons. One is that the fainting associated with heat exhaustion can be dangerous because the victim may be operating machinery or controlling an operation that should not be left unattended; moreover, the victim may be injured when he or she faints. Also, the signs and symptoms seen in heat exhaustion are similar to those of heat stroke. It is a medical emergency. Workers suffering from heat exhaustion should be removed from the hot environment and given fluid replacement. They should also be encouraged to get adequate rest.
3. *Heat cramps* are usually caused by performing hard physical labor in a hot environment. These cramps have been attributed to an electrolyte imbalance caused by sweating. It is important to understand that cramps can be caused by both too much and too little salt. Cramps appear to be caused by the lack of water replenishment. Because sweat is a hypotonic solution (±0.3 percent NaCl), excess salt can build up in the body if the water lost through sweating is not replaced. Thirst cannot be relied on as a guide to the need for water; instead, water must be taken every 15 to 20 minutes in hot environments. Under extreme conditions, such as working for 6 to 8 hours in heavy protective gear, a loss of sodium may occur. Recent studies have shown that drinking commercially available carbohydrate-electrolyte replacement liquids is effective in minimizing physiological disturbances during recovery.
4. In *heat collapse* (fainting), the brain does not receive enough oxygen because blood pools in the extremities. As a result, the exposed individual may lose consciousness. This reaction is similar to that of heat exhaustion and does not affect the body's heat balance. However, the onset of heat

collapse is rapid and unpredictable. To prevent heat collapse, the worker should gradually become acclimatized to the hot environment.

5. *Heat rashes* are the most common problem in hot work environments. Prickly heat manifests as red papules and usually appears in areas where the clothing is restrictive. As sweating increases, these papules give rise to a prickling sensation. Prickly heat occurs on skin that is persistently wetted by unevaporated sweat; and heat rash papules may become infected if they are not treated. In most cases, heat rashes will disappear when the affected individual returns to a cool environment.
6. A factor that predisposes an individual to *heat fatigue* is lack of acclimatization. The use of a program of acclimatization and training for work in hot environments is advisable. The signs and symptoms of heat fatigue include impaired performance of skilled sensorimotor, mental, or vigilance jobs. There is no treatment for heat fatigue except to remove the heat stress before a more serious heat-related condition develops.

Workload assessments should be made under conditions of high temperature and heavy workload. The workload category is determined by averaging metabolic rates in kilocalories used per hour (kcal/hr) for the tasks and then ranking them:

- Light work: up to 200 kcal/hr
- Medium work: 200–350 kcal/hr
- Heavy work: 350–500 kcal/hr

Some examples of activities and the workload involved are as follows:

- Light handwork: writing, hand knitting
- Heavy handwork: typewriting
- Heavy work with one arm: hammering in nails (shoemaker, upholsterer)
- Light work with two arms: filing metal, planing wood, and raking a garden
- Moderate work with the body: cleaning a floor, beating a carpet
- Heavy work with the body: railroad track laying, digging, debarking trees

The most commonly used temperature index is the Wet Bulb Globe Temperature (WBGT). The WBGT reading should be for continuous all-day or several hours of exposures and then should be averaged over a 60-minute period. Intermittent exposures should be averaged over a 120-minute period. These averages should be calculated using the following formulae:

1. For indoor and outdoor conditions with no solar load, WBGT is calculated as

$$\text{WBGT} = 0.7\ \text{NWB} + 0.3\ \text{GT}$$

2. For outdoors with a solar load, WBGT is calculated as

$$\text{WBGT} = 0.7\ \text{NWB} + 0.2\ \text{GT} + 0.1\ \text{DB}$$

where WBGT = Wet Bulb Globe Temperature Index, NWB = Nature Wet-Bulb Temperature, DB = Dry-Bulb Temperature, and GT = Globe Temperature.

Exposure limits are valid for employees wearing light clothing. They must be adjusted for the insulation from clothing that impedes sweat evaporation and other body cooling mechanisms. Ventilation, air cooling, fans, shielding, and insulation are the five major types of engineering controls used to reduce heat stress in hot work environments. Heat reduction can also be achieved using power assists and tools that reduce the physical demands placed on a worker. However, for this approach to be successful, the metabolic effort required for the worker to use or operate these devices must be less than the effort required without them. Another method is to reduce the effort necessary to operate power assists. The worker should be allowed to take frequent rest breaks in a cooler environment.

The human body can adapt to heat exposure to some extent. This physiological adaptation is called acclimatization. After a period of acclimatization, the same activity will produce fewer cardiovascular demands. The worker will sweat more efficiently (causing better evaporative cooling), and thus will more easily be able to maintain normal body temperatures. A properly designed and applied acclimatization program decreases the risk of heat-related illnesses. Such a program basically involves exposing employees to work in a hot environment for progressively longer periods. NIOSH (The National Institute for Occupational Safety and Health) stated that, for workers who have had previous experience in jobs where heat levels are high enough to produce heat stress, the regimen should be 50 percent exposure on day 1, 60 percent on day 2, 80 percent on day 3, and 100 percent on day 4. For new workers who will be similarly exposed, the regimen should be 20 percent on day 1, with a 20 percent increase in exposure each additional day.

Cool (50°F to 60°F) water or any cool liquid (except alcoholic beverages) should be made available to workers to encourage them to drink small amounts frequently (e.g., one cup every twenty minutes). Ample supplies of liquids should be placed close to the work area. Although some commercial replacement drinks contain salt, this is not necessary for acclimatized individuals because most people add enough salt to their summer diets.

Engineering controls, such as general ventilation, are used to dilute hot air with cooler air (generally cooler air that is brought in from the outside). This technique clearly works better in cooler climates than in hot ones. A permanently installed ventilation system usually handles large areas or entire buildings. Portable or local exhaust systems may be more effective or practical in smaller areas. Air treatment/air cooling differs from ventilation because it reduces the air temperature of the air by removing heat (and sometimes humidity) from the air. Air conditioning is a method of air cooling but it is expensive to install and operate. An alternative to air conditioning is the use of chillers to circulate cool water through heat exchangers, over which air from the ventilation system is then passed; chillers are more efficient in cooler climates or in dry climates where evaporative cooling can be used. Local air cooling can be effective in reducing air temperature in specific areas. Two methods have been used successfully in industrial settings. One type, cool rooms, can be used to enclose a specific workplace or to offer a recovery area near hot jobs. The second type is a portable blower with a built-in air chiller. The main advantage of a blower, aside from portability, is minimal set-up time. Another way to reduce heat stress is to increase airflow or convection using fans, etc. in the work area (as long as the air temperature is less than the worker's skin temperature). Changes in air speed can help workers stay cooler by increasing both the convective heat exchange (the exchange between the skin surface and the surrounding air) and the rate of evaporation. Because this method does not actually cool the air, any increase in air speed must impact the worker directly to be effective.

Heat conduction methods include insulating the hot surface that generates the heat and changing the surface itself. Simple engineering controls, such as shields, can be used to reduce radiant heat (i.e., heat coming from hot surfaces within the worker's line of sight). Surfaces that exceed 95°F are sources of infrared radiation that can add to the worker's heat load. Flat black surfaces absorb heat more than smooth, polished ones. Having cooler surfaces surrounding the worker assists in cooling because the worker's body radiates heat toward those surfaces. With some sources of heat radiation, such as heating pipes, it is possible to use both insulation and surface modifications to achieve a substantial reduction in radiant heat. Instead of reducing heat radiation from the source, shielding can be used to interrupt the path between the source and the worker. Polished surfaces make the best barriers, although special glass or metal mesh surfaces can be used if visibility is a problem. Shields should be located so that they do not interfere with air flow, unless they are also being used to reduce convective heating. The reflective surface of the shield should be kept clean to maintain its effectiveness.

Administrative controls, such as training, are the key to good work practices. Unless all employees understand the reasons for using new or changing old work practices, the chances of such a program succeeding are greatly reduced. NIOSH (1986) states that a good heat stress training program should include at least the following components:

1. Knowledge of the hazards of heat stress
2. Recognition of predisposing factors, danger signs, and symptoms
3. Awareness of first aid procedures for, and the potential health effects of, heat stroke
4. Employee's responsibilities in avoiding heat stress
5. Dangers of using drugs, including therapeutic ones, and alcohol in hot work environments
6. Use of protective clothing and equipment
7. Purpose and coverage of environmental and medical surveillance programs and the advantages of worker participation in such programs

Hot jobs should be scheduled for the cooler part of the day, and routine maintenance and repair work in hot areas should be scheduled for the cooler seasons of the year. Other administrative controls are to reduce the physical demands of work. For example, reduce the amount of excessive lifting or digging with heavy objects; provide recovery areas, such as air-conditioned enclosures and rooms; use shifts, such as early morning, cool part of the day, or night work; use intermittent rest periods with water breaks; use relief workers; use worker pacing; assign extra workers; and limit worker occupancy, or the number of workers present, especially in confined or enclosed spaces.

Some types of personal protective clothing can be employed, such as reflective clothing, which can vary from aprons and jackets to suits that completely enclose the worker from neck to feet. Such clothing can stop the skin from absorbing radiant heat. However, because most reflective clothing does not allow air exchange through the garment, the reduction in radiant heat must more than offset the corresponding loss in evaporative cooling. For this reason, reflective clothing should be worn as loosely as possible. In situations where radiant heat is high, auxiliary cooling systems can be used under the reflective clothing.

Also, commercially available ice vests, although heavy, may accommodate as many as seventy-two ice packets, which are usually filled with water. Carbon dioxide (dry ice) can also be used as a coolant. The cooling offered by ice packets lasts only 2 to 4 hours at moderate to heavy heat loads, and frequent replacement is necessary. However, ice vests should not encumber the worker and thus permit maximum mobility. Cooling with ice is also relatively inexpensive.

Wetted clothing is another simple and inexpensive personal cooling technique. It is effective when reflective or other impermeable protective clothing is worn. The clothing may be wetted terry cloth coveralls or wetted two-piece, or whole-body, cotton suits. This approach to auxiliary cooling can be quite effective under conditions of high temperature and low humidity, where evaporation from the wetted garment is not restricted. Water-cooled garments range from a hood, which cools only the head, to vests and long johns, which offer partial or complete body cooling. Use of this equipment requires a battery-driven circulating pump, liquid-ice coolant, and a container. Although this system has the advantage of allowing wearer mobility, the weight of the components limits the amount of ice that can be carried and thus reduces the effective use time. The heat transfer rate in liquid cooling systems may limit their use to low-activity jobs; even in such jobs, their service time is only about 20 minutes per pound of cooling ice. To keep outside heat from melting the ice, an outer insulating jacket should be an integral part of these systems.

Circulating air is the most highly effective, as well as the most complicated, personal cooling system. By directing compressed air around the body from a supplied air system, both evaporative and convective cooling are improved. The greatest advantage occurs when circulating air is used with impermeable garments or double-cotton overalls.

One type, used when respiratory protection is also necessary, forces exhaust air from a supplied-air hood (bubble hood) around the neck and down inside an impermeable suit. The air then escapes through openings in the suit. One problem with this system is the limited mobility of workers whose suits are attached to an air hose. Another is that of getting air to the work area itself. These systems should therefore be used in work areas where workers are not required to move around much or to climb.

The weight of a self-contained breathing apparatus (SCBA) increases stress on a worker, and this stress contributes to overall heat stress. Chemical protective clothing such as totally encapsulating chemical protection suits will also add to the heat stress problem.

Care must be taken to not increase existing heat loads. The major initiative is to reduce the heat stress that exists because the physical problems that workers face due to heat exposure can be deadly.

IONIZING RADIATION

Ionizing radiation has always been a mystery to most people. Actually, much more is known about ionizing radiation than the hazardous chemicals that constantly bombard the workplace. After all, there are only four types of radiation (α-particles, β-particles, γ-rays, and neutrons), whereas there are thousands of chemicals. There are instruments that can detect each type of radiation and provide an accurate dose-received value. This is not so for chemicals, where the best that we could hope for in a real-time situation is a detection of the presence of a chemical and not what the chemical is. With radiation detection instruments, the boundaries of contamination can be detected and set, while detecting such boundaries for chemicals is nearly impossible except for a solid.

It is possible to maintain a lifetime dose for individuals exposed to radiation. Most workers wear a personal dosimeter that provides a check on the levels of exposure. The same is impossible for chemicals where no standard unit of measurement, such as the roentgen equivalent in man (rem), exists for radioactivity in chemicals. The health effects of specific doses are well known: for example, 20 to 50 rems—causes minor changes in blood, 60 to 120 rems—vomiting occurs but no long-term illness, or 5,000 to 10,000 rems—certain death within 48 hours. Certainly radiation can be dangerous, but exposures can usually be controlled by one or a combination of three factors: distance, time, and/or shielding. Certainly distance is the best because the amount of radiation from a source drops off quickly as a factor of the inverse square of the distance; for example, at 8 feet away, the exposure is 1/64 of the radiation emanating from the source. As for time, many radiation workers are only allowed to stay in a radiation area for a limited length of time, and then they must leave that area. Shielding often conjures up lead plating or lead suits (similar to when x-rays are taken by a physician or dentist). Wearing a lead suit may seem appropriate but the weight alone can be prohibitive. Lead shielding can be used to protect workers from γ-rays (similar to x-rays). Once they are emitted, they could pass through anything in their path and continue on their way, unless a lead shield is thick enough to protect the worker.

For β-particles, aluminum foil will stop their penetration. Thus, a protective suit will prevent β-particles from reaching the skin, where they can burn and cause surface contamination. α-Particles can enter the lungs and cause the tissue to become electrically charged (ionized). Protection from α-particles can be obtained with the use of air-purifying respirators with proper cartridges to filter out radioactive particles. Neutrons are found around the core of a nuclear reactor and are absorbed by both water and the material in the control rods of the reactor. If a worker is not close to the core of the reactor, then no exposure can occur.

Ionizing radiation is a potential health hazard. The areas where potential exposure can occur are usually highly regulated, posted, and monitored on a continuous basis (see Figure 23.8). There is a maximum yearly exposure permitted. Once

FIGURE 23.8 Radiation symbol used as a warning of its presence.

that maximum has been reached, a worker can have no more exposure. The generally used number is 5 rems per year. This is fifty times higher than the U.S. Environmental Protection Agency recommends for the public on a yearly basis. The average public exposure is supposed be no more than 0.1 rems per year. Five rems has been employed for many years and seems to provide reasonable protection for workers. Exposure to radiation should be considered serious because overexposure can lead to serious health problems or even death.

LASERS

The term "laser" is an acronym for Light Amplification by Stimulated Emission of Radiation. Light can be produced by atomic processes that generate laser light. A laser consists of an optical cavity, a pumping system, and an appropriate lasing medium. It is a form of nonionizing radiation.

The optical cavity contains the media to be excited and mirrors to redirect the produced photons back along the same general path. The pumping system uses photons from another source, such as a xenon gas flash tube (optical pumping), to transfer energy to the media. It may use electrical discharge within the pure gas or gas mixture media (collision pumping), or it may rely on the binding energy released in chemical reactions to raise the media to the metastable or lasing state. The laser medium can be a solid (state), gas, dye (in liquid), or semiconductor. Lasers are commonly designated by the type of lasing material employed.

Solid-state lasers have lasing material distributed in a solid matrix [e.g., the ruby or neodymium-YAG (yttrium aluminum garnet) laser]. The neodymium-YAG laser emits infrared light at 1.064 micrometers, while gas lasers [helium and helium-neon (HeNe) are the most common gas lasers] have a primary output of a visible red light. CO_2 lasers emit energy in the far-infrared, 10.6 micrometers, and are used for cutting hard materials. Excimer lasers (the name is derived from the terms excited and dimers) use reactive gases such as chlorine and fluorine mixed with inert gases such as argon, krypton, or xenon. When electrically stimulated, a pseudomolecule or dimer is produced and when lased it produces light in the ultraviolet range. Dye lasers use complex organic dyes like rhodamine 6G in liquid solution or suspension as lasing media. They are tunable over a broad range of wavelengths. Semiconductor lasers, sometimes called diode lasers, are solid-state lasers. These electronic devices are generally very small and use low power. They may be built into larger arrays (e.g., the writing source in some laser printers or compact disk players).

TABLE 23.2
Major Uses of Lasers

Alignment	Laboratory instruments
Annealing	Interferometry
Balancing	Metrology:
Biomedical:	Plasma diagnostics
Cellular research	Spectroscopy
Dental	Velocimetry
Diagnostics	Special photography
Dermatology	Scanning microscopy
Ophthalmology	Military:
Surgery	Distance ranging
Communications	Rifle simulation
Construction:	Weaponry
Alignment	Nondestructive training
Ranging	Scanning
Surveying	Sealing
Cutting	Scribing
Displays	Soldering
Drilling	Welding
Entertainment	
Heat treating	
Holography	
Information handling:	
Copying	
Displays	
Plate making	
Printing	
Reading	
Scanning	
Typesetting	
Videodisk (playing and making)	
Marking	

The wavelength output from a laser depends on the medium being excited. The uses of a laser are quite varied, as can be seen by the major use categories in Table 23.2 (see Figure 23.9).

Some of the potential hazards that can be expected include those associated with compressed gases, cryogenic materials, toxic and carcinogenic materials, noise, explosion hazard, radiation, electricity, and flammability.

The types of biological or health effects that might occur are eye injury, thermal injury, skin hyperpigmentation, and erythema or carcinogenesis (see Table 23.3).

Lasers and laser systems are assigned one of four broad Classes (I to IV), depending on the potential for causing biological damage. Class I cannot emit laser radiation at known hazardous levels [typically continuous wave (cw): 0.4 W at visible wavelengths]. Users of Class I laser products are generally exempt from radiation hazard

FIGURE 23.9 Using a laser as a level.

TABLE 23.3
Summary of Basic Biological Hazards of Light

Wavelengths	Hazards to Eyes	Hazards to Skin
Photobiological Spectral Domain	Eye Effects	Skin effects
Ultraviolet C (0.200–0.280 μm)	Photokeratitis	Erythema (sunburn) Skin cancer
Ultraviolet B (0.280–315 μm)	Photokeratitis	Accelerated skin aging Increased pigmentation
Ultraviolet A (0.315–0.400 μm)	Photochemical UV cataract	Pigment darkening Skin burn
Visible (0.400–0.780 μm)	Photochemical and thermal retinal injury	Photosensitive reactions Skin burn
Infrared A (0.780–1.400 μm)	Cataract, retinal burns	Skin burn
Infrared B (1.400–3.00 μm)	Corneal burn Aqueous flare IR cataract	Skin burn
Infrared C (3.00–1000 μm)	Corneal burn only	Skin burn

controls during operation and maintenance. Because lasers are not classified on beam access during service, most Class I industrial lasers consist of a higher class (high-power) laser enclosed in a properly interlocked and labeled protective enclosure. In some cases, the enclosure may be a room (walk-in protective housing) that requires a means to prevent operation when operators are inside the room.

Class IA is a special designation based on a 1,000-second exposure and applies only to lasers that are not intended for viewing, such as a supermarket laser scanner. The upper power limit of Class IA is 4.0 mW. The emission from a Class IA laser is defined such that the emission does not exceed the Class I limit for emission duration of 1,000 seconds.

Class IIA lasers are low-power visible lasers that emit above Class I levels but at a radiant power not above 1 mW. The concept is that the human aversion reaction to bright light will protect a person. Only limited controls are specified.

Class IIIA lasers are intermediate-power lasers (cw: 1 to 5 mW) and only hazardous for intrabeam viewing. Some limited controls are usually recommended.

Class IIIB lasers are moderate-power lasers (cw: 5 to 500 mW or the diffuse reflection limit, whichever is lower). In general, Class IIIB lasers will not be a fire hazard, nor are they generally capable of producing a hazardous diffuse reflection. Specific controls are recommended.

Class IV lasers are high-power lasers (cw: 500 mW or the diffuse reflection limit) and are hazardous to view under any condition (directly or diffusely scattered), and are a potential fire hazard and a skin hazard. Significant controls are required of Class IV laser facilities.

Accident data on laser usage have shown that Class I, Class II, Class IA, and Class IIIA lasers are normally not considered hazardous from a radiation standpoint unless illogically used. Direct exposure on the eye by a beam of laser light should always be avoided with any laser, no matter how low the power.

There are some uses of Class IIIB and Class IV lasers where the entire beam path may be totally enclosed, the beam path is confined by design to significantly limit access, and where the beam path is totally open. In each case, the controls required vary as follows:

- Enclosed (total) beam path
- Limited open beam path

Protective equipment (e.g., eye protection, temporary barriers, clothing and/or gloves, respirators, etc.) is recommended, for example, only if the hazard analysis indicated a need, or if the Safe Operating Procedure (SOP) requires periods of beam access such as during setup or infrequent maintenance activities. Temporary protective measures for service can be handled in a manner similar to the service of any embedded Class IV laser.

When the entire beam path from a Class IIIB or Class IV laser is not sufficiently enclosed or baffled to ensure that radiation exposures do not exceed the maximum permissible exposure, a laser-controlled area is required. During periods of service, a controlled area may be established on a temporary basis. Two controls are required for both Class IIIB and Class IV installations: (1) posting with appropriate laser warning signs, and (2) operated by qualified and authorized personnel.

Control measures recommended for Class IIIB and required for Class IV lasers are as follows:

- Supervision directly by an individual knowledgeable in laser safety: entry of any noninvolved personnel requires approval.
- A beam stop of an appropriate material must be used to terminate all potentially hazardous beams.
- Use diffusely reflecting materials near the beam, where appropriate: appropriate laser protective eyewear must be provided to all personnel within the laser controlled area.

- The beam path of the laser must be located and secured above or below eye level for any standing or seated position in the facility.
- All windows, doorways, open portals, etc. of an enclosed facility should be covered or restricted to reduce any escaping laser beams below appropriate ocular Maximum Permissible Exposure (MPE) level, and require storage or disabling of lasers when not in use.

In addition, there are specific controls required at the entryway to a Class IV laser controlled area. These can be summarized as follows:

- All personnel entering a Class IV area shall be adequately trained and provided with proper laser protective eyewear.
- All personnel shall follow all applicable administrative and procedural controls.
- All Class IV area and entryway controls shall allow rapid entrance and exit under all conditions.
- The controlled area shall have a clearly marked "Panic Button" (a non-lockable disconnect switch) that allows rapid deactivation of the laser.

Class IV areas also require some form of area and entryway controls such as

- Non-defeatable entryway controls
- Defeatable entryway controls
- Procedural entryway controls
- Entryway warning systems

Administrative and procedural controls are standard operating procedures, alignment procedures, limitations on spectators, and protective equipment. Protective equipment for laser safety generally means eye protection in the form of goggles or spectacles, clothing, barriers, and other devices designed for laser protection.

Engineering controls are normally designed and built into the laser equipment to provide for safety. In most instances, these are included on the equipment (i.e., provided by the laser manufacturer). Some of these controls are protective housing, master switch control, optical viewing system safety, beam stop or attenuation, laser activation warning system, service access panels, protective housing interlock requirements, and a remote interlock connector.

Lasers emit beams of coherent radiation of a single color or wavelength, in contrast to conventional light sources, which produce incoherent radiation over a range of wavelengths. The laser is made up of light waves that are nearly parallel to each other, all traveling in the same direction. Atoms are pumped to a higher energy state and are stimulated to de-excite to a lower energy level, and in the process release coherent radiation that is collimated and gives off radiation to produce the coherent laser beam.

Because the laser is highly collimated (i.e., has a small divergence angle), it has a high energy density and a large number of photons in a narrow beam. Direct viewing of the laser source or its reflections should be avoided. The work area should contain no reflective surfaces (such as mirrors or highly polished furniture) as even a reflected laser beam can be hazardous. Suitable shielding to contain the laser beam should be provided. OSHA covers protection against laser hazards in its construction regulations.

The eye is the organ most vulnerable to injury induced by laser energy. The reason for this is the ability of the cornea and lens to focus the collimated laser beam on a small spot on the retina. The fact that infrared radiation of certain lasers may not be visible to the naked eye contributes to the potential hazard. Lasers generating in the ultraviolet range of the electromagnetic spectrum can produce corneal burns rather than retinal damage because of the way the eye handles ultraviolet light. Other factors that have a bearing on the degree of eye injury induced by laser light are

1. Pupil size (the smaller the pupil diameter, the less the amount of laser energy permitted to the retina)
2. Ability of the cornea and lens to focus the incident light on the retina
3. Distance from the source of energy to the retina
4. Energy and wavelength of the laser
5. Pigmentation of the eye of the subject
6. Place of the retina where the light is focused
7. Divergence of the laser light
8. Presence of scattering media in the light path

LEAD

Overexposure to lead is one of the most common overexposures found in industry. It is also a major potential public health risk. Lead poisoning is the leading environmentally induced illness in children. At greatest risk are children under the age of six because they are undergoing rapid neurological and physical development. In general populations, lead may be present at hazardous concentrations in food, water, and air. Sources include paint, urban dust, and folk remedies.

Lead is commonly added to industrial paints because of its characteristic of resisting corrosion. Industries with particularly high potential exposures include construction work involving welding, cutting, brazing, abrasive surface blasting, etc., on lead painted surfaces; most smelter operations either as a trace contaminant or as a major product; secondary lead smelters, where lead is recovered from batteries; radiator repair shops; and firing ranges. Oral ingestion may represent a major route of exposure in contaminated workplaces. Lead soldering usually does not represent an inhalation risk because controlling the temperature of lead below 900°F (melting temperature = 621°F) is effective in controlling lead fuming. Most exposures occur with inorganic lead. Organic (tetraethyl and tetramethyl) lead, which was added to gasoline up until the late 1970s, is not commonly encountered. Organic forms may be absorbed through the skin, while inorganic forms cannot. Inorganic lead is not metabolized, but is directly absorbed, distributed, and excreted. The rate depends on its chemical and physical form and on the physiological characteristics of the exposed person (e.g., nutritional status and age). Once in the blood, lead is distributed primarily among the person's blood, soft tissue (kidney, bone marrow, liver, and brain), and mineralizing tissue (bones and teeth). Absorption via the gastrointestinal

(GI) tract following ingestion is highly dependent upon the presence of and levels of calcium, iron, fats, and proteins.

The precautions for working with or around lead are very similar to the requirements for asbestos. Although asbestos is more of a respiratory hazard while lead is more of a digestive hazard, once lead enters the bloodstream, it can cause kidney and neurological problems. Lead contained in the bones remains there for nearly the life of the individual. OSHA has a PEL of 50 g/m^3 during an 8-hour period or when a level of 50 g/dl of blood is detected. Once workers are leaded to this level, they must be removed from work until the blood lead level decreases to 40 g/dl of blood. There are many other requirements regarding lead exposure found in 29 CFR 1910.1025.

Lead contamination is easily spread. Thus, precautions must be taken to decontaminate anything leaving the regulated area, including work clothing and shoes. Workers should wear respirators in areas where dust or vapors can become airborne and disposable outer garments to prevent the spread of lead to other areas, including their own homes. Lead easily enters the digestive track. This reinforces the need for washing hands prior to eating or smoking. Eating or smoking should never occur in a regulated area. Lead exposure can lead to serious illness or even death.

NOISE-INDUCED HEARING LOSS

Occupational exposure to noise levels in excess of the current OSHA standards places hundreds of thousands of workers at risk for developing material hearing impairment, hypertension, and elevated hormone levels. Workers in some industries (i.e., construction, oil and gas well drilling and servicing) are not fully covered by the current OSHA standards and lack the protection of an adequate hearing conservation program. Occupationally induced hearing loss continues to be one of the leading occupational illnesses in the United States and is often called a "silent epidemic." OSHA has designated this issue as a priority for rulemaking action to extend hearing conservation protection, provided in the general industry standard, to the construction industry and other uncovered industries.

According to the U.S. Bureau of the Census, Statistical Abstract of the United States, there are more than 7.2 million workers employed in the construction industry (6 percent of all employment). The National Institute for Occupational Safety and Health's (NIOSH) National Occupational Exposure Survey (NOES) estimates that 421,000 construction workers are exposed to noise above 85 dBA. NIOSH estimates that 15 percent of workers exposed to noise levels of 85 dBA or higher will develop permanent hearing impairment.

Research demonstrates that construction workers are regularly overexposed to noise. The extent of daily exposure to noise in the construction industry depends on the nature and duration of the work. For example, rock drilling—up to 115 dBA, chain saw—up to 125 dBA, abrasive blasting—105 to 112 dBA, heavy equipment operation—95 to 110 dBA, demolition—up to 117 dBA, and needle guns—up to 112 dBA. Exposure to 115 dBA is permitted for a maximum of 15 minutes for an 8-hour workday. No exposure above 115 dBA is permitted. Traditional dosimetric measurements may substantially underestimate noise exposure levels for construction workers because short-term peak exposures, which may be responsible for acute and chronic effects, can be lost in lower, full-shift time-weighted average measurements.

There are a variety of control techniques, documented in the literature, to reduce overall worker exposure to noise. Such controls reduce the amount of sound energy released by the noise source, divert the flow of sound energy away from the receiver, or protect the receiver from the sound energy reaching him or her. For example, types of noise controls include proper maintenance of equipment, revised operating procedures, equipment replacements, acoustic shields and barriers, equipment redesign, enclosures, administrative controls, and personal protective equipment.

Under OSHA's general industry standard, feasible administrative and engineering controls must be implemented whenever employee noise exposures exceed 90 dBA [8-hour time-weighted average (TWA)]. In addition, an effective hearing conservation program (including specific requirements for monitoring noise exposure, audiometric testing, audiogram evaluation, hearing protection for employees with a standard threshold shift, training, education, and recordkeeping) must be made available whenever employee exposures equal or exceed an 8-hour TWA sound level of 85 dBA (29 CFR 1910.95). Similarly, under the construction industry standard, the maximum permissible occupational noise exposure is 90 dBA (8-hour TWA), and noise levels in excess of 90 dBA must be reduced through feasible administrative and engineering controls. However, the construction industry standard includes only a general minimum requirement for hearing conservation and lacks the specific requirements for an effective hearing conservation program included in the general industry standard (29 CFR 1920.95). NIOSH and the American Conference of Governmental Industrial Hygienists (ACGIH) have also recommended exposure limits (NIOSH: 85 dBA TWA, 115 dBA ceiling; ACGIH: 85 dBA).

Noise, or unwanted sound, is one of the most pervasive occupational health problems. It is a by-product of many industrial processes. Sound consists of pressure changes in a medium (usually air), caused by vibration or turbulence. These pressure changes produce waves emanating away from the turbulent or vibrating source. Exposure to high levels of noise causes hearing loss and may cause other harmful health effects as well. The extent of damage depends primarily on the intensity of the noise and the duration of the exposure. Noise-induced hearing loss can be temporary or permanent. Temporary hearing loss results from short-term exposures to noise, with normal hearing returning after a period of rest. Generally, prolonged exposure to high noise levels over a period of time gradually causes permanent damage.

OSHA's hearing conservation program is designed to protect workers with significant occupational noise exposures from suffering material hearing impairment, even if they are subject to such noise exposures over their entire working lifetimes. The following summarizes the required component of OSHA's hearing conservation program:

1. The hearing conservation program requires employers to monitor noise exposure levels in a manner that will accurately identify employees who are exposed to noise at or above 85 decibels (dBA) averaged over eight working hours, or an 8-hour TWA. That is, employers must monitor all employees whose noise exposure is equivalent to or greater than a noise

exposure received in 8 hours where the noise level is constantly 85 dBA. The exposure measurement must include all continuous, intermittent, and impulsive noise within an 80-dBA to 130-dBA range and must be taken during a typical work situation. This requirement is performance oriented as it allows employers to choose the monitoring method that best suits each individual situation. Monitoring should be repeated when changes in production, process, or controls increase noise exposure. Such changes may mean that additional employees need to be monitored, and/or their hearing protectors may no longer provide adequate attenuation. Under this program, employees are entitled to observe monitoring procedures, and they must be notified of the results of exposure monitoring. The method used to notify employees is left to the discretion of the employers. Instruments used for monitoring employee exposures must be carefully checked or calibrated to ensure that the measurements are accurate. Calibration procedures are unique to specific instruments. Employers have the duty to ensure that measuring instruments are properly calibrated. They may find it useful to follow the manufacturer's instructions to determine when and how extensively to calibrate.

2. Audiometric testing not only monitors the sharpness and acuity of an employee's hearing over time, but also provides an opportunity for employers to educate employees about their hearing and the need to protect it. The employer shall establish and maintain an audiometric testing program (see Figure 23.10). The important elements of an audiometric testing program include baseline audiograms, annual audiograms, training, and follow-up procedures. Audiometric testing must be made available at no cost to all

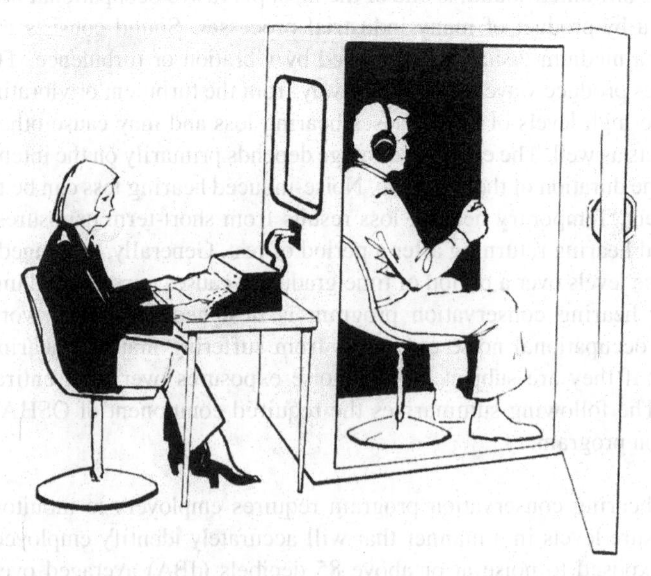

FIGURE 23.10 Example of a testing booth for conducting hearing test. (*Source:* Courtesy of the National Mine Health and Safety Academy.)

employees who are exposed to an action level of 85 dBA or above, measured as an 8-hour TWA. The audiometric testing program follow-up should indicate whether the employer's hearing conservation program is preventing hearing loss. A licensed or certified audiologist (specialist involved in evaluation of the hearing function), an otolaryngologist (physician specializing in the diagnosis and treatment of disorders of the ear, nose, and throat), or a physician must be responsible for the program. Both professionals and trained technicians may conduct audiometric testing. The professional in charge of the program need not be present when a qualified technician conducts tests. The professional's responsibilities include overseeing the program and the work of the technicians, reviewing problem audiograms, and determining whether referral is necessary. The employee needs a referral for further testing when test results are questionable or when problems of a medical nature are suspected. If additional testing is necessary or if the employer suspects a medical pathology of the ear is caused or aggravated by the wearing of hearing protectors, the employee shall be referred for a clinical audiological evaluation or otological exam, as appropriate. There are two types of audiograms required in the hearing conservation program: baseline and annual audiograms.

3. The baseline audiogram is the reference audiogram against which future audiograms are compared. Baseline audiograms must be provided within 6 months of an employee's first exposure at or above an 8-hour TWA of 85 dBA. An exception is the use of mobile test vans to obtain audiograms. In these instances, baseline audiograms must be completed within 1 year after an employee's first exposure to workplace noise at or above a TWA of 85 dBA. Employees, however, must be fitted with, issued, and are required to wear hearing protectors for any period exceeding 6 months after their first exposure until the baseline audiogram is obtained. Baseline audiograms taken before the effective date of the hearing conservation program (April 7, 1983) are acceptable baselines if the professional supervisor determines that the audiogram is valid. Employees should not be exposed to workplace noise for 14 hours preceding the baseline test; however, appropriate hearing protectors can serve as a substitute for this requirement and can be worn during this time period.
4. Annual audiograms must be conducted within 1 year of the baseline. It is important to test hearing on an annual basis to identify deterioration in hearing ability so that protective follow-up measures can be initiated before hearing loss progresses. Annual audiograms must be routinely compared to baseline audiograms to determine whether the audiogram is valid and to determine whether the employee has lost hearing ability, that is, if a standard threshold shift (STS) has occurred. An STS is an average shift in either ear of 10 dBA or more at 2,000, 3,000, and 4,000 hertz. An averaging method of determining STS was chosen because it diminished the number of persons falsely identified as having STS and who were later shown not to have had a change in hearing ability. Additionally, the method is sensitive enough to identify meaningful shifts in hearing early on.

5. If an STS is identified, employees must be fitted or refitted with adequate hearing protectors, shown how to use them, and required to wear them. Employees must be notified within 21 days from the time the determination is made that their audiometric test results show an STS. Some employees with an STS may need to be referred for further testing if the professional determines that their test results are questionable, or if they have an ear problem of a medical nature that is thought to be caused or aggravated by wearing hearing protectors. If the suspected medical problem is not thought to be related to wearing hearing protection, employees must be informed that they should see a physician. If subsequent audiometric tests show that the STS identified on a previous audiogram is not persistent, employees whose exposure to noise is less than a TWA of 90 dBA may discontinue wearing hearing protectors. An annual audiogram may be substituted for the original baseline audiogram if the professional supervising the program determines that the employee's STS is persistent. The original baseline audiogram, however, must be retained for the length of the employee's employment. This substitution will ensure that the same shift is not repeatedly identified. The professional also may decide to revise the baseline audiogram if an improvement in hearing occurs. This will ensure that the baseline reflects actual hearing thresholds to the extent possible. Audiometric tests must be conducted in a room meeting specific background levels and with calibrated audiometers that meet American National Standard Institute (ANSI) specifications of SC-1969.
6. Hearing protectors must be available to all workers exposed to 8-hour TWA noise levels of 85 dBA or above. This requirement will ensure that employees have access to protectors before they experience a loss in hearing. Hearing protectors must be worn by (1) employees for any period exceeding 6 months from the time they are first exposed to 8-hour TWA noise levels of 85 dBA or above—this must be done until they receive their baseline audiograms in situations where baseline audiograms are delayed because it is inconvenient for mobile test vans to visit the workplace more than once a year; (2) employees who have incurred standard threshold shifts since these workers have demonstrated that they are susceptible to noise; and (3) employees exposed over the permissible exposure limit of 90 dBA over an 8-hour TWA. Employees should decide, with the help of a person who is trained in fitting hearing protectors, which size and type of protectors are most suitable for their working environment. Protectors selected should be comfortable to the wearer and offer sufficient attenuation to prevent hearing loss. Hearing protectors must adequately reduce the severity of the noise level for each employee's work environment (see Figure 23.11). The employer must reevaluate the suitability of the employee's present protectors whenever there is a change in working conditions that may cause the hearing protector being used to be inadequate. If workplace noise levels increase, employees must be given more effective protectors. The protector must reduce employee exposures to at least 90 dBA and to 85 dBA when an STS has already occurred in the worker's hearing. Employees must be

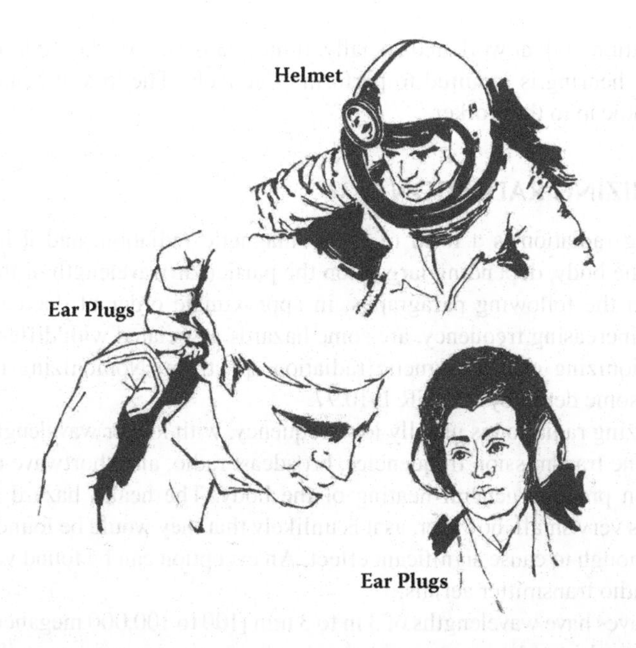

FIGURE 23.11 Examples of hearing protectors for workers. (*Source:* Courtesy of the National Mine Health and Safety Academy.)

shown how to use and care for their protectors and must be supervised on the job to ensure that they continue to wear them correctly.

7. Employee training is very important. When workers understand the reasons for their hearing conservation program's requirements and the need to protect their hearing, they will be better motivated to participate actively in the program and to cooperate by wearing their protectors and taking audiometric tests. Employees exposed to TWAs of 85 dBA and above must be trained, at least annually, in the effects of noise; the purpose, advantages, and disadvantages of various types of hearing protectors; the selection, fit, and care of protectors; and the purpose and procedures of audiometric testing. The training program may be structured in any format, with different portions conducted by different individuals and at different times, as long as the required topics are covered.
8. Noise exposure measurement records must be kept for 2 years. Records of audiometric test results must be maintained for the duration of employment of the affected employee. Audiometric test records must include the name and job classification of the employee, the date, the examiner's name, the date of the last acoustic or exhaustive calibration, measurements of the background sound pressure levels in audiometric test rooms, and the employee's most recent noise exposure measurement.

Sometimes the loss of hearing due to industrial noise is called the "silent epidemic." Because this type of hearing loss is not correctable by either surgery or the use of hearing aids, it certainly is a monumental loss to the worker. It distorts communication both at work and socially. It may cause the worker to lose his or her job if acute hearing is required to perform effectively. The loss of hearing is definitely a handicap to the worker.

NONIONIZING RADIATION

Nonionizing radiation is a form of electromagnetic radiation, and it has varying effects on the body, depending largely on the particular wavelength of the radiation involved. In the following paragraphs, in approximate order of decreasing wavelength and increasing frequency, are some hazards associated with different regions of the nonionizing electromagnetic radiation spectrum. Nonionizing radiation is covered in some detail by 29 CFR 1910.97.

Nonionizing radiation is usually low frequency, with longer wavelengths, including powerline transmission frequencies, broadcast radio, and shortwave radio. Each of these can produce general heating of the body. The health hazard from these radiations is very small, however, as it is unlikely that they would be found in intensities great enough to cause significant effect. An exception can be found very close to powerful radio transmitter aerials.

Microwaves have wavelengths of 3 m to 3 mm [100 to 100,000 megahertz (MHz)]. They are found in radar, communications, some types of cooking, and diathermy applications. Microwave intensities may be sufficient to cause significant heating of tissues. The effect is related to wavelength, power intensity, and time of exposure. Generally, longer wavelengths produce greater penetration and temperature rise in deeper tissues than shorter wavelengths. However, for a given power intensity, there is less subjective awareness to the heat from longer wavelengths than there is to the heat from shorter wavelengths because absorption of longer wavelength radiation takes place beneath the body's surface.

An intolerable rise in body temperature, as well as localized damage to specific organs, can result from an exposure of sufficient intensity and time. In addition, flammable gases and vapors may ignite when they are inside metallic objects located in a microwave beam. Power intensities for microwaves are given in units of milliwatts per square centimeter (mW/cm^2), and areas having a power intensity of over 10 mW/cm^2 for a period of 0.1 hours or longer should be avoided.

Radio frequency (RF) and microwave (MW) radiation are electromagnetic radiation in the frequency range 3 kilohertz (kHz) to 300 gigahertz (GHz). Usually, MW radiation is considered a subset of RF radiation, although an alternative convention treats RF and MW radiation as two separate spectral regions. Microwaves occupy the spectral region between 300 GHz and 300 MHz, while RF or radio waves include 300 MHz to 3 kHz. RF/MW radiations are nonionizing in that there is insufficient energy [less than 10 electron volts (eV)] to ionize biologically important atoms. The primary health effects of RF/MW energy are considered thermal. The absorption of RF/MW energy varies with frequency. Microwave frequencies produce a skin effect (a person can literally sense their skin starting to feel warm). RF radiation may penetrate the body and be absorbed in deep body organs without the skin being affected, which can warn an individual of danger. A great deal of research has turned up other nonthermal effects. All the standards of Western countries have, so far, based their

TABLE 23.4
Recommended Ratios of Illumination for Various Tasks

Conditions	Luminance Ratios	
	Office	Industrial
Between tasks and adjacent darker surroundings	3:1	3:1
Between tasks and adjacent lighter surroundings	1:3	1:3
Between tasks and more remote darker surfaces	5:1	20:1
Between tasks and more remote lighter surfaces	1:5	1:20
Between luminaires (or windows, skylights) and surfaces adjacent to them	20:1	NC[a]
Anywhere within normal field of view	40:1	NC[a]

[a] NC means not controllable in practice.

exposure limits solely on preventing thermal problems. In the meantime, research continues. Use of RF/MW radiation includes aeronautical radios, citizen's band (CB) radios, cellular phones, cooking of foods in microwave ovens, heat sealers, vinyl welders, high-frequency welders, induction heaters, flow solder machines, communications transmitters, radar transmitters, ion implant equipment, microwave drying equipment, sputtering equipment, glue curing, and power amplifiers.

Infrared radiation does not penetrate below the superficial layer of the skin, so that its only effect is to heat the skin and the tissues immediately below it. Except for thermal burns, the health hazard upon exposure to low-level conventional infrared radiation sources is negligible.

Visible radiation, which is about midway in the electromagnetic spectrum, is important because it can affect both the quality and accuracy of work. Good lighting conditions generally result in increased product quality with less spoilage and increased production. Lighting should be bright enough for easy vision and directed so that it does not create glare. Illumination levels and brightness ratios recommended for manufacturing and service industries are found in Table 23.4

Ultraviolet radiation in industry may be found around electrical arcs, and such arcs should be shielded by materials opaque to the ultraviolet. The fact that a material may be opaque to ultraviolet has no relation to its opacity to other parts of the spectrum. Ordinary window glass, for instance, is almost completely opaque to the ultraviolet in sunlight; at the same time, it is transparent to visible light waves. A piece of plastic, dyed a deep red-violet, may be almost entirely opaque in the visible part of the spectrum and transparent in the near-ultraviolet. Electric welding arcs and germicidal lamps are the most common, strong producers of ultraviolet in industry. The ordinary fluorescent lamp generates a good deal of ultraviolet inside the bulb, but it is essentially all absorbed by the bulb and its coating.

The most common exposure to ultraviolet radiation is from direct sunlight, and a familiar result of overexposure—one that is known to all sunbathers—is sunburn. Most everyone is also familiar with certain compounds and lotions that reduce the effects of the sun's rays, but many are unaware that some industrial materials, such as cresols, make the skin especially sensitive to ultraviolet rays. So much so that after having been exposed to cresols, even a short exposure in the sun usually results in severe sunburn.

Nonionizing radiation, although perceived not to be as dangerous as ionizing radiation, does have its share of adverse health effects accompanying it.

VIBRATION

Vibrating tools and equipment at frequencies between 40 and 90 hertz can cause damage to the circulatory and nervous systems. Care must be taken with low frequencies, which have the potential to put workers at risk for vibration injuries. One of the most common cumulative trauma disorders (CTDs) resulting from vibration is Raynaud's Syndrome. Its most common symptoms are intermittent numbness and tingling in the fingers, skin that turns pale, ashen, and cold, and eventual loss of sensation and control in the fingers and hands. Raynaud's comes about by use of vibrating hand tools such as palm sanders, planers, jack-hammers, grinders, and buffers. When such tools are required for a job, an assessment should be made to determine if any other method(s) can be used to accomplish the desired task. If not, other techniques, such as time/use limitations, alternating workers, or other such administrative actions, should be considered to help reduce the potential for a vibration-induced CTD. The damage caused by vibrating tools can be reduced by

- Using vibration dampening gloves
- Purchasing low vibration tools and equipment
- Putting antivibration material on handles of existing tools
- Reducing length of exposure
- Changing the actual work procedure if possible
- Using balanced and dampening tools and equipment
- Rotating workers to decrease exposure time
- Decreasing the pace of the job as well as the speed of tools or equipment

Individuals subject to whole-body vibration have experienced visual problems; vertebral degeneration; breathing problems; motion sickness; pains in the abdomen, chest, and jaw; backache; joint problems; muscle strain; and problems with their speech. Although there are still many questions regarding vibration, it is definite that physical problems can transpire from exposure to vibration.

WORKPLACE STRESS

All of us need a certain amount of stress to keep us alert or just to keep us going. How much stress is good and how much is bad is always the question. Because so many individual differences exist for workers, it is nearly impossible to predict what is too much stress or too little stress. Some individuals thrive on stress while others have adverse bodily reactions to stress. No exacting measurement can predict stress levels. Those instruments that are used are often based upon very subjective factors and judgments.

In the workplace, some factors have been shown to create stress for some workers. They are the lack of control on the job, the failure to fit into the workplace social structure, excessive demands of a job, or severe time expectations. Other job characteristics that contribute to workplace stress are

- Jobs with high demand and low control
- Repetitious or hurried-pace jobs
- Low-skill jobs
- Jobs requiring interaction with demanding colleagues
- High-vigilance jobs
- Jobs with nontraditional work schedules or substantial overtime

The lack of job security has become very real with so much reorganizing and downsizing transpiring in the business community. Career pressures including increased responsibility are viewed as stressors by many workers. Some workers have jobs that are too complex for their talents, abilities, or skills. Last, but by no means unimportant, is the physical environment such as noise, temperature extremes, lighting, space, and odors. Other work environment stressors are

- Risk of workplace violence
- High or conflicting expectations
- Cronyism or favoritism
- Various types of harassment (bullying, sexual, verbal, etc.)

Employers must take workplace stress seriously when they have warning signs such as

- Worker have little to say about how they do their job
- Workers who express feelings of being rushed
- Productivity is decreasing
- Employees are frequently sick or increase in work injuries
- Anxiety due to changes in organization, cuts, or loss of benefits
- More displays of anger or frustration
- Employees express a sense of being stressed

These types of stressors may manifest themselves in actual physical disorders such as gastrointestinal disorders. Most health conditions develop or are exacerbated by stressors. Many workers have complained and filed work-related claims for all types of illnesses where stress is considered a contributing factor in the illness or disability.

Most workplaces have instituted employee assistance programs to assist workers in coping with workplace stress. Other action employers can take include

- Give workers greater control and say over their jobs.
- Realize stressor at home can affect work.
- Improve communications and relationships.

- Institute social support actions such as social events.
- Value the workforce any way possible.
- Encourage personal health and fitness, stress management classes, and flexible work time when possible.
- Tell employees the truth.

Employers must ensure that supervisors are trained in the effective use of these programs. Worker complaints of stress should not be taken lightly. Employees must take a role by

- Accepting and adjusting to change
- Taking control when they can
- Using time effectively
- Trying to have a sense of humor and laughing off as many issues as possible
- Becoming active (exercise is an excellent stress reliever)
- Being well rested when arriving at work
- Trying to take short breaks from issues causing stress
- Getting professional help if needed

Stress is a part of everyday life. All workplaces have a degree of stress that may or may not be an issue. If stress is hampering workplace function or performance, it may need attention and problem solving. Stress has been discovered to be the basis of many serious workplace events in recent history.

REFERENCES

United States Department of Health and Human Services. http://www.healthypeople.gov/2020.

United States Department of Labor, Occupational Safety and Health Administration. Subject Index. "Internet." April 1999. Available: http://www.osha.gov.

United States Department of Labor, Occupational Safety and Health Administration. General Industry Digest (OSHA 2201). Washington, D.C.: Government Printing Office, 1995.

United States Department of Labor, Occupational Safety and Health Administration. Title 29 Code of Federal Regulations 1910. Washington, D.C.: Government Printing Office, 1999.

United States Department of Labor, Occupational Safety and Health Administration. Title 29 Code of Federal Regulations 1926. Washington, D.C.: Government Printing Office, 1999.

24 Controls and Personal Protective Equipment

INTRODUCTION

This chapter discusses the requirements for personal protective equipment (PPE). It also addresses other control mechanisms and procedures. Engineering controls should be the primary method used to eliminate or minimize hazard exposure in the workplace. Administrative controls must be set in motion prior to the use of PPE. When such controls are not practical or applicable, PPE shall be employed to reduce or eliminate personnel exposure to hazards. PPE will be provided, used, and maintained when it has been determined that its use is required and that such use will lessen the likelihood of occupational injuries and/or illnesses. All personal protective clothing and equipment should be of a safe design and appropriate for the work to be performed. Only those items of protective clothing and equipment that meet National Institute for Occupational Safety and Health (NIOSH) or American National Standards Institute (ANSI) standards are to be procured or accepted for use.

The Occupational Safety and Health Administration (OSHA) requires employers to protect their employees from workplace hazards such as machines, work procedures, and hazardous substances that can cause injury or illness. The preferred way to do this is through engineering controls, predetermined safe work practices, and administrative controls; but when these controls are not feasible or do not provide sufficient protection, an alternative or supplementary method of protection should be provided workers in the form of PPE and the know-how to use it properly.

This chapter will help employers be in compliance with OSHA's general PPE requirements, but it is not a substitute for OSHA standards requiring PPE (Title 29, Code of Federal Regulations [CFR] 1910.132, Subpart I). This standard requires employers to establish general procedures for a PPE program, to give employees necessary protective equipment, and to train them to use it properly. Although not specifically directed to construction and the maritime industry, the information, methods, and procedures in this guide are also applicable to, and will help in compliance with, OSHA's general PPE requirements for the construction industry at 29 CFR 1926.95 and for the maritime industry at 29 CFR 1915.152.

PPE includes a variety of devices and garments to protect workers from injuries or hazardous exposures. PPE designed to protect eyes, face, head, ears, feet, hands and arms, and the whole body is available. PPE includes such items as goggles, face shields, safety glasses, hard hats, safety shoes, gloves, safety vests, earplugs, earmuffs, respirators, and suits for full body protection.

CONTROLLING HAZARDS

Many ways to control hazards have been used over the years but usually these can be broken down into five primary approaches. The preferred ways to do this are through engineering controls, awareness devices, predetermined safe work practices, administrative controls, and management controls. When these controls are not feasible or do not provide sufficient protection, an alternative or supplementary method of protection is to provide workers with personal protective equipment (PPE) and the know-how to use it properly.

ENGINEERING CONTROLS

When a hazard is identified in the workplace, every effort should be made to eliminate it so that employees are not harmed. Elimination may be accomplished by designing or redesigning a piece of equipment or process. This could be the installation of a guard on a piece of machinery that prevents workers from contacting the hazard. The hazard can be engineered out of the operation. Another way to reduce or control the hazard is to isolate the process, such as in the manufacture of vinyl chloride used to make such items as plastic milk bottles, where the entire process becomes a closed circuit. The results are no one being exposed to vinyl chloride gas, which is known to cause cancer. Thus, any physical controls that are put in place are considered the best approach from an engineering perspective. Keep in mind that you are a consumer of products. Thus, at times you can leverage the manufacturer to implement safeguards or safety devices on products that you are looking to purchase. Let your vendor do the engineering for you or do not purchase their product. This may not always be a viable option. To summarize the engineering controls that can be used, the following may be considered:

- Substitution
- Elimination
- Ventilation
- Isolation (see Figure 24.1)
- Process or design change

AWARENESS DEVICES

Awareness devices are linked to the senses. They are warning devices that can be heard and seen. They act as alerts to workers but create no type of physical barrier. They are found in most workplaces and carry with them a moderate degree of effectiveness. Such devices are

- Backup alarms
- Warning signals, both audible and visual
- Warning signs

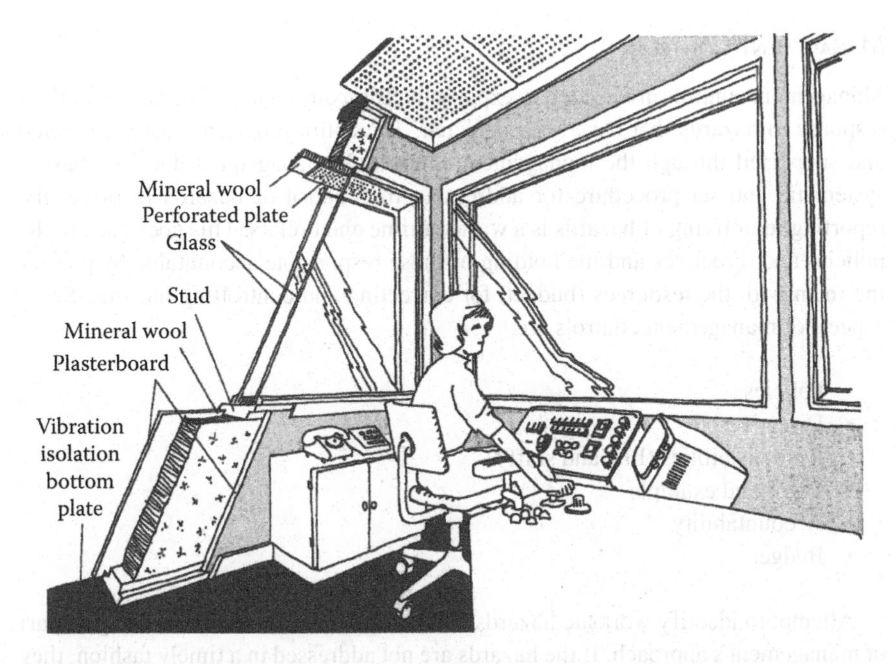

FIGURE 24.1 Typical isolation booth for a machine or equipment operator. (Courtesy of the Occupational Safety and Health Administration.)

Work Practices

Work practices concern the ways in which a job task or activity is performed. This may mean that you create a specific procedure for completing the task or job. It may also mean that you implement special training for a job or task. It also presupposes that you might require inspection of the equipment or machinery prior to beginning work or when a failure has occurred. An inspection should be done prior to restarting the process or task. It may also require that a lockout/tagout procedure be used to create a zero potential energy release.

Administrative Controls

A second approach is to control the hazard through administrative directives. This may be accomplished by rotating workers, which allows you to limit their exposure, or having workers only work in areas when no hazards exist during that part of their shift. This applies particularly to chemical exposures and repetitive activities that could result in ergonomics-related incidents. Examples of administrative controls include

- Requiring specific training and education
- Scheduling off-shift work
- Worker rotation

Management Controls

Management controls are needed to express the company's view of hazards and their response to hazards that have been detected. The entire program must be directed and supported through the management controls. If management does not have a systematic and set procedure for addressing the control of hazards in place, the reporting/identifying of hazards is a waste of time and dollars. This goes back to the policies and directives and the holding of those responsible accountable by providing them with the resources (budget) for correcting and controlling hazards. Some aspects of management controls are

- Policies
- Directives
- Responsibilities (line and staff)
- Vigor and example
- Accountability
- Budget

Attempt to identify worksite hazards; addressing them should be an integral part of management's approach. If the hazards are not addressed in a timely fashion, they will not be identified or reported. If dollars become the main reason for not fixing or controlling hazards, you will lose the motivation of your workforce to identify or report them.

PERSONAL PROTECTIVE EQUIPMENT

PPE includes a variety of devices and garments to protect workers from injuries. PPE is designed to protect eyes, face, head, ears, feet, hands and arms, and the whole body. It includes such items as goggles, face shields, safety glasses, hard hats, safety shoes, gloves, vests, earplugs, earmuffs, and suits for full body protection (see Figure 24.2).

In November 2007, OSHA announced a final rule regarding employer-paid personal protective equipment. Under the rule, all PPE, with a few exceptions, will be provided at no cost to the employee. OSHA anticipates that this rule will have substantial safety benefits that will result in more than 21,000 fewer occupational injuries per year.

Employees exposed to safety and health hazards may need to wear PPE to be protected from injury, illness, and death caused by exposure to those hazards. This final rule will clarify who is responsible for paying for PPE, which OSHA anticipates will lead to greater compliance and potential avoidance of thousands of workplace injuries each year.

The final rule contains a few exceptions for ordinary safety-toed footwear, ordinary prescription safety eyewear, logging boots, and ordinary clothing and weather-related gear. The final rule also clarifies OSHA's requirements regarding payment for employee-owned PPE and replacement PPE.

Level A Protection
Totally encapsulating vapor-tight suit with full-facepiece SCBA or supplied-air respirator.

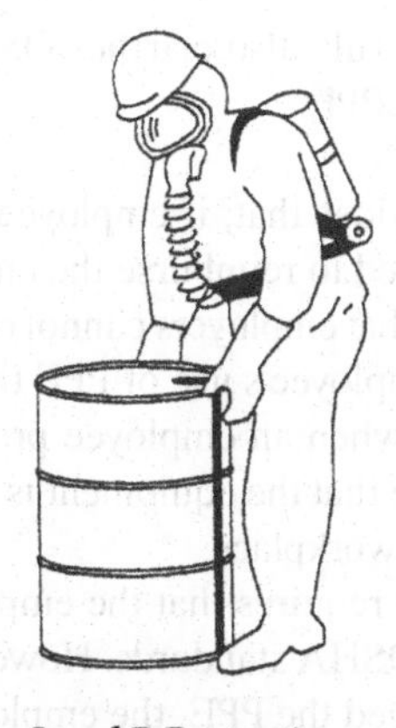

Level B Protection
Totally encapsulating suit does not have to be vapor-tight. Same level of respiratory protection as Level A.

Level C Protection
Full-face canister air purifying respirator. Chemical protective suit with full body coverage.

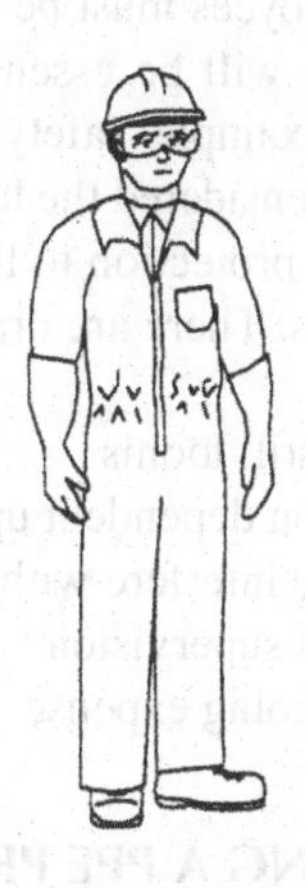

Level D Protection
Basic work uniform, i.e., longsleeve converalls, gloves, hardhat, boots, faceshield or goggles.

FIGURE 24.2 Levels of protection using a variety of PPE. (Courtesy of the National Institute for Occupational Safety and Health.)

The final rule does not create new requirements regarding what PPE employers must provide. It does not require payment for uniforms, items worn to keep clean, or other items that are not PPE.

The final rule contains exceptions for certain ordinary protective equipment, such as safety-toe footwear, prescription safety eyewear, everyday clothing and weather-related gear, and logging boots.

The final rule also clarifies OSHA's intent regarding employee-owned PPE, and replacement PPE:

- It provides that, if employees choose to use PPE they own, employers will not need to reimburse the employees for the PPE. The standard also makes clear that employers cannot require employees to provide their own PPE, and the employee's use of PPE they already own must be completely voluntary. Even when an employee provides his or her own PPE, the employer must ensure that the equipment is adequate to protect the employee from hazards at the workplace.
- It also requires that the employer pay for replacement PPE used to comply with OSHA standards. However, when an employee has lost or intentionally damaged the PPE, the employer is not required to pay for its replacement.

The final rule requires employers to pay for almost all personal protective equipment that is required by OSHA's general industry, construction, and maritime standards. Employers already pay for approximately 95 percent of these types of PPE.

When employees must be present and engineering or administrative controls are not feasible, it will be essential to use PPE as an interim control and not a final solution. For example, safety glasses may be required in the work area. Too often, PPE usage is considered the last thing to do in the scheme of hazard control. PPE can provide added protection to the employee even when the hazard is being controlled by other means. There are drawbacks to the use of PPE, including

- Hazard still looms
- Protection dependent upon the worker using PPE
- PPE may interfere with performing task and productivity
- Requires supervision
- Is an ongoing expense

ESTABLISHING A PPE PROGRAM

A PPE program sets out procedures for selecting, providing, and using PPE as part of an organization's routine operation. A written PPE program is easier to establish and maintain than a company policy and easier to evaluate than an unwritten one. To develop a written program, you should consider the following:

1. Identify steps taken to assess potential hazards in every employee workspace and in workplace operating procedures.
2. Identify appropriate PPE selection criteria.
3. Identify how you will train employees on the use of PPE, including
    a. What PPE is necessary?
    b. When is PPE necessary?
    c. How to properly inspect PPE for wear and damage.
    d. How to properly put on and adjust the fit of PPE.

   e. How to properly take off PPE.
   f. The limitations of the PPE.
   g. How to properly care for and store PPE.
4. Identify how you will assess employee understanding of PPE training.
5. Identify how you will enforce proper PPE use.
6. Identify how you will provide for any required medical examinations.
7. Identify how and when to evaluate the PPE program.

HAZARD ASSESSMENT

Recent regulatory requirements make hazard analysis part of the PPE selection process. Hazard analysis procedures should be used to assess the workplace to determine if hazards are present, or are likely to be present, which may necessitate the use of PPE. As part of this assessment, the employees' work environment should be examined for potential hazards that are likely to present a danger to any part of their bodies. If it is not possible to eliminate workers' exposure or potential exposure to the hazard through the efforts of engineering controls, work practices, and administrative controls, then the proper PPE must be selected, issued, and worn. The checklist found in Figure 24.3 may be of assistance in conducting a hazard analysis.

EYE AND FACE PROTECTION

Workers must be provided with eye protection whenever they are exposed to potential eye injuries during their work if work practices or engineering controls do not eliminate the risk of injury. Some causes of eye injuries include the following:

1. Dust and other flying particles, such as metal shavings or wool fibers
2. Molten metal that might splash
3. Acids and other caustic liquid chemicals that might splash
4. Blood and other potentially infectious body fluids that might splash, spray, or splatter
5. Intense light such as that created by welding arcs and lasers

Eye protection must shield against the specific hazard(s) encountered in the workplace. It must be reasonably comfortable to wear. Eye protection must not restrict vision or movement. It must be durable, easy to clean and disinfect, and must not interfere with the function of other required PPE. In addition, the American National Standards Institute (ANSI) has issued standard requirements for the design, construction, testing, and use of protective devices for eyes and face. OSHA requires that all protective eyewear purchased for employees meets the requirements of ANSI Z87.1-1989 for devices purchased after July 5, 1994, and ANSI Z87.1-1968 for devices purchased before that date.

Prescription eyeglasses designed for ordinary wear do not provide the level of protection necessary to protect against workplace hazards. Special care must be taken

Personal Protective Equipment Program

Hazard Assessment Certification Form

Date:	Location:
Assessment Conducted By:	
Specific Tasks Performed at This Location:	

Hazard Assessment and Selection of Personal Protective Equipment

I. Overhead Hazards

- Hazards to consider include:
- Suspended loads that could fall
- Overhead beams or loads that could be hit against
- Energized wires or equipment that could be hit against
- Employees work at elevated site who could drop objects on others below
- Sharp objects or corners at head level

Hazards Identified:

Head Protection

Hard Hat:	Yes	No
If yes, type: ☐ **Type A** (impact and penetration resistance, plus low-voltage electrical insulation) ☐ **Type B** (impact and penetration resistance, plus high-voltage electrical insulation) ☐ **Type C** (impact and penetration resistance)		

II. Eye and Face Hazards

- Hazards to consider include:
- Chemical splashes
- Dust
- Smoke and fames
- Welding operations
- Lasers/optical radiation
- Projectiles

Hazards Identified:

Eye Protection

Safety glasses or goggles	Yes	No
Face shield	Yes	No

III. Hand Hazards

- Hazards to consider include:
- Chemicals
- Sharp edges, splinters, etc.

FIGURE 24.3 PPE hazard analysis checklist. (*Source:* Courtesy of the Occupational Safety and Health Administration.)

- Temperature extremes
- Biological agents
- Exposed electrical wires
- Sharp tools, machine parts, etc.
- Material handling

Hazards Identified:

Hand Protection

Gloves	Yes	No
☐ Chemical resistant ☐ Temperature resistant ☐ Abrasion resistant ☐ Other (Explain)		

IV. Foot Hazards

- Hazards to consider include:
- Heavy materials handled by employees
- Sharp edges or points (puncture risk)
- Exposed electrical wires
- Unusually slippery conditions
- Wet conditions
- Construction/demolition

Hazards Identified:

Foot Protection

Safety shoes	Yes	No
Types: ☐ Toe protection ☐ Metatarsal protection ☐ Puncture resistant ☐ Electrical insulation ☐ Other (Explain)		

V. Other Identified Safety and/or Health Hazards:

Hazard	Recommended Protection

I certify that the above inspection was performed to the best of my knowledge and ability, based on the hazards present on ______________________.

____
(Signature)

FIGURE 24.3 (continued)

when choosing eye protection for employees who wear eyeglasses with corrective lenses. Consider choosing the following:

1. Prescription spectacles, with side shields and protective lenses meeting the requirements of ANSI Z87.1 that also correct the individual employee's vision
2. Goggles that can fit comfortably over corrective eyeglasses without disturbing the alignment of the eyeglasses
3. Goggles that incorporate corrective lenses mounted behind protective lenses

Protective eyewear must be provided to employees who wear contact lenses and are exposed to potential eye injury. Eye protection provided to these employees may also incorporate corrective eyeglasses. Thus, if an employee must wear eyeglasses in the event of contact lens failure or loss, he or she will still be able to use the same protective eyewear.

Often called safety spectacles, protective eyewear is made with safety frames constructed of metal and/or plastic and fitted with either corrective or plano impact-resistant lenses. They come with and without side shields, but most workplace operations require side shields. Impact-resistant spectacles can be used for moderate impact from particles produced by such jobs as carpentry, woodworking, grinding, and scaling. Side shields protect against particles that might enter the eyes from the side. Side shields are made of wire mesh or plastic. Eye-cup type side shields provide the best protection.

Each type of goggle is designed for specific hazards. Generally, goggles protect eyes, eye sockets, and the facial area immediately surrounding the eyes from impact, dust, and splashes. Some goggles fit over corrective lenses (see Figure 24.4). The specific use or application for each type of protective eyewear found in Figure 24.4 is shown in Table 24.1 which is to be used in conjunction with Figure 24.4.

Welding shields are constructed of vulcanized fiber or fiberglass and fitted with a filtered lens; these protective devices are designed for the specific hazards associated with welding. Welding shields protect employees' eyes from burns caused by infrared or intense radiant light, and they protect the face and eyes from flying sparks, metal spatter, and slag chips produced during welding, brazing, soldering, and cutting. The intensity of light or radiant energy produced by welding, cutting, or brazing operations varies according to a number of factors, including the task producing the light, the electrode size, and the arc current. Table 24.2 shows the minimum protective shade for a variety of welding, cutting, and brazing operations. To protect employees who are exposed to intense radiant energy, begin by selecting a shade too dark to see the welding zone. Then try lighter shades until one is found that allows a sufficient view of the welding zone without going below the minimum protective shade.

Laser safety goggles provide a range of protection against the intense concentrations of light produced by lasers. The type of laser safety goggles chosen will depend on the equipment and operating conditions in the workplace. Laser safety goggles specifically designed to protect employees' eyes from the specific intensity of light produced by the laser should be provided. The level of protection will vary according

1. **Goggles**, Flexible Fitting, Regular Ventilation
2. **Goggles**, Flexible Fitting, Hooded Ventilation
3. **Goggles**, Cushioned Fitting, Rigid Body
4. **Spectacles**, Metal Frame, Without Sideshields
5. **Spectacles**, Plastic Frame, With Sideshields
6. **Spectacles**, Metal-Plastic Frame, With Flat Fold Sideshields
7. **Welding Goggles**, Eyecup Type, Tinted Lenses

7a. **Chipping Goggles**, Eyecup Type, Clear Safety Lenses (not illustrated)

8. **Welding Goggles**, Eyecup Type, Tinted Plate Lens

8a. **Chipping Goggles**, Coverspec Type, Clear Safety Lenes (not illustrated)

9. **Welding Goggles**, Coverspec Type, Tinted Plate Lens
10. **Faceshield** (Available with Plastic or Mesh Window, Tinted/Transparent)
11. **Welding Helmets**

*Theses are also available without sideshields for limited use requiring only frontal protection.

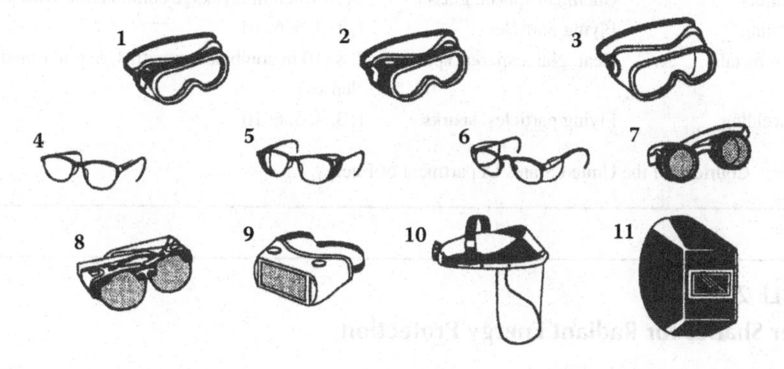

FIGURE 24.4 Types of protective eyewear. (Courtesy of the United States Department of Energy.)

to the level of radiation emitted by the laser. If employees are exposed to laser beams, then the maximum power density or intensity that the lasers can produce must be determined. Based on this knowledge, the lenses that will protect against this maximum intensity must be selected. Table 24.3 shows the minimum optical density of lenses required for various laser intensities. Employers with lasers emitting radiation between two measures of power density (or light blocking capability) must provide lenses that offer protection against the higher of the two intensities. Every pair of safety goggles, intended for use with laser beams, must bear a label with the following information:

- The laser wavelengths for which they are intended to be used
- The optical density of those wavelengths
- The visible light transmission

Face shields are transparent sheets of plastic extended from the brow to below the chin across the entire width of the employee's head. Some are polarized for glare protection. Choose face shields to protect employees' faces from nuisance dusts

TABLE 24.1
Function of Various Types of Protective Eyewear

Operation	Hazards	Recommended Protectors: (see Figure 21-2)
Acetylene-burning Acetylene-cutting Acetylene-welding	Sparks, harmful rays, molten metal, flying particles	7, 8, 9
Chemical handling	Splash, acid burns, fumes	2, 10 (For severe exposure add 10 over 2)
Chipping	Flying particles	1, 3, 4, 5, 6, 7A, 8A
Electric (arc) welding	Sparks, intense rays, molten metal	9, 11 (11 in combination with 4, 5, 6 in tinted lenses advisable)
Furnace operations	Glare, heat, molten metal	7, 8, 9 (For severe exposure add 10)
Grinding-light	Flying particles	1, 3, 4, 5, 6, 10
Grinding-heavy	Flying particles	1, 3, 7A, 8A (For severe exposure add 10)
Laboratory	Chemical splash, glass	2 (10 when in breakage combination with 4, 5, 6)
Machining	Flying particles	1, 3, 4, 5, 6, 10
Molten metals	Heat, glare, sparks, splash	7, 8 (10 in combination with 4, 5, 6 in tinted lenses)
Spot welding	Flying particles, sparks	1, 3, 4, 5, 6, 10

Source: Courtesy of the United States Department of Energy.

TABLE 24.2
Filter Shades for Radiant Energy Protection

Welding Operation	Shade Number
Shielded metal-arc welding 1/18-, 3/32-, 1/8-, 5/32-inch-diameter electrodes	10
Gas-shielded arc welding (nonferrous) 1/16-, 3/32-, 1/8-, 5/32-inch-diameter electrodes	11
Gas-shielded arc welding (ferrous) 1/16-, 3/32-, 1/8-, 5/32-inch-diameter electrodes	12
Shielded metal-arc welding 3/16-, 7/32-, 1/4-inch-diameter electrodes	12
5/16-, 3/8-inch-diameter electrodes	12
Atomic hydrogen welding	10–14
Carbon-arc welding	14
Soldering	2
Torch brazing	3 or 4
Light cutting, up to 1 inch	3 or 4
Medium cutting, 1 to 6 inches	4 or 5
Heavy cutting, over 6 inches	5 or 6
Gas welding (light), up to 1/8 inch	4 or 5
Gas welding (medium), 1/8 to 1/2 inch	5 or 6
Gas welding (heavy), over 1/2 inch	6 or 8

Source: Courtesy of the United States Department of Energy.

TABLE 24.3
Criteria for Laser Safety Glasses

Intensity, CW Maximum Power Density (watts/cm³)	Optical Density (O.D.)	Attenuation Factor
10^{-2}	5	10^{5}
10^{-1}	6	10^{6}
1.0	7	10^{7}
10^{+1}	8	10^{8}

Source: Courtesy of the United States Department of Energy.

and potential splashes or sprays of hazardous liquids. Face shields do not protect employees from impact hazards. However, face shields in combination with goggles or safety spectacles to protect against impact hazards, even in the absence of dust or potential splashes, provide for additional protection beyond that offered by goggles or spectacles alone.

Each kind of protective eyewear is designed to protect against specific hazards. By completing the hazard assessment of the workplace outlined in the previous section, you will be able to identify the specific workplace hazards that pose a threat to employees' eyes and faces. Train employees to use and care for the eye protection provided. Employees must know how to clean their eye protectors. Allow time at the end of their shifts to do the following:

- Disassemble goggles or spectacles.
- Thoroughly clean all parts with soap and warm water.
- Carefully rinse off all traces of soap.
- Replace all defective parts.

Occasionally disinfect protective eyewear. To do so, after cleaning, the following can be done:

1. Immerse and swab all parts for ten minutes in a germicidal solution.
2. Remove all parts from the solution and hang in a clean place to air dry at room temperature or with heated air.
3. Do not rinse the parts after submerging them in the disinfectant. Rinsing will remove the germicidal residue that remains after drying.
4. Ultraviolet disinfecting and spray-type disinfecting solutions may also be used after washing.

If the goggles or spectacles do not need to be individually designed to incorporate an employee's corrective lenses and the eyewear is disinfected between uses by different employees, more than one employee may use the same set of protective eyewear.

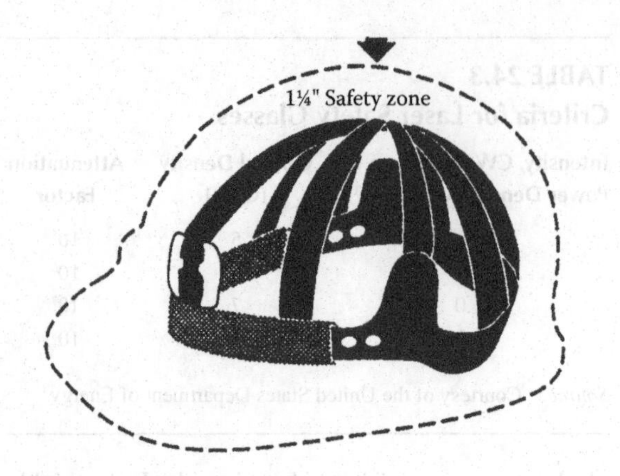

FIGURE 24.5 Head protection suspension clearance. (Courtesy of the United States Department of Energy.)

HEAD PROTECTION

Employers must provide head protection for employees if objects might fall from above and strike them on the head, if they might bump their heads against fixed objects such as exposed pipes or beams, or if they work near exposed electrical conductors. In general, protective helmets, or hard hats, should

- Resist penetration by objects.
- Absorb the shock of a blow.
- Be water resistant and slow burning.
- Come with instructions explaining proper adjustment and replacement of the suspension and headband.

Hard hats require a hard outer shell and a shock-absorbing lining. The lining should incorporate a headband and straps that suspend the shell from 1 to 1¼ inches (2.54 to 3.18 centimeters) away from the user's head (see Figure 24.5). This design provides shock absorption during impact and ventilation during wear. As with devices designed to protect eyes, the design, construction, testing, and use of protective helmets must meet standards established by ANSI. Protective helmets purchased after July 5, 1994, must comply with ANSI Z89.1-1986 and they fall into Type I under the new ANSI Z89.1-1997. Type II helmets must meet new requirements for (1) impact resistance from blows to the front, back, sides, and top of the head; (2) off-center penetration resistance; and (3) chin strap retention. The old classes listed below have been changed to Class G (General), Class E (Electrical), and Class C (Conductive-no electrical protection). Hard hats are divided into three industrial classes:

1. Class A now Class G: These helmets are for general service. They provide good impact protection but limited voltage protection. They are used mainly in mining, building construction, shipbuilding, lumbering, and manufacturing.

2. Class B now Class E: Choose Class B helmets if employees are engaged in electrical work. They protect against falling objects and high-voltage shocks and burns.
3. Class C now Class C: Designed for comfort, these lightweight helmets offer limited protection. They protect workers from bumping against fixed objects but do not protect against falling objects or electric shock.

Look at the inside of any protective helmet being considered for employees, and a label should be seen showing the manufacturer's name, the ANSI standard it meets, and its class. Figure 24.5 shows the basic design of hard hats. Each kind of protective helmet is designed to protect against specific hazards. By completing the previous hazard assessment outlined, identify the specific workplace hazards that pose a threat to employees' heads.

Issuing appropriate head protection to employees is a major first step but employers must make sure that the hard hats continue to provide sufficient protection for employees. Do this by training employees in the proper use and maintenance of hard hats, including daily inspection of them. If employees identify any defects (e.g., the suspension system shows signs of deterioration such as cracking, tearing, or fraying), then remove the hard hats from service. Also remove protective headwear if the suspension system no longer holds the shell from 1 to 1¼ inches (2.54 to 3.18 centimeters) away from the employee's head; if the brim or shell is cracked, perforated, or deformed; or if the brim or shell shows signs of exposure to heat, chemicals, ultraviolet light, or other radiation, which may appear as loss of surface gloss, chalking, or flaking (a sign of advanced deterioration).

In specific situations, a chin strap must be provided for the protective helmets worn by employees. For example, this type of helmet must be used to protect employees working at elevated levels, whether in an aerial lift, at the edge of a pit, or around helicopters. The chin straps should be designed to prevent the hard hats from being bumped off the employees' heads.

Long hair (longer than 4 inches) can be drawn into machine parts such as chains, belts, rotating devices, suction devices, and blowers. Hair may even be drawn into machines otherwise guarded with mesh. Although employees are not required to cut their hair, they must be required to cover and protect their hair with bandanas, hair nets, turbans, soft caps, or the like. These items, however, must not themselves present a hazard.

Paints, paint thinners, and some cleaning agents can weaken the shell of the hard hat and may eliminate electrical resistance. Consult the helmet manufacturer for information on the effects of paint and cleaning materials on their hard hats. Keep in mind that paint and stickers can also hide signs of deterioration in the hard hat shell. Limit their use.

Ultraviolet light and extreme heat, such as that generated by sunlight, can reduce the strength of hard hats. Therefore, employees should not store or transport hard hats on the rear-window shelves of automobiles or otherwise in direct sunlight. It is thus very important that workers be trained to maintain and care for their head protection. Instruct employees to clean their protective helmets periodically by immersing them for 1 minute in hot (approximately 140°F, or 60°C) water and detergent, scrubbing, and rinsing in clear hot water. Training should communicate both the importance of wearing head protection and taking proper care of it.

FOOT AND LEG PROTECTION

The employer must provide foot and leg protection if a workplace hazard assessment reveals potential dangers to these parts of the body. Some of the potential hazards that might be identified include the following:

- Heavy objects such as barrels or tools that might roll onto or fall on employees' feet
- Sharp objects such as nails or spikes that might pierce the soles or uppers of ordinary shoes
- Molten metal that might splash onto feet or legs
- Hot or wet surfaces
- Slippery surfaces

The type of foot and leg protection provided to employees will depend on the specific workplace hazards identified and the specific parts of the feet or legs exposed to potential injury. Safety footwear must meet minimum compression and impact performance standards and testing requirements established by ANSI. Protective footwear purchased after July 5, 1994, must meet the requirements of ANSI Z41-1991. Protective footwear bought before that date must comply with ANSI Z41-1967. There are many forms of foot and leg protection to choose from (see Figure 24.6).

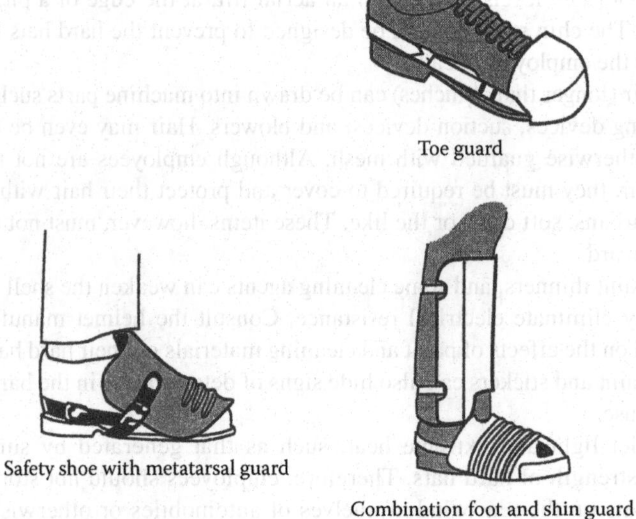

FIGURE 24.6 Examples of foot and leg protection. (Courtesy of United States Department of Energy.)

Leggings are used to protect the lower legs and feet from heat hazards such as molten metal or welding sparks. Safety snaps allow for quick removal of the leggings. Metatarsal guards are made of aluminum, steel, fiber, or plastic; these guards may be strapped to the outside of shoes to protect the instep area from impact and compression. Toe guards may be made of steel, aluminum, or plastic. They fit over the toes of regular shoes and protect only the toes from impact and compression hazards. Combination foot and shin guards may be used in combination with toe guards when greater protection is needed.

Sturdy safety shoes should have impact-resistant toes and heat-resistant soles to protect against hot work surfaces common in roofing, paving, and hot metal industries. The metal insoles of some safety shoes protect against puncture wounds. Safety shoes may also be designed to be electrically conductive to prevent the buildup of static electricity in areas with the potential for explosive atmospheres, or nonconductive to protect workers from workplace electrical hazards. All safety shoes must comply with the ANSI standard(s) mentioned above. In addition, depending on the types of worker exposures, there may be a need to provide specially designed safety shoes such as conductive or electrical-hazard safety shoes.

Electrically conductive shoes protect against the buildup of static electricity. Essentially, these shoes ground the employees who are wearing them. Employees working in explosive and hazardous locations such as explosives manufacturing facilities or grain elevators must wear conductive shoes to reduce the risk of static electricity buildup on an employee's body that could produce a spark and cause an explosion or fire. During training, employees must be instructed not to use foot powder or wear socks made of silk, wool, or nylon with conductive shoes. Foot powder insulates and retards the conductive ability of the shoes. Silk, wool, and nylon produce static electricity. Conductive shoes are not general-purpose shoes and must be removed upon completion of the tasks for which they are required. Employees exposed to electrical hazards must never wear conductive shoes.

Electrical hazard, safety-toe shoes are nonconductive and will prevent employees' feet from completing an electrical circuit to the ground. They can protect employees against open circuits of up to 600 volts in dry conditions. Electrical hazard, safety-toe shoes should be used in conjunction with other insulating equipment and precautions to reduce or eliminate the potential for employees' bodies or parts of their bodies to provide a path for hazardous electrical energy. Note: Nonconductive footwear must not be used in explosive or hazardous locations; in such locations, electrically conductive shoes are required. Train employees to recognize that the insulating protection of electrical hazard, safety-toe shoes may be compromised if

- The shoes are wet.
- The rubber soles are worn through.
- Metal particles become embedded in the soles or heels.
- Other parts of the employees' bodies come into contact with conductive, grounded items.

In addition to insulating employees' feet from the extreme heat of molten metal, foundry shoes prohibit hot metal from lodging in shoe eyelets, tongues, or other parts. These snug-fitting leather or leather-substitute shoes have leather or rubber soles and rubber heels. In addition, all foundry shoes must have built-in safety toes.

HAND AND ARM PROTECTION

If the workplace hazard assessment reveals that employees risk injury to their hands and arms, and engineering and work practice controls do not eliminate such hazards, then hand and arm protection must be provided to employees. The injuries that may need to be guarded against in the workplace include burns, bruises, abrasions, cuts, punctures, fractures, amputations, or chemical exposures.

For many workplace operations, machine guards such as point-of-operation guards will be sufficient. For example, install a barrier that makes it impossible for employees to put their hands at the point where a table saw blade makes contact with the wood it cuts. For other hazardous operations, employers may be able to institute work procedures that eliminate the risk of injury to employees' hands or arms. When such measures fail to eliminate the hazard, however, protective gloves will be the primary means of protecting employees' hands. When the risk of injury includes the arm, protective sleeves, often attached to the gloves, may be appropriate.

There is no one kind of glove that universally protects the worker from hand injury hazards. The nature of the hazard(s) and the operation to be performed will determine the selection of gloves. The variety of potential occupational hand injuries may make selecting the appropriate pair of gloves more difficult than choosing other protective equipment. Take care to choose gloves designed for the particular circumstances of the particular workplace. Gloves are made from a wide variety of materials and are designed for virtually every workplace hazard. In general, however, they may be divided into four groups:

1. Durable work gloves made of metal mesh, leather, or canvas
2. Fabric and coated fabric gloves
3. Chemical and liquid resistant gloves
4. Insulating rubber gloves

Metal Mesh, Leather, or Canvas Gloves

Sturdy gloves made from metal mesh, leather, or canvas provide protection against cuts, burns, and sustained heat. Leather gloves protect against sparks, moderate heat, blows, chips, rough objects, and some low voltage electrical shocks (see Figure 24.7). Welders, in particular, need the durability of higher-quality leather gloves.

Aluminized gloves are usually used for welding, furnace, and foundry work because they provide reflective and insulating protection against heat. Aluminized gloves require an insert made of synthetic materials that protects against heat and cold. Aramid fiber gloves are made from synthetic material that protects against heat and cold. Many glove manufacturers use aramid fiber to make gloves that are cut- and abrasive-resistant and wear well.

Several manufacturers make gloves with other synthetic fabrics that offer protection against heat and cold. In addition to protection against temperature extremes,

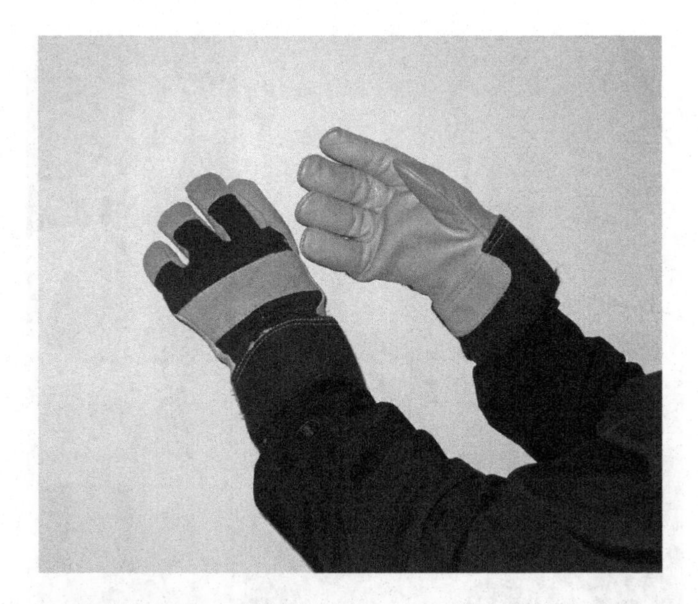

FIGURE 24.7 Example of leather gloves.

gloves made with other synthetic materials are cut- and abrasive-resistant (see Figure 24.8) and may withstand some diluted acids. These materials do not stand up well against alkalis and solvents.

Fabric and Coated Fabric Gloves

These gloves are made of cotton or other fabric to provide varying degrees of protection. Fabric gloves can protect against dirt, slivers, chaffing, and abrasion. They do not provide sufficient protection, however, to be used with rough, sharp, or heavy materials. However, adding a plastic coating to some fabric gloves strengthens them and makes them effective protection for a variety of tasks.

Coated fabric gloves made of cotton flannel with napping on one side work well for many tasks. By coating the un-napped side with plastic, fabric gloves are transformed into general-purpose hand protection offering slip-resistant qualities. These gloves are used for tasks ranging from handling bricks and wire rope to handling chemical containers in laboratory operations. When selecting gloves to protect against chemical exposure hazards, always check with the manufacturer (or review the manufacturer's product literature) to determine the gloves' effectiveness against the specific chemicals and conditions in the workplace.

Chemical- and Liquid-Resistant Gloves

Gloves made of rubber (latex, nitrile, or butyl), plastic, or synthetic rubber-like material (such as Neoprene) protect workers from burns, irritation, and dermatitis caused by contact with oils, greases, solvents, and other chemicals. The use of rubber gloves also reduces the risk of exposure to blood and other potentially infectious substances.

FIGURE 24.8 Example of a cut-resistant glove.

Butyl rubber gloves protect against nitric acid, sulfuric acid, hydrofluoric acid, red fuming nitric acid, rocket fuels, and peroxide. Highly impermeable to gases, chemicals, and water vapor, butyl rubber gloves also resist oxidation and ozone corrosion. In addition, they resist abrasion and remain flexible at low temperatures.

Natural latex or rubber gloves are comfortable to wear, and the pliability of latex gloves as well as their protective qualities make them a popular general-purpose glove. In addition to resisting abrasions caused by sandblasting, grinding, and polishing, these gloves protect workers' hands from most water solutions of acids, alkalis, salts, and ketones. When selecting hand protection, employers should be aware that latex gloves have caused allergic reactions in some individuals and thus may not be appropriate for all employees. Hypoallergenic gloves, glove liners, and powderless gloves are possible alternatives for individuals who are allergic to latex gloves.

Neoprene gloves have good pliability, finger dexterity, high density, and tear resistance, which protect against hydraulic fluids, gasoline, alcohols, organic acids, and alkalis.

Nitrile rubber gloves are sturdy and provide protection from chlorinated solvents such as trichloroethylene and perchloroethylene. Although intended for jobs requiring dexterity and sensitivity, nitrile gloves stand up to heavy use even after prolonged exposure to substances that cause other gloves to deteriorate. In addition, nitrile gloves resist abrasions, punctures, snags, and tears.

Appendix F provides an example of a glove selection chart. Most manufacturers of gloves have developed their own glove selection chart that should be used when purchasing their brands of gloves.

BODY PROTECTION

Employers must provide body protection for employees if they are threatened with bodily injury of one kind or another while performing their jobs, and if engineering, work practices, and administrative controls have failed to eliminate such hazards. Workplace hazards that could cause bodily injury include intense heat; splashes of hot metals and other hot liquids; impacts from tools, machinery, and materials; cuts; hazardous chemicals; contact with potentially infectious materials such as blood; or radiation.

As with all protective equipment, protective clothing is available to protect against specific hazards. Personal protective equipment must be provided for the parts of the body exposed to possible injury. Depending on hazards in the workplace, employees may need to be provided with one or more of the following: vests, jackets, aprons, coveralls, surgical gowns, or full body suits. If the hazard assessment indicates the need for full body protection against toxic substances or harmful physical agents, employers must

- Inspect the clothing carefully.
- Ensure proper fit.
- Ensure that protective clothing functions properly for its intended use.

Protective clothing comes in a variety of materials, each suited to particular hazards. Conduct a hazard assessment to identify the sources of any possible bodily injury. Install any feasible engineering controls and institute work practice controls to eliminate the hazards. If the possibility of bodily injury still exists, provide protective clothing constructed of material that will protect against the specific hazards in your workplace. Materials for protective clothing include the following:

1. Paper-like fiber disposable suits made of this material provide protection against dust and splashes.
2. Treated wool and cotton protective clothing adapts well to changing workplace temperatures and is comfortable as well as fire resistant. Treated cotton and wool clothing protects against dust, abrasions, and rough and irritating surfaces.
3. Duck, which is a closely woven cotton fabric, protects employees against cuts and bruises while they handle heavy, sharp, or rough materials.
4. Leather protective clothing is often used to protect against dry heat and flame.

5. Rubber, rubberized fabrics, neoprene, and plastics protective clothing protect against certain acids and other chemicals.

The purpose of chemical protective clothing and equipment is to shield or isolate individuals from the chemical, physical, and biological hazards that may be encountered during hazardous materials operations. During chemical operations, it is not always apparent when exposure occurs. Many chemicals pose invisible hazards and offer no warning properties.

It is important that protective clothing users realize that no single combination of protective equipment and clothing is capable of protecting against all hazards. Thus, protective clothing should be used in conjunction with other protective methods. For example, engineering or administrative controls to limit chemical contact with personnel should always be considered an alternative measure for preventing chemical exposure. The use of protective clothing can itself create significant wearer hazards, such as heat stress, physical and psychological stress, in addition to impaired vision, mobility, and communication. In general, the greater the level of chemical protective clothing, the greater the associated risks to workers. For any given situation, equipment and clothing should be selected that provide an adequate level of protection. Overprotection as well as under-protection can be hazardous and should be avoided.

Protective Clothing Applications

Protective clothing must be worn whenever the wearer faces potential hazards arising from chemical exposure. Some examples include

- Emergency response
- Chemical manufacturing and process industries
- Hazardous waste site cleanup and disposal (see Figure 24.9)
- Asbestos removal and other particulate operations
- Agricultural application of pesticides

Within each application, there are several operations that require chemical protective clothing. For example, in emergency response, the following activities dictate chemical protective clothing use:

- Site survey
- Rescue
- Spill mitigation
- Emergency monitoring
- Decontamination

Selection of Protective Clothing

The approach in selecting personal protective clothing must encompass an "ensemble" of clothing and equipment items that are easily integrated to provide both an

FIGURE 24.9 Worker on a hazardous waste site with protective ensemble.

appropriate level of protection and still allow one to carry out activities involving chemicals. In many cases, simple protective clothing by itself may be sufficient to prevent chemical exposure, such as wearing gloves in combination with a splash apron and face shield (or safety goggles).

The following is a checklist of components that may form the chemical protective ensemble:

- Protective clothing (suit, coveralls, hoods, gloves, boots)
- Respiratory equipment (SCBA, combination SCBA/SAR, air purifying respirators)
- Cooling system (ice vest, air circulation, water circulation)
- Communications device
- Head protection
- Eye protection
- Ear protection
- Inner garment
- Outer protection (overgloves, overboots, flashcover)

Ensemble Selection Factors

- Specific chemical hazards
- Physical environment
- Duration of exposure
- Protective clothing or equipment available

Classification of Protective Clothing

Personal protective clothing includes the following:

- Fully encapsulating suits
- Nonencapsulating suits
- Gloves, boots, and hoods
- Firefighters' protective clothing
- Proximity, or approach clothing
- Blast or fragmentation suits
- Radiation-protective suits

Be aware that different materials will protect against different chemicals and physical hazards. When chemical or physical hazards are present, check with the clothing manufacturer to make sure that the material selected will provide protection against the specific chemical or physical hazards in the workplace for which they are procured (see Table 24.4).

HEARING PROTECTION

Determining the need to provide hearing protection can be tricky. Employees' exposure to excessive noise depends upon a number of factors:

- How loud is the noise, as measured in decibels (dB)?
- What is the duration of each employee's exposure to the noise?
- Do employees move between separate work areas with different noise levels?
- Is noise generated from one source or multiple sources?

Usually, the louder the noise, the shorter the exposure time before employers must provide hearing protection. For instance, employees may be exposed to a noise level of 90 dBA for 8 hours per day before hearing protection is required for them. Suppose, however, that the noise level reaches 115 dBA in the workplace. Then the employer must provide hearing protection if employees' anticipated exposure exceeds 15 minutes.

Noises are considered continuous if the interval between occurrences of the maximum noise level is 1 second or less. Noises not meeting this definition are considered impact or impulse noises. Exposure to impact or impulse noises (loud momentary explosions of sound) must not exceed 140 dBA. Examples of impact or impulse noises may include the noise from a powder-actuated nail gun, the noise from a punch press, or the noise from drop hammers.

TABLE 24.4
Types of Protective Clothing for Full Body Protection

Types of Protective Clothing for Full Body Protection Description	Type of Protection	Use Considerations
Fully encapsulating suit One-piece garment. Boots and gloves may be integral, attached and replaceable, or separate.	Protects against splashes, dust gases, and vapors.	Does not allow body heat to escape. May contribute to heat stress in wearer, particularly if worn in conjunction with a closed-circuit SCBA; a cooling garment may be needed. Impairs worker mobility, vision, and communication.
Nonencapsulating suit Jacket, hood, pants or bib overalls, and one-piece coveralls.	Protects against splashes, dust, and other materials but not against gases and vapors. Does not protect parts of head or neck.	Do not use where gas-tight or pervasive splashing protection is required. May contribute to heat stress in wearer. Tape-seal connections between pant cuffs and boots and between gloves and sleeves.
Aprons, leggings, and sleeve protectors Fully sleeved and gloved apron. Separate coverings for arms and legs. Commonly worn over nonencapsulating suit.	Provides additional splash protection of chest, forearms, and legs.	Whenever possible, should be used over a nonencapsulating suit to minimize potential heat stress. Useful for sampling, labeling, and analysis operations. Should be used only when there is a low probability of total body contact with contaminants.
Firefighters' protective clothing Gloves, helmet, running or bunker coat, running or bunker pants (NFPA No. 1971, 1972, 1973, and boots (1974).	Protects against heat, hot water, and some particles. Does not protect against gases and vapors, or chemical permeation or degradation. NFPA Standard No. 1971 specifies that a garment consists of an outer shell, an inner liner, and a vapor barrier with a minimum water penetration of 25 lb/in.^{2} (1.8 kg/cm^2) to prevent passage of hot water.	Decontamination is difficult. Should not be worn in areas where protection against gases, vapors, chemical splashes, or permeation is required.
Proximity garment (approach suit) One- or two-piece overgarment with boot covers, gloves, and hood of aluminized nylon or cotton fabric. Normally worn over other protective clothing, firefighters' bunker gear, or flame-retardant coveralls.	Protects against splashes, dust, gases, and vapors.	Does not allow body heat to escape. May contribute to heat stress in wearer, particularly if worn in conjunction with a closed-circuit SCBA; a cooling garment may be needed. Impairs worker mobility, vision, and communication.

continued

TABLE 24.4 (continued)
Types of Protective Clothing for Full Body Protection

Types of Protective Clothing for Full Body Protection Description	Type of Protection	Use Considerations
Blast and fragmentation suit Blast and fragmentation vests and clothing, bomb blankets, and bomb carriers.	Provides some protection against very small detonations. Bomb blankets and baskets can help redirect a blast.	Does not provide for hearing protection.
Radiation-contamination protective suit Various types of protective clothing designed to prevent contamination of the body by radioactive particles.	Protects against alpha and beta particles. Does *not* protect against gamma radiation.	Designed to prevent skin contamination. If radiation is detected on site, consult an experienced radiation expert and evacuate personnel until the radiation hazard has been evaluated.
Flame/fire retardant coveralls Normally worn as an undergarment.	Provides protection from flash fires.	Adds bulk and may exacerbate heat stress problems and impair mobility

Source: Courtesy of the Occupational Safety and Health Administration.

As with other types of hazards, the employer must implement feasible engineering controls and work practices before resorting to PPE such as earplugs or earmuffs. If engineering and work practice controls do not lower employee exposure to workplace noise to acceptable levels, then employees must be provided with appropriate PPE.

If employees move from location to location and the noise level is different in each location, or if the noise level in an area changes throughout the day (e.g., equipment turns on and off), then an "equivalent noise factor" must be calculated to determine whether employers must provide hearing protection. This means measuring the noise level at each location in which the employee works. For each location, divide the actual time the employee spends there by the permissible duration for the noise at the measured level found in Table 24.5. Add all the results from dividing C/T. If the total is greater than 1, employers must implement engineering controls, work practices, or provide hearing protection to exposed employees. The formula for calculating this exposure is as follows:

$$F_e = (C_1 / T_1) + (C_2 / T_2) + \ldots (C_n / T_n)$$

where F_e is the equivalent noise factor, C is the period of actual noise exposure at an essentially constant level at each location in which the employee works, and T is the permissible duration of noise exposure at an essentially constant noise level from Table 24.5.

Plain cotton does not effectively protect against occupational noise. There are several products to choose from that are effective in protecting employees' hearing.

TABLE 24.5
Permissible Noise Exposure Limits

Duration per Day (hours)	Sound Level dBA Slow Response
8	90
6	92
4	95
3	97
2	100
1-1/2	102
1	105
½	110
1/4 or less	115

Source: Courtesy of United States Department of Energy.

FIGURE 24.10 Workers wearing ear muffs as hearing protection.

Single-use earplugs made of waxed cotton, foam, or fiberglass wool are self-forming and, when properly inserted, work as well as most molded earplugs. Preformed and molded earplugs can be reusable or at times single use/disposable; this type of plug must be individually fitted by a professional. Nondisposable plugs should be cleaned after each use.

Earmuffs require a perfect seal around the ear (see Figure 24.10). Glasses, long sideburns, long hair, and facial movements such as chewing may reduce the protective value of earmuffs. It is possible to purchase special earmuffs designed for use with eyeglasses or beards.

Hearing protectors only reduce the amount of noise that gets through to the ears. The amount of this reduction is referred to as attenuation. Attenuation differs according to the type of hearing protection used and how well it fits. Hearing protectors must be capable of achieving the attenuation needed to reduce the employee's noise exposure to within the acceptable limits. OSHA's regulation in Appendix B of 29 CFR 1910.95, Occupational Noise Exposure, describes methods for estimating the attenuation of a particular hearing protector based on the device's noise reduction rating (NRR). Manufacturers of hearing protection devices must report the device's NRRs on the product packaging.

Train your employees to use hearing protection. Once hearing protection and training in how to use it has been provided to employees, its effectiveness should be monitored. If employees are exposed to occupational noise at or above 85 dBA averaged over an 8-hour period, then a hearing conservation program must be instituted that includes regular testing of employees' hearing by qualified professionals. The OSHA occupational noise standard, at 29 CFR 1910.95, provides the requirements for a hearing conservation program.

RESPIRATORY PROTECTION

In the respiratory protection program, hazard assessment and selection of proper respiratory PPE is conducted in the same manner as for other types of PPE. To control those occupational diseases caused by breathing air contaminated with harmful dusts, fogs, fumes, mists, gases, smokes, sprays, or vapors, the primary objective shall be to prevent atmospheric contamination. This should be accomplished as far as feasible by accepted engineering control measures (e.g., enclosure or confinement of the operation, general and local ventilation, and substitution of less toxic materials). When effective engineering controls are not feasible, or while they are being instituted, appropriate respirators must be used. (See OSHA Standards Respiratory Protection, 29 CFR 1910.134.)

Management must develop a respiratory program, then implement it and ensure that workers follow its requirements. OSHA requires that voluntary use of respirators, when not required by the company, must be controlled as strictly as under required circumstances. To prevent violations of the Respiratory Protection Standard, employees are not allowed voluntary use of their own or company-supplied respirators of any type. The exception is employees whose only use of respirators involves the voluntary use of filtering (nonsealing) face pieces (dust masks).

Evaluations of the workplace are necessary to ensure that the written respiratory protection program is being properly implemented. This includes consulting with employees to ensure that they are using the respirators properly. Evaluations must be conducted as necessary to ensure that the provisions of the current written program are being effectively implemented and that it continues to be effective.

Program evaluation includes discussions with employees required to use respirators to assess the employees' views on program effectiveness and to identify any problems. Any problems that are identified during this assessment must be corrected. Factors to be assessed include, but are not limited to

1. Respirator fit (including the ability to use the respirator without interfering with effective workplace performance)

2. Appropriate respirator selection for the hazards to which the employee is exposed
3. Proper respirator use under the workplace conditions the employee encounters
4. Proper respirator maintenance

The company needs to retain written information regarding medical evaluations, fit testing, and the respirator program. This information will facilitate employee involvement in the respirator program, assist the company in auditing the adequacy of the program, and provide a record for compliance determinations by OSHA.

Effective training for employees who are required to use respirators is essential. The training must be comprehensive, understandable, and recur annually and more often if necessary. Training will be provided prior to requiring the employee to use a respirator in the workplace. The training must ensure that each employee can demonstrate knowledge of at least the following:

1. Why the respirator is necessary and how improper fit, usage, or maintenance can compromise the protective effect of the respirator
2. Limitations and capabilities of the respirator
3. How to use the respirator effectively in emergency situations, including situations in which the respirator malfunctions
4. How to inspect, put on and remove, use, and check the seals of the respirator
5. What the procedures are for maintenance and storage of the respirator
6. How to recognize medical signs and symptoms that may limit or prevent the effective use of respirators

Retraining shall be conducted annually when changes in the workplace occur or the type of respirator renders previous training obsolete. Inadequacies in the employee's knowledge or use of the respirator indicates that the employee has not retained the requisite understanding or skill, and if another situation arises in which retraining appears necessary, the employer must proceed as soon as possible with retraining to ensure safe respirator use.

Training will be conducted by approved instructors. Training is divided into the following sections:

1. Classroom instruction:
   a. Overview of the company's respiratory protection program and OSHA's standard
   b. Respiratory protection safety procedures
   c. Respirator selection
   d. Respirator operation and use
   e. Why respirator is necessary
   f. How improper fit, usage, or maintenance can compromise the protective effect
   g. Limitations and capabilities of the respirator
   h. How to use the respirator effectively in emergency situations, including respirator malfunctions

i. How to inspect, put on and remove, use, and check the seals of the respirator
j. What the procedures are for maintenance and storage of the respirator
k. How to recognize medical signs and symptoms that may limit or prevent the effective use of respirators
l. Change-out schedule and procedure for air purifying respirators

2. Fit testing:
a. Hands-on respirator training
b. Respirator inspection
c. Respirator cleaning and sanitizing
d. Recordkeeping
e. Respirator storage
f. Respirator fit check
g. Emergencies

Basic Respiratory Protection Safety Procedures

1. Only authorized and trained employees may use respirators. Those employees may use only the respirator that they have been trained on and properly fitted to use.
2. Only physically qualified employees may be trained and authorized to use respirators. A preauthorization and annual certification by a qualified physician will be required and maintained. Any changes in an employee's health or physical characteristics will be reported to the occupational health department and will be evaluated by a qualified physician.
3. Only the proper prescribed respirator or self-containing breathing apparatus (SCBA) may be used for the job or work environment. Air cleansing respirators may be worn in work environments when oxygen levels are between 19.5 percent and 23.5 percent and when the appropriate air cleansing canister, as determined by the manufacturer and approved by NIOSH for the known hazardous substance, is used. SCBAs will be worn in oxygen-deficient environments (below 19.5 percent oxygen).
4. Employees working in environments where a sudden release of a hazardous substance is likely must wear an appropriate respirator for that hazardous substance. For example, employees working in an ammonia compressor room would have an ammonia APR respirator on their person.
5. Only SCBAs can be used in oxygen-deficient environments, or environments with an unknown hazardous substance, or unknown quantity of a known hazardous substance or any environment that is determined "Immediately Dangerous to Life or Health" (IDLH).
6. Employees with respirators loaned on "permanent check-out" must be responsible for their sanitation, proper storage, and security. Respirators damaged by normal wear must be repaired or replaced by the company when returned.
7. The last employee using a respirator or SCBA available for general use must be responsible for its proper storage and sanitation. Monthly and after each

use, every respirator must be inspected with documentation to ensure its availability for use.

8. All respirators must be located in a clean, convenient, and sanitary location.
9. In the event that employees must enter a confined space, or work in environments with hazardous substances that would be dangerous to life or health, each employee should wear the required SCBA for the environment. Employees are to follow the established emergency response program for hazardous materials and/or confined space entry program when applicable.
10. Management must establish and maintain surveillance of jobs and workplace conditions and monitor the degree of employee exposure or stress in order to maintain the proper procedures and provide the necessary PPE.
11. Management must establish and maintain safe operating procedures for the safe use of PPE. There must be strict enforcement and disciplinary action for failure to follow all general and specific safety rules or standard operating procedures for general PPE use. Procedures must be maintained as an attachment to the respiratory protection program. Standard operating procedures for PPE use, under emergency response situations, must be maintained as an attachment to the emergency response program.

Respirator User Policies

Adherence to the following guidelines will help ensure the proper and safe use of respiratory equipment. Employees must wear only the respirator they have been instructed to use. For example, they should not wear an SCBA if they have been assigned and fitted for a respirator. Each employee must wear the correct respirator for the particular hazard. Some situations, such as chemical spills or other emergencies, may require a higher level of protection than the respirator can handle. Also, the proper cartridge must be matched to the hazard (a cartridge designed for dusts and mists will not provide protection against chemical vapors). The employee must check the respirator for a good fit before each use. Positive and negative fit checks should be conducted. He or she must check the respirator for deterioration before and after use, being sure to not use a defective respirator. Employees must recognize indications that cartridges and canisters are at their end of service. If in doubt, they must change the cartridges or canisters before using the respirator; then each must practice moving and working while wearing the respirator in order to get used to it. Require employees to clean the respirator after each use, thoroughly dry it, and place the cleaned respirator in a sealable plastic bag. Also, have them store respirators carefully in a protected location away from excessive heat, light, and chemicals.

Selection of Respirators

The company is to evaluate the respiratory hazard(s) in each workplace, identify relevant workplace and user factors, and base respirator selection on these factors. Included are estimates of employee exposures to respiratory hazard(s) and an identification of the contaminant's chemical state and physical form. This selection includes appropriate protective respirators for use in IDLH atmospheres, and limits

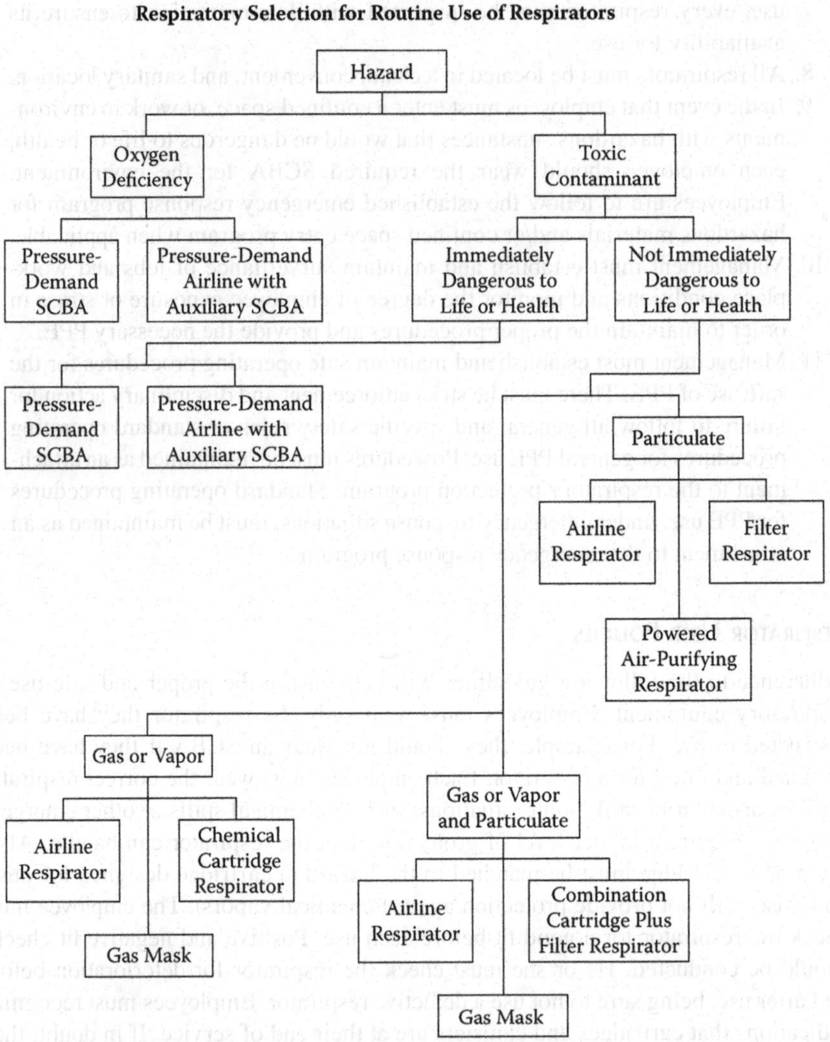

Respiratory Selection for Routine Use of Respirators

Hazard

Oxygen Deficiency

Pressure-Demand SCBA

Pressure-Demand Airline with Auxiliary SCBA

Toxic Contaminant

Immediately Dangerous to Life or Health

Pressure-Demand SCBA

Pressure-Demand Airline with Auxiliary SCBA

Not Immediately Dangerous to Life or Health

Particulate

Airline Respirator

Filter Respirator

Powered Air-Purifying Respirator

Gas or Vapor

Airline Respirator

Chemical Cartridge Respirator

Gas Mask

Gas or Vapor and Particulate

Airline Respirator

Combination Cartridge Plus Filter Respirator

Gas Mask

FIGURE 24.11 Respirator selection flow chart. (Courtesy of the Occupational Safety and Health Administration.)

the selection and use of air-purifying respirators (APR). All selected respirators are to be NIOSH-certified (see Figure 24.11).

Filter classifications are marked on the filter or filter package. These classifications are N-Series: Not Oil Resistant (approved for non-oil particulate contaminants such as dust, fumes, or mists not containing oil); R-Series: Oil Resistant (approved for all particulate contaminants, including those containing oil such as dusts, mists, or fumes—time restriction of 8 hours when oils are present); and P-Series: Oil Proof (approved for all particulate contaminants including those containing oil such as dust, fumes, or mists). (See manufacturer's time use restrictions on packaging.)

FIGURE 24.12 Example of a self-contained breathing apparatus (SCBA). (Courtesy of the National Institute for Occupational Safety and Health.)

Respirators used in IDLH atmospheres are a full-facepiece pressure demand SCBA certified by NIOSH for a minimum service life of 30 minutes, or a combination full-facepiece pressure demand supplied-air respirator (SAR) with an auxiliary self-contained air supply. Respirators provided only for escape from IDLH atmospheres shall be NIOSH-certified for escape from the atmosphere in which they will be used (see Figure 24.12).

The respirators selected shall be adequate to protect the health of the employee and ensure compliance with all other OSHA statutory and regulatory requirements, under routine and reasonably foreseeable emergency situations. The respirator selected shall be appropriate for the chemical state and physical form of the contaminant.

Identification of Filters and Cartridges

All filters and cartridges shall be labeled and color-coded with the NIOSH approval label. The label must not be removed and must remain legible. A change schedule for filters and canisters has been developed to ensure that these elements of the respirators remain effective.

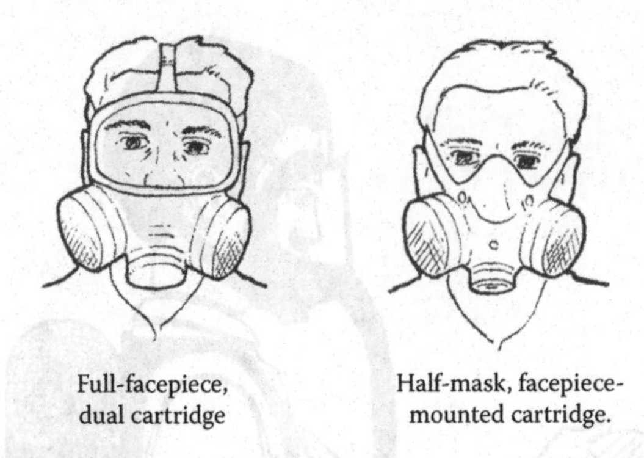

FIGURE 24.13 Example of an air-purifying respirator. (Courtesy of the National Institute for Occupational Safety and Health.)

Respirator Filter and Canister Replacement

An important part of the respiratory protection program includes identifying the useful life of canisters and filters used on air-purifying respirators (see Figure 24.13). Each filter and canister shall be equipped with an end-of-service-life indicator (ESLI) certified by NIOSH for the contaminant. If there is no ESLI appropriate for conditions, a change schedule for canisters and cartridges that is based on objective information or data will ensure that canisters and cartridges are changed before the end of their service life.

Filter and Cartridge Change Schedule

Stocks of spare filters and cartridges shall be maintained to allow immediate change when required or desired by the employee. Cartridges shall be changed based on the most limiting factor below:

- Prior to expiration date
- Manufacturer's recommendations for use and environment
- After each use
- When requested by employee
- When contaminate odor is detected
- When restriction to airflow has occurred as evidenced by an increased effort by the user to breathe normally

Cartridges shall remain in their original sealed packages until needed for immediate use. Filters shall be changed based on the most limiting factor below:

- Prior to expiration date
- Manufacturer's recommendations for the specific use and environment
- When requested by employee

- When contaminate odor is detected
- When restriction to air flow has occurred as evidenced by increase effort by user to breathe normally
- When discoloring of the filter media is evident

Respiratory Protection Schedule by Job and Working Condition

The company maintains a respiratory protection schedule by job and working condition. This schedule is provided to each authorized and trained employee. The schedule provides the following information on job and working conditions: work location, hazards present, type of respirator or SCBA required, type of filter or canister required, location of respirator or SCBA, and filter or cartridge change-out schedule.

The schedule should be reviewed and updated at least annually and whenever any changes are made in the work environments, machinery, equipment, or processes or if different respirator models are introduced or existing models are removed.

Permanent respirator schedule assignments are for the watch person who engages in welding and will be provided with their own company-provided dust-mist-fume filter APR. This respirator will be worn during all welding operations.

Physical and Medical Qualifications

Records of medical evaluations must be retained and made available in accordance with 29 CFR 1910.1020. The medical evaluation is required based upon the fact that the use of a respirator may place a physiological burden on employees. The amount of burden varies with the type of respirator worn, the job and workplace conditions in which the respirator is used, and the medical status of each employee. The company must provide a medical evaluation to determine the employee's ability to use a respirator before the employee is fit tested or required to use the respirator in the workplace.

Medical evaluation procedures require that each employee is provided with a medical questionnaire by a designated occupational healthcare provider. The company shall ensure that a follow-up medical examination is provided for an employee who gives a positive response to any question among questions in Part B of the questionnaire or whose initial medical examination demonstrates the need for a follow-up medical examination. The follow-up medical examination shall include any medical tests, consultations, or diagnostic procedures that the physician deems necessary to make a final determination.

The medical questionnaire and examinations shall be administered confidentially during the employee's normal working hours or at a time and place convenient to the employee. The medical questionnaire shall be administered in a manner that ensures that the employee understands its content. The company shall provide the employee with an opportunity to discuss the questionnaire and examination results with the physician.

The following information must be provided to the physician before the physician makes a recommendation concerning an employee's ability to use a respirator:

1. The type and weight of the respirator to be used by the employee
2. The duration and frequency of respirator use (including its use for rescue and escape)
3. The expected physical work effort
4. Additional protective clothing and equipment to be worn
5. Temperature and humidity extremes that may be encountered
6. Any supplemental information provided previously to the physician explaining that an employee need not be provided a subsequent medical evaluation if the information and the physician remain the same
7. A copy of the written respiratory protection program and a copy of the OSHA Standard 1910.134

In determining the employee's ability to use a respirator, the company must obtain from the physician a written recommendation regarding the employee's ability to use the respirator. The recommendation shall provide only the following information: any limitations on respirator use related to the medical condition of the employee, or relating to the workplace conditions in which the respirator will be used, including whether or not the employee is medically able to use the respirator; the need, if any, for follow-up medical evaluations; and a statement that the physician has provided the employee with a copy of the physician's written recommendations. If the respirator is a negative pressure respirator and if the physician finds a medical condition that may place the employee's health at increased risk when the respirator is used, then the company shall provide a powered air purifying respirator (PAPR). If the physician's medical evaluation finds that the employee can use such a respirator, or if a subsequent medical evaluation finds that the employee is medically able to use a negative pressure respirator, then the company is no longer required to provide a PAPR.

Additional Medical Evaluations

At a minimum, the company must provide additional medical evaluations that comply with the requirements of this regulation if

1. An employee reports medical signs or symptoms that are related to their ability to use a respirator.
2. A physician, supervisor, or the respirator program administrator should inform the company that an employee needs to be re-evaluated.
3. Information from the respiratory protection program, including observations made during fit testing and program evaluation, indicates a need for employee re-evaluation.
4. A change occurs in workplace conditions (e.g., physical work effort, protective clothing, or temperature) that may result in a substantial increase in the physiological burden placed on an employee.

Respirator Fit Testing

Before an employee is required to use any respirator with a negative or positive pressure tight-fitting face piece, the employee must be fit tested with the same make, model, style, and size of respirator that will be used. The company must ensure that an employee using a tight-fitting face piece respirator is fit tested prior to initial use of the respirator, whenever a different respirator face piece (size, style, model or make) is used, and at least annually thereafter.

A qualitative fit test (QLFT) uses an agent that can be detected if there is a leak in the seal of the respirator mask. A quantitative fit test (QNFT) actually measures the ratio of the agent or contaminant outside the respirator mask and the amount that leaks into the mask through the seal on the face piece. The company must establish a record of the QLFT or QNFT administered to employees, including

1. The name or identification of the employee tested
2. Type of fit test performed
3. Specific make, model, style, and size of respirator tested
4. Date of test
5. The pass/fail results for the QLFT or fit tests factor and strip chart recording or other recording of the test results for QNFT

Additional fit tests must be conducted whenever the employee reports, or the company, physician, supervisor, or program administrator makes visual observations of, changes in an employee's physical condition that could affect respirator fit. Such conditions include, but are not limited to, facial scarring, dental changes, cosmetic surgery, or an obvious change in body weight.

If after passing a QLFT or QNFT the employee notifies the company, program administrator, supervisor, or physician that the fit of the respirator is unacceptable, the employee shall be given a reasonable opportunity to select a different respirator face piece and to be retested.

Types of Fit Tests

The fit test shall be administered using an OSHA-accepted QLFT or QNFT protocol. These protocols and procedures are contained in Appendix A of OSHA Standard 1910.134. The QLFT may only be used to fit test negative pressure air-purifying respirators that must achieve a fit factor of 100 or less. If the fit factor, as determined through an OSHA-accepted QNFT protocol, is equal to or greater than 100 for tight-fitting half face pieces, or equal to or greater than 500 for tight-fitting full face pieces, then that respirator has passed the QNFT. Fit testing of tight-fitting atmosphere-supplying respirators and tight-fitting powered air-purifying respirators shall be accomplished by performing quantitative or qualitative fit testing in the negative pressure mode, regardless of the mode of operation (negative or positive pressure) used for respiratory protection. Qualitative fit testing of these respirators shall be accomplished by temporarily converting the respirator user's actual face piece into a negative pressure respirator with appropriate filters, or by using an identical negative pressure air-purifying respirator face piece with the same sealing surfaces as a surrogate for the atmosphere-supplying or powered air-purifying respirator face piece.

Quantitative fit testing of these respirators must be accomplished by modifying the face piece to allow sampling inside the face piece in the breathing zone of the user, midway between the nose and mouth. This requirement shall be accomplished by installing a permanent sampling probe onto a surrogate face piece, or by using a sampling adapter designed to temporarily provide a means of sampling air from inside the face piece.

Any modifications to the respirator face piece for fit testing must be completely removed, and the face piece restored to NIOSH approved configuration, before that face piece can be used in the workplace. Fit test records shall be retained for respirator users until the next fit test is administered. Written materials required to be retained must be made available upon request to affected employees.

Respirator Operation and Use

Respirators will only be used following the respiratory protection safety procedures established in this program. The operation and use manual for each type of respirator must be maintained by the respiratory program administrator and be available to all qualified users.

Surveillance, by the direct supervisor, shall be maintained of work area conditions and degree of employee exposure or stress. When there is a change in work area conditions or degree of employee exposure or stress that may affect respirator effectiveness, the company must re-evaluate the continued effectiveness of the respirator.

For continued protection of respirator users, the following general use rules apply:

1. Users shall not remove respirators while in a hazardous environment.
2. Respirators are to be stored in sealed containers out of harmful atmospheres.
3. Store respirators away from heat and moisture.
4. Store respirators such that the sealing area does not become distorted or warped.
5. Store respirators such that the face piece is protected.

To ensure face piece seal protection, the company should not permit respirators with tight-fitting face pieces to be worn by employees who have facial hair that comes between the sealing surface of the face piece and the face or that interferes with valve function. The same applies to employees who have any condition that interferes with the face-to-face piece seal or valve function.

If an employee wears corrective glasses or goggles or other personal protective equipment, the company shall ensure that such equipment is worn in a manner that does not interfere with the seal of the face piece to the face of the user.

Continuing Effectiveness of Respirators

The company must ensure that all employees leave the respirator use area to wash their faces and respirator face pieces as necessary to prevent eye or skin irritation associated with respirator use. If they detect vapor or gas breakthrough, changes in breathing resistance, or leakage of the face piece, they must leave the area as quickly as possible. They must also leave the area to replace the respirator or the filter, cartridge, or canister elements.

If the employee detects vapor or gas breakthrough, changes in breathing resistance, or leakage of the face piece, the company must replace or repair the respirator before allowing the employee to return to the work area.

Procedures for IDLH Atmospheres

For all IDLH atmospheres, the company shall ensure that one employee or, when needed, more than one employee is located outside the IDLH atmosphere. Visual, voice, or signal line communication must be maintained between the employee(s) in the IDLH atmosphere and the employee(s) located outside the IDLH atmosphere. The employee(s) located outside the IDLH atmosphere shall be trained and equipped to provide effective emergency rescue. The company or designee must be notified before the employee(s) located outside the IDLH atmosphere enter the IDLH atmosphere to provide emergency rescue. The company or designee authorized to do so by the company, once notified, shall provide necessary assistance appropriate to the situation. Employee(s) located outside the IDLH atmospheres must be equipped with

1. Pressure demand or other positive pressure SCBAs, or a pressure demand or other positive pressure supplied-air respirator with auxiliary SCBA.
2. Appropriate retrieval equipment for removing the employee(s) who enter(s) these hazardous atmospheres where retrieval equipment would contribute to the rescue of the employee(s) and would not increase the overall risk resulting from entry or an equivalent means for rescue where retrieval equipment is not required.

Cleaning and Disinfecting

The company shall provide each respirator user with a respirator that is clean, sanitary, and in good working order. The company must ensure that respirators are cleaned and disinfected using the standard operating procedure (SOP) under these conditions:

1. Respirators issued for the exclusive use of an employee shall be cleaned and disinfected as often as necessary to be maintained in a sanitary condition.
2. Respirators issued to more than one employee must be cleaned and disinfected before being worn by different individuals.
3. Respirators maintained for emergency use shall be cleaned and disinfected after each use.
4. Respirators used in fit testing and training must be cleaned and disinfected after each use.
5. Cleaning and storage of respirators assigned to specific employees are the responsibility of the employee.

Respirator Inspection

All respirators/SCBAs, available for both general use and permanent check-out, must be inspected after each use and at least monthly. Should any defects be noted, the respirator/SCBA shall be taken to the program administrator. Damaged respirators must be either repaired or replaced. Inspection of respirators loaned on permanent check-out status is the responsibility of the trained employee. Respirators shall be inspected as follows:

1. All respirators used in routine situations must be inspected before each use and during cleaning.
2. All respirators maintained for use in emergency situations shall be inspected at least monthly and in accordance with the manufacturer's recommendations, and must be checked for proper function before and after each use.
3. Emergency escape-only respirators shall be inspected before being carried into the workplace for use.

Respirator inspections include the following: a check of respirator function, tightness of connections, and the condition of the various parts including, but not limited to, the face piece, head straps, valves, connecting tube, cartridges, canisters, and filters. A check of elastomeric parts for pliability and signs of deterioration must also be performed.

Self-contained breathing apparati shall be inspected monthly. Air and oxygen cylinders must be maintained in a fully charged state and shall be recharged when the pressure falls to 90 percent of the manufacturer's recommended pressure level. The company must determine that the regulator and warning devices function properly.

For emergency-use respirators, additional requirements apply. Certify the respirator by documenting the date the inspection was performed, the name (or signature) of the person who performed the inspection, the findings, required remedial action, and a serial number or other means of identifying the inspected respirator. Provide this information on a tag or label that is attached to the storage compartment for the respirator. It must be kept with the respirator, or included in inspection reports stored as paper or electronic files. This information shall be maintained until replaced following a subsequent certification.

Respirator Storage

All respirators shall be stored to protect them from damage, contamination, dust, sunlight, extreme temperatures, excessive moisture, and damaging chemicals. They must be packed or stored to prevent deformation of the face piece and exhalation valve. Emergency respirators shall be kept accessible to the work area, stored in compartments or in covers that are clearly marked as containing emergency respirators, and stored in accordance with any applicable manufacturer instructions.

Respirator Repairs

Respirators that fail an inspection or are otherwise found to be defective must be removed from service to be discarded, repaired, or adjusted in accordance with the following procedures:

1. Repairs or adjustments to respirators must be made only by persons appropriately trained to perform such operations and shall use only the respirator manufacturer's NIOSH-approved parts designed for the respirator.
2. Repairs must be made according to the manufacturer's recommendations and specifications for the type and extent of repairs to be performed.
3. Reducing and admission valves, regulators, and alarms shall be adjusted or repaired only by the manufacturer or a technician trained by the manufacturer.

Respirators can be used effectively and safely if the mandates of the regulation found in 29 CFR 1910.134 are followed. Employers should be certain that their respirator program is working effectively if their intent is to prevent occupational injuries and illnesses.

SUMMARY

Many factors must be considered when selecting PPE to protect employees from workplace hazards. With all of the types of operations that can present hazards and all of the types of PPE available to protect the different parts of a worker's body from specific types of hazards, this selection process can be confusing and at times overwhelming. Because of this, OSHA requires that employers implement a PPE program to help employers systematically assess the hazards in the workplace and select the appropriate PPE that will protect workers from those hazards. As part of a PPE program, employers must do the following:

1. Assess the workplace to identify equipment, operations, chemicals, and other workplace components that could harm employees.
2. Implement engineering controls and work practices to control or eliminate these hazards to the extent feasible.
3. Select the appropriate types of PPE to protect your employees from hazards that cannot be eliminated or controlled through engineering controls and work practices.
4. Inform employees why the PPE is necessary and when it must be worn.
5. Train employees how to use and care for the selected PPE and how to recognize PPE deterioration and failure.
6. Require employees to wear the selected PPE in the workplace.

The basic information presented here attempts to establish and illustrate a logical, structured approach to hazard assessment and PPE selection and use. These steps must be followed in order to prevent occupational injuries and illnesses.

REFERENCES

United States Department of Energy. OSH Technical Reference Manual. Washington, D.C.: U.S. Department of Energy, 1993.

United States Department of Labor, Occupational Safety and Health Administration. Assessing the Need for Personal Protective Equipment: A Guide for Small Business Employers (OSHA 3151). Washington, D.C.: U.S. Department of Labor, 1997.

25 Safety Hazards

INTRODUCTION

There are such a myriad of safety hazards facing employers and workers within the workplace that it is difficult to select the most important ones and thus err by leaving out ones that others believe are significant. Although great detail cannot be provided in one chapter, it is the intent of this chapter to provide enough information to set the tone regarding the need to recognize and identify safety hazards. If detailed information is needed, then other references can be sought by the reader.

ABRASIVE BLASTING (29 CFR 1910.94 AND 1910.244)

Abrasive blasting cleaning is in wide use in many industries and presents both a physical hazard due to the force of particles being propelled, and a health hazard depending upon what material (e.g., silica) is used to abrade the article.

All abrasive blasting cleaning nozzles are to be equipped with an operating valve that must be held open manually and should operate as a deadman switch or positive-pressure control. A support should be provided upon which the nozzle can be mounted when not in use. All abrasive blasting cleaning enclosures must be exhaust ventilated in such a way that a continuous inward flow of air will be maintained at all openings in the enclosure during the blasting operation.

ABRASIVE WHEEL EQUIPMENT/GRINDERS (29 CFR 1910.212, 1910.215, AND 1910.243)

Grinding wheels rotate at high speeds and have the potential to break apart (explode). Thus, abrasive wheel machinery and portable power tools should only be used on machines provided with safety guards, with the following exceptions:

1. Wheels used for internal work while within the work being ground.
2. Mounted wheels, used in portable operations, two inches (five centimeters) and smaller in diameter.
3. Type 16, 17, 18, 18R, and 19 cones, plugs, and threaded-hole pot balls where the work offers protection.

Abrasive wheel machinery and portable power tool safety guards must cover the spindle end, nut, and flange projections, except

1. Safety guards on all operations where the work provides a suitable measure of protection to the operator may be so constructed that the spindle end, nut, and outer flange are exposed.

FIGURE 25.1 Example of a very unsafe abrasive wheel.

2. Where the nature of the work is such as to entirely cover the side of the wheel, the side covers of the guard may be omitted.
3. The spindle end, nut, and outer flange may be exposed on machines designed as portable saws.

Work rests are to be adjusted so that they are no more than 1/8 inch from the abrasive wheel. Abrasive wheel safety guards for bench and floor stands and for cylindrical grinders should not expose the grinding wheel periphery more than 65 degrees above the horizontal plane of the wheel spindle. The protecting member must be adjustable for variations in wheel size so that the distance between the wheel periphery and the adjustable tongue (tongue guard) or end of the peripheral member at the top shall never exceed 1/4 inch. Machines designed for a fixed location must be securely anchored to prevent movement, or designed in such a manner that in normal operation they will not move (see Figure 25.1).

Other precautions include the wearing of goggles or face shields when grinding. Each grinder must have an individual on and off control switch and be effectively grounded. Before new abrasive wheels are mounted, they should be visually inspected and ring tested. Dust collectors and powered exhausts must be provided on grinders used in operations that produce large amounts of dust, and splashguards must be mounted on grinders that use coolant in order to prevent the coolant from reaching employees.

AIR RECEIVERS (29 CFR 1910.169)

Air receivers have safety requirements similar to compressors and must be operated and lubricated in accordance with the manufacturer's recommendations. Every receiver must be equipped with a pressure gauge and with one or more automatic, spring-loaded safety valves. All safety valves must be tested frequently and at regular intervals to determine whether they are in good operating condition. The total relieving capacity of the safety valve must be capable of preventing pressure in the receiver from exceeding the maximum allowable working pressure of the receiver by more than 10 percent.

Every air receiver must be provided with a drainpipe and valve at the lowest point for the removal of accumulated oil and water. Compressed air receivers are to be periodically drained of moisture and oil. The inlet of air receivers and piping systems must be kept free of accumulated oil and carbonaceous materials.

AISLES AND PASSAGEWAYS (29 CFR 1910.17, 1910.22, AND 1910.176)

Aisles and passageways must be free from debris and kept clear for travel. Where mechanical material handling equipment is used, sufficient safe clearance must be allowed for aisles, at loading docks, through doorways, and wherever turns or passage must be made. Aisles and passageways used by mechanical equipment shall be kept clear and in good repair with no obstructions across or in aisles that could create hazards. All permanent aisles and passageways should be appropriately marked. Areas where workers are walking should be covered or guardrails should be provided to protect workers from the hazards of open pits, tanks, vats, ditches, etc.

BELT SANDING MACHINES (29 CFR 1910.213)

Belt sanding machines used for woodworking must be provided with guards at each nip point where the sanding belt runs onto a pulley, and the unused run of the sanding belt must be shielded to prevent accidental contact.

CHAINS, CABLES, ROPES, AND HOOKS (29 CFR 1910.179 AND 1910.180)

Rigging apparati used for lifting and material handling, such as hooks, wire rope, and chains, must be visually inspected daily and monthly with a full, written, dated, and signed report of the condition and kept in a file and readily available to designated personnel (see Figure 25.2).

Hoist ropes on crawler, locomotive, and truck cranes must be free from kinks or twists and should be never wrapped around the load.

U-bolts and rope clips used on hoist ropes for overhead and gantry cranes must be installed so that the U-bolt is in contact with the dead end (short or non-load-carrying end) of the rope. Rope clips must be installed in accordance with the manufacturer's recommendation. All nuts on newly installed clips must be tightened after one hour of use.

COMPRESSORS AND COMPRESSED AIR (29 CFR 1910.242)

Because compressed air is supplied by a compressor, great care must be taken to ensure that certain types of equipment are operating in a safe manner. Safety

FIGURE 25.2 Rigging sling attached to crane's hook.

devices for a compressed air system should be checked frequently. Compressors should be equipped with pressure relief valves and pressure gauges. The air intakes must be installed and equipped so as to ensure that only clean, uncontaminated air enters the compressor. Air filters installed on the compressor intake facilitate this. Before any repair work is done on the pressure system of a compressor, the pressure must be bled off and the system locked out. All compressors must be operated and lubricated in accordance with the manufacturer's recommendations. Signs are to be posted to warn of the automatic starting feature of compressors. The belt drive system must be totally enclosed to provide protection from any contact.

No worker should direct compressed air toward another person, and employees are prohibited from using highly compressed air for cleaning purposes. If compressed air is used for cleaning off clothing, the pressure must be reduced to less than 10 pounds per square inch (psi). When using compressed air for cleaning, employees must wear protective chip guarding eyewear and personal protective equipment.

Safety chains or other suitable locking devices must be used at couplings of high-pressure hose lines where a connection failure would create a hazard. Before compressed air is used to empty containers of liquid, the safe working pressure of the container must be checked.

When compressed air is used with abrasive blast cleaning equipment, the operating valve must be held open manually. As compressed air is used to inflate auto tires, a clip-on chuck and an inline regulator preset to 40 psi are required. It should be prohibited to use compressed air to clean up or move combustible dust because such action could cause the dust to be suspended in the air and cause a fire or explosion hazard.

COMPRESSED GAS CYLINDERS (29 CFR 1910.101 AND 1910.253)

Compressed gas cylinders have exploded and become airborne. There is a lot of stored energy in a compressed gas cylinder, which is why they should be handled with great care and respect. A collar or valve protection device such as a recess is required to protect cylinders with a water weight capacity greater than 30 pounds. Cylinders should be legibly marked to clearly identify the gas contained within the cylinder. Compressed gas cylinders must be stored in areas that are protected from external heat sources, such as flame impingement, intense radiant heat, electric arcs, and high temperature lines. Inside buildings, cylinders must be stored in well-protected, well-ventilated, dry locations away from combustible materials by 20 feet. Also, the in-plant handling, storage, and utilization of all compressed gases in cylinders, portable tanks, rail tank cars, or motor vehicle cargo tanks must be in accordance with Compressed Gas Association pamphlet P-1-1965.

Cylinders must be located or stored in areas where they will not be damaged by passing or falling objects or subject to tampering by unauthorized persons. Cylinders must be stored or transported in a manner to prevent them from creating a hazard by tipping, falling, or rolling and stored 20 feet away from highly combustible materials (see Figure 25.3). Where a cylinder is designed to accept a valve protection cap, caps must to be in place with the exception of when the cylinder is in use or is connected for use.

Cylinders containing liquefied fuel gas should be stored or transported in a position so that the safety relief device is always in direct contact with the vapor space in the cylinder. All valves must be closed off before a cylinder is moved, when the cylinder is empty, and at the completion of each job. Low-pressure fuel-gas cylinders should be checked periodically for corrosion, general distortion, cracks, or any other defect that might indicate a weakness or render it unfit for service.

COMPRESSED GASES (29 CFR 1910.101, 1910.102, 1910.103, 1910.104, 1910.106, AND 1910.253)

There are several hazards associated with compressed gases, including oxygen displacement, fires, explosions, toxic effects from certain gases, as well as the physical hazards associated with pressurized systems. Special storage, use, and handling precautions are necessary in order to control these hazards. There are specific safety requirements for many of the compressed gases (e.g., acetylene, hydrogen, nitrous oxide, and oxygen).

FIGURE 25.3 Compressed gas cylinder in a well on a truck to protect it.

Acetylene cylinders must be stored and used in a vertical, valve-end-up position only. Under no conditions should acetylene be generated, piped (except in approved cylinder manifolds), or utilized at a pressure in excess of 15 psi (103 kPa gauge pressure) or 30 psi (206 kPa absolute). The use of liquid acetylene is prohibited. The in-plant transfer, handling, and storage of acetylene in cylinders must be in accordance with the Compressed Gas Association in pamphlet C-1.3-1959.

Hydrogen containers must comply with one of the following:

- They must be designed, constructed, and tested in accordance with the appropriate requirements of ASME's Boiler and Pressure Vessel Code, Section VIII Unfired Pressure Vessels (1968).
- They must be designed, constructed, tested, and maintained in accordance with United States Department of Transportation specifications and regulations.

Hydrogen systems must be located so that they are readily accessible to delivery equipment and authorized personnel. They must be located above ground, not beneath electric power lines, and not near flammable liquid piping or piping of other flammable gases. Permanently installed containers must be provided with substantial noncombustible supports on firm noncombustible foundations.

Nitrous oxide piping systems for the in-plant transfer and distribution of nitrous oxide must be designed, installed, maintained, and operated in accordance with the Compressed Gas Association pamphlet G-8.1-1964.

Oxygen cylinders in storage must be separated from fuel-gas cylinders or combustible materials (especially oil or grease) by a minimum distance of 20 feet or by a noncombustible barrier at least 5 feet high having a fire-resistance rating of 1/2 hour.

CONFINED SPACES (29 CFR 1910.146)

Many workplaces contain spaces that are considered confined because their configurations hinder the activities of employees who must enter, work in, and exit them. For example, employees who work in process vessels generally must squeeze in and out through narrow openings and perform their tasks while cramped or contorted. OSHA uses the term "confined space" to describe such spaces. In addition, there are many instances where employees who work in confined spaces face increased risk of exposure to serious hazards. In some cases, confinement itself poses an entrapment hazard. In other cases, confined space work keeps employees closer to hazards, such as asphyxiating atmospheres or the moving parts of machinery. OSHA uses the term "permit-required confined space" or permit space) to describe those spaces that both have "confined space" and have elements that pose health or safety hazards (see Figure 25.4).

Improper entry into confined spaces has resulted in approximately 200 lives lost each year. Confined spaces are those of adequate size and shape to allow a person to enter but have limited openings for workers to enter and exit and are not designed for continuous human occupancy. Examples of confined spaces are storage tanks, silos, pipelines, manholes, and underground utility vaults.

Confined spaces should be thoroughly emptied of any corrosive or hazardous substances, such as acids or caustics, before entry. All lines to a confined space containing inert, toxic, flammable, or corrosive materials must have the valve off and blanked or disconnected and separated before entry. All impellers, agitators, or other moving parts and equipment inside confined spaces should be locked out and tagged out if they present a hazard.

In evaluating confined space accidents in the past, certain scenarios seem to continuously occur. These included the failure to recognize an area as a confined space; failure to test, evaluate, and monitor for hazardous atmospheres; failure to train workers regarding safe entry; and failure to establish rescue procedures. At times, either natural or mechanical ventilation is enough to make the space safe, which can be provided prior to confined space entry. Appropriate atmospheric tests must be performed to check for oxygen deficiency, explosive concentrations, and toxic substances in the confined space before entry. The atmosphere inside the confined space must be frequently tested or continuously monitored during the conduct of work.

With the promulgation of the "Permit-Required Confined Space Entry" standard (29 CFR 1910.146), all these failures are addressed. First, a written and signed permit is required prior to entry if the space contains or has the potential to contain a hazardous atmosphere (e.g., oxygen deficient, flammable, or toxic); if it contains materials

FIGURE 25.4 Confined space entry.

that could engulf the entrant; if the space's configuration could cause the entrant to become entrapped or asphyxiated by converging walls; or if it contains any other recognized serious safety (e.g., electrical) or health (e.g., pathogen bacteria) hazard.

This regulation requires that all permit-required spaces be identified, evaluated, and controlled. Procedures for entry must exist. Appropriate equipment and training for authorized entrants must be provided. Entry supervisors and attendants must be trained and present, and a written/signed entry permit should exist prior to entry. Trained and available rescue personnel shall (over one-half of the deaths in confined spaces are rescuers) be available. The space must have posted warning signs and barriers erected, and personal protective equipment and rescue equipment must be provided.

It is important to know that work in confined spaces can be accomplished safely by paying attention to these key regulatory requirements prior to entry. These regulations help employers identify, evaluate/test, and monitor, train, and plan for rescue.

CONTAINERS AND PORTABLE TANK STORAGE (29 CFR 1910.106)

See flammable and combustible liquids.

CONTROL OF HAZARDOUS ENERGY SOURCES [LOCKOUT/TAGOUT] (29 CFR 1910.147)

Lockout/tagout deals with preventing the release of energy from machines, equipment, and electrical circuits that are perceived to be de-energized. OSHA estimates that compliance with the lockout/tagout standard will prevent about 120 fatalities and approximately 28,000 serious and 32,000 minor injuries each year. Approximately 39 million general industry workers are to be protected from accidents during maintenance and servicing of equipment under this ruling.

The standard for the control of hazardous energy sources (lockout/tagout) covers servicing and maintenance of machines and equipment in which the unexpected energizing or start-up of the machines or equipment or the release of stored energy can cause injury to employees. The rule generally requires that energy sources for equipment be turned off or disconnected, and that the switch either be locked or labeled with a warning tag. About 3 million workers actually servicing equipment face the greatest risk. These include craft workers, machine operators, and laborers. OSHA's data show that packaging and wrapping equipment, printing presses, and conveyors account for a high proportion of the accidents associated with lockout/tagout failures. Typical injuries include fractures, lacerations, contusions, amputations, and puncture wounds, with the average lost time for injuries running at 24 days.

Agriculture, maritime, and construction employers are not covered under standard 29 CFR 1910.147. Also, the generation, transmission, and distribution of electric power by utilities, and work on electric conductors and equipment, are excluded. The general requirements under this ruling require employers to

1. Develop an energy control program.
2. Use locks when equipment can be locked out (see Figure 25.5).
3. Ensure that new equipment or overhauled equipment can accommodate locks.
4. Employ additional means to ensure safety when tags, rather than locks, are installed by using an effective tagout program.
5. Identify and implement specific procedures (generally in writing) for the control of hazardous energy, including preparation for shutdown, equipment isolation, lockout/tagout application, release of stored energy, and verification of isolation.
6. Institute procedures for the release of lockout/tagout, including machine inspection, notification and safe positioning of employees, and the removal of a lockout/tagout device.
7. Obtain standardized locks and tags that indicate the identity of the employee using them and that are of sufficient quality and durability to ensure their effectiveness.
8. Require that each lockout/tagout device be removed by the employee who applied the device.
9. Conduct inspections of energy control procedures at least annually.
10. Train employees in specific energy control procedures with training reminders as part of the annual inspection of these control procedures.

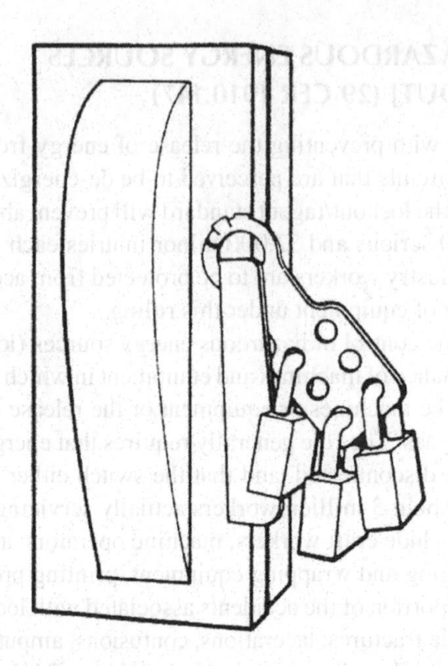

FIGURE 25.5 Using gang lock device to prevent turning on electrical power.

11. Adopt procedures to ensure safety when equipment must be tested during servicing, when outside contractors are working at the site, when multiple lockouts are needed for the crew servicing equipment, and when shifts or personnel change occur.

Excluded from coverage are normal production operations, including repetitive, routine, and minor adjustments that would be covered under OSHA's machine guarding standard. Exclusion includes work on cord- and plug-connected electrical equipment when it is unplugged, and the employee working on the equipment has complete control over the plug. Hot tap operations, involving gas, steam, water, or petroleum products, are excluded when the employer can show that continuity of service is essential, shutdown is impractical, and documented procedures are to be followed to provide proven effective protection for employees.

In summary, all machinery or equipment capable of movement must be de-energized or disengaged and locked out during cleaning, servicing, adjusting, and setting-up operations. Where the power-disconnecting means for equipment does not also disconnect the electrical control circuit, the appropriate electrical enclosures must be identified. A means should be provided to ensure that the control circuit can also be disconnected and locked out. The locking out of control circuits, in lieu of locking out main power disconnects, must be prohibited. All equipment control valve handles must be provided with a means for locking out. Lockout procedures require that stored energy (e.g., mechanical, hydraulic, air, etc.) be released or blocked before equipment is locked out for repairs. Appropriate employees must be provided with individually keyed personal safety locks and are expected to keep personal control of their key(s) while they have safety locks in use. It should be required that only the employee exposed to the hazard place or remove the safety lock. Employees must check the safety of the lockout by attempting a start-up after making sure no one is exposed. Employees must be instructed to always push the control circuit stop button immediately after checking the safety of the lockout. A means must be provided to identify any or all employees who are working on locked-out equipment by their locks or accompanying tags. A sufficient number of accident-prevention signs or tags and safety padlocks must be provided for any reasonably foreseeable repair emergency. When machine operations, configuration, or size requires the operator to leave his or her control station to install tools or perform other operations, and that part of the machine could move if accidentally activated, then that element must be separated, locked, or blocked out. In the event that equipment or lines cannot be shut down, locked out, and tagged, a safe job procedure must be established and rigidly followed.

CRANE, DERRICK, AND HOIST SAFETY (29 CFR 1910.179, 1910.180, AND 1910.181)

Moving large, heavy loads is crucial to today's manufacturing and construction industries. Much technology has been developed for these operations, including careful training and extensive workplace precautions. There are significant safety issues to consider, both for the operators of the diverse "lifting" devices and for workers in proximity to them.

More than 250,000 crane operators, a very large but undetermined number of other workers, and the general public are at risk of serious and often fatal injury due to accidents involving cranes, derricks, hoists, and hoisting accessories. There are approximately 125,000 cranes in operation today in the construction industry, as well as an additional 80,000 to 100,000 in general and maritime industries. According to the U.S. Bureau of Labor Statistics' (BLS) Census of Fatal Occupational Injuries, seventy-nine fatal occupational injuries were related to cranes, derricks, hoists, and hoisting accessories in 1993. OSHA's analysis of crane accidents in general industry and construction identified an average of seventy-one fatalities each year.

The BLS identified eight fatal injuries in 1993 among crane and tower crane operators; this corresponds to a risk of more than one death per thousand workers (1.4) over a working lifetime of 45 years. According to the 1987 BLS supplementary data system (twenty-three states reporting), over 1,000 construction injuries were reported to involve cranes and hoisting equipment. However, underreporting of crane-related injuries and fatalities, due to misclassification and a host of other factors, masks the true magnitude of the problem.

OSHA's analysis identified the major causes of crane accidents to include boom or crane contact with energized power lines (nearly 45 percent of the cases) under the hook lifting device, overturned cranes, dropped loads, boom collapse, crushing by the counter weight, failure to use outriggers, falls, and rigging failures.

Some cranes are neither maintained properly nor inspected regularly to ensure safe operation. Crane operators often do not have the necessary qualifications to operate each piece of equipment safely, and the operators' qualifications required in the existing regulations may not provide adequate guidance to employers. The issues of crane inspection/certification and crane operator qualifications and certification

FIGURE 25.6 Cranes find many uses in the workplace.

must be further examined. In fact, OSHA's crane standards for construction, general industry, and maritime have not been updated since 1971 and rely heavily on outdated 1968 American National Standards Institute (ANSI) consensus standards.

These types of heavy lifting equipment are very sophisticated and time has only increased their complexity. Thus, only those with unique experience and training should be allowed to operate and maintain these types of lifting devices. When failure of equipment or operator occurs, the outcome is often catastrophic. Cranes, hoists, and derricks should be inspected regularly and well maintained to preclude their failure during performance (see Figure 25.6).

DIP TANKS CONTAINING FLAMMABLE OR COMBUSTIBLE LIQUID (29 CFR 1910.108)

Dip tanks with more than 150 gallons or 10 square feet in liquid surface area must be equipped with a properly trapped overflow pipe leading to a safe location outside the building. No open flames, spark-producing devices, or heated surfaces having a temperature sufficient to ignite vapors in any flammable vapor should exist. Areas in the vicinity of dip tanks must be kept clear of combustible stock as is practical and must be kept entirely free of combustible debris.

All dip tanks exceeding 150 gallons of flammable liquid capacity or having a liquid surface area exceeding 4 square feet must be protected with at least one of the following automatic extinguishing facilities: water spray system, foam system, carbon dioxide system, dry chemical system, or automatic dip tank cover.

DOCKBOARDS (29 CFR 1910.30)

Dockboards must be strong enough to carry the load imposed upon them. Portable dockboards must be anchored or equipped with devices that will prevent their slipping. Handholds should exist on dockboards that provide a safe and effective means of being able to handle them. Railroad cars should be provided with mechanisms that would prevent movement while dockboards are being used.

DRINKING WATER (29 CFR 1910.141)

Potable drinking water must be provided in all places of employment. Potable drinking water dispensers must be designed, constructed, and serviced to ensure sanitary conditions, capable of being closed, and have a tap. Open containers such as barrels, pails, or tanks for drinking water from which the water must be dipped or poured, whether or not they are fitted with a cover, are prohibited. A common drinking cup is not allowed.

ELECTRICAL (29 CFR 1910.303, 1910.304, 1910.305, 1910.331, AND 1910.333)

Electricity is accepted as a source of power without much thought to the hazards encountered. Some employees work with electricity directly, as is the case with engineers, electricians, and people who work with wiring, such as overhead lines, cable harnesses, or circuit assemblies. Others, such as office workers and salespeople, work with electricity indirectly. Approximately 700 workers are electrocuted each year, with many workers suffering injuries such as burns and cuts.

OSHA's electrical standards address the government's concern that electricity has long been recognized as a serious workplace hazard. Beyond burns and cuts, employees are subject to such dangers as electric shock, electrocution, fires, and explosions. The objective of these standards is to minimize such potential hazards by specifying design characteristics of safety approaches and designs in the use of electrical equipment and systems.

Electrical equipment must be free from recognized hazards that are likely to cause death or serious physical harm to employees. Flexible cords and cables (extension cords) should be protected from accidental damage. Unless specifically permitted, flexible cords and cable may not be used as a substitute for the fixed wiring of a structure, where attached to building surfaces; where concealed or they run through holes in walls, ceilings, or floors; or where they run through doorways, windows, or similar openings. Flexible cords must be connected to devices and fittings so that strain relief is provided that will prevent force from being directly transmitted to joints or terminal screws.

A grounding electrode conductor must be used for a grounding system to connect both the equipment grounding conductor and the grounded circuit conductor to the grounding electrode. Both the equipment grounding conductor and the grounding electrode conductor must be connected to the ground circuit conductor on the supply side of the service disconnecting means or on the supply side of the system disconnecting means or over-current devices if the system is separately derived. For ungrounded service-supplied systems, the equipment grounding conductor must be connected to the grounding electrode conductor at the service equipment. The path to ground from circuits, equipment, and enclosures will be permanent and continuous.

Electrical equipment shall be free from recognized hazards that are likely to cause death or serious physical harm. Each disconnecting means shall be legibly marked to indicate its purpose, unless it is located so the purpose is evident. Listed or labeled equipment must be used or installed in accordance with any instructions included in the listing or labeling. Unused openings in cabinets, boxes, and fittings must be effectively closed.

Safety-related work practices must be employed to prevent electric shock or other related injuries resulting from either direct or indirect electrical contact when work is performed near or on equipment of circuits that are or may be energized. Electrical safety-related work practices cover both qualified persons (those who have training in avoiding the electrical hazards of working on or near exposed energized parts) and unqualified persons (those with little or no such training) (see Figure 25.7).

There must be written lockout and/or tagout procedures. Overhead power lines must be de-energized and grounded by the owner or operator of the lines or other protective measures must be provided before work is started. Protective measures, such as guarding or insulating the lines, must be designed to prevent employees from contacting the lines.

Unqualified employees and mechanical equipment must be at least 10 feet away from overhead power lines. If the voltage exceeds 50,000 volts, the clearance distance should be increased 4 inches for each 10,000 volts. OSHA requires portable ladders to have nonconductive side rails if used by employees who would be working where they might contact exposed energized circuit parts. Also, conductors must be spliced or joined either with devices identified for such use, or by brazing, welding, or soldering with a fusible alloy or metal. All splices, joints, and free ends of conductors must be covered with insulation equivalent to that of the conductor or with an insulating device suitable for the purpose.

All portable electrical tools and equipment must be grounded or of double insulated type. Electrical appliances such as vacuum cleaners, polishers, and vending machines must be grounded. Extension cords being used must have a grounding conductor. Multiple plug adapters are prohibited. Ground-fault circuit interrupters should be installed on each temporary 15- or 20-ampere, 120-volt AC circuit at locations where construction, demolition, modifications, alterations, or excavations are

FIGURE 25.7 A qualified electrician applying his trade.

being performed. All temporary circuits must be protected by suitable disconnecting switches or plug connectors at the junction with permanent wiring. If electrical installations in hazardous dust or vapor areas exist, they need to meet the National Electrical Code (NEC) for hazardous locations. In wet or damp locations, the electrical tools and equipment must be appropriate for this use or all hazardous locations or otherwise protected. The location of electrical power lines and cables (overhead, underground, under floors, or other side of walls) must be determined before digging, drilling, or similar work begins.

All energized parts of electrical circuits and equipment must guard against accidental contact by approved cabinets or enclosures. Sufficient access and working space must be provided and maintained about all electrical equipment to permit ready and safe operations and maintenance.

Low-voltage protection must be provided in the control device of motor-driven machines or equipment that could cause (probable) injury from inadvertent starting. Each motor disconnecting switch or circuit breaker must be located within sight of the motor control device. Each motor located within sight of its controller, or the controller disconnecting means that is capable of being locked in the open position, must have a separate disconnecting means installed in the circuit within sight of the motor.

FIGURE 25.8 Possible electrocution requires prior emergency preparedness. (*Source:* Courtesy of the Occupational Safety and Health Administration.)

All employees should be required to report, as soon as practical, any obvious hazard to life or property observed in connection with electrical equipment or lines (see Figure 25.8). Employees must be instructed to make preliminary inspections or appropriate tests to determine what conditions exist before starting work on electrical equipment or lines. Employees who regularly work on or around energized electrical equipment or lines should be instructed in the cardiopulmonary resuscitation (CPR) method.

ELEVATED SURFACES (29 CFR 1910.23)

Elevated surfaces present a real potential for injuries from both falls and objects and falling from above. Surfaces elevated more than 30 inches above the floor or ground must be provided with standard guardrails (Figure 25.9). Elevated surfaces (beneath which people or machinery could be exposed to falling objects) must be provided with standard 4-inch toeboards. Material on elevated surfaces must be piled, stacked, or racked in a manner that prevents them from tipping, falling, collapsing, rolling, or spreading.

EMERGENCY ACTION PLANS (29 CFR 1910.38)

The lack of an emergency action and fire plan is a safety hazard to workers. Workers must have knowledge of how to react and what is expected of them. Training and practice drills must take place. An emergency action plan, to ensure employee safety in the event of fire or other emergencies, must be prepared in writing and reviewed with affected employees. The plan should include the escape procedures and routes, critical plant operations, employee accounting, following an emergency evacuation

FIGURE 25.9 Falls from elevated surfaces can be deadly. (*Source:* Courtesy of the Occupational Safety and Health Administration.)

plan, rescue and medical duties, means of reporting emergencies, and persons to be contacted for information or clarification. Employers must also apprise employees of fire hazards associated with the materials and processes to which they are exposed.

Not every employer is required to have an emergency action plan (EAP). The standards that require such plans include the following:

- Process Safety Management of Highly Hazardous Chemicals, 1910.119
- Fixed Extinguishing Systems, General, 1910.160
- Fire Detection Systems, 1910.164
- Grain Handling, 1910.272
- Ethylene Oxide, 1910.1047
- Methylenedianiline, 1910.1050
- 1,3 Butadiene, 1910.1051

If the employer has ten or fewer employees, then he or she can communicate the plan orally instead of by the required written plan that must be maintained in the workplace and available to all workers. When required, employers must develop emergency action plans that

- Describe the routes for workers to use and procedures to follow.
- Account for all evacuated employees.
- Remain available for employee review.
- Include procedures for evacuating disabled employees.
- Address evacuation of employees who stay behind to shut down critical plant equipment.
- Include preferred means of alerting employees to a fire emergency.
- Provide for an employee alarm system throughout the workplace.
- Require an alarm system that includes voice communication or sound signals such as bells, whistles, or horns.

- Make the evacuation signal known to employees.
- Ensure emergency training.
- Require the employer to review the plan with new employees and with all employees whenever the plan is changed.

In addition, employers must designate and train employees to assist in a safety and orderly evacuation of other employees. Employers must also review the EAP with each employee covered when the following occur:

- Plan is developed or an employee is assigned initially to a job
- Employee's responsibilities under the plan change
- Plan is changed

Prior to the development of an EAP, it is necessary to perform a hazard assessment to determine potentially toxic materials and unsafe conditions. For information on chemicals, the manufacturer or supplier can be contacted to secure Material Safety Data Sheet (MSDS).

EXIT DOORS (29 CFR 1910.36)

Doors that are required to serve as exits must be designed and constructed so that the way of exit travel is obvious and direct. Any windows that could be mistaken for exit doors must be made inaccessible by means of barriers or railings. Exit doors must be able to open in the direction of exit travel without the use of a key or any special knowledge or effort when the building is occupied. Revolving, sliding, or overhead doors should be prohibited from serving as a required exit door. Where panic hardware is installed on a required exit door, it must allow the door to open by applying a force of 15 pounds or less in the direction of the exit traffic. Doors in cold storage rooms must be provided with an inside release mechanism that will release the latch and open the door even if it is padlocked or otherwise locked on the outside.

Where exit doors open directly onto any street, alley, or other area where vehicles may be operated, there must be adequate barriers and warnings provided to prevent employees from stepping into the path of traffic. Doors that swing in both directions and are located between rooms where there is frequent traffic are to be provided with viewing panels in each door (see Figure 25.10).

EXITS AND EXIT ROUTES (29 CFR 1910.36)

Every workplace must have enough exits suitably located to enable everyone to get out of the facility quickly. Considerations include the type of structure, the number of persons exposed, the fire protection available, the type of industry involved, and the height and type of construction of the building or structure. In addition, fire doors must not be blocked or locked when employees are inside. Delayed opening of fire doors, however, is permitted when an approved alarm system is integrated into the fire door design. Exit routes from buildings must be free of obstructions and properly marked with exit signs. See 29 CFR Part 1910.36 for details about all requirements.

FIGURE 25.10 Exit doors that swing in the direction of emergency travel.

An exit route is a continuous and unobstructed path of exit travel from any point within a workplace to a place of safety. An exit route consists of threes parts:

1. Exit access is the portion of the route that leads to an exit.
2. Exit is the portion of an exit route that is generally separated from other areas to provide a protected way of travel to the exit discharge.
3. Exit discharge is the part of the exit route that leads directly outside or to a street, walkway, refuge area, public way, or open space with access to the outside.

Normally, a workplace must have at least two exit routes to permit prompt evacuation of their employees and other facility occupants during an emergency. More than two exits are required if the number of employees, the size of the facility, or the arrangement of the workplace will not allow a safe evacuation. Exit routes must be located as far away as practical from each other in case one is blocked by fire or smoke.

Requirements for Exits

The requirements for exits are as follow:

- Exits must be separated from the workplace by fire-resistant materials; that is, a 1-hour fire-resistance rating if the exit connects three or fewer stories and a 2-hour fire-resistance rating if the exit connects more than three floors.
- Exits can have only those openings necessary to allow access to the exit from occupied areas of the workplace or to the exit discharge. Openings must be protected by a self-closing, approved fire door that remains closed or automatically closes in an emergency.

FIGURE 25.11 Example of an exit sign.

- Always keep the line-of-sight to exit signs clearly visible.
- Install "EXIT" signs using plainly legible letters (see Figure 25.11).

Safety Features for Exit Routes

Exits routes should have special safety features such as the following:

- Keep exit routes free of explosives or highly flammable furnishings and other decorations.
- Arrange exit routes so employees will not have to travel toward a high-hazard area unless the path of travel is effectively shielded from the high-hazard area.
- Ensure that exit routes are free and unobstructed by materials, equipment, locked doors, or dead-end corridors.
- Provide lighting for exit routes adequate for employees with normal vision.
- Keep exit route doors free of decorations or signs that obscure their visibility of exit route doors.
- Post signs along the exit access indicating the direction of travel to the nearest exit and exit discharge if that direction is not immediately apparent.
- Mark doors or passages along an exit access that could be mistaken for an exit "Not an Exit" or with a sign identifying its use (such as "Closet").
- Renew fire-retardant paints or solutions when needed.
- Maintain exit routes during construction, repairs, and alterations.

Design and Construction Requirements

There are specific design and construction requirements for exits and exit routes that include the following:

- Exit routes must be permanent parts of the workplace.
- Exit discharges must lead directly outside or to a street, walkway, refuge area, public way, or open space with access to the outside.

- Exit discharge areas must be large enough to accommodate people likely to use the exit route.
- Exit route doors must unlock from the inside. They must be free of devices or alarms that could restrict use of the exit route if the device or alarm fails.
- Exit routes can be connected to rooms only by side-hinged doors, which must swing out in the direction of travel if the room may be occupied by more than 50 people.
- Exit routes must support the maximum permitted occupant load for each floor served, and the capacity of an exit route may not decrease in the direction of exit route travel to the exit discharge.
- Exit routes must have ceilings at least 7 feet, 6 inches high.
- An exit access must be at least 28 inches wide at all points. Objects that project into the exit must not reduce its width.

EXPLOSIVES AND BLASTING AGENTS (29 CFR 1910.109)

The outcome of mishandling explosives is disastrous and a very unforgiving process. Those individuals who have been specifically trained in the use and handling of explosive are not expected to make mistakes; these are unacceptable and usually end in deaths.

All explosives must be kept in approved magazines. Stored packages of explosives are to be laid flat with the top side facing up. Black powder, when stored in magazines with other explosives, must be stored separately.

Smoking, matches, open flames, spark-producing devices, and firearms (except firearms carried by guards) are not permitted inside 50 feet of magazines. The land surrounding magazines must be kept clear of combustible materials for a distance of at least 25 feet. Combustible materials cannot be stored within 25 feet of a magazine. The manufacturers of explosives and pyrotechnics must meet the requirements of OSHA's Process Safety Management Highly of Hazardous Chemicals (29 CFR 1910.119) standard.

FAN BLADES (29 CFR 1910.212)

When the periphery of the blades of a fan is less than 7 feet above the floor or working level, the blades must be guarded. The guard must have an opening no larger than ½ inch.

FALL PROTECTION (29 CFR 1910.23 AND 1910.66 APPENDIX I)

In 1999, slips, trips, and falls (STFs) occurred to over 1 million workers in the United States and 17,000 died as a result. It is estimated that 3.8 million disabling injuries occur each year in the workforce. About 15 percent if all accidents are STFs. The cost for a disabling injury is approximately $28,000. It appears that 17 percent (969) of deaths come from falls from elevations.

It is well known that falls to same levels occur most often in industrial settings. These falls to same level result in many injuries, but falls from heights (elevation) result in more serious injuries and are one of the leading causes of deaths. Workers working at heights often find that gravity never takes a day off. Falls from heights occur so quickly that seldom does a worker or anyone have a chance to react quickly enough to prevent the fall or recover from the loss of balance.

Fall from heights occur because of inattention, slipping, tripping, and loss of balance. These types of falls often are the result from a ladder, scaffold, or roof. Each worker on a walking working surface with unprotected sides or edges should be protected from falling to a lower level by the use of guardrails, safety nets, personal fall arrest systems, or the equivalent. Regulations require fall protection if a worker is working above certain heights. The specific regulations and their height requirements are as follows:

- 29 CFR 1910 General Industry 4 feet
- 29 CFR 1915 Shipyards 5 feet
- 29 CFR 1926 Construction Standards 6 feet

Falls in the workplace are the leading cause of death to workers; this also includes workers within the Service Industry. Thus, the need for fall protection in workplaces when workers are performing their jobs is definitely a requirement. Fall protection must be provided so that workers can concentrate on the job tasks without any fear of falling. Requirements relating to fall protection, as described in this chapter, do not apply to scaffolds, cranes and derricks, ladders and stairways, or electrical power transmission and distribution, each of which has its own requirements. The major components of fall protection, described herein, are for installation, construction, and proper use of body harnesses and belts, lanyards, and lifelines, and the requirements for the training of fall protection.

It is the employer's responsibility to determine if the walking or working surfaces on which his or her employees are to work have the strength and structural integrity to support employees safely. Employees are allowed to work on those surfaces only when the surfaces have the requisite strength and structural integrity. Any time a worker is on a walking or working surface (horizontal and vertical surface), or an edge with an unprotected side or edge that is 4 feet or more above a lower level, that worker must be protected from falling by using guardrail systems, safety net systems, or personal fall arrest systems. If the employer can demonstrate that it is not feasible or creates a greater hazard to use these systems, the employer shall develop and implement a fall protection plan to protect the worker.

Each worker on a walking or working surface 4 feet or more above a lower level where edges exist must be protected from falling by using a guardrail system, safety net system, or personal fall arrest system. If a guardrail system is chosen to provide the fall protection, and a controlled access zone for the edge work, the control line may be used in lieu of a guardrail along the edge that parallels the leading edge.

Workers in a hoist area must be protected from falling 4 feet or more to lower levels by using guardrail systems or personal fall arrest systems. If guardrail systems (chains, gates, or guardrails), or portions thereof, are removed to facilitate the hoisting operation (e.g., during landing of materials), and the worker must lean through the access opening or out over the edge of the access opening (to receive or guide equipment and materials, for example), that employee must be protected from fall hazards by using a personal fall arrest system.

Each worker on a walking or working surface must be protected from tripping in, or stepping into or through holes (including skylights), and from objects falling through holes (including skylights), by using covers.

Also, workers on ramps, runways, and other walkways must be protected from falling 4 feet or more to lower levels by using guardrail systems; workers at the edge of an excavation 4 feet or more in depth must be protected from falling by using guardrail systems, fences, or barricades.

Workers less than 4 feet above dangerous equipment must be protected from falling into or onto the dangerous equipment by the use of guardrail systems or equipment guards. When workers are 4 feet or more above dangerous equipment, they must be protected from fall hazards by using guardrail systems, personal fall arrest systems, or safety net systems.

Workers engaged in roofing activities on low-slope roofs, with unprotected sides and edges 4 feet or more above lower levels, must be protected from falling by using guardrail systems, safety net systems, or personal fall arrest systems; or, they must be protected by using a combination of a warning line system and guardrail system, a warning line system and a safety net system, a warning line system and a personal fall arrest system, or a warning line system and a safety monitoring system. When on a steep roof with unprotected sides and edges 4 feet or more above lower levels, workers must be protected from falling by using guardrail systems with toeboards, safety net systems, or personal fall arrest systems.

Each employee working on, at, above, or near wall openings (including those with chutes attached), where the outside bottom edge of the wall opening is 4 feet or more above lower levels and the inside bottom edge of the wall opening is less than 39 inches above the walking or working surface, must be protected from falling by using a guardrail system, safety net system, or personal fall arrest system.

When workers are exposed to falling objects, the employer must have each employee wear a hard hat and implement one of the following measures:

- Erect toeboards, screens, or guardrail systems to prevent objects from falling from higher levels.
- Erect a canopy structure and keep potential fall objects far enough from the edge of the higher level so that those objects would not go over the edge if they were accidentally displaced.
- Barricade the area to which objects could fall; prohibit employees from entering the barricaded area; and keep objects that may fall far enough away from the edge of a higher level so that those objects would not go over the edge if they were accidentally displaced.

The most common fall protection systems are

- Guardrails
- Safety nets
- Personal fall arrest systems

- Other fall protection systems:
  - Horizontal and vertical lifelines
  - Ladder climbing devices
  - Work positioning and travel restraint systems
  - Warning line systems
  - Aerial lift equipment/work platforms
  - Raising/lowering devices
- Covers

Personal Fall Arresting System

Effective January 1, 1998, body belts are not acceptable as part of a personal fall arrest system. Note: The use of a body belt in a positioning device system is acceptable. The full body harness is now the order of the day. After all, fallen workers could only hang about 1 minute, 30 seconds until they pass out from pooling of blood in their extremities and at times they actually slid from the belt and fell the remaining distance to the ground or surface. Rescues could not be performed in such a short time.

The belt itself often caused injuries to the back, etc. when workers fell. The full body harness, similar to a parachute harness, allows for a worker to hang in it upon a fall for some 30 minutes before it starts to become uncomfortable and rescue can usually transpire in that amount of time. The full body harness is actually a piece of equipment with a tag containing its serial number and other information for tracking purposes. The requirement for full body harness was viewed by workers as being impractical, uncomfortable, or cumbersome to work while wearing it. At worksites today, it does not pose these problems and is an important part of a fall protection system (see Figure 25.12).

Falls are a common accident in the workplace but can be prevented by taking time to plan when workers must be at heights. Some companies have gone to a philosophy of 100 percent fall protection so that no worker is left unprotected even while working on a ladder where no fall protection device is usually required. When used, improved fall protection devices have shown a marked decrease in falls from elevation that end in injury or death. Each year the newspapers have a picture of two or three workers hanging in body harnesses from the side of a building where a two-point scaffold has failed. Once rescued, the only thing hurt may be their pride. When properly used, fall protection has proven itself a part of good safety practices and good business. To check compliance with the OSHA requirements for a personal fall arrest system, regulation 29 CFR 1910.66 Appendix I provides more detail.

FIRE PROTECTION (29 CFR 1910.157)

Workplace fires and explosions kill 200 and injure more than 5,000 workers each year. In 1995, more than 75,000 workplace fires cost businesses more than $2.3 billion. Fires wreak havoc among workers and their families and destroy thousands of businesses each year, putting people out of work and severely impacting their livelihoods. The human and financial tolls underscore the serious nature of workplace fires.

FIGURE 25.12 Worker wearing a personal fall arrest system.

The OSHA standards require employers to provide proper exits, firefighting equipment, and employee training to prevent fire deaths and injuries in the workplace. Each workplace building must have at least two means of escape remote from each other that can be used in a fire emergency. Fire doors must not be blocked or locked to prevent emergency use when employees are within the buildings. Delayed opening of fire doors is permitted when an approved alarm system is integrated into the fire door design. Exit routes from buildings must be clear, free of obstructions, and properly marked with signs designating exits from the building.

Each workplace building must have a full complement of the proper type of fire extinguisher for the fire hazards present—except when the employer wishes to have employees evacuate instead of fighting small fires. Employees expected or anticipated to use fire extinguishers must be instructed on the hazards of fighting fires, how to properly operate the fire extinguishers available, and what procedures to follow in alerting others to the fire emergency. Only approved fire extinguishers are permitted to be used in the workplace, and they must be kept in good operating condition. Proper maintenance and inspection of this equipment are required of each employer (Figure 25.13).

Training of all employees in what must be done in an emergency is required. Where the employer wishes to evacuate employees instead of having them fight

FIGURE 25.13 Fire extinguisher in close proximity to the alarm box.

small fires, there must be written emergency plans and employee training for proper evacuation. Emergency action plans are required that describe the routes to use and procedures for employees to follow. Also, procedures for accounting for all evacuated employees must be part of the plan. The written plan must be available for employee review. Where needed, special procedures for helping physically impaired employees must be addressed in the plan; also, the plan must include procedures for those employees who must remain behind temporarily to shut down critical plant equipment before evacuating.

The preferred means of alerting employees to a fire emergency must be part of the plan. An employee alarm system must be available throughout the workplace complex and must be used for emergency alerting for evacuation. The alarm system may be voice communication or sound signals such as bells, whistles, or horns. Employees must know the evacuation signal.

Employers need to implement a written fire prevention plan to complement the fire evacuation plan. This should minimize the frequency of evacuation. Stopping unwanted fires from occurring is the most efficient way to handle them. The written plan shall be available for employee review. Housekeeping procedures for storage and cleanup of flammable materials and flammable waste must be included in the plan. Recycling of flammable waste such as paper is encouraged; however, handling and packaging procedures must be included in the plan. Procedures must be in place for controlling workplace ignition sources such as smoking, welding, and burning. Heat-producing equipment such as burners, heat exchangers, boilers, ovens, stoves, fryers, etc. must be properly maintained and kept clean of accumulations of flammable residues; flammables must not be stored close to these pieces of equipment. All employees must be apprised of the potential fire hazards of their job and the procedures called for in the employer's fire prevention plan. The plan shall be reviewed with all new employees when they begin their job and with all employees when the plan is changed.

Properly designed and installed fixed fire suppression systems enhance fire safety in the workplace. Automatic sprinkler systems throughout the workplace are among the most reliable firefighting means. The fire sprinkler system detects the fire, sounds an alarm, and puts the water where the fire and heat are located. Automatic fire suppression systems require proper maintenance to keep them in serviceable condition. When it is necessary to take a fire suppression system out of service while business continues, the employer must temporarily substitute a fire watch of trained employees standing by to respond quickly to any fire emergency in the normally protected area. The fire watch must interface with the employer's fire prevention plan and emergency action plan. Signs must be posted in areas protected by total flooding fire suppression systems that use agents that are a serious health hazard, such as carbon dioxide, Halon 1211, etc. Such automatic systems must be equipped with area pre-discharge alarm systems to warn employees of the impending discharge of the system and allow time to evacuate the area. There must be an emergency action plan to provide for the safe evacuation of employees within the protected area. Such plans should be part of the overall evacuation plan for the workplace facility.

The local fire department must be well acquainted with your facility, its location, and specific hazards. The fire alarm system must be certified, as required, and tested at least annually. Interior standpipes must be inspected regularly. Outside private fire hydrants must be flushed at least once a year and on a routine preventive maintenance schedule. All fire doors and shutters must be in good operating condition, unobstructed, and protected against obstructions, including their counterweights.

FLAMMABLE AND COMBUSTIBLE LIQUIDS (29 CFR 1910.106)

Flammable liquids must be kept in covered containers or tanks when not actually in use (see Figure 25.14). The quantity of flammable or combustible liquid that may

FIGURE 25.14 Label for a flammable liquid.

be located outside an inside storage room or storage cabinet in any one fire area of a building cannot exceed

- 25 gallons of Class IA liquids in containers
- 120 gallons of Class IB, IC, II, or III liquids in containers
- 660 gallons of Class IB, IC, II, or III liquids in a single portable tank

Flammable and combustible liquids are to be drawn from or transferred into containers within buildings only through a closed piping system, from safety cans, by means of a device drawing through the top, or by gravity through an approved self-closing valve. The transfer of liquids by means of air pressure is prohibited. No more than 60 gallons of Class I or Class II liquids or no more than 120 gallons of Class III liquids may be stored in a storage cabinet. Inside storage rooms for flammable and combustible liquids must be constructed to meet the required fire-resistant ratings and wiring specifications for their uses.

Outside storage areas must be graded so as to divert spills away from buildings or other exposures, or be surrounded with curbs at least 6 inches high with appropriate drainage to a safe location for accumulated liquids. The areas shall be protected against tampering or trespassing, where necessary, and shall be kept free of weeds, debris, and other combustible material not necessary to their storage.

Adequate precautions must be taken to prevent the ignition of flammable vapors. Sources of ignition include, but are not limited to, open flames, lightning, smoking, cutting and welding, hot surfaces, frictional heat, static, electrical and mechanical sparks, spontaneous ignition (including heat-producing chemical reactions), and radiant heat.

Class I liquids are not to be dispensed into containers unless the nozzle and container are electrically interconnected. All bulk drums of flammable liquids are to be grounded and bonded to containers during dispensing.

FLAMMABLE AND COMBUSTIBLE MATERIALS

Combustible scrap, debris, and waste materials (oily rags, etc.) stored in covered metal receptacles must be removed promptly from the worksite. Proper storage must be practiced to minimize the risk of fire, including spontaneous combustion. Fire extinguishers must be selected and provided for the types of materials in areas where they are to be used. "NO SMOKING" rules should be enforced in areas involving the storage and use of hazardous materials.

FLOORS [GENERAL CONDITIONS] (29 CFR 1910.22 AND 1920.23)

All floor surfaces must be kept clean, dry, and free from protruding nails, splinters, loose boards, holes, or projections. Where wet processes are used, drainage must be maintained and false floors, platforms, mats, or other dry standing places must be provided where practical.

In every building or other structure, or part thereof, used for mercantile, business, industrial, or storage purposes, the loads approved by the building official must be marked on plates of approved design and supplied and securely affixed by the owner of the building, or their duly authorized agent, in a conspicuous place in each space to which they relate. Such plates must not be removed or defaced but, if lost, removed, or defaced, shall be replaced by the owner or his or her agent.

Every stairway and ladderway floor opening must be guarded by standard railings with standard toeboards on all exposed sides except at the entrance. For infrequently used stairways, the guard may consist of a hinged cover and removable standard railings. The entrance to ladderway openings must be guarded to prevent a person from walking directly into the opening.

Every hatchway and chute floor opening must be guarded by a hinged floor opening cover equipped with standard railings to leave only one exposed side, or a removable railing with toeboards on not more than two sides and a fixed standard railing with toeboards on all other exposed sides.

Every floor hole into which a person can accidentally walk must be guarded by either a standard railing with standard toeboard on all exposed sides, or by a floor hole cover that is hinged in place. While the cover is not in place, the floor hole must be attended or protected by a removable standard railing.

Every open-sided floor, platform, or runway 4 feet or more above the adjacent floor or ground level must be guarded by a standard railing with toeboards on all open sides, except where there is an entrance to a ramp, stairway, or fixed ladder. Runways not less than 18 inches wide used exclusively for special purposes may have the railing on one side omitted where operating conditions necessitate. Regardless of height, open-sided floors, walkways, platforms, or runways above or adjacent to dangerous equipment must be guarded with a standard railing and toeboard.

FORKLIFT TRUCKS [POWERED INDUSTRIAL TRUCKS] (29 CFR 1910.178)

The American Society of Mechanical Engineers (ASME) defines a powered industrial truck as a mobile, power-propelled truck used to carry, push, pull, lift, stack, or tier materials. Powered industrial trucks are also commonly known as forklifts, pallet trucks, rider trucks, fork trucks, or lift trucks. Each year, tens of thousands of forklift-related injuries occur in U.S. workplaces. Injuries usually involve employees being struck by lift trucks or falling while standing or working from elevated pallets and tines. Many employees are injured when lift trucks are inadvertently driven off loading docks or when the lift falls between a dock and an unchocked trailer. Most incidents also involve property damage, including damage to overhead sprinklers, racking, pipes, walls, machinery, and other equipment. Unfortunately, a majority of employee injuries and property damage can be attributed to lack of procedures, insufficient or inadequate training, or lack of safety-rule enforcement.

If at any time, a powered industrial truck is found to be in need of repair, defective, or in any way unsafe, the truck must be taken out of service until it has been restored to a safe operating condition. High-lift rider trucks must be equipped with substantial overhead guards unless operating conditions do not permit. Fork trucks must be equipped with vertical-load backrest extensions when the types of loads present a hazard to the operators. Each industrial truck must have a warning backup alarm, whistle, gong, or other device that can be clearly heard above the normal noise in the areas where operated. The brakes of trucks must be set and wheel chocks placed under the rear wheels to prevent the movement of trucks, trailers, or railroad cars while loading or unloading.

Only trained and authorized operators are permitted to operate a powered industrial truck. Training must be done for the industrial truck being operated by the operator. Methods must be devised to train and evaluate operators in the safe operation of powered industrial trucks.

FUELING (29 CFR 1910.178, 1910.180, AND 1910.181)

It is prohibited to fuel an internal combustion engine with a flammable liquid while the engine is running. Fueling operations must be done in such a manner that the likelihood of spillage will be minimal. When spillage occurs during fueling operations, the spilled fuel must be washed away completely, evaporated, or other measures taken to control vapors before restarting the engine. Fuel tank caps must be replaced and secured before starting the engine.

Fueling hoses must be of a type designed to handle the specific type of fuel. It is prohibited to handle or transfer gasoline in open containers. No open lights, open flames, sparking, or arcing equipment is allowed during fueling or transfer of fuel operations, and no smoking should be permitted.

HAND TOOLS (29 CFR 1910.242)

Hand and power tools are a common part of our everyday lives and are present in nearly every industry. These tools help us easily perform tasks that otherwise would be difficult or impossible. However, these simple tools can be hazardous and have the potential to cause severe injuries when used or maintained improperly. Special attention toward hand and power tool safety is necessary in order to reduce or eliminate these hazards (see Figure 25.15).

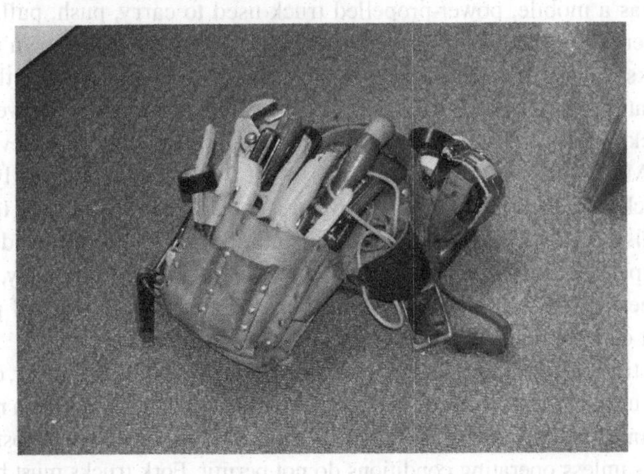

FIGURE 25.15 A tool pouch of hand tools.

Hand tools are nonpowered. They include anything from axes to wrenches. The greatest hazards posed by hand tools result from misuse and improper maintenance. Some examples include using a screwdriver as a chisel may cause the tip of the screwdriver to break and fly off, hitting the user or other employees. If a wooden handle on a tool such as a hammer or an axe is loose, splintered, or cracked, the head of the tool may fly off and strike the user or another worker. A wrench must not be used if its jaws are sprung, because it might slip. Impact tools such as chisels, wedges, or drift pins are unsafe if they have mushroomed heads. The head might shatter on impact, sending sharp fragments flying.

The employer is responsible for the safe condition of all tools and equipment used by employees but the employees have the responsibility for properly using and maintaining such tools. Employers should caution employees that saw blades, knives, or other tools be directed away from aisle areas and other employees working in close proximity. Knives and scissors must be sharp. Dull tools can be more hazardous than sharp ones. Appropriate personal protective equipment (e.g., safety goggles, gloves, etc.) should be worn due to hazards that may be encountered while using portable power tools and hand tools.

Safety requires that floors be kept as clean and dry as possible to prevent accidental slips with or around dangerous hand tools. Around flammable substances, sparks produced by iron and steel hand tools can be dangerous ignition sources. Where this hazard exists, spark-resistant tools made from brass, plastic, aluminum, or wood will provide a degree of safety.

Employees who use hand and power tools and who are exposed to the hazards of falling, flying, abrasive, and splashing objects, or exposed to harmful dusts, fumes, mists, vapors, or gases, must be provided with the particular specific personal protective equipment necessary to protect them from the hazard.

HOIST AND AUXILIARY EQUIPMENT (29 CFR 1910.179)

Each overhead electric hoist must be equipped with a limit device to stop the hook at its highest and lowest point of safe travel. Each hoist must automatically stop and hold any load up to 125 percent of its rated load if its actuating force is removed. The rated load of each hoist must be legibly marked and visible to the operator. Stops must be provided at the safe limits of travel for a trolley hoist. The controls of each hoist should be plainly marked to indicate the direction of travel or motion.

Close-fitting guards or other suitable devices must be installed on the hoist to ensure that hoist ropes are maintained in the sheave groves. All hoist chains and ropes must be of sufficient length to handle the full range of movement of the application, while still maintaining two full wraps on the drum at all times. Nip points or contact points between hoist ropes and sheaves that are permanently located within 7 feet of the floor, ground, or working platform must be guarded. No chains or rope slings that are kinked or twisted should be used, and never use the hoist rope or chain wrapped around the load as a substitute for a sling. The operator must be instructed to avoid carrying loads over people.

FIGURE 25.16 Poor housekeeping impedes travel and presents falling hazards.

HOUSEKEEPING (29 CFR 1910.22)

Housekeeping is, by far, one of the easiest preventive methods to implement and tends to have the most immediate impact upon the workplace. Poor housekeeping leads to a myriad of hazards from slips, trips, and falls to fire hazards. Housekeeping is a portion of a hazard removal program in which everyone in the workplace can take an active role. Because of this, it will impact more than just accident prevention; it will also increase morale and productivity (see Figure 25.16).

HYDRAULIC POWER TOOLS (29 CFR 1910.217)

Hydraulic power tools must use only approved fire-resistant fluid that retains its operating characteristics at the most extreme temperatures to which it will be exposed. The manufacturer's recommended safe operating pressure for hoses, valves, pipes, filters, and other fittings must not be exceeded.

JACKS (29 CFR 1910.244)

All jacks (lever and ratchet jacks, screw jacks, and hydraulic jacks) must have a device that stops them from jacking up too high. Also, the manufacturer's load limit must be permanently marked in a prominent place on the jack and should not be exceeded. A jack should never be used to support a lifted load. Once the load has been lifted, it must immediately be blocked up. Use wooden blocking under the base if necessary to make the jack level and secure. If the lift surface is metal, place a 1-inch-thick hardwood block or equivalent between it and the metal jack head to reduce the danger of slippage. To set up a jack, make certain of the following:

- The base rests on a firm level surface.
- The jack is correctly centered.
- The jack head bears against a level surface.
- The lift force is applied evenly.

Proper maintenance of jacks is essential for safety. All jacks must be inspected before each use and lubricated regularly. If a jack is subjected to an abnormal load or shock, it should be thoroughly examined to make sure it has not been damaged. Hydraulic jacks exposed to freezing temperatures must be filled with adequate antifreeze liquid.

LADDERS, FIXED (29 CFR 1910.27)

A fixed ladder must be able to support at least two loads of 250 pounds each, concentrated between any two consecutive attachments. Fixed ladders also must support added anticipated loads caused by ice buildup, winds, rigging, and impact loads resulting from using ladder safety devices. Fixed ladders must be used at a pitch no greater than 90 degrees from the horizontal, as measured from the back side of the ladder (see Figure 25.17).

FIGURE 25.17 An example of a fixed ladder.

Individual-rung/step ladders must extend at least 42 inches above an access level or landing platform either by the continuation of the rung spacing as horizontal grab bars or by providing vertical grab bars that must have the same lateral spacing as the vertical legs of the ladder rails. Each step or rung of a fixed ladder must be able to support a load of at least 250 pounds applied in the middle of the step or rung.

The minimum clear distance between the sides of individual-rung/step ladders and between the side-rails of other fixed ladders must be 16 inches. The rungs of individual-rung/step ladders must be shaped to prevent slipping off the end of the rungs. The rungs and steps of fixed metal ladders manufactured after March 15, 1991, must be corrugated, knurled, dimpled, coated with skid-resistant material, or treated to minimize slipping. The minimum perpendicular clearance between fixed ladder rungs, cleats, and steps and any obstruction behind the ladder must be 7 inches, except for the clearance of an elevator-pit ladder, which must be 4½ inches. The minimum perpendicular clearance between the centerline of fixed ladder rungs, cleats, steps, and any obstruction on the climbing side of the ladder must be 30 inches. If obstructions are unavoidable, clearance may be reduced to 24 inches, provided a deflection device is installed to guide workers around the obstruction. The step-across distance between the center of the steps or rungs of fixed ladders and the nearest edge of a landing area must be no less than 7 inches and no more than 12 inches. A landing platform must be provided if the step-across distance exceeds 12 inches. Fixed ladders without cages or wells must have at least a 15-inch clear width to the nearest permanent object on each side of the centerline of the ladder.

Fixed ladders must be provided with cages, wells, ladder safety devices, or self-retracting lifelines where the length of climb is less than 24 feet but the top of the ladder is at a distance greater than 24 feet above lower levels. If the total length of the climb on a fixed ladder equals or exceeds 24 feet, the following requirements must be met: Fixed ladders must be equipped with either (1) ladder safety devices, (2) self-retracting lifelines and rest platforms at intervals not to exceed 150 feet, or (e) a cage or well, and multiple ladder sections, each ladder section not to exceed 50 feet in length. These ladder sections must be offset from adjacent sections, and landing platforms must be provided at maximum intervals of 50 feet.

The side-rails of through- or side-step fixed ladders must extend 42 inches above the top level or landing platform served by the ladder. Parapet ladders must have an access level at the roof if the parapet is cut to permit passage through it; if the parapet is continuous, the access level is the top of the parapet. Steps or rungs for through-fixed-ladder extensions must be omitted from the extension; and the extension of side-rails must be flared to provide between 24 and 30 inches of clearance between side-rails. When safety devices are provided, the maximum clearance distance between side-rail extensions must not exceed 36 inches.

Cages must not extend less than 27 inches or more than 30 inches from the centerline of the step or rung, and must not be less than 27 inches wide. The inside of the cage must be clear of projections. Horizontal bands must be fastened to the side-rails of rail ladders or directly to the structure, building, or equipment for individual-rung ladders. Horizontal bands must be spaced at intervals not more than 4 feet apart, measured from centerline to centerline. Vertical bars must be on the inside of the horizontal bands and must be fastened to them. Vertical bars must be spaced at intervals not more than 9½ inches, measured centerline to centerline.

The bottom of the cage must be between 7 and 8 feet above the point of access to the bottom of the ladder. The bottom of the cage must be flared not less than 4 inches between the bottom horizontal band and the next higher band. The top of the cage must be a minimum of 42 inches above the top of the platform or the point of access at the top of the ladder. Provisions must be made for access to the platform or other points of access.

Wells must completely encircle the ladder and be free of projections. The inside face of the well on the climbing side of the ladder must extend between 27 and 30 inches from the centerline of the step or rung. The inside width of the well must be at least 30 inches. The bottom of the well above the point of access to the bottom of the ladder must be between 7 and 8 feet.

All safety devices must be able to withstand, without failure, a drop test consisting of a 500-pound weight dropping 18 inches. All safety devices must permit the worker to ascend or descend without continually having to hold, push, or pull any part of the device, leaving both hands free for climbing. All safety devices must be activated within 2 feet after a fall occurs, and limit the descending velocity of an employee to 7 feet per second or less. The connection between the carrier or lifeline and the point of attachment to the body harness must not exceed 9 inches in length.

Mountings for rigid carriers must be attached at each end of the carrier, with intermediate mountings spaced along the entire length of the carrier, to provide the necessary strength to stop workers' falls. Mountings for flexible carriers must be attached at each end of the carrier. Cable guides for flexible carriers must be installed with spacing between 25 and 40 feet along the entire length of the carrier, to prevent wind damage to the system. The design and installation of mountings and cable guides must not reduce the strength of the ladder. Side-rails and steps or rungs for side-step fixed ladders must be continuous in extension.

Fixed ladders with structural defects (such as broken or missing rungs, cleats, or steps, broken or split rails, or corroded components) must be withdrawn from service until repaired. Defective fixed ladders are considered withdrawn from use when they are (1) immediately tagged with "Do Not Use" or similar language, (2) marked in a manner that identifies them as defective, or (3) blocked, for example, with a plywood attachment that spans several rungs.

LADDERS, PORTABLE (29 CFR 1910.25 AND 1910.26)

Non-self-supporting and self-supporting portable ladders must support at least four times the maximum intended load; extra heavy-duty type. The ability of a self-supporting ladder to sustain loads must be determined by applying the load to the ladder in a downward vertical direction. The ability of a non-self-supporting ladder to sustain loads must be determined by applying the load in a downward vertical direction when the ladder is placed at a horizontal angle of 75.5 degrees (see Figure 25.18).

When portable ladders are used for access to an upper landing surface, the side-rails must extend at least 3 feet above the upper landing surface. When such an extension is not possible, the ladder must be secured, and a grasping device such as a

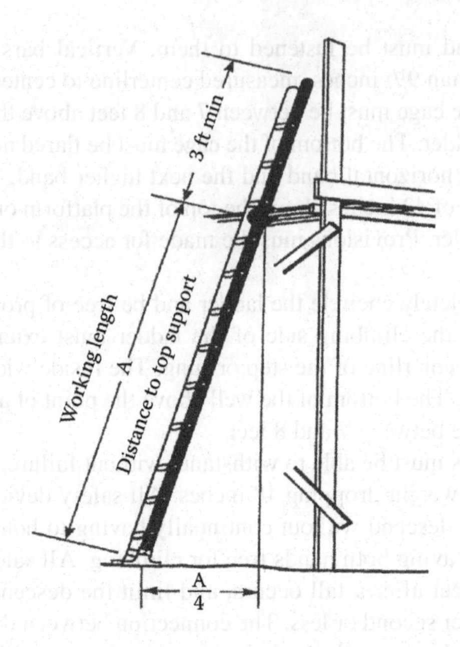

FIGURE 25.18 Safe use of portable ladder. (*Source:* Courtesy of the Occupational Safety and Health Administration.)

grab rail must be provided to assist workers in mounting and dismounting the ladder. An extension ladder must not deflect under a load that would cause the ladder to slip off its supports.

Ladders must be maintained free of oil, grease, and other slipping hazards. Ladders must not be loaded beyond the maximum intended load for which they were built or beyond their manufacturer's rated capacity. Ladders must be used only for the purpose for which they were designed. Non-self-supporting ladders must be used at an angle where the horizontal distance from the top support to the foot of the ladder is approximately one-quarter of the working length of the ladder. Wood job-made ladders with spliced side-rails must be used at an angle where the horizontal distance is one-eighth the working length of the ladder.

Ladders must be used only on stable and level surfaces unless secured to prevent accidental movement. Ladders must not be used on slippery surfaces unless secured or provided with slip-resistant feet to prevent accidental movement. Slip-resistant feet must not be used as a substitute for the care in placing, lashing, or holding a ladder upon slippery surfaces. Ladders placed in areas such as passageways, doorways, or driveways, or where they can be displaced by workplace activities or traffic, must be secured to prevent accidental movement or a barricade must be used to keep traffic or activities away from the ladder. The area around the top and bottom of the ladders must be kept clear.

The top of a non-self-supporting ladder must be placed with two rails supported equally unless it is equipped with a single support attachment. Ladders must not be moved, shifted, or extended while in use. Ladders must have nonconductive side-rails if they are used where the worker or the ladder could contact exposed energized electrical equipment.

A double-cleated ladder, or two or more ladders, must be provided when ladders are the only way to enter or exit a work area having twenty-five or more employees, or when a ladder serves simultaneous two-way traffic. Ladder rungs, cleats, and steps must be parallel, level, and uniformly spaced when the ladder is in position for use. Rungs, cleats, and steps of portable and fixed ladders (except as provided below) must not be spaced less than 10 inches apart, or more than 14 inches apart, along the ladder's side-rails. Rungs, cleats, and steps of step stools must not be less than 8 inches apart, nor more than 12 inches apart, between center lines of the rungs, cleats, and steps.

Ladders must be inspected by a competent person for visible defects on a periodic basis and after any incident that could affect their safe use. Single-rail ladders must not be used. When ascending or descending a ladder, the worker must face the ladder. Each worker must use at least one hand to grasp the ladder when climbing. A worker on a ladder must not carry any object or load that could cause him or her to lose balance and fall. The top or top step of a stepladder must not be used as a step. Cross bracing on the rear section of stepladders must not be used for climbing unless the ladders are designed and provided with steps for climbing on both front and rear sections.

Ladders must not be tied or fastened together to create longer sections unless they are specifically designed for such use. A metal spreader or locking device must be provided on each stepladder to hold the front and back sections in an open position when the ladder is being used. Two or more separate ladders used to reach an elevated work area must be offset with a platform or landing between the ladders, except when portable ladders are used to gain access to fixed ladders.

Ladder components must be surfaced to prevent injury from punctures or lacerations and prevent snagging of clothing. Wood ladders must not be coated with any opaque covering, except for identification or warning labels, which may be placed only on one face of a side-rail.

Portable ladders with structural defects (such as broken or missing rungs, cleats, or steps, broken or split rails, corroded components, or other faulty or defective components) must immediately be marked defective or tagged with "Do Not Use" or similar language and withdrawn from service until repaired. Ladder repairs must restore the ladder to a condition meeting its original design criteria before the ladder is returned to use.

Under these provisions of the OSHA standard, employers must provide a training program for each employee using ladders and stairways. The program must enable each employee to recognize hazards related to ladders and stairways and to use proper procedures to minimize these hazards. For example, employers must ensure that each employee is trained by a competent person in the following areas, as applicable:

1. The nature of fall hazards in the work area
2. The correct procedures for erecting, maintaining, and disassembling the fall protection systems to be used

3. The proper construction, use, placement, and care in handling all stairways and ladders
4. The maximum intended load-carrying capacities of ladders used

LUNCH ROOMS (29 CFR 1910.141)

Employees are not to consume food or beverages in toilet rooms or in any area exposed to a toxic material. A covered receptacle of corrosion-resistant or disposable material is to be provided in the lunch areas for disposal of waste food. The cover may be omitted when sanitary conditions can be maintained without the use of a cover.

MACHINE GUARDING (29 CFR 1910.212 AND 1910.219)

Moving machine parts have the potential for causing severe workplace injuries, such as crushed fingers or hands, amputations, burns, and blindness, just to name a few. Safeguards are essential for protecting workers from these needless and preventable injuries. Machine guarding and related machinery violations continuously rank among the top ten OSHA citations issued. In fact, "Machine Guarding: General Requirements" (29 CFR 1910.212) and "Mechanical Power Transmission" (29 CFR 1910.219) were the number three and number eight, respectively, OSHA violations for the manufacturing sector from October 2006 to September 2007, with 2,258 and 1,476 federal citations issued, respectively. Mechanical power presses have also become an area of increasing concern. In April 1997, OSHA launched a National Emphasis Program on Mechanical Power Presses (CPL 2-1.24). This program targets industries that have high amputation rates and includes both education and enforcement efforts.

Machine guarding must be provided to protect employees in the machine area from hazards such as those created by point-of-operation, nip points, rotating parts, flying chips, and sparks (see Figure 25.19). The guard itself should not create an accident hazard. The point-of-operation guarding device must be designed to prevent operators from having any part of their body in the danger zone during operating cycles. Special supplemental hand tools for placing and removing material must be designed so that operators can perform without their hands being in the danger zone. Some of the machines that usually require point-of-operation guarding are guillotine cutters, shears, alligator shears, power presses, milling machines, power saws, jointers, portable power tools, and forming rolls and calendars.

Machines designed for a fixed location must be securely anchored to prevent walking or moving, or designed in such a manner that they will not move during normal operation. Sufficient clearance, provided around and between machines, facilitates safer operations, setup, servicing, material handling, and waste removal.

It is also good practice to have a training program to instruct employees on safe methods of machine operation. Supervision should ensure that employees are following safe machine operating procedures. A regular program of safety inspection of machinery and equipment should exist that will ensure all machinery and equipment is kept clean and properly maintained.

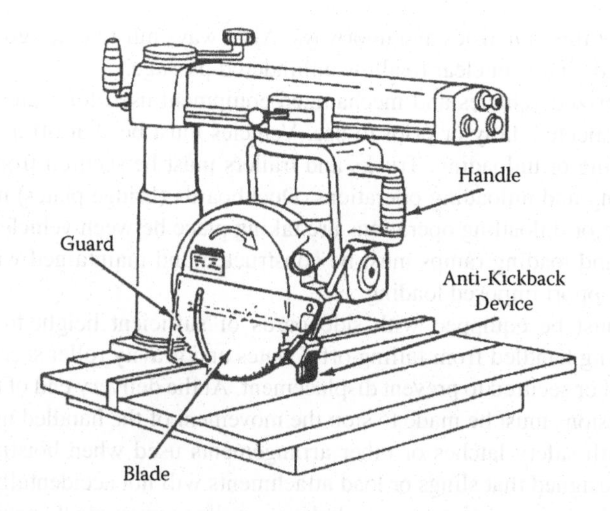

FIGURE 25.19 A self-adjusting guard for the blade of a radial saw.

As a part of safeguarding, a power shut-off switch should be within reach of the operator's position at each machine, and the electric power to each machine must be locked out for maintenance, repair, or security. Noncurrent-carrying metal parts of electrically operated machines must be bonded and grounded. Foot-operated switches must be guarded or arranged in order to prevent accidental actuation by personnel or falling objects. Any manually operated valves and switches controlling the operation of equipment and machines must be clearly identified and readily accessible. Provisions must exist to prevent machines from automatically starting when power is restored after a power failure or shutdown. All emergency stop buttons shall be colored red.

All exposed moving chains and gears must be properly guarded. All pulleys and belts that are within 7 feet of the floor or working level must be properly guarded. Fan blades must be protected with a guard having openings no larger than ½ inch when operating within 7 feet of the floor.

Splash guards should be mounted on machines that use coolant to prevent the coolant from reaching employees. Revolving drums, barrels, and containers must be guarded by an enclosure that is interlocked with the drive mechanism, so that revolution cannot occur unless the guard enclosures are in place, and so guarded. Machine guarding is an integral part of machine safety and should be addressed without question based upon the number of injuries that transpire.

MACHINERY, FIXED (29 CFR 1910.212)

See machine guarding.

MATERIAL HANDLING (29 CFR 1910.176)

Some form of material handling transpires in almost all workplaces. Material handling is one of the leading causes of occupational injuries, whether it is manual lifting or the use of lifting equipment. Great care must be taken to ensure safe clearance for equipment through aisles and doorways. Aisle ways must be designated, permanently marked, and kept clear to allow unhindered passage.

Any motorized vehicles and mechanized equipment used for material handling should be inspected daily or prior to use. Vehicles must be shut off and brakes set prior to loading or unloading. Trucks and trailers must be secured from movement during loading and unloading operations. Dockboards (bridge plates) must be used when loading or unloading operations are taking place between vehicles and docks. Dockplates and loading ramps must be constructed and maintained with sufficient strength to support imposed loading.

Chutes must be equipped with sideboards of sufficient height to prevent the materials being handled from falling off. Chutes and gravity roller sections must be firmly placed or secured to prevent displacement. At the delivery end of the rollers or chutes, provisions must be made to stop the movement of the handled materials.

Hooks with safety latches or other arrangements used when hoisting materials must be so designed that slings or load attachments will not accidentally slip off the hoist hooks. Securing chains, ropes, chokers, or slings must meet specifications for the job to be performed. When hoisting material or equipment, provisions shall be made to ensure that no one will be passing under suspended loads.

Containers of combustibles or flammables, when stacked while being moved, are always separated by dunnage sufficient to provide stability. Material safety data sheets must be available to employees handling hazardous substances.

Material handling is one hazard in the workplace that impacts everyone. Almost every worker is exposed to the potential hazards associated with material handling at some period during their work history.

MECHANICAL POWER PRESSES (29 CFR 1910.217)

OSHA has identified power presses for special emphasis based upon the number of amputations that occur on a routine bases. The employer must provide and ensure the usage of point-of-operation guards exist or the use of properly applied and adjusted point-of-operation devices. These tools help prevent entry of hands or fingers into the point-of-operation transpires. Accidents occur when operators reach through, over, under, and around the guard or protective device during operations performed on a mechanical power press. This requirement shall not apply when the point-of-operation opening is 1/4 inch or less. On these and similarly functioning machines, hand and foot operations must be provided with guards to prevent inadvertent initiation of the press.

All dies should be stamped with the tonnage and stroke requirements or be otherwise recorded and readily available to the die setter. The employer is to provide and enforce the use of safety blocks whenever dies are being adjusted or repaired in the press. Brushes, swabs, or other tools must be provided for lubrication, so that employees cannot reach into the point of operation.

Presence-sensing devices may not be used to initiate the slide motion except when used in total conformance with paragraph (h) of 29 CFR 1910.217, which requires certification of the control system. Any machine using full-revolution clutches must incorporate a single-stroke mechanism. A main disconnect capable of being locked in the off position must be provided with every power press control system.

To ensure safe operating conditions and to maintain a record of inspections and maintenance work, the employer must establish a program of regular inspections of the power presses to include the date, serial number of the equipment, as well as the signature of the inspector. OSHA requires that all point-of-operation injuries be reported within 30 days to OSHA's Office of Standards and Development or the state agency administering an occupational safety and health program.

MOTOR VEHICLE SAFETY

According to the Bureau of Labor Statistics, more than 2,000 deaths a year result from occupational motor vehicle incidents, more than 30 percent of the total annual number of fatalities from occupational injuries. These deaths include driver and passenger deaths in highway crashes, farm equipment accidents, and industrial vehicle incidents, as well as pedestrian fatalities. There are no specific OSHA standards concerning workplace motor vehicle safety; however, most occupational fatalities occur on public highways where there are seat belt requirements and traffic laws. OSHA issued a Notice of Proposed Rulemaking in July 1990 for a standard that would have required seat belt use and driver awareness programs. Nothing has transpired regarding this proposal.

PNEUMATIC TOOLS (29 CFR 1910.243)

Pneumatic tools are powered by compressed air and include chippers, drills, hammers, and sanders. There are several dangers encountered in the use of pneumatic tools. The main one is the danger of getting hit by one of the tool attachments or by some kind of fastener the worker is using with the tool. Eye protection is required and face protection is recommended for employees working with pneumatic tools. Noise is another hazard. Working with noisy tools, such as jackhammers, requires proper, effective use of hearing protection.

When using pneumatic tools, employees must check to see that the tools are fastened securely to the hose to prevent them from becoming disconnected. A short wire, or positive locking device, attaching the air hose to the tool will serve as an added safeguard. A safety clip or retainer must be installed to prevent attachments, such as chisels on a chipping hammer, from being unintentionally shot from the barrel. Screens must be set up to protect nearby workers from being struck by flying fragments around chippers, riveting guns, staplers, or air drills.

Compressed air guns should never be pointed toward anyone, and users should never "dead-end" them against themselves or anyone else.

PORTABLE (POWER OPERATED) TOOLS AND EQUIPMENT (29 CFR 1910.243)

Tools are such a common part of our lives that it is difficult to remember that they may pose hazards. All tools are manufactured with safety in mind but, tragically, serious accidents often occur before steps are taken to search out, avoid, or eliminate

FIGURE 25.20 Examples of power tools.

tool-related hazards. In the process of removing or avoiding hazards, workers must learn to recognize the hazards associated with specific types of tools and the safety precautions necessary to prevent those hazards (see Figure 25.20). All hazards involved in the use of power tools can be prevented by following five basic safety rules:

- Keep all tools in good condition with regular maintenance.
- Use the right tool for the job.
- Examine each tool for damage before use.
- Operate tools according to the manufacturer's instructions.
- Provide and use the proper protective equipment.

Employees and employers have a responsibility to work together to establish safe working procedures. If a hazardous situation is encountered, it should be brought to the attention of the proper individual immediately.

Power tools can be hazardous when improperly used. There are several types of power tools, based on the power source they use: electric, pneumatic, liquid fuel, hydraulic, and powder-actuated. Employees should be trained in the use of all tools (not just power tools). They should understand the potential hazards as well as the safety precautions to prevent hazards from occurring. The following general precautions should be observed by power tool users:

1. Never carry a tool by its cord or hose.
2. Never yank the cord or the hose to disconnect it from its receptacle.
3. Keep cords and hoses away from heat, oil, and sharp edges.
4. Disconnect tools when not in use, before servicing, and when changing accessories such as blades, bits, and cutters.

5. All observers should be kept at a safe distance away from the work area.
6. Secure work with clamps or a vise, freeing both hands to operate the tool.
7. Avoid accidental starting. The worker should not hold a finger on the switch button while carrying a plugged-in tool.
8. Tools should be maintained with care. They should be kept sharp and clean for the best performance. Follow instructions in the user's manual for lubricating and changing accessories.
9. Be sure to keep a secure footing and maintain good balance.
10. Proper apparel should be worn. Loose clothing, ties, or jewelry can become caught in moving parts.
11. All portable electric tools that are damaged must be removed from use and tagged "Do Not Use."

Hazardous moving parts of a power tool must be safeguarded. For example, belts, gears, shafts, pulleys, sprockets, spindles, drums, flywheels, chains, or other reciprocating, rotating, or moving parts of equipment must be guarded if such parts are exposed to contact by employees. Guards, as necessary, should be provided to protect the operator and others from the following:

- Point-of-operation
- In-running nip points
- Rotating parts
- Flying chips and sparks

Safety guards must never be removed when a tool is being used. For example, portable circular saws must be equipped with guards. An upper guard must cover the entire blade of the saw. A retractable lower guard must cover the teeth of the saw, except when it makes contact with the work material. The lower guard must automatically return to the covering position when the tool is withdrawn from the work.

The following hand-held powered tools must be equipped with a momentary contact "on-off" control switch: drills; tappers; fastener drivers; horizontal, vertical, and angle grinders with wheels larger than 2 inches in diameter; disc and belt sanders; reciprocating saws; saber saws; and other similar tools. These tools may also be equipped with a lock-on control, provided that turnoff can be accomplished by a single motion of the same finger or fingers that turn it on.

The following hand-held powered tools may be equipped with only a positive "on-off" control switch: platen sanders, disc sanders with discs 2 inches or less in diameter, grinders with wheels 2 inches or less in diameter, routers, planers, laminate trimmers, nibblers, shears, scroll saws and jigsaws with blade shanks 1/4 inch wide or less.

Other hand-held powered tools such as circular saws having a blade diameter greater than 2 inches, chain saws, and percussion tools without means for positive accessory holding must be equipped with a constant pressure switch that will shut off the power when the pressure is released.

Employees using electric tools must be aware of several dangers; the most serious is the possibility of electrocution. Among the chief hazards of electric-powered tools are burns and slight shocks that can lead to injuries or even heart failure. Under certain conditions, even a small amount of current can result in fibrillation of the heart and eventual death. A shock can also cause the user to fall off a ladder or other elevated work surface.

To protect the user from shock, tools must either have a three-wire cord with ground, be grounded, be double insulated, or be powered by a low-voltage isolation transformer. Three-wire cords contain two current-carrying conductors and a grounding conductor. One end of the grounding conductor connects to the tool's metal housing. The other end is grounded through a prong on the plug. Anytime an adapter is used to accommodate a two-hole receptacle, the adapter wire must be attached to a known ground. The third prong should never be removed from the plug. Double insulation is more convenient. The user and the tools are protected in two ways: by normal insulation from the wires inside the tool, and by a housing that cannot conduct electricity to the operator in the event of a malfunction. These general practices should be followed when using electric tools:

- Electric tools should be operated within their design limitations.
- Gloves and safety footwear are recommended during the use of electric tools.
- When not in use, tools should be stored in a dry place.
- Electric tools should not be used in damp or wet locations.
- Work areas should be well lighted.

Powered abrasive grinding, cutting, polishing, and wire buffing wheels create special safety problems because they may throw off flying fragments. Before an abrasive wheel is mounted, it should be inspected closely and sound- or ring-tested to ensure that it is free from cracks or defects. To test, wheels should be tapped gently with a light nonmetallic instrument. If the wheel sounds cracked or dead, it could fly apart in operation and therefore must not be used. A robust and undamaged wheel will give a clear metallic tone or "ring." To prevent the wheel from cracking, the user should be sure it fits freely on the spindle. The spindle nut must be tightened enough to hold the wheel in place, without distorting the flange. Follow the manufacturer's recommendations. Care must be taken to ensure that the spindle wheel does not exceed the abrasive wheel specifications. Due to the possibility of a wheel disintegrating (exploding) during start-up, employees should never stand directly in front of the wheel as it accelerates to full operating speed. Portable grinding tools must be equipped with safety guards to protect workers not only from the moving wheel surface, but also from flying fragments in case of breakage. In addition, when using a powered grinder,

- Always use eye protection.
- Turn off the power when not in use.
- Never clamp a hand-held grinder in a vise.

POWDER-ACTUATED TOOLS (29 CFR 1910.243)

Powder- or explosive-actuated tools operate like a loaded gun and should be treated with the same respect and precautions. In fact, they are so dangerous that they must be operated only by specially trained employees. All employees who operate powder-actuated tools must be trained in their use and carry a valid operator's card. Safety precautions to remember include the following:

1. These tools should not be used in an explosive or flammable atmosphere.
2. Before using the tool, the worker should inspect it to determine that it is clean, that all moving parts operate freely, and that the barrel is free from obstructions.
3. The tool should never be pointed at anybody.
4. The tool should not be loaded unless it is to be used immediately. A loaded tool should not be left unattended, especially where it would be available to unauthorized persons.
5. Hands should be kept clear of the barrel end. To prevent the tool from firing accidentally, two separate motions are required for firing: one to bring the tool into position and another to pull the trigger. The tools must not be able to operate until they are pressed against the work surface with a force of at least 5 pounds greater than the total weight of the tool.
6. If a powder-actuated tool misfires, the employee should wait at least 30 seconds and then try firing it again. If it still will not fire, the user should wait another 30 seconds so that the faulty cartridge is less likely to explode, and then carefully remove the load. The bad cartridge should be put in water.
7. Suitable eye and face protection are essential when using a powder-actuated tool.
8. The muzzle end of the tool must have a protective shield or guard centered perpendicularly on the barrel to confine any flying fragments or particles that might otherwise create a hazard when the tool is fired. The tool must be designed so that it will not fire unless it has this kind of safety device.
9. All powder-actuated tools must be designed for varying powder charges so that the user can select a powder level necessary to do the work without excessive force.
10. If the tool develops a defect during use, it should be tagged and taken out of service immediately until it is properly repaired.

When using powder-actuated tools to apply fasteners, there are some precautions to consider. Fasteners must not be fired into material that would let them pass through to the other side. The fastener must not be driven into materials such as brick or concrete any closer than 3 inches to an edge or corner. In steel, the fastener must not come any closer than 1/2 inch from a corner or edge. Fasteners must not be driven into very hard or brittle materials that might chip, splatter, or make the fastener ricochet. An alignment guide must be used when shooting a fastener into an existing hole. A fastener must not be driven into a spalled area caused by an unsatisfactory fastening.

Each powder-actuated tool must be stored in its own locked container when not being used. A sign at least 7 inches by 10 inches with boldface type reading "POWDER- OR EXPLOSIVE-ACTUATED TOOL IN USE" should be posted conspicuously when the tool is being used. Powder-actuated tools are to be left unloaded until they are actually ready for use. Powder-actuated tools must be inspected for obstructions or defects each day before use. Their operators should have and use appropriate personal protective equipment such as hard hats, safety goggles, safety shoes, and ear protectors.

POWERED PLATFORMS FOR BUILDING MAINTENANCE (29 CFR 1910.66)

All completed building maintenance equipment installation must be inspected and tested in the field before being placed in service. A similar inspection and test must be made following any major alteration to an existing installation. No hoist is to be subjected to a load in excess of 125 percent of its rated load.

Structural supports, tie-downs, tie-in guides, anchoring devices, and any affected parts of a building included in the installation must be designed by or under the direction of a registered, experienced engineer. Exterior installations must be capable of withstanding prevailing climatic conditions. The building installation must provide safe access to, and egress from, the equipment and sufficient space to conduct necessary maintenance.

Affected parts of the building must have the capacity to sustain all the loads imposed by the equipment. The affected parts of the building must be designed to allow the equipment to be used without exposing employees to a hazardous condition.

Repairs or major maintenance of those building portions that provide support for the suspended equipment must not affect the capability of the building to meet the requirements of this standard.

The equipment power circuit must be an independent electrical circuit that is to remain separate from all other equipment within or on the building, other than the power circuits used for hand tools that will be used in conjunction with the equipment. If the building is provided with an emergency power system, the equipment power circuit may also be connected to this system.

Equipment installations must be designed by or under the direction of a registered experienced professional engineer. The design is to provide for a minimum live load or 250 pounds for each occupant of a suspended or supported platform. Equipment that is exposed to wind when not in service must be designed to withstand forces generated by winds of at least 100 mph at 30 feet above grade and, when in service, must be able to withstand forces generated by wind of at least 50 mph at all elevations.

Each suspended unit component, except suspension ropes and guardrail systems, must be capable of supporting at least four times the maximum intended live load applied or transmitted to that component.

POWER TRANSMISSION EQUIPMENT GUARDING (29 CFR 1910.219)

All belts, pulleys, sprockets and chains, flywheels, shafting and shaft projections, gears, couplings, other rotating or reciprocating parts, or any movable portion within 7 feet of the floor or working platform must be effectively guarded.

All guards for inclined belts must conform to the standards for construction of horizontal belts, and shall be arranged in such a manner that a minimum clearance of 7 feet is maintained between the belt and floor at any point outside the guard. Where both runs of horizontal belts are 7 feet or less from the floor or working surface, the guard is to extend at least 15 inches above the belt or to a standard height. An exception is that where both runs of a horizontal belt are 42 inches or less from the floor, the belt must be fully enclosed by guards made of expanded metal, perforated or solid sheet metal, wire mesh on a frame of angle iron, or iron pipe securely fastened from the floor to the frame of the machine.

Flywheels located so that any part is 7 feet or less above the floor or work platform must be guarded with an enclosure of sheet, perforated, or expanded metal, or woven wire. Flywheels protruding through a working floor must be entirely enclosed by a guardrail and toeboards.

Gears, sprocket wheels, and chains must be enclosed unless they are more than 7 feet above the floor, or the mesh points are guarded. Couplings with bolts, nuts, or setscrews extending beyond the flange of the coupling must be guarded by a safety sleeve.

PRESSURE VESSELS (29 CFR 1910.106, 1910.216, AND 1910.217)

Generally, a pressure vessel is a storage tank or vessel that has been designed to operate at pressures above 15 pounds per square inch (psi). Recent inspections of pressure vessels have shown that there are a considerable number of cracked and damaged vessels in workplaces. Cracked and damaged vessels can result in leakage or rupture failures. Potential health and safety hazards of leaking vessels include poisonings, suffocations, fires, and explosion hazards. Rupture failures can be much more catastrophic and can cause considerable damage to life and property. The safe design, installation, operation, and maintenance of pressure vessels in accordance with the appropriate codes and standards are essential to worker safety and health.

Pressure vessel design, construction, and inspection is referenced in the ASME Boiler and Pressure Vessel Code, 1968, and current 1910.106(b), 1910.217(b)(12), 1910.261(a)(3), and OSHA Technical Manual CPL 2-2.208, Chapter 10 (Pressure Vessel Guidelines). These set the guidelines for pressure vessel safety. Two consequences result from a complete rupture: (1) blast effects can happen due to a sudden expansion of the pressurized fluid, and (2) fragmentation damage and injury can result if vessel rupture occurs.

In the event of leakage failure, the hazard consequences can range from no effect to very serious effects. Suffocation or poisoning can occur, depending on the nature of the contained fluid, if the leakage occurs into a closed space. Fire and explosion can cause physical hazards when a flammable fluid ignites. Chemical and thermal burns can result from contact with process liquids.

Most of the pressure or storage vessels in service in the United States have been designed and constructed in accordance with one of the following two pressure vessel design codes:

1. The ASME Code, or Section VIII of the ASME (American Society of Mechanical Engineers) "Boiler and Pressure Vessel Code"
2. The API Standard 620 or the American Petroleum Institute Code, which provides rules for lower pressure vessels not covered by the ASME Code

Information should be readily available that identifies the specific pressure and storage vessels being assessed, and that provides general information about the vessel. This information should include the following items:

1. Current owner of the vessel
2. Vessel location, including the original location and current location if it has been moved
3. Vessel identification, including the manufacturer's serial number and the National Board number if registered with NB
4. Manufacturer identification, including the name and address of manufacturer and the authorization or identification number of the manufacturer
5. Date of manufacture of the vessel
6. Data report for the vessel (ASME U-1 or U-2, API 620 form, or other applicable report)
7. Date vessel was placed in service
8. Interruption dates if not in continuous service

Information giving the history of conditions under which the vessel or tank has been kept and operated will be helpful in safety assessment. Information should be kept on the following items:

1. Fluids handled, including type and composition, temperature, and pressures
2. Type of service (e.g., continuous, intermittent, or irregular)
3. Significant changes in service conditions (e.g., changes in pressures, temperatures, fluid compositions) and the dates of the changes
4. Vessel's history of alterations, rerating, and repairs performed and any date(s) of changes or repairs

Information about inspections performed on the vessel or tank and the results obtained, which will assist in the safety assessment, should include the following items:

1. Inspection(s) performed:
   a. Type, extent, and dates
2. Examination methods:
   a. Preparation of surfaces and welds
   b. Techniques used (visual, magnetic particle, penetration test, radiography, ultrasonic).
3. Qualifications of personnel:
   a. ASNT (American Society for Nondestructive Testing) levels or equivalent of examining employees and supervisory personnel

4. Inspection results and reports:
   a. Report form used (NBIC NB-7, API 510, or other)
   b. Summary of type and extent of damage or cracking
   c. Disposition (no action, delayed action, or repaired)

Information acquired from the above items is not adaptable to any kind of numerical ranking for quantitative safety assessment purposes. However, the information can reveal the owner or user's apparent attention to good practice, careful operation, regular maintenance, and adherence to the recommendations and guidelines developed for susceptible applications. If the assessment indicates cracking and other serious damage problems, it is important that the inspector obtain qualified technical advice and opinion. After all, safety and health are of primary concern regarding pressure vessels.

RAILINGS (29 CFR 1910.23)

OSHA's general requirements apply to all stair rails and handrails. Stairways having four or more risers, or rising more than 30 inches in height—whichever is less—must have at least one handrail. A stair rail must also be installed along each unprotected side or edge. When the top edge of a stair rail system also serves as a handrail, the height of the top edge must be no more than 37 inches or less than 36 inches from the upper surface of the stair rail to the surface of the tread. Winding or spiral stairways must have a handrail to prevent using areas where the tread width is less than 6 inches. Stair rails installed after March 15, 1991, must not be less than 36 inches in height (see Figure 25.21).

Midrails, screens, mesh, intermediate vertical members, or equivalent intermediate structural members must be provided between the top rail and stairway steps to the stair rail system. Midrails, when used, must be located midway between the top of the

FIGURE 25.21 Example of a stair rail and handrail.

stair rail system and the stairway steps. Screens or mesh, when used, must extend from the top rail to the stairway step and along the opening between top rail supports.

Intermediate vertical members, such as balusters, when used, must not be more than 19 inches apart. Other intermediate structural members, when used, must be installed so that there are no openings of more than 19 inches wide.

Handrails and the top rails of the stair rail systems must be able to withstand, without failure, at least 200 pounds of weight applied within two inches of the top edge in any downward or outward direction, at any point along the top edge. The height of handrails must not be more than 37 inches nor less than 30 inches from the upper surface of the handrail to the surface of the tread. The height of the top edge of a stair rail system used as a handrail must not be more than 37 inches nor less than 36 inches from the upper surface of the stair rail system to the surface of the tread.

Stair rail systems and handrails must be surfaced to prevent injuries such as punctures or lacerations and to keep clothing from snagging. Handrails must provide an adequate handhold for employees to grasp to prevent falls. The ends of stair rail systems and handrails must be built to prevent dangerous projections, such as rails protruding beyond the end posts of the system. Temporary handrails must have a minimum clearance of 3 inches between the handrail and walls, stair rail systems, and other objects. Unprotected sides and edges of stairway landings must be provided with standard 42-inch guardrail systems.

SAWS, PORTABLE CIRCULAR (29 CFR 1910.243)

All portable, power-driven circular saws (except those used for cutting meat) having a blade diameter greater than 2 inches must be equipped with guards above and below the baseplate or shoe. The upper guards shall cover the saw to the depth of the teeth, except for the minimum arc required to permit the baseplate to be tilted for bevel cuts. The lower guard is to cover the saw to the depth of the teeth, except for the minimum required to allow proper retraction and contact with the work. When the tool is withdrawn from the work, the lower guard is then automatically returned to the covering position.

SCAFFOLDS (29 CFR 1910.28)

Analysis of 1986 BLS data to support OSHA's scaffolding standard estimates that, of the 500,000 injuries and illnesses occurring in the construction industry annually, 10,000 are related to scaffolds. In addition, of the estimated 900 occupational fatalities occurring annually, at least 80 are associated with work on scaffolds. Seventy-two percent of the workers injured in scaffold accidents, covered by the BLS study, were attributed to either the planking or support giving way, or to the employee slipping or being struck by a falling object. Plank slippage was the most commonly cited cause.

All scaffolds and their supports must be capable of holding the load they are designed to carry with a safety factor of at least 4. All planking must be of scaffold grade, as recognized by grading rules for the species of wood used. The maximum permissible spans for 2-inch by 9-inch or wider planks are shown in the following table (see Table 25.1).

TABLE 25.1
Permissible Plank Spans for Scaffolds

Maximum Permissible Span Using Full-Maximum Intended Load (lb per sq ft)	Maximum Permissible Span Using Thickness, Undressed Lumber (feet)	Normal Thickness Lumber (feet)
25	10	8
50	8	6
756		Not Applicable

The maximum permissible span for a 1¼-inch by 9-inch or wider plank for full thickness is 4 feet, with medium loading of 50 pounds per square foot. Scaffold planks must extend over their supports not less than 6 inches or more than 18 inches. Scaffold planking must overlap a minimum of 12 inches or secured from movement.

SKYLIGHTS (29 CFR 1910.23)

Over a number of years, many workers have fallen through skylights after assuming they were designed to support human weight. This misconception has resulted in many deaths and injuries. Every skylight floor opening and hole is to be guarded by a standard skylight screen or a fixed standard railing on all exposed sides.

SPRAY-FINISHING OPERATIONS (29 CFR 1910.107)

In conventional dry-type spray booths, over-spray dry filters or rolls, if installed, are to conform to the following. The spraying operations, except electrostatic spraying, must ensure an average air velocity over the open face of the booth of not less than 100 feet per minute. Electrostatic spraying operations may be conducted with an air velocity of not less than 60 feet per minute, depending on the volume of finishing material being applied, its flammability, and explosive characteristics. Visible gauges, or audible alarm, or pressure-activated devices must be installed to indicate or ensure that the required air velocity is maintained. Filter pads must be inspected after each period of use, and clogged filter pads discarded and replaced. Inspections are to ensure proper replacement of filter media.

Spray booths must be installed so that all portions are readily accessible for cleaning. A clear space of not less than 3 feet on all sides must be kept free from storage or combustible construction. Space within the spray booth on the downstream and upstream sides of filters must be protected with approved automatic sprinklers. There shall be no open flame- or spark-producing equipment present in any spraying area or within 20 feet thereof, unless separated by a partition. Electrical wiring and equipment not subject to deposits of combustible residues, but located in a spraying area, must be explosion proof.

The quantity of flammable or combustible liquids kept in the vicinity of spraying operations must be the minimum required for operations and should ordinarily not exceed a supply for one day or one shift. Bulk storage of portable containers of flammable or combustible liquids must be in a separate building detached from other important buildings or cut off in a standard manner. Whenever flammable or combustible liquids are transferred from one container to another, both containers must be effectively bonded and grounded to prevent the discharge of sparks or static electricity.

All spraying areas must be kept as free from the accumulation of deposits of combustible residues as practical, with cleaning conducted daily if necessary. Scraper, spuds, or other tools used for cleaning purposes must be of nonsparking materials. Residue scrapings and debris contaminated with residue must be immediately removed from the premises. "No smoking" signs in large letters on contrasting color background shall be conspicuously posted in all spraying areas and paint storage rooms.

Adequate ventilation must be ensured before spray operations begin. Mechanical ventilation must be provided when spraying operations are performed in enclosed areas. When mechanical ventilation is provided during spraying operations, it must be so arranged that it will not circulate the contaminated air.

STAIRS, FIXED INDUSTRIAL (29 CFR 1910.23 AND 1910.24)

Fixed stairways must be provided for access from one structure to another where operations necessitate regular travel between levels and for access to operating platforms at any equipment that requires attention routinely during operations. Fixed stairs must also be provided where access to an elevation is daily, or at each shift where such work may expose employees to harmful substances or for purposes for which carrying of tools or equipment by hand is normally required. Spiral stairways are not permitted, except for special limited usage and secondary access situations where it is not practical to provide a conventional stairway.

Every flight of stairs having four or more risers must be provided with a standard railing on all open sides. Handrails must be provided on at least one side of closed stairways, preferably on the right side descending. Fixed stairways must have a minimum width of 22 inches. Stairs must be constructed so the riser height and tread width are uniform throughout and do not vary by more than 1/4 inch. Other general requirements include

1. A stairway or ladder must be provided at all worker points of access where there is a break in elevation of 19 inches or more and no ramp, runway, embankment, or personnel hoist is provided.
2. When there is only one point of access between levels, it must be kept clear to permit free passage by workers. If free passage becomes restricted, a second point of access must be provided and used.
3. Where there are more than two points of access between levels, at least one point of access must be kept clear.
4. Stairways must be installed at least 30 degrees and no more than 50 degrees—from the horizontal.
5. Where doors or gates open directly onto a stairway, a platform must be provided that extends at least 20 inches beyond the swing of the door.

6. All stairway parts must be free of dangerous projections such as protruding nails.
7. Slippery conditions on stairways must be corrected.

Extreme care must be taken by workers when ascending and descending a set of stairs. Many serious injuries and even fatalities occur when workers slip and fall on stairways.

STORAGE (29 CFR 1910.176)

Stored materials stacked in tiers must be stacked, blocked, interlocked, and limited in height so that they are secure against sliding or collapse (see Figure 25.22). Storage areas must be kept free from accumulation of materials that constitute hazards from tripping, fire, explosion, or pest harborage. Vegetation control must be exercised when necessary. Where mechanical handling equipment is used, sufficient safe clearance must be allowed for aisles, at loading docks, through doorways, and whenever turns or passage must be made.

TANKS, OPEN-SURFACE (29 CFR 1910.94)

All employees working in and around open-surface tank operations must be instructed as to the hazards of their respective jobs and in the personal protection and first-aid procedures applicable to these hazards.

On open-surface tanks where ventilation is used to control potential exposure to employees, ventilation must be adequate to reduce the concentration of the air

FIGURE 25.22 Example of adequately stacked materials.

contaminant to the degree that a hazard to employees does not exist. Whenever there is a danger of splashing, the employees shall be required to wear either tight-fitting chemical goggles or an effective face shield.

There shall be a supply of clean, cold water near each tank containing liquid that may be harmful to the skin if splashed on the worker's body. The water pipe shall be provided with a quick opening valve and at least 48 inches of hose not smaller than 3/4 inch. Alternatively, deluge shower and eye flushes must be provided.

TIRE INFLATION

Because tires have the potential to release a large amount of energy if they explode, and they can cause injury during a faulty inflation procedure, care must be taken to follow safety procedures. Tires are mounted and/or inflated on drop center wheels. Therefore, each repair shop must have a safe practice procedure posted and enforced. When tires are mounted and/or inflated on wheels with split rims and/or retainer rings, a safe practice procedure must also be posted and enforced. Each tire inflation hose should have a clip-on chuck with at least 24 inches of hose between the chuck and an in-line hand valve and gauge. The tire inflation control valve must automatically shut off so that airflow stops when the valve is released. Employees should be strictly forbidden from taking a position directly over or in front of a tire that is being inflated. A tire restraining device, such as a cage, rack, or other effective means, must be used while inflating tires mounted on split rims, or rims using retainer rings.

TOEBOARDS (29 CFR 1910.23)

Toeboards are used to protect workers from being struck by objects falling from elevated areas. While railings protect floor openings, platforms and scaffolds must be equipped with toeboards whenever people can pass beneath the open side, wherever there is moving machinery, and wherever there is equipment with which falling material could cause a hazard. A standard toeboard must be at least 4 inches in height and may be of any substantial material, either solid or open, with openings not to exceed 1 inch in greatest dimension (see Figure 25.23).

TOILETS (29 CFR 1910.141)

Water closets are to be provided according the following: 1 to 15 persons, one facility; 16 to 35 persons, two facilities; 36 to 35 persons, three facilities; 56 to 80 persons, four facilities; 81 to 110 persons, five facilities; 111 to 150 persons, six facilities; over 150 persons, one for each additional 40 persons. Where toilet rooms will be occupied by no more than one person at a time and can be locked from inside, separate rooms for each sex need not be provided.

Each water closet is to occupy a separate compartment with a door and walls or partitions between fixtures sufficiently high to ensure privacy. Wash basins (lavatories) must be provided in every place of employment. Lavatories must

FIGURE 25.23 Example of toeboards in use on a guardrail system.

have hot, cold, or tepid running water, hand soap or equivalent, and hand towels, blowers, or equivalent.

The previous requirements do not apply to mobile crews or normally unattended locations, as long as employees working at these locations have transportation immediately available to nearby toilet facilities.

TRANSPORTING EMPLOYEES AND MATERIALS

All employees who operate vehicles on public thoroughfares must have valid operator licenses. When seven or more employees are regularly transported in a van, bus, or truck, the operator's license must be appropriate for the class of vehicle being driven. Each van, bus, or truck used regularly to transport employees must be equipped with an adequate number of seats and seat belts. When employees are transported by truck, provisions must be provided to prevent their falling from the vehicle. Vehicles used to transport employees must be equipped with lamps, brakes, horns, mirrors, windshields, turn signals, and be in good repair. Transport vehicles must be provided with handrails, steps, stirrups, or similar devices, so placed and arranged that employees can safely mount or dismount.

Employee transport vehicles must be equipped at all times with at least two warning reflectors as well as a fully charged fire extinguisher in good condition with at least a four B: C rating. Cutting tools or tools with sharp edges are not to be carried in passenger compartments of employee transport vehicles and are to be placed in closed boxes or containers that are secured in place. Employees are prohibited from riding on top of any load that can shift, topple, or otherwise become unstable.

WALKING/WORKING SURFACES (29 CFR 1910.21 AND 1910.22)

Slips, trips, and falls constitute the majority of general industry accidents. They cause 15 percent of all accidental deaths, and are second only to motor vehicles as a cause of fatalities. The OSHA standards for walking and working surfaces apply to all permanent places of employment, except where only domestic, mining, or agricultural work is performed.

Wet working/walking surfaces must be covered with nonslip materials. All spilled materials must be cleaned up immediately. Any holes in the floor, sidewalk, or other walking surface, before being repaired properly, must be covered or otherwise made safe. All aisles and passageways must be kept clear and marked as appropriate. There must be safe clearance for walking in aisles where motorized or mechanical handling equipment is being operated.

Materials or equipment should be stored in such a way that sharp projections will not interfere with the walkway. Changes of direction or elevations should be readily identifiable. There should be adequate headroom provided for the entire length of any aisle or walkway.

WELDING, CUTTING, AND BRAZING (29 CFR 1910.251, 1910.252, 1910.253, 1910.254, AND 1910.255)

Welding, cutting, and brazing are hazardous activities that pose a unique combination of both safety and health risks to more than 500,000 workers in a wide variety of industries. The risk from fatal injuries alone is more than four deaths per thousand workers over a working lifetime. An estimated 562,000 employees are at risk of exposure to chemical and physical hazards of welding, cutting, and brazing. Fifty-eight deaths from welding and cutting incidents, including explosions, electrocutions, asphyxiation, falls, and crushing injuries, were reported by the Bureau of Labor Statistics in 1993.

There are numerous health hazards associated with exposure to fumes, gases, and ionizing radiation formed or released during welding, cutting, and brazing, including heavy metal poisoning, lung cancer, metal fume fever, flash burns, and others. These risks vary, depending upon the type of welding materials and welding surfaces.

Only authorized and trained personnel are permitted to use welding, cutting, or brazing equipment. Operators must have a copy of the appropriate operating instructions, which they are directed to follow. Only approved apparati (e.g., torches, regulators, pressure reducing valves, acetylene generators, and manifolds) can be used.

Compressed gas cylinders must be regularly examined for obvious signs of defects, such as deep rusting or leakage, and care should be used in handling and storing cylinders, safety valves, and relief valves to prevent damage. Precautions must be taken to prevent the mixture of air or oxygen with flammable gases, except at a burner or in a standard torch. Cylinders must be kept away from sources of heat. Cylinders must be kept away from elevators, stairs, and gangways. Cylinders are prohibited from being used as rollers or supports. Empty cylinders must be appropriately marked, and their valves closed. Signs reading "DANGER—NO SMOKING, MATCHES, OR OPEN LIGHTS" or the equivalent shall be posted. Cylinders, cylinder valves, couplings, regulators, hoses, and apparatus must be kept free of oily or greasy substances. Do not drop or strike cylinders. Unless secured on special trucks, regulators must be removed and valve-protection caps put in place before moving cylinders. Liquefied gases must be stored and shipped valve-end up with valve covers in place.

Provisions must be made to never crack a fuel gas cylinder valve near sources of ignition. Before a regulator is removed, the valve must be closed and gas must be released from the regulator. Red is used to identify the acetylene (and other fuel-gas) hose, green is for the oxygen hose, and black is for inert gas and the air hose. Pressure-reducing regulators are used only for gas and the pressures for which they are intended.

When the object to be welded cannot be moved and fire hazards cannot be removed, shields must be used to confine heat, sparks, and slag. A fire watch should be assigned when welding or cutting is performed in locations where a serious fire might develop. When welding is done on metal walls, precautions should be taken to protect combustibles on the other side. Before hot work is begun, used drums, barrels, tanks, and other containers should be thoroughly cleaned so that no substances remain that could explode, ignite, or produce toxic vapors.

Arc welding and cutting operations must be shielded by noncombustible or flame-retardant screens that will protect employees and other persons working in the vicinity from the direct rays of the arc. Arc welding and cutting cables must be of the completely insulated, flexible type, capable of handling the maximum current requirement of the work in progress. Cables in need of repair should not be used. When the welder or cutter has occasion to leave work or to stop work for any appreciable length of time, or when the welding or cutting machine is to be moved, the power supply switch to the equipment should be shut off. All ground return cables and all arc welding and cutting machine grounds must be in accordance with regulatory requirements. Ground connections must be made directly to the material being welded. The open circuit (no-load) voltage of arc welding and cutting machines must be as low as possible and not in excess of the recommended limits.

Under wet conditions, automatic controls for reducing no-load voltage should be used. Electrodes must be removed from the holders when not in use. It is required that electric power to the welder shut off when no one is in attendance. Suitable fire extinguishing equipment must be available for immediate use. The welder shall be forbidden to coil or loop welding electrode cable around his or her body. Wet machines must be thoroughly dried and tested before being used, thereby decreasing the risk of electrical shock or electrocution (see Figure 25.24).

It is required that eye protection helmets, hand shields, and goggles meet appropriate standards. Employees exposed to hazards created by welding, cutting, or brazing operations must be protected with personal protective equipment and clothing.

WOODWORKING MACHINERY (29 CFR 1910.213)

Suffice it to say, there would not be an OSHA regulation on woodworking machinery if an extreme hazard did not exist for these fast, sharp, and powerful pieces of machinery. For this reason, specific precautions must be undertaken to prevent injuries.

FIGURE 25.24 A welder performing arc welding.

All woodworking machinery—such as table saws, swing saws, radial saws, band saws, jointers, tenoning machines, boring and mortising machines, shapers, planers, lathes, sanders, veneer cutters, and other miscellaneous woodworking machinery—shall be enclosed or guarded, except that part of the blade doing the actual cutting, to protect the operator and other employees from hazards inherent to the operation.

Power control devices must be provided on each machine to make it possible for the operator to cut off the power to the machine without leaving his or her position at the point-of-operation. Power controls and operating controls should be located within easy reach of the operator while at his or her regular work location, making it unnecessary for the operator to reach over the cutter to make adjustments. This does not apply to constant pressure controls used only for setup purposes.

Restarting machinery in operations where injury to the operator might result because motors were to restart after power failures, provision must be made to prevent machines from automatically restarting upon restoration of power.

Band saw blades must be enclosed and guarded except for the working portion of the blade between the bottom of the guide rolls and the table. Band saw wheels must be fully encased. The outside periphery of the enclosure must be solid. The front and back should be either solid or wire mesh or perforated metal.

Circular table saws should have a hood over the portion of the saw above the table mounted so that the hood will automatically adjust itself to the thickness of, and remain in contact with, the material being cut. Circular table saws should have a spreader aligned with the blade, spaced no more than ½ inch behind the largest blade mounted in the saw. A spreader in connection with grooving, dadoing, or rabbeting is not required. Circular table saws used for ripping must have non-kickback fingers or dogs.

Inverted swing or sliding cut-off saws shall be provided with a hood that will cover the part of the saw that protrudes above the top of the table or material being cut.

Radial saws must have an upper guard that completely encloses the upper half of the saw blade. The sides of the lower exposed portion of the blade must be guarded by a device that will automatically adjust to the thickness of and remain in contact with the material being cut. Radial saws used for ripping need to have non-kickback fingers or dogs. Radial saws must have an adjustable stop to prevent the forward travel of the blade beyond the position necessary to complete the cut in repetitive operations. Radial saws must be installed so that the cutting head will return to the starting position when released by the operator.

Rip saws must have a spreader aligned with the blade and be no thinner than the blade. A spreader in connection with grooving, dadoing, or rabbeting is not required. Rip saws shall have non-kickback fingers or dogs.

Self-feed circular saws' feed rolls and blades must be protected by a hood or guard to prevent the hand of the operator from coming into contact with the in-running rolls at any point.

Swing or sliding cut-off saws must be provided with a hood that will completely enclose the upper half of the saw. Swing or sliding cut-off saws must be provided with limit stops to prevent the saws from extending beyond the front or back edges of the table. Swing or sliding cut-off saws must be provided with an effective device to return the saw automatically to the back of the table when released at any point of its travel.

Woodworking machinery must be provided with the safeguards previously discussed in order to protect those using them from injuries such as due to laceration, cuts, and amputation. The use of protective equipment should be employed to further protect workers with this type of machinery.

WORKPLACE VIOLENCE

Workplace violence has emerged as an important safety and health issue in today's workplace. Its most extreme form, homicide, is the second leading cause of fatal occupational injury in the United States. Nearly 1,000 workers are murdered, and 1.5 million are assaulted in the workplace each year. According to the BLS Census of Fatal Occupational Injuries (CFOI), there were 709 workplace homicides in 1998, accounting for 12 percent of the total 6,026 fatal work injuries in the United States. Environmental conditions associated with workplace assaults have been identified, and control strategies implemented in a number of work settings. OSHA has developed guidelines and recommendations to reduce worker exposures to this hazard but is not initiating rulemaking at this time.

According to the Department of Justice's National Crime Victimization Survey (NCVS), assaults and threats of violence against Americans at work number almost 2 million a year. The most common type of workplace violent crime was simple assault, with an average of 1.5 million a year. There were 396,000 aggravated assaults, 51,000 rapes and sexual assaults, 84,000 robberies, and 1,000 homicides. Again, according to the NCVS, retail sales workers were the most numerous victims, with

330,000 being attacked each year. They were followed by police, with an average of 234,200 officers victimized. The risk rate for various occupations per 1,000 workers was as follows:

- Police officers—306
- Private security guards—218
- Taxi drivers—184
- Prison guards—117
- Bartenders—91
- Mental health professionals—80
- Gas station attendants—79
- Convenience, liquor store clerks—68
- Mental health custodial workers—63
- Junior high/middle school teachers—57
- Bus drivers—45
- Special education teachers—41
- High school teachers—29
- Elementary school teachers—16
- College teachers—3

Factors that may increase a worker's risk for workplace assault, as identified by the National Institute for Occupational Safety and Health (NIOSH), are

- Contact with the public
- Exchange of money
- Delivery of passengers, goods, or services
- Having a mobile workplace such as a taxicab or police cruiser
- Working with unstable or volatile persons in health care, social services, or criminal justice settings
- Working alone or in small numbers
- Working late at night or during early morning hours
- Working in high-crime areas
- Guarding valuable property or possessions
- Working in community-based settings

The Occupational Safety and Health Administration's (OSHA's) response to the problem of workplace violence in certain industries has been the production of OSHA's guidelines and recommendations to those industries for implementing workplace violence prevention programs. In 1996, OSHA published "Guidelines for Preventing Workplace Violence for Health Care and Social Service Workers." In 1998, OSHA published "Recommendations for Workplace Violence Prevention Programs in Late-Night Retail Establishments." The guidelines and recommendations are based on OSHA's Safety and Health Program Management Guidelines and contain four basic elements:

1. Management's commitment and employee involvement may include simply clear goals for worker security in smaller sites or a written program for larger organizations.
2. Worksite analysis involves identifying high-risk situations through employee surveys, workplace walkthroughs, and reviews of injury/illness data.
3. Hazard prevention and control calls for designing engineering, administrative, and work practice controls to prevent or limit violent incidents.

4. Training and education ensure that employees know about potential security hazards and ways to protect themselves and their co-workers.

Although not exhaustive, OSHA's guidelines and recommendations include policies, procedures, and corrective methods to help prevent and mitigate the effects of workplace violence. Engineering controls remove hazards from the workplace or create a barrier between the worker and specific hazards.

Administrative and work practice controls affect the way jobs or tasks are performed. Some recommended engineering and administrative controls include

1. Physical barriers such as bullet-resistant enclosures, pass-through windows, or deep service counters
2. Alarm systems, panic buttons
3. Convex mirrors, elevated vantage points, clear visibility of service and cash register areas
4. Bright and effective lighting
5. Adequate staffing
6. Arrange furniture to prevent entrapment
7. Cash-handling controls, use of drop safes
8. Height markers on exit doors
9. Emergency procedures to use in case of robbery
10. Training in identifying hazardous situations and appropriate responses in emergencies
11. Video surveillance equipment and closed-circuit TV
12. Establish liaison with local police

Post-incident response and evaluation are essential to an effective violence prevention program. All workplace violence programs should provide treatment for victimized employees and employees who may be traumatized by witnessing a workplace violence incident. Several types of assistance can be incorporated into the post-incident response, including

- Trauma-crisis counseling
- Critical incident stress debriefing
- Employee assistance programs to assist victims

Workplace homicides are the second leading cause of death in the workplace. Homicide is the number-one cause of death for women on the job. Although workplace murders appear to be declining somewhat, they still represented 15 percent of all deaths, that is, more than 900 workers who went to work but never came home, and 80 percent of them died at the hands of robbers or other criminals. Almost half of workplace homicides occur in the retail industry, where those working late at night are particularly vulnerable. Employers and employees will want to examine their operations from a variety of perspectives—from work practices to physical barriers to employee training. The options they select to reduce the risk of violence will depend upon their individual circumstances.

For example, a gas station may find pass-through windows with bullet-resistant glass, increased lighting inside the station and over the pumps, and clearing windows of signs to permit an unobstructed view for police officers in the street to be useful measures. A convenience store might use video surveillance equipment, combined with an alarm system, convex mirrors in the store, and drop safes to foil would-be thieves. A liquor store in a high crime area might increase staffing levels at night and restrict customer access to only one door. A facility that backs up to a wooded area might increase lighting at the rear of the store, lock rear doors at night, and limit deliveries to daytime hours.

All these facilities might find it helpful to train employees in emergency procedures to use in case of violence. Employees may also benefit from training in how to handle aggressive or abusive customers. OSHA's recommendations are not a fixed formula, but rather a listing of common-sense strategies and practices that can help stop thieves. By making cash more difficult to get, store owners will discourage potential criminals and protect innocent employees.

SUMMARY

Although this not a complete description of all potential hazards regulated by OSHA, it is a fair representation that will provide the reader with a quick reference source to start and guide a more in-depth research of safety hazards. The final word on regulations and control of safety hazards is found in Part 29 of the Code of Federal Regulations.

REFERENCES

United States Department of Health and Human Services, National Institute for Occupational Safety and Health. Criteria for a Recommended Standard; Welding, Brazing and Thermal Cutting, Publication 88-110. Washington, D.C.: U.S. Government Printing Office, 1988.

United States Department of Health and Human Services, National Institute for Occupational Safety and Health. NIOSH Current Intelligence Bulletin 57. Violence in the Workplace: Risk Factors and Prevention Strategies, Washington, D.C.: U.S. Department of Health and Human Services, 1996.

United States Department of Labor, Bureau of Labor Statistics. *National Census of Fatal Occupational Injuries*, Washington, D.C.: U.S. Department of Labor, 1998.

United States Department of Labor, Occupational Safety and Health Administration. Draft Report by OSHA's Crane Safety Task Group. Washington: U.S. Department of Labor, September 1990.

United States Department of Labor, Occupational Safety and Health Administration. General Industry Digest (OSHA 2201). Washington, D.C.: U.S. Department of Labor, 1995.

United States Department of Labor, Occupational Safety and Health Administration. 29 Code of Federal Regulations 1910. Washington, D.C.: U.S. Department of Labor, 1999.

United States Department of Labor, Occupational Safety and Health Administration. 29 Code of Federal Regulations 1926. Washington, D.C.: U.S. Department of Labor, 1999.

United States Department of Labor, Occupational Safety and Health. Guidelines for Pressure Vessel Safety Assessment, Instruction Pub 8-1.5. Washington, D.C.: U.S. Department of Labor, 1989.

Warchol, G. *Workplace Violence, 1992–96*. National Crime Victimization Survey. (Report No. NCJ-168634). Washington, D.C.: 1998.

26 Conclusion

OVERVIEW

Accident prevention has been addressed in many ways in this book. Suffice it to say that an organized approach is going to serve your best interest. As with anything we attempt in life, whether it be starting a business or accomplishing a goal, a plan is a must to achieve positive outcomes. In writing this book an attempt to cover all the bases was made. Thus, an attempt was made to provide you with the tools to accomplish your plans for a safer and healthier workplace and a place where workers go home the same way they came to work—uninjured and not sick.

Employers must address and manage accident prevention, addressing the people issues, analyzing accidents/incidents, using tested accident prevention techniques, and looking at specific situations and potential hazards that may exist. It will pay off to use the information and guidance herein.

In an ideal situation, one would expect the following to be an integral part of an accident/incident prevention effort in the workplace:

Management leadership and employee participation would include

- Visible management leadership provides the motivating force for an effective safety and health program. Site safety and health issues would regularly be included on agendas of management operations meetings. Management clearly demonstrates—by involvement, support, and example—the primary importance of safety and health for everyone on the worksite. Performance is consistent and sustained or has improved over time.
- Employee participation provides the mechanism through which workers identify hazards, recommend and monitor abatement, and otherwise participate in their own protection.

  Workers and their representatives participate fully in development of the safety and health program and conduct of training and education. Workers participate in inspections and audits, program reviews conducted by management or third parties, and collection of samples for monitoring purposes, and have necessary training and education to participate in such activities. Employers encourage and authorize employees to stop activities that present potentially serious safety and health hazards.
- Implementation means tools, provided by management, including budget, information, personnel, assigned responsibility, adequate expertise and authority, means to hold responsible persons accountable (line accountability), and program review procedures.

All tools necessary to implement a good safety and health program are more than adequate and effectively used. The management safety and health representative has expertise appropriate to facility size and process, and has access to professional advice when needed. Safety and health budgets and funding procedures are reviewed periodically for adequacy.

- Contractor safety in an effective safety and health program protects all personnel on the worksite, including the employees of contractors and subcontractors. It is the responsibility of management to address contractor safety. The site's safety and health program ensures protection of everyone employed at the worksite, that is, regular full-time employees, contractors, and temporary and part-time employees.

Workplace analysis would include the following:

- Survey and hazard analysis in an effective, proactive safety and health program will seek to identify and analyze all hazards. In large or complex workplaces, components of such analysis are the comprehensive survey and analyses of job hazards and changes in conditions. Regular surveys, including documented comprehensive workplace hazard evaluations, are conducted by certified safety and health professionals or professional engineers, etc. Corrective action is documented and hazard inventories are updated. Hazard analysis is integrated into the design, development, implementation, and changing of all processes and work practices.
- Inspections are conducted to identify new or previously missed hazards and failures in hazard controls; an effective safety and health program will include regular site inspections. Inspections are planned and overseen by certified safety or health professionals. Statistically valid random audits of compliance with all elements of the safety and health program are conducted. Observations are analyzed to evaluate progress.
- A reliable hazard reporting system enables employees, without fear of reprisal, to notify management of conditions that appear hazardous and to receive timely and appropriate responses. Management responds to reports of hazards in writing within specified time frames. The workforce readily identifies and self-corrects hazards; they are supported by management when they do so.

Accident and record analysis would include the following:

- Accident investigation is an effective part of a program that provides for investigation of accidents and "near-miss" incidents, so that their causes, and the means for their prevention, are identified. All loss-producing accidents and "near-misses" are investigated for root causes by teams or individuals that include trained safety personnel and employees.
- Data analysis in an effective program will analyze injury and illness records for indications of sources and locations of hazards, and jobs that experience higher numbers of injuries. By analyzing injury and illness trends over

time, patterns with common causes can be identified and prevented. All levels of management and the workforce are made aware of results of data analyses and resulting preventive activity. External audits of accuracy of injury and illness data, including review of all available data sources, are conducted. Scientific analysis of health information, including nonoccupational databases, is included where appropriate in the program.

Hazard prevention and control result from the use of the previous components and the following:

- *Hazard control:* Workforce exposure to all current and potential hazards should be prevented or controlled using engineering controls wherever feasible and appropriate, work practices and administrative controls, and personal protective equipment (PPE).

  Hazard controls are fully in place and continually improved upon based on workplace experience and general knowledge. Documented reviews of needs are conducted by certified health and safety professionals or professional engineers, etc.
- *Maintenance:* An effective safety and health program will provide for facility and equipment maintenance, so that hazardous breakdowns are prevented. There is a comprehensive safety and preventive maintenance program that maximizes equipment reliability.
- *Medical Programs:* An effective safety and health program will include a suitable medical program where it is appropriate for the size and nature of the workplace and its hazards. Healthcare providers are on-site for all production shifts and are involved in hazard identification and training. Healthcare providers periodically observe the work areas and activities and are fully involved in hazard identification and training.

Emergency response is an example of planning for the expected and unexpected by insuring that the following exist:

- *Emergency preparedness:* There should be appropriate planning, training and drills, and equipment for response to emergencies. Note that in some facilities, the employer plan is to evacuate and call the fire department. In such cases, only applicable items suggested in the next text should be considered. A designated emergency response team with adequate training is on-site. All potential emergencies have been identified. Plan is reviewed by the local fire department. Plan and performance are reevaluated at least annually and after each significant incident. Procedures for terminating an emergency response condition are clearly defined.
- First aid or emergency care should be readily available to minimize harm if an injury or illness occurs. Personnel trained in advanced first aid or emergency medical care are always available on-site. In larger facilities a healthcare provider is on-site for each production shift.

Safety and health training should cover the safety and health responsibilities of all personnel who work at the site or affect its operations. It is most effective when incorporated into other training about performance requirements and job practices. It should include all subjects and areas necessary to address the hazards at the site. Knowledgeable persons conduct safety and health training that is scheduled, assessed, and documented. Training covers all necessary topics and situations, and includes all persons working at the site (hourly employees, supervisors, managers, contractors, and part-time and temporary employees). Employees participate in creating site-specific training methods and materials. Employees are trained to recognize inadequate responses to reported program violations. A retrievable recordkeeping system provides for appropriate retraining, makeup training, and modifications to training as the result of evaluations.

In all actuality, what probably will exist is something somewhere between a good effort and the ideals previously espoused.

Managers must demonstrate commitment by developing a good safety and health program and allocating resources specifically for safety. No safety and health program should be developed by management in a vacuum. The best way to gain commitment and ownership is to actively involve all supervisors and employees in the safety and health program development and implementation process. This approach will make it truly effective.

Supervisors must assume the key role and model safety and health directives in the workplace. They must understand that they are responsible for the safety and health of those who work with them. These same supervisors should be held accountable for their safety performance.

Safety and health programs must go beyond management's commitment. Managers and supervisors must make an effort to ensure that hazards are identified and reported when observed. Those hazards then must be removed from the workplace or, at the very least, controlled, to mitigate the potential danger. If employees are not trained in hazard recognition and safe work procedures, they cannot be expected to perform their work in a safe manner.

The motivation to perform safely is critical to safe performance once the worker is trained. If an employee senses that the environment is not safe, if it is not conducive to their working in a safe manner, and if their efforts toward safety are not recognized, then even trained workers will not be motivated to perform safely. Employees will not follow leadership that it views as only superficial with regard to safety and health on the job.

Employees must feel motivated and comfortable about reporting hazards as a part of an accident prevention effort of the company. In like manner, accidents and injuries should be reported so that the facts relevant to causes of events can be determined without becoming a fault-finding witch hunt. Fault-finding certainly will not foster an honest search for the root causes and possible prevention of the occurrence and reoccurrence of various types of serious occupational events.

When an accident occurs, the questions to ask are: Who? What? When? Where? Why? How? You will want to know what failed and what allowed the unsafe condition or act to exist. The answer is called the root cause of the event. Your final question should ask: How can I prevent this from occurring again?

Each accident/incident should be methodically analyzed using an accident investigation/root cause analysis approach. Because many root cause analysis methods exist, it will be the investigator's responsibility to select the appropriate analysis approach (e.g., barrier analysis). Use of proper accident/incident investigation methods and tracking will lead to intervention, which will successfully prevent further occurrence of these occupational accidents and incidents.

The identification of hazards forces you to analyze your past history of accidents. Data to examine should include the OSHA 300 log, workers' compensation claims, and records of noninjury accidents. Recognition of hazards requires that inadequate guards and protective devices; defective tools, equipment, and materials; inadequate working, storage, and traveling space; inadequate warning systems; fire and explosion hazards; substandard housekeeping; hazardous atmospheric conditions; excessive noise; radiation exposure; and inadequate illumination or ventilation be identified in order to control hazards. If you cannot recognize these hazards, then nothing can be done to prevent them from causing an accident or incident.

The recognition of hazards is of primary importance. After workers and supervisors have been trained regarding hazard identification, their understanding of Occupational Safety and Health regulations will help identify jobs, equipment and machinery, areas, and industry processes that have exhibited hazards in the past. It will be your responsibility to identify the potential for hazards, or the existence of hazards within your workplace. Some tools used for identification have been presented in the form of hazard hunts, job hazard analysis, and job safety observations, while other techniques provide the means to prevent and control existing hazards.

The prevention and control of hazards can be accomplished in many ways. The primary way is through the reporting of hazards by employees and the correction of hazards by supervisors or other company qualified personnel. Second, employers can use preventive maintenance programs, special emphasis programs, training, workplace audits, engineering controls, administrative controls, and, as a last resort, personal protective equipment.

It takes a concerted effort to address the safety and health issues that are causing your occupational injuries and illnesses. This certainly means that many of you will need to try out the accident/incident techniques that you have not used previously, or at least try that with which you have had no familiarity. This means that you may have to develop new programs, implement new ways of addressing people issues, change your approach to safety and health, or admit that you need help with your safety and health initiative. Then get the help you need.

If you fail to address the pillage and carnage that lurk in damaged equipment and occupationally related injuries and illnesses, then only one word describes you—"negligent." This may seem harsh but safety and health issues do not cure themselves or disappear. If you turn a blind eye to them, they may become even worse. They can easily exacerbate, and you can end up with a death or catastrophe. This is a step beyond negligent and should be considered a criminal act. To neglect your responsibility to provide a safe and healthy workplace for workers is inexcusable.

Remember that you are not working on your own. Books like this provide guidance and direction on how to address occupational safety and health problems. This book lays the foundation for a safer and healthier workplace for your workforce.

Appendix A: Written Safety and Health Program*

MANAGEMENT'S COMMITMENT

Safety is a management function which requires management's participation in planning, setting objectives, organizing, directing, and controlling the program. Management's commitment to safety and health is evident in every decision the company makes and every action this company takes. Therefore, the management of *Name of Company* assumes total responsibility for implementing and ensuring the effectiveness of this safety and health program. The best evidence of our company's commitment to safety and health is this written program, which will be fully implemented on each companys facility or project.

ASSIGNING RESPONSIBILITY

The individual assigned with the overall responsibility and authority for implementing this safety and health program is *Name of Individual*, Safety Director. Management fully supports the safety director and will provide the necessary resources and leadership to ensure the effectiveness of this safety and health program.

The safety director will supplement this written safety and health program by:

- Establishing workplace objectives and safety recognition programs
- Working with all government officials during accident investigations and safety inspections
- Maintaining safety and individual training records
- Encouraging reporting of unsafe conditions and promoting a safe workplace (some of these responsibilities will be delegated to supervisors for implementation)

SAFETY AND HEALTH POLICY STATEMENT

Example of Statement for All Employees

To all Employees: *Name of Company* is committed to providing a safe and healthful workplace that is free from recognized hazards. The safety and health of our employees is one of the highest priorities of the *Name of Company*. It is the policy

* Excerpted from Reese, C. D., Moran, J. B., and Lapping, K. *Model Construction Safety and Health Program*. Washington, D.C.: Laborers' Health and Safety Fund of North America, 1993.

of this company that accident prevention will be given primary importance in all phases of operation and administration. Therefore, management has developed this safety and health program to reduce injuries and illnesses that are so prevalent in our company.

The effectiveness of this program depends upon the cooperation and communication of management officials, supervisors and employees. Everyone must be capable of recognizing hazards in the workplace and understand their role. Each supervisor will make the safety and health of all employees an integral part of his/her regular management functions. In addition, each employee will adhere to established company safety rules and procedures. Participation of all employees is essential in order to ensure the effectiveness of this program.

Management will make every effort to provide adequate safety training to employees prior to allowing an employee to begin work. Employees in doubt about how to do a job or task safely are required to ask a qualified person for assistance. Employees must report all injuries and unsafe conditions to management as soon as possible so that corrective measures can be taken to prevent future accidents.

Please read this safety and health program and follow the safe work procedures described. Safety is everyone's business and everyone (management officials, supervisors, employees) will be held accountable for participating in this program.

Please think Safety and always work Safely.

____

President

COMPANY SAFETY GOALS AND OBJECTIVES

On each *Name of Company* project jobsite, or facility the site superintendent or manager will be accountable to management for the successful achievement of targeted company safety and health goals. *Name of Company's* safety and health goals are:

- Zero fatalities or serious injuries.
- Reduce injuries, lost workday accidents, and workers' compensation claims.
- Prevention of damage or destruction to company property or equipment.
- Increased productivity through reduction of injuries.
- Reduced workers' compensation costs.
- Enhance company's image by working safely.
- Keep safety a paramount part of the workers' daily activities.
- Recognize and reward safe work practices.
- Improve morale and productivity.

SAFETY ENFORCEMENT POLICY

Whenever a violation of safety rules occurs, the following enforcement policy will be implemented:

FIRST OFFENSE—Employee received verbal warning and proper instruction pertaining to the specific safety violation. (A notation of the violation may be made and placed in the employee's personnel file.)
SECOND OFFENSE—Written warning with a copy placed in the employee's personnel file.
THIRD OFFENSE—Receipt of two (2) written reprimands in any twelve (12) month period may result in suspension.
FOURTH OFFENSE— Employee is dismissed from employment.

The company reserves the right to terminate immediately any employee who acts unsafely on* Name of Company *jobsites.

Responsibilities for safety and health include the establishment and maintenance of an effective communication system between management officials, supervisors, and workers. To this end, all personnel are responsible for assuring that their messages are received and understood by the intended receiver.

Specific safety and health responsibilities for *Name of Company* personnel are as follows:

Management Officials

Active participation in, and support of, safety and health programs are essential. Therefore, all management officials of the *Name of Company* will display their interest in safety and health matters at every opportunity. At least one manager (as designated) will participate in safety and health meetings, accident investigations, and worksite inspections. Each manager will establish realistic goals for accident reduction in his/her area of responsibility and will establish the necessary implementing instructions for meeting the goals. Goals and implementing instructions must be within the framework established by this document. Incentives may be included as a part of implementing instructions.

Supervisors

The safety and health of the employees are primary responsibilities of the supervisors. To accomplish this obligation, supervisors will:

1. Assure that all safety and health rules, regulations, policies, and procedures are understood by conducting pre-work safety orientations with all workers and reviewing rules as the job or conditions change or when individual workers show a specific need.
2. Require the proper care and use of all necessary personal protective equipment to protect workers from hazards.
3. Identify and eliminate job hazards expeditiously through hazard analysis procedures.
4. Receive and take initial action on employee suggestions, awards, or disciplinary measures.

5. Conduct supervisor/worker meetings the first five minutes of each work shift to discuss safety matters and work plans for the workday.
6. Train employees (both new and experienced) in the safe and efficient methods to accomplish each job or task.
7. Review accident trends and establish prevention measures.
8. Attend safety meetings and actively participate in the proceedings.
9. Participate in accident investigations and safety inspections.
10. Promote employee participation in this safety and health program.
11. Actively follow the progress of injured workers and display an interest in their rapid recovery and return to work.

Employees

Safety is a management responsibility; however, each employee is expected, as a condition of employment for which he/she is paid, to work in a manner which will not inflict self-injury or cause injury to fellow workers. Each employee must understand that responsibility for his/her own safety is an integral job requirement. Each employee of *Name of Company* will:

- Observe and comply with all safety rules and regulations that apply to his/her trade.
- Report all on-the-job accidents and injuries to his/her supervisor immediately.
- Report all equipment damage to his/her supervisor immediately.
- Follow instructions and ask questions of his/her supervisor when in doubt about any phase of his/her operation.
- Report all unsafe conditions or situations that are potentially hazardous.
- Operate only equipment or machinery that he/she is qualified to operate. When in doubt, ask for directions.
- Know what emergency telephone numbers to call in case of fire and/or personal injury.
- Help to maintain a safe and clean work area.
- Talk with management at any reasonable time concerning problems that affect his/her safety or work conditions.

The most important part of making this program effective is the individual employee. Without your cooperation, the most stringent program can be ineffective. Protect yourself and your fellow workers by following the rules. Remember: Work safely so that you can return home each day the same way you left. Your family needs you and this company needs you! ***Don't Take Chances—Think Safety First!***

COMPETENT/QUALIFIED PERSONS

The Occupational Safety and Health Administration's (OSHA) Construction Standards (29 CFR 1926) require every employer to designate competent persons to conduct frequent and regular inspections of the job site, materials, and equipment.

To comply with OSHA competent/qualified person requirements, each project will have a project competent person capable of identifying existing and predictable hazards and having authority to take prompt corrective measures to eliminate them. This individual may designate other competent persons to perform certain tasks, such as supervising scaffold erection.

Competent/qualified persons will be designated for each project and listed on the company's Safety and Health Competent Person Assignments Form. This form will be completed and displayed at all operations requiring the presence of a competent/qualified person. The form should be updated and replaced as necessary to reflect current designated competent/qualified persons and their area of expertise and responsibility.

The core of an effective safety and health program is hazard identification and control. Periodic inspections and procedures for correction and control provide methods of identifying existing or potential hazards in the workplace, and eliminating or controlling them. The hazard control system provides a basis for developing safe work procedures, and injury and illness prevention training. If hazards occur or recur, this reflects a breakdown in the hazard control system.

This written safety and health program establishes procedures and responsibilities for the identification and correction of workplace hazards. The following activities will be used by this company to identify and control workplace hazards:

- Jobsite Inspections
- Accident Investigation
- Safety and Health Committee

JOBSITE SAFETY INSPECTIONS

Safety inspections of the jobsite will occur periodically every (*Insert Frequency*), when conditions change, or when a new process or procedure is implemented. These inspections should focus on the identification and correction of potential safety, health, and fire hazards. Individuals should use the site evaluation worksheet when conducting jobsite safety inspections. In addition, the "safe work" procedures should be reviewed by personnel conducting safety inspections of the jobsite.

As part of this safety and health program, the work procedures for each company worksite or facility will:

- Identify "high hazard" areas of operation and determine inspection priorities.
- Establish inspection responsibilities and schedules.
- Develop an administrative system to review, analyze, and take corrective action on inspection findings.

ACCIDENT INVESTIGATION

All accidents will be investigated to determine causal factors and prevent future recurrences of similar accidents. A written report of investigation findings will be prepared by each injured employee's supervisor and submitted to management for review. Written reports for accidents resulting in fatalities or serious injuries will also be submitted to company attorneys.

Whenever an accident is reported, the supervisor of the injured worker(s) should respond to the scene of the accident as soon as possible and complete the supervisor's accident report. All witnesses should be interviewed privately as soon as possible after the accident. If possible, the supervisor should interview the worker(s) at the scene of the accident so that events leading up to the accident can be re-enacted.

Photographs should be taken as soon as possible after the accident and include the time and date taken.

Supervisors are required to submit accident investigation reports that answer the questions: who, what, when, where and how:

- Who was involved? The investigation report should identify the injured worker(s) name and occupation.
- What happened? The investigation report should describe the accident, the injury sustained, eyewitnesses, the date, time, and location of the accident.
- Why did the accident occur? All the facts surrounding the accident should be noted here, including, but not limited to, the following.
    - What caused the situation to occur?
    - Was/were the worker(s) qualified to perform the function involved in the accident?
    - Were they properly trained?
    - Were operating procedures established for the task involved?
    - Were procedures followed, and if not, why not?
    - Where else might this or a similar situation exist, and how can it be avoided?
    - What should be done? Methods for preventing future accidents of a similar nature should be identified.
- What has been done? A follow-up report will be completed by the site safety representative to determine if the suggested action was implemented, and if so, whether similar accidents were prevented as a result of such implementation.

SAFETY AND HEALTH COMMITTEE

Each *Name of Company* worksite will establish a safety and health committee to assist with implementation of this program and the control of identified hazards. The Safety and Health Committee will be comprised of employees and management representatives. The committee should meet regularly, but not less than once a month. Written minutes from safety and health committee meetings will be available and posted on the project bulletin board for all employees to see.

The safety and health committee will participate in periodic inspections to review the effectiveness of the safety program and make recommendations for improvement of unsafe and unhealthy conditions. This committee will be responsible for monitoring the effectiveness of this program. The committee will review safety inspection and accident investigation reports and, where appropriate, submit suggestions for further action. The committee will also, upon request from OSHA, verify abatement action taken by *Name of Company* in response to safety and health citations.

Objectives of Labor/Management Safety and Health Committee

1. Reduce accidents through a cooperative effort to identify and eliminate as many unsafe conditions and acts as possible.
2. Promote employee training in areas of recognition, avoidance, and prevention of workplace hazards.
3. Encourage employee participation in the company safety and health program.
4. Establish a line of communication for the worker to voice his/her concern(s) on existing or potential hazards and receive positive feedback.
5. Develop a mechanism which enables workers to provide suggestions on how to improve safety and health on the jobsite.
6. Provide a forum for joint labor-management cooperation on safety and health issues in the workplace.

Functions of Labor/Management Safety and Health Committee

1. Involve workers in problem-solving.
2. Examine accident and injury statistics and set safety objectives.
3. Communicate accident prevention information to the workforce.
4. Review reports of recent accidents.
5. Identify and correct hazardous conditions and practices.
6. Assist in identifying the causes of hazards.
7. Regularly review minutes of previous meetings to ensure that action has been taken.

MONTHLY PROJECT OR PRODUCTION SAFETY MEETING

A monthly safety meeting will be conducted on each worksite/jobsite to provide affected parties with relevant information concerning existing or potential worksite hazards, corrective actions, and/or abatement. Minutes from these meetings should be recorded and a copy sent to the corporate safety office. The following parties should attend these monthly safety meetings:

- Company President/CEO or designated representative
- Middle manager
- Supervisor
- Foreman or lead person
- Safety and health representative

All employees from managers to workers will receive safety education and training through all phases of work performed by *Name of Company*. The following safety education and training practices will be implemented and enforced at all company jobsites.

NEW-HIRE SAFETY ORIENTATION

New employees or current employees who are transferred from another project job or facility must attend specific new-hire safety orientation. This program provides each employee the basic information about the *Name of Company* worksite safety and health rules, federal and state OSHA standards, and other applicable safety rules and regulations. Employee attendance is mandatory prior to working on the construction project. The site superintendent will record attendance using the New Hire Safety Orientation Form and maintain a file documenting all workers who attend new-hire safety orientation.

The worksite/jobsite safety orientation program will introduce new employees to:

- Company Safety and Health Program and Policy
- The worksite/jobsite and the employee's role within it
- Hazard communication requirements
- Emergency procedures
- Location of first aid stations, fire extinguishers, telephone, lunchroom, washroom, and parking
- Site-specific hazards
- Safety and health responsibilities
- Reporting of injuries and hazardous conditions
- Use of personal protective equipment
- Tool handling and storage
- Review of each safety and health rule applicable to the job
- Introduction to safety and health representative(s)
- Introduction to supervisor
- Site tour or map where appropriate

Management understands that a new employee can absorb only so much information in the first few days. Therefore, each new employee will be paired with a veteran employee who can reinforce the new employee's training while, at the same time, raising the safety awareness of the experienced "buddy."

SUPERVISOR TRAINING

The supervisor/foreman is responsible for the prevention of accidents for tasks under his/her direction, as well as thorough accident prevention and safety training for employees he/she supervises. Therefore, all supervisors/foremen will receive training so that they have a sound theoretical and practical understanding of the following:

- The site-specific safety program
- OHSAct and applicable regulations
- OSHA Hazard Communication standard
- Site emergency response plan
- First aid and CPR
- Accident and injury reporting and investigation procedures

- Hazard assessment in their areas of expertise, and topics appropriate for toolbox talks
- OSHA recordkeeping requirements
- Communication techniques

In addition to the training requirements described above, managers will receive additional training on, but not limited to, the following topics:

- Implementation and monitoring of the company's safety program
- Personnel selection techniques
- Jobsite planning
- Contractor supervision
- Worksite documents
- OSHA recordkeeping requirements

SAFETY BULLETIN BOARD

A safety bulletin board will be located on each worksite/jobsite where it will be visible to all employees. The bulletin board will contain information such as:

- Safety and health committee meeting minutes
- Safety promotions/awards
- Safety meeting dates and times
- OSHA 300A Form (February 1 to April 30 of each year)
- Available safety training
- Safety inspection findings
- Emergency phone numbers
- Any violation cited by OSHA
- "It's the Law" OSHA Job Safety and Health poster (OSHA 3165)

Additional items may be posted with management's approval.

SAFETY TALKS

Supervisors/foremen will conduct weekly work group sessions, also known as safety talks/toolbox meetings each ________________ immediately prior to start of work. These safety talks may be held more frequently depending on the circumstances (i.e., fatality, injury, new operations, etc.). The supervisor/foreman will provide appropriate materials (handouts, audio/visual aids, etc.) to discussion leaders in advance of each meeting. Discussion leaders will be selected for each meeting by the supervisor/foreman.

These weekly meetings should not exceed 15 minutes. Active employee participation and a question-and-answer session are recommended during each meeting.

Meetings will be scheduled whenever new operations are introduced into the workplace to ensure that all employees are familiar with the safe job procedures and requirements for performing the job safely.

Employee attendance at a toolbox meeting must be recorded on the Employee Training Record Form. If discussion at the meeting identifies a suspected safety or health hazard, a copy of the form must be forwarded to the site superintendent.

Various types of reports are necessary to meet OSHA recordkeeping requirements, insurance carriers, and other government regulatory agencies. Additionally, some clients may require additional site recordkeeping requirements.

The *Name of Company* has established uniform recordkeeping procedures for all company worksites/jobsites to measure the overall safety and health performance of each project.

OSHA RECORDS

The Occupational Safety and Health Administration (OSHA) requires *Name of Company* to record and maintain injury and illness records. These records are used by management to evaluate the effectiveness of this safety and health program. The safety director shall be responsible for following the OSHA recordkeeping regulations listed below:

- Obtain a report on every injury or illness requiring medical treatment.
- Record each injury or illness on the OSHA form "Log of Work-Related Injuries and Illnesses" (Form 300).
- Prepare a supplementary record of the occupational injuries and illnesses within seven day of the incident on an Employer's "Injury or Illness Incident Report" (Supplementary Record, Form 301).
- Prepare the "Summary of Work-Related Injuries and Illnesses" OSHA Form 300A, post it no later than February 1, and keep it posted where employees can see it until April 30 of each year as well as provide copies as required or requested.
- Maintain these records in company files for five years.

MEDICAL/EXPOSURE RECORDS

Medical/exposure records will be maintained for 30 years from the time of the end of an employee's employment unless a different retention period is specified by a specific standard. These records are confidential information and will remain in the custody of the safety director. Information from an employee's medical record will only be disclosed to the employee or his/her designated representative after written consent from the employee.

All employees will be informed by posted notice of the existence, location, and availability of medical/exposure records at the time of initial employment and at least annually thereafter. *Name/Title of Individual* is responsible for maintaining and providing access to these records.

TRAINING RECORDS

Training records will be maintained in each employee's personnel file and available for review upon request. Experience indicates that supervisors/foremen who receive basic first aid and CPR training are much more safety conscious and usually have better crew safety performance records. Therefore, all field supervisory personnel will be required to attend basic first aid and CPR training unless they possess a valid first aid and CPR card issued in their name.

Each *Name of Company* facility/worksite/jobsite will have adequate first aid supplies and certified, trained personnel available for the treatment of personnel injured on the job. It is also imperative that all treatments be documented in the first aid log.

Prompt medical attention should be sought for any serious injury or if there is doubt of an employee's condition.

FIRST AID SUPPLIES

First aid supplies will be available and in serviceable condition at all company worksites. Items which must be kept sterile in the first aid kit shall be contained in individual packaging. All first aid kits will contain, but not be limited to, the following items:

- 1 Pkg—Adhesive bandages, 1" (16 per pkg)
- 1 Pkg—Bandage compress, 4" (1 per pkg)
- 1 Pkg—Scissors and tweezers (1 each per pkg)
- 1 Pkg—Triangular bandage, 40" (1 per pkg)
- 1 Pkg—Antiseptic soap or pads (3 per pkg)

MEDICAL SERVICES

Each *Name of Company* worksite will have medical services available either on the worksite or at a location nearby. Emergency phone numbers will be posted on the jobsite for employees to call in the event of an injury or accident on the worksite. Nurses will be available from _____ a.m. until _____ p.m. to respond to medical emergencies. First aid will be available from the *Name of Fire Department* at all other times.

JOBSITE FIRST AID LOG

A first aid log should be maintained in the *Name of Company* first aid facility. This log should reflect the following information:

- Injured employee's name
- Immediate supervisor
- Date and time of injury
- Nature of the injury
- Injured employee's job or craft
- Treatment rendered and disposition of employee (returned to work or sent for medical attention)

EMERGENCY PROCEDURES

All employees will be provided with the locations of the first aid stations on each worksite/jobsite. Instructions for using first aid equipment are located in each station. In the event of an emergency, employees should contact any supervisor or individual who is trained in first aid. Supervisors and employees trained in first aid will be visible by a first aid emblem on their hard hat or jacket.

Fire

Fire is one of the most hazardous situations encountered on a worksite/jobsite because of the potential for large losses. Prompt reaction to, and rapid suppression of, any fire is essential. *Name of Company* will develop a fire protection program for each worksite/jobsite. The program shall provide for effective firefighting equipment to be available without delay and designed to effectively meet all fire hazards as they occur. In addition, each fire protection program shall require that:

- All firefighting equipment be conspicuously located and readily available at all times.
- All firefighting equipment be inspected and maintained in operating condition.
- All fire protection equipment be inspected no less than once monthly with documentation maintained for each piece of equipment inspected.
- Discharged extinguishers or damaged equipment be immediately removed from service and replaced with operable equipment.
- All supervisors and employees seek out potential fire hazards and coordinate their abatement as rapidly as possible.
- Each individual assigned safety responsibilities receive the necessary training to properly recognize fire hazards, inspect and maintain fire extinguishers, and the proper use of each.
- A trained and equipped firefighting brigade be established, as warranted by the project, to assure adequate protection to life.

Evacuation

Some emergencies may require company personnel to evacuate the worksite/jobsite. In the event of an emergency which requires evacuation from the workplace, all employees are required to go the area adjacent to the project that has been designated as the "safe area." The safe area for this project is located: *Description of location.* Employers should have a written evacuation plan and it should be drilled or practiced.

SUMMARY

This is a sample written safety and health program and should be used as a template that would be added to or revised to meet the specific needs of a company. No one example like this one would meet the needs of all companies.

Appendix B: Accident Investigation Forms (Revised)*

* Courtesy of the Mine Health and Safety Administration.

INITIAL INVESTIGATION INFORMATION

ACCIDENT I.D. No.: ______________________

ACCIDENT INVESTIGATOR: ______________________ PHONE: ____________

SPECIAL INVESTIGATOR: ______________________ PHONE: ____________

WITNESSES TO ACCIDENT: ______________________

NAME	TITLE

DATE OF ACCIDENT: ______________ TIME OF ACCIDENT: ________ AM/PM

DATE OF INVESTIGATION: ______________________

MINE RESCUE INVOLVED: ☐ YES ☐ NO

RESPONSIBLE PERSON AT ACCIDENT SITE: ______________________

NAME	TITLE

OFFICIAL RESPONSIBLE FOR EQUIPMENT: ______________________

If Accident Occurred Off Property or in an Office, Describe Location and Identify Owner/Operator:

____

____

____

____

____

____

INITIAL ACCIDENT CLASSIFICATION

From the Information Received Regarding the Accident, Write a Short Summary (Maximum—50 Words) of What Occurred. Using All of the Descriptors Below Which Apply:

____

____

____

____

____

Check the Appropriate Descriptors Below. Finalize **AFTER** Investigation is Complete. List all Secondary Descriptors in Order of Importance.

- ☐ Electrical
- ☐ Exploding Vessels Under Pressure
- ☐ Explosives and Breaking Agents
- ☐ Falling, Rolling, Sliding Rock or Material
- ☐ Fire
- ☐ Handling Material
- ☐ Hand Tools (Not Powered)
- ☐ Non-Powered Haulage
- ☐ Powered Haulage
- ☐ Hoisting
- ☐ Ignition or Explosion of Gas or Dust
- ☐ Machinery
- ☐ Slip or Fall of Person
- ☐ Stepping or Kneeling on Object
- ☐ Striking or Bumping
- ☐ Power Tools
- ☐ Welding
- ☐ Other: _______________

Primary Descriptor: _______________

Secondary Descriptors: _______________

____

____

____

____

IMMEDIATE SUPERVISOR DATA

NAME: ____

(FIRST) *(MIDDLE)* *(LAST)*

EXPERIENCE AS A SUPERVISOR: _______ MONTHS

TOTAL EXPERIENCE: _______ MONTHS

HEALTH AND SAFETY COURSES/TRAINING RECEIVED	DATE:

TIME LAST PRESENT AT SCENE: ______________________________

MINUTES PRIOR TO ACCIDENT: ______________________________

WERE INSTRUCTIONS RELATED TO ACCIDENT GIVEN? ☐YES ☐NO

WHAT INSTRUCTIONS WERE ISSUED? ______________________________

____

____

TIME OF LAST CONTACT WITH PRINCIPAL IN ACCIDENT: ______________

MINUTES PRIOR TO ACCIDENT: ______________________________

WHERE, IF NOT AT SCENE: ______________________________

WAS SUPERVISOR AWARE OF UNSAFE PRACTICES OR CONDITIONS?

☐YES ☐NO

WHAT WARNINGS WERE GIVEN ABOUT UNSAFE PRACTICES OR CONDITIONS? ______

____

DID SOMEONE OTHER THAN THE FOREMAN/SUPERVISOR DIRECT WORK OF THE PRINCIPAL IN ACCIDENT?

☐YES ☐NO NAME: ____________________ TITLE: ____________________

HIS/HER EXPERIENCE IN TASK IN WHICH PRINCIPAL WAS INJURED: _______MONTHS

WHAT DIRECTION DID HE/SHE GIVE VICTIM? ______________________________

____

PRINCIPAL PERSON DATA

NAME: ______________________ SOCIAL SECURITY NO.: ________

(LAST ONLY) *(LAST FOUR DIGITS)*

Company Number: ______________________

DATE OF BIRTH: SEX: ______ HEIGHT: ______(IN.) WEIGHT: ______(LB)

______/______/______ (from records) (from records)

JOB TITLE OR CLASSIFICATION: ______________________

EXPERIENCE: ______________________

Total: ______ Years ______ Mos. ______ With Company: ______ Years ______ Mos. ______

At This Operation: ______ Years ______ Mos. ______ In Job Classification: ______ Years Mos. ______

In Task in Which Accident Occurred: ______ Years ______ Mos. ______

Physical Impairments or Limitations: (describe): ______________________

____

____

Eye Sight: Left Eye 20/ ______ Right Eye 20/ ______

Primary & Secondary Education (Last Year Completed): ______________________

College (Last Year Completed): ______________________

Technical/Vocational Schooling: ______________________ Years ______ Months ______

Vocational Area(s): ______________________

Accident History Past Five Years (All Employees): ______________________

____

____

Safety Violation History (Last Two Years): ______________________

____

____

Disciplinary Actions Last Two Years: ______________________

____

____

Hours Worked (Here & Elsewhere) in 48 Prior to Accident: ______ Hours

120 Prior to Accident: ______ Hours

Sleep in Last 48 Hours: ______ Hours Time Since Last Slept: ______ Hours

PRINCIPAL PERSON DATA

Medication in Use at Time of Accident: (Prescription and Non-Prescription)

____

____

____

Degree of Injury: Fatal ______ Permanent Disability ______

Temporary Disability ______ Minor Injury ______ No Injury ______

Date Returned to Regular Job at Full Capacity: ____/____/____

Number of Days Away from Work: ________________

Number of Days of Restricted Work Activity: ________________

Injury(s) or Illness From Accident (As Defined by Principal, Witness or Investigators): ______

____

____

____

Is It Known, From the Principal Directly, or Through Acquaintances, That the Events Below Occurred in the Principal's Life during the Week (Approximately) Prior to the Accident? Check Those Which Apply. Explain Briefly in Space Provided. Also Note Other "Stressors" Than Those Listed Which Apply.

- ☐ Death of Spouse, Child or Living Partner
- ☐ Death of Other Family Member
- ☐ Divorce or Separation
- ☐ Jail Term
- ☐ Personal Injury or Illness, On Job or Off
- ☐ Marriage, Marital Reconciliation, or New Living Partner
- ☐ Retirement Date Set and Imminent
- ☐ Change in Health of Family Member
- ☐ Pregnancy, Self or Intimate Acquaintance
- ☐ Gain of New Family Member
- ☐ Change in Financial State
- ☐ Death of Close Friend
- ☐ Change in Number of Arguments with Spouse
- ☐ Foreclosure of Mortgage or Loan
- ☐ Change in Responsibilities in Union
- ☐ Son or Daughter Leaving Home
- ☐ Trouble with Relatives
- ☐ Outstanding Personal Achievement at Work or Elsewhere
- ☐ Wife Began or Stopped Work or Being Partner
- ☐ Change in Living Conditions

PRINCIPAL PERSON DATA

☐ Trouble with Supervisor or Co-worker
☐ Change in Work Hours or Conditions, Including Shift Change
☐ Change in Residence
☐ Change in Church Activities
☐ Change in Social Activities
☐ Change in Eating Habits
☐ Vacation Just Completed or Imminent
☐ Minor Violations of the Law

Medical Data:

Physical Measurements:

1. Height _______ in.
2. Eve Height, Standing _______ in.
3. Shoulder Height _______ in.
4. Knee Height _______ in.
5. Crotch Height _______ in.
6. Functional Reach _______ in.
7. Shoulder Width _______ in.

Right Handed _______

Left Handed _______

Weight: _______ lb

Medical Examiner: ____

Medical Data:

Physical Measurements:

PRINCIPAL PERSON DATA

For Fatals: CAUSE OF DEATH: ______________________________

DATE OF DEATH: ______________ Time of Death: ______________

State of Health at Time of Accident: ______________________________

____

INJURIES FROM ACCIDENT: ______________________________

____

Blood Chemistry (Must Include Alcohol and Carbon Monoxide): ______________

Attach Photographs of Victim and Close-Up Photograph of Injury:

Check ( ) When Complete.

Attach Medical Examiner's Report if Available.

PRINCIPAL PERSON TRAINING DATA

Certifications:

☐ Foremen/Supervisor
☐ Competent Person
☐ Hoisting Engineer
☐ Electrical
☐ Energized Surface High Voltage Lines

Training Programs Completed: ________________ Date: ________________

☐ Introduction
☐ Emergency
☐ First Aid
☐ Task Training
☐ Other (Detail) ________________

Company Training Completed (O.J.T., Formal Courses, Etc.):

TRAINING	OJT/FORMAL	INSTRUCTOR	DATES

Did Principal in Accident Have Task Training Specifically Related to Accident?

☐ Yes ☐ No When: ________________ By Whom? ________________

How Was the Training Given? ________________

____

Who Was Responsible for the Determination That the Principal Was Adequately Trained to Perform Regular Job and Task Being Performed at the Time of the Accident? Experience/Testing Required?

Name	Title	Experience/Test

Do Workers at This Company Perform This Job Exactly as They Were Trained?

☐ Yes ☐ No

If "No," Explain: ________________

____

____

GENERAL ACCIDENT DATA

TIME WORK SHIFT STARTED: _______AM/PM

TIME PRINCIPAL(S) BEGAN WORK ON JOB IN WHICH ACCIDENT OCCURRED: ______AM/PM

EXACT TASK AND ACTIVITY WHEN ACCIDENT OCCURRED: ____________________

WHAT DIRECTLY INFLICTED THE INJURIES (IF ANY): ____________________

MACHINERY AND TOOLS INVOLVED:

TYPE	MANUFACTURER	MODEL NO.	SERIAL NO.

Description of Damage to Equipment and Estimate Cost to Repair/Replace in Manhours (detail):

Warning Time From First Sign of Imminent Danger to:

Injury Producing Event ____________________

First Available Avoidance Activity ____________________

Were Principal(s) or Helper(s) Working Under Pressure to Complete Job Rapidly?

☐ Yes ☐ No If "Yes," Explain: ____________________

Were Proper and Safe Tools Available for the Job? ☐ Yes ☐ No

Were They Used? ☐ Yes ☐ No If "No," Explain: ____________________

GENERAL ACCIDENT DATA

WAS A COMPANY OR GOVERNMENT SAFETY RULE VIOLATED BY THE PRINCIPAL OR OTHER INDIVIDUAL(S) INVOLVED IN THE ACCIDENT?

☐ YES ☐ NO

CITE THE RULE ______________________________ CHECK ( ) WHEN COMPLETED.

DESCRIBE HOW THE RULE WAS VIOLATED: ______________________________

____

What Were Helpers or Co-Workers Doing at the Time of the Accident? ______________

____

____

____

Were These Normal Activities? ☐ Yes ☐ No If "No," Explain: ______________

____

____

Was First Aid Given to Principal(s)? ☐ Yes ☐ No

By Certified Person? ☐ Yes ☐ No

What First Aid Was Administered? ______________________________

____

Was Principal Accompanied to Emergency Transportation, Hospital, or Doctor?

☐ Yes ☐ No By Whom? ______________________________

Did Principal Complete the Shift Prior to Seeking Medical Attention?

☐ Yes ☐ No

How Was Principal Assigned to Job?

Bid for Job—Regular Job ☐

Assigned by Foreman ☐

Assigned by Foreman, Temporary or Replacement ☐

How Much Production Was Lost as a Result of the Accident?

_______ tons, or _______ production time hours.

Is There Any Reason to Question the Accuracy of the Accident Description Given by Principal(s)?

Superviory or Witnesses? ☐ Yes ☐ No

Describe Reasons: ______________________________

____

GENERAL ACCIDENT DATA

Temperature at Accident Site: _______ °F	Visibility: (estimate) _______ ft
_______ °C	(consider normal dust, smoke, haze conditions for area of the accident)

Lighting:

- ☐ Caplight
- ☐ Approved Machine Lighting
- ☐ Mercury Vapor Overhead
- ☐ Quartz Iodine Overhead
- ☐ Fluorescent Overhead
- ☐ Incandescent Overhead
- ☐ Sunlight
- ☐ Other: ____________________

Entry Dimensions: _______ Height: _______ inches Width: _______ inches

Weather (surface accidents):

Clear _______ Cloudy _______ Rain _______ Sleet _______ Snow _______ Fog _______

Wind Speed: _______ mph Wind Direction: _______

Ground Condition: Wet? ☐ Yes ☐ No Ice Covered? ☐ Yes ☐ No

Noise Level Interference with Normal Conversation? ☐ Yes ☐ No

Source: ____________________

Chemical Spills? ☐ Yes ☐ No

Chemical	Amount

Accident Environment:

Housekeeping: ☐ Good ☐ Poor

☐ Leaking Storage Tanks	
☐ Leaking Machinery	
☐ Wires/Cords	
☐ Improper Tool Storage	
☐ Improper Supply Storage	
☐ Waste Accumulation	
☐ Poor Roadway Maintenance	
☐ Other (Specify)	

GENERAL ACCIDENT DATA

Communications Available:

- ☐ 2 Way Radio
- ☐ Radio Paging
- ☐ Dial Telephone
- ☐ Paging Telephone

Distance from Accident Site _______ ft

Delay in Emergency Action Caused by Communication System, From Time of Accident: _______ minutes

Because of: Location ☐ System Failure ☐ Incorrect Use ☐

Hard Hat Worn? ☐ Yes ☐ No If Involved in Accident, Did It Fail? ☐ Yes ☐ No

If Fainted, Describe How: ________________________________

____

Manufacturer: ________________________ Model: ________________

Gloves Worn? ☐ Yes ☐ No

Type of Gloves: Leather ☐ Rubber ☐ Cotton ☐ Other ☐ (Specify) ____________

Condition of Gloves if Involved in Accident: ☐ Good ☐ Poor

Relationship to Accident or to Injury: ________________________________

____

Type of Shoes or Boots Worn: Leather ☐ Rubber ☐ Man-made materials ☐

Hard toe: ☐ Yes ☐ No High Top (above ankle): ☐ Yes ☐ No

Type of Sole: Leather ☐ Hard Rubber ☐ Soft Rubber ☐ Plastic ☐

Relationship to Accident or to Injury: ________________________________

____

Hearing Protection Worn? ☐ Yes ☐ No

Type of Hearing Protection: Individual Ear Plugs ☐ Protective Caps or Muffs ☐

Manufacturer: ________________________ Model: ________________

Properly Worn? ☐ Yes ☐ No

Relationship to Accident or Injury: ________________________________

____

Eye Protection Used? Safety Glasses: ☐ Yes ☐ No

Prescription: ☐ Yes ☐ No

Side Protection: ☐ Yes ☐ No Failed? ☐ Yes ☐ No

GENERAL ACCIDENT DATA

If Failed, Describe How: ______________________________

Relationship to Accident or Injury: ______________________________

____

Respirator Used? ☐ Yes ☐ No

Portable Breathing Apparatus? ☐ Yes ☐ No

If Involved in Accident or Injury, Describe How: ______________________________

____

Relationship to Accident or Injury: ______________________________

____

Manufacturer: ____________________ Model: ____________________

Personal Flotation Device Worn at Time of Accident? ☐ Yes ☐ No

Manufacturer: ____________________ Model: ____________________

Relationship to Accident or Injury: ______________________________

____

Safety Harness and Lifeline in Use at Time of Accident? ☐ Yes ☐ No

Type: ____________________ Condition: ____________________

Relationship to Accident or Injury: ______________________________

____

ACCIDENT DATA, ELECTRICAL

(1) Voltage of circuit involved: _______ Volts _______ AC _______ DC

(2) Voltage to which principal(s) exposed: _______ Volts

(3) Type of supply circuit: _______

(4) Provide diagram of circuit involved: Check _______ when completed.

(5) Type of conductor: _______

(6) Size of conductor: _______ Volts

Insulation rating: _______ Volts

(7) Condition of conductors/protection: _______ Good _______ Poor

Describe condition of failed or deteriorated components: _______

(8) Ground fault trip value (3 phase only): _______

(9) Grounding: Resistor: _______ Ohms frame grounding type: _______

Ground fault relays type: _______ Delay setting: _______

(10) Did lack of grounding contribute to the accident? ☐ Yes ☐ No

(11) Condition of building/equipment/floor: _______ Dry _______ Wet

Describe drainage, level, irregularities, material: _______

(12) Was interrupted circuit tagged? ☐ Yes ☐ No Locked-out? ☐ Yes ☐ No

____

(13) Were power switches properly labeled? ☐ Yes ☐ No

If not, use diagram to identify labels needed.

(14) Was circuit protection device provided? ☐ Yes ☐ No

Was it overridden? ☐ Yes ☐ No

(15) Did the accident result from a faulty splice? ☐ Yes ☐ No

Was it a temporary splice? ☐ Yes ☐ No Describe splice and equipment used:

____

____

Manufacturer: _______

(16) Did design of equipment contribute to accident? ☐ Yes ☐ No

How? (operation, maintenance, inherent safety): _______

____

(17) Was poor maintenance a factor? ☐ Yes ☐ No If so, describe:

____

____

(18) Did test equipment contribute to accident? ☐ Yes ☐ No If so, type:

_______ Voltmeter _______ AC _______ DC _______ Ammeter

_______ AC _______ DC _______ Other (specify) _______

Manufacturer: _______ Model: _______

If person testing was other than victim, was he qualified to perform task? ☐ Yes ☐ No

ACCIDENT DATA, ELECTRICAL

(19) Approval/permissibility ________

Equipment	Number	Agency

(20) If hand tool involved: Type: ________

Condition: ________ Good ________ Poor Insulation: ________ Single ________ Double

Manufacturer: ________ Model Number: ________

(21) Rotating machinery and electrical tools:

Motors/Generators	Nameplate Data	Protection
AC	HP	Fuse Amperage:
DC	KW	Circuit Breaker:
Phases		Amperage:
		Starter/Contactor:
		O.L. Amperage:

(22) Electrical Equipment:

Power Center Nameplate Data	Grounding Transformers
Manufacturer	Zig Zag
Input Voltage	Other
AC	
DC	
KVA	
Power Transformer Nameplate Data	**Overcurrent Transformers**
Manufacturer:	Manufacturer:
As connected,	Type: Bushing Bar
Primary Voltage:	Window/Doughnut
Delta Wye	CT Ratio: /
As connected,	
Secondary Voltage:	Current Relays
Delta Wye	Manufacturer:
Phase: Single	Instantaneous
Three	Trip Setting
KVA:	Sustained
Type: Oil filled	Overload Trip
Dry	Time Delay
	Setting

ACCIDENT DATA, ELECTRICAL

Electrical Equipment (cont'd.)

Circuit Breakers, Fuses	Cables
Manufacturer:	Manufacturer:
Type:	Type:
Setting:	Approval number:
Volts:	Ampacity:
Ground fault:	Trailing cable:
Trip value:	Power cable:
CT ratio:	Insulation type:
	AWG size:

ACCIDENT DATA, EXPLODING VESSELS UNDER PRESSURE

(1) What failed? ______________________________

____

(2) Manufacturer: ______________ Model: ______________

Serial number: ______________ Age: _______ Years

Failed pressure: __________ PSI

(3) Rated capacity: __________ PSI

How determined? ______________________________

(4) Working fluid/gas: _______ Water _______ Air _______ Oil _______ Inert Gas
_______ Combustible gas _______ Other

(Specify ______________________________

If fluid, obtain sample for analysis. Check _______ when completed.

(5) Date failed element last tested: _______/_______/_______

Certified by: ______________ Approval number: ______________

(6) Modifications made to manufactured form: ______________________________

______________ Original part ______________ Replacement part

(7) Did design of part on equipment contribute to accident? ☐ Yes ☐ No

How? ______________________________

____

____

(8) Pinches involved? ☐ Yes ☐ No If so, describe: ______________________________

____

(9) Prior damage: ☐ Yes ☐ No Describe: ______________________________

(10) Did a maintenance problem contribute to accident? ☐ Yes ☐ No

How? ______________________________

____

(11) Fire present? ☐ Yes ☐ No Temperature in working area _______ °F

(12) How did injury occur? ______________________________

____

ACCIDENT DATA, EXPLOSIVES AND BREAKING AGENT

(1) Type of explosive or agent:

Age: _______

(2) Manufacturer: _________________________ Brand Name: ____________________________

(3) Quantity involved in accident? ___________ Ignition source ___________________________

(4) Was explosive or agent properly stored? ☐ Yes ☐ No If not, describe:

____

____

(5) Fuse: Electric _______ Thermal _______

(6) Exactly how was energy release initiated? ____

____

____

____

(7) Were there deviations from normal use procedures? ☐ Yes ☐ No

If so, describe: ____

____

____

____

____

ACCIDENT DATA — FALLING, ROLLING, OR SLIDING ROCK OR MATERIAL OF ANY KIND

(1) Material involved: ____________ Particle size: ____________

(2) Weight: _______ lb Length: _______ in. Width: _______ in. Height: _______ in.

(3) Were accident victims aware of falling, rolling, or sliding material hazard?

☐ Yes ☐ No What, if any, precautions had been taken to prevent fall, roll, or slide or to remove hazard by planned fall, roll, or slide?

____

____

____

____

____

(4) Vertical distance of fall, roll, or slide: _______ feet

Angle of fall, roll, or slide: _______ degrees

Normal angle of repose of material: _______ degrees

(5) Was fall, roll, or slide induced by: Use of explosives ____________

Action of mobile machine or drill: ____________

Weather conditions: ____________ Watering procedures: ____________

Explain: ____________

____

____

(6) Was any falling object protection structure (FOPS) involved in the accident?

☐ Yes ☐ No Manufacturer: ____________ Standards applicable: ____________

How did it perform intended function? ____________

(7) Describe condition after accident: ____________

____

____

Describe geological conditions: ____________

____

____

____

ACCIDENT DATA, FIRE

(1) What burned? ______________________________

(2) What was ignition source? ______________________________

(3) How long did fire burn? _______ minutes _______ hours

(4) How much burned? (Estimate each material, its amount, and describe the materials remaining after the fire was extinguished.)

(5) How was the fire detected?

Air monitoring _______ Observation/contact _______

Other detection system: Describe: ______________________________

(6) Was toxic gas present? ☐ Yes ☐ No

Type	Amount	How Determined

(7) Locate the fire on a mine map and the locations of all persons affected by it.

Indicate escape route used. Check _______ (when completed).

(8) Describe ventilation changes for miner protection. (Fan changes, stoppings, barricades, etc.)

Draw on mine map. Check _______ (when completed).

____

____

(9) How was fire extinguished? (Identify agent, number of extinguishers used.)

____

____

(10) Identify extinguishers that failed to function. Give reason, if known. Recover for analysis.

(Exhibit # ______________________________)

____

(11) Were protective shelters used? ☐ Yes ☐ No If used, did they function properly?

☐ Yes ☐ No

If not, explain: ______________________________

____

ACCIDENT DATA, HANDLING MATERIAL

(1) Material being handled: ________________________________

(2) Material: _______ Loose _______ Bagged _______ Boxed _______ Solid Item

(3) Weight: _______ lb Length: _______ in. Width: _______ in. Height: _______ in.

(4) Activity at time of accident (describe purpose of action, distance material to be removed, etc.):

____

____

(5) Was activity a normal requirement of the job? ☐ Yes ☐ No If so, was activity being done as trained? ☐ Yes ☐ No

According to company safe job procedures? ☐ Yes ☐ No If not as trained, nor according to safe job procedures, explain how and why:

____

____

(6) Tool or aid used? ☐ Yes ☐ No If used, describe: ________________________________

Was tool or aid used correctly? ☐ Yes ☐ No If not, explain:

____

____

____

(7) Action: _______ Lifting _______ Pulling _______ Pushing _______ Carrying _______ Shoveling _______ Other (describe): ________________________________

(8) Injury resulted from: _______ Improper lifting, pulling, or pushing _______ Dropped material _______ Overload _______ Pinching _______ Other (describe):

____

____

____

____

ACCIDENT DATA, HAND TOOLS

(1) Type of tool: ____________________

(2) Manufacturer: ____________ Model: ____________ Size: ________

(3) _______ Left Handed _______ Right Handed Tool Weight: _______ oz./lb

(4) Type of handle: _______ Wood _______ Rubber _______ Plastic _______ Metal

(5) Was handle slippery? ☐ Yes ☐ No If so, why? ____________

(6) Was tool designed for the job? ☐ Yes ☐ No If not, explain (screwdriver being used for punch or chisel, etc.):

____

(7) Was there a material failure of tool? ☐ Yes ☐ No If so, describe: ____________

____

(8) Condition of tool prior to the accident: _______ Good _______ Poor Describe: ________

____

(9) Describe exactly how tool inflicted injury: ____________

____

____

____

(10) If electrical shock was involved, was tool insulated? ☐ Yes ☐ No

Complete Accident Data, Electrical Form if appropriate. Recover tool for analysis.

ACCIDENT DATA, NON-POWERED HAULAGE

(1) Type of equipment:

_______ Ground rail (track)	_______ Suspended rail (track)	_______ Chute
_______ Wheelbarrow	_______ Skid	_______ Other
_______ Hand Track	_______ Dolly	(explain)

(2) Manufacturer: _______________ Model: _______________ Age: _______

(3) Number of wheel on transport device: _______________

(4) Weight: Empty lb _______ At time of accident _______ lb

(5) Condition (list defective or excessively worn components): _______________

____

(6) Did maintenance problem contribute to accident? ☐ Yes ☐ No If so, how? (broken welds, brake failure, etc.):

____

(7) Did equipment have adequate handholds? ☐ Yes ☐ No If not, describe:

____

(8) Did design of equipment contribute to the accident? ☐ Yes ☐ No

If so, how? _______________

____

(9) Was the equipment moving? ☐ Yes ☐ No _______ Forward/Backward

Was the victim facing direction of travel? ☐ Yes ☐ No

(10) Route incline at point to accident (% grade, slope in degree or feet of rise per 100 feet): _______

____

(11) Condition of road bed: _______ Smooth _______ Irregular _______ Slippery

Describe: _______________

(12) Were road edges guarded? ☐ Yes ☐ No

(13) Describe approved safe job procedures for work being done at time of accident:

____

____

(14) How was injury inflicted? _______________

____

(15) Sketch position of victim(s) with respect to transport device at time of accident (give dimensions and include track clearance).

ACCIDENT DATA, POWERED HAULAGE

(1) Type of equipment: ____________________

(2) Manufacturer: ____________________ Model: ____________________

Serial Number: ____________________ Age: __________

(3) Approval/permissibility numbers: ____________________ Agency: __________

(4) Condition of equipment: _______ Good _______ Poor If poor, describe:

____

____

(5) Last scheduled maintenance: __________ (Attach copy of maintenance records, if available.)

Last unscheduled repair: __________ (Attach copy of repair records, if available.)

(6) Was maintenance factor involved in the accident? ☐ Yes ☐ No

If so, describe: ____________________

(7) Last inspection: __________ By __________ Title __________

(8) Modification related to accident (from manufacturer's configuration):

☐ Yes ☐ No Describe modification and by whom done: __________

____

(9) Was modification unique to this operation? ☐ Yes ☐ No

Was modification unique to this piece of equipment? ☐ Yes ☐ No

(10) Was equipment failure involved in the accident? ☐ Yes ☐ No If so, explain: __________

____

(11) Primary power type: _______ Electric _______ Pneumatic _______ Hydraulic

_______ Diesel/gas

(12) Was equipment operating at design power? ☐ Yes ☐ No

If over power/pressure, note amount: ____________________

(13) Was equipment operating within design load? ☐ Yes ☐ No If not, how much was it overloaded? _______ lb

(14) Was equipment provided with "emergency shut-off" control? ☐ Yes ☐ No

Was it used? ☐ Yes ☐ No Did it work properly? ☐ Yes ☐ No

If not, explain: ____________________

ACCIDENT DATA, POWERED HAULAGE

(15) Were guards missing? ☐ Yes ☐ No If missing, specify: ____________

____

Were guards damaged? ☐ Yes ☐ No If damaged, describe: ____________

____

(16) Machine voltage and type of frame grounding for equipment: ____________

(17) If a rollover, were ROPS and/or FOPS provided? ☐ Yes ☐ No

Did ROPS/FOPS perform intended function? ☐ Yes ☐ No If not, explain: ________

______________________________ Condition after accident: ____________

Standard: ______________________ Manufacturer: ____________

(18) If a rollover, was the vehicle operating with a high center of gravity? (high bucket, high load, etc.)

☐ Yes ☐ No If so, explain: ______________________________

____

(19) Was the vehicle moving? ☐ Yes ☐ No ______ Forward ______ Backward

Was the operator facing direction of travel? ☐ Yes ☐ No

(20) Was the victim(s) at a normal work/operating station? ☐ Yes ☐ No

If not, describe position at time of accident: ______________________________

____

(21) Did the operator station design contribute to the accident? ☐ Yes ☐ No If so, provide photo or sketch of station.

______ Vision restrictions ______ Motion/size ______ Control location

______ Other (explain) ______________________________

(22) Sketch equipment. Note dimensions, victim(s) position before and after accident, and track clearance. ______ Check when completed.

(23) Condition of road bed: ______ Smooth ______ Irregular ______ Slippery

Describe: ______________________________

(24) Route incline at point of accident (% grade, slope in degrees or feet of rise per 100 feet): ______

(25) Were road bed edges guarded? ☐ Yes ☐ No

____

ACCIDENT DATA, HOISTING

(1) Type of hoist: ____________________

(2) Manufacturer: ____________ Model: ____________ Age: ________

(3) Last inspected: ____________ By: ____________

(4) Rated capacity: _______ lb Size of rope: _______ strands × _______ wires
Diameter of rope: _______ in. Date rope placed in service: ____________

(5) Angle of lift: ____________________

(6) Type of emergency braking: ____________________

(7) Was hoist overloaded? ☐ Yes ☐ No If so, by how much _______ lb

(8) Were all guards in place? ☐ Yes ☐ No If not, describe unguarded areas: ____________

____

(9) Was a maintenance problem involved? ☐ Yes ☐ No If so, describe: ____________

____

(10) Was there a mechanical/physical failure? ☐ Yes ☐ No If so, collect sample of failed parts for analysis. If rope failure involved, test must be completed for strength. Check _______ when completed.

(11) Did the design of the hoist contribute to the accident? ☐ Yes ☐ No.
If so, explain: ____________________

____

(12) Was there a trained hoist person on duty? ☐ Yes ☐ No
Name: ____________________ Date of last training ____________
Last physical exam: ____________ Was he operating the hoist at the time of the accident?
☐ Yes ☐ No
If not, identify the person operating hoist: ____________________

____

Experience of person operating hoist: ____________________

(13) Was the hoist person alone? ☐ Yes ☐ No If not, identify person(s) with operator:

____

(14) Were communications being maintained at time of accident? ☐ Yes ☐ No If not, explain:

____

ACCIDENT DATA, HOISTING

Communications system: _______ telephone _______ two-way radio _______ bells
other (describe): ______________________________

Was there a communications error? ☐ Yes ☐ No If so, describe: ______________

____

____

____

(15) Past accident and maintenance history. When records have been reviewed or copied, check _______

(16) Sketch hoist and load. Indicate damage on sketch, give dimensions and depth of shaft, depths at which accident began:

ACCIDENT DATA, MACHINERY

(1) Type of machinery: ______

(2) Manufacturer: ______ Model: ______

Serial Number: ______ Age: ______

(3) Power source: Electric ______ Propane ______ Diesel ______ Gasoline ______

Air ______ Hydraulic ______ Other ______ Motor Driven H.P. ______

Component driven H.P. ______

(4) Approval/Permissibility: Number: ______ Agency: ______

(5) Modified (from manufacturer's configuration)? ☐ Yes ☐ No

If so, describe: ______

(6) Was modification unique to mine? ☐ Yes ☐ No

Was modification unique to this place of equipment? ☐ Yes ☐ No

(7) Were all guards in place? ☐ Yes ☐ No If not, explain: ______

(8) Note condition and estimate the readability of gauges related directly in the accident sequence:

______

(9) Last inspection date: ______ By: ______

Title: ______ Report: ______

(10) Was a material failure involved in the accident? ☐ Yes ☐ No

If so, describe: ______

(11) Was maintenance involved in the failure? ☐ Yes ☐ No If so, how?

______

Attach maintenance records if available.

(12) Are operation instructions/warnings related to accident visible on the machine ☐ Yes ☐ No

Describe: ______

(13) Determine if equipment was operated in an overpowered manner (too high air pressure, etc.)

☐ Yes ☐ No

How much? ______

ACCIDENT DATA, MACHINERY

(14) Was machine being worked past rated capacity? ☐ Yes ☐ No

Rated capacity? _______ Actual load: _______________

(15) Did machine lighting contribute to accident? ☐ Yes ☐ No

If so, describe: _______________

(16) Did machine design contribute to accident? ☐ Yes ☐ No

Was design of operator station a factor in accident? ☐ Yes ☐ No

_______ Control placement _______ Physical size restraints or excess

_______ Instrumentation _______ Vision limitations

Explain: _______________

(17) If rollover, was machine operating with a high center of gravity?

☐ Yes ☐ No If so, explain: _______________

____

(18) Was safety seat belt provided? ☐ Yes ☐ No Was it worn? ☐ Yes ☐ No

(19) Was equipment provided with ROPS or FOPS? ☐ Yes ☐ No

Manufacturer: _______________

If rollover, did ROPS perform intended function? ☐ Yes ☐ No

If not, explain: _______________

____

Did ROPS or FOPS contribute to injury? ☐ Yes ☐ No If so, explain:

____

____

Condition of ROPS/FOPS after accident? _______________

____

(20) Sketch the area of the accident, the machine, and detail controls if they contributed to accident. Show position of injured victim(s) with respect to machine.

ACCIDENT DATA, SLIP OR FALL OF PERSON

(1) Slipped on what?/Fell from what? ____________________

Fell onto what?/Fell how far? ____________________

(2) Activity when slip or fall occurred: ____________________

____

(3) Location within mine/plant/office: ____________________

____

(4) Equipment or machine involved? ☐ Yes ☐ No Type:

Manufacturer: ____________________

Model: ____________________ Area of machine involved: ____________

(5) Surface: _______ Concrete floor _______ Ground _______ Steel plate

_______ Metal grate _______ Other Describe: ____________________

Surface condition: ____________________

(6) Depth of irregularities: ____________________

Sketch accident site.

(7) Does lighting at accident site reveal surface conditions? ☐ Yes ☐ No

If so, explain: ____________________

____

(8) Were foot and hand holds adequate? ☐ Yes ☐ No If not, explain:

____

____

ACCIDENT DATA, STEPPING OR KNEELING ON OBJECT

(1) What was stepped on? ______

What was knelt on? ______

(2) Was the stepping or kneeling a normal part of task performance?

☐ Yes ☐ No

(3) Was task being done according to safe job procedures in mine?

☐ Yes ☐ No

If not, explain: ______

(4) Explain what happened to victim after stepping or kneeling action: ______

______

______

______

(5) Have other accidents happened in the same location and in approximately the same way?

☐ Yes ☐ No If so, explain: ______

______

(6) Has victim had other accidents of this kind not previously reported? ______

☐ Yes ☐ No If yes, explain: ______

______

(7) Sketch position of object and victim after accident.

(8) Describe condition and construction of walkway or floor: ______

______

______

______

ACCIDENT DATA, STRIKING OR BUMPING

(1) What bumped or struck victim(s)? ____________________

What did victim(s) bump or strike? ____________________

(2) Where was the object or person bumped or struck located relative to the victim(s)? ____________________

(3) Was it marked? ☐ Yes ☐ No If yes, describe: ____________________

(4) Was the lighting in the area adequate? ☐ Yes ☐ No If not, explain:

____

(5) What inflicted the injury, if any? ____________________

(6) Was victim(s) performing task according to safe operating procedures of mine? ☐ Yes ☐ No

If not, explain: ____________________

(7) Have other accidents happened in the same location and in approximately the same way?

☐ Yes ☐ No

If so, explain: ____________________

(8) Has victim(s) had other accidents of this kind, not reported previously?

☐ Yes ☐ No If so, explain: ____________________

ACCIDENT DATA, CONVEYORS

(1) Type of conveyor: Apron _______ Belt _______ Drag chain _______

Screw _______ Light _______ Vibrating _______ Bucket _______

Pneumatic _______ Roller _______

(2) Manufacturer: _____________________ Model: _____________________ Age: _________

(3) Length of conveyor: _______ ft Horsepower: _______

(4) If a bridge conveyor: Length _______ ft

Position with respect to feeding machine: _____________________

____

Location in workplace: _____________________

(5) If extensible belt: Total designed length _______ ft

Extension at time of accident _______ ft

(6) Design carrying capacity: _______ tons/minute _______ tons/hour

(7) Incline of conveyor at accident site (percent grade, incline in degrees, or slope in feet per 100 ft):

____

(8) Is conveyor system equipped with automatic braking? ☐ Yes ☐ No

(9) Is conveyor equipped with automatic temperature sensors? ☐ Yes ☐ No

(10) Is conveyor equipped with automatic shut-off? ☐ Yes ☐ No

(11) Are lube points extended away from moving elements of conveyor and to points where oiler has firm footing? ☐ Yes ☐ No If not, explain:

____

____

(12) Is conveyor adequately guarded? ☐ Yes ☐ No If not, describe:

____

____

(13) Describe belt cleaning system (scraper, plow roller, brush, etc.):

____

(14) State tools and machines normally used to clean up spillage: _____________________

(15) Are stop/start controls installed along belt conveyor? ☐ Yes ☐ No

Condition: _____________________

Distance between controls: _______ ft

ACCIDENT DATA, BINS, HOPPERS, AND SURGE PILES

(1) Type: Bin or hopper _______ Surge piles _______

(2) Bin or hopper size: Design capacity _______ tons

Contained at time of accident: _______ tons

Top dimensions: _______

Manufacturer: _______

(3) Surge pile size: Height _______ in./ft Diameter _______ in./ft

(4) Material involved: Type (describe): _______

Size: _______

(5) Was feeder operating at time of accident? ☐ Yes ☐ No

(6) Type of feeder: Conveyor _______ Front end loader _______ Dragline _______

Dozer _______ Shovel _______ Other (specify) _______

(7) Was feeder guarded? ☐ Yes ☐ No If not, explain: _______

____

(8) Could feeder operator see victim(s)? ☐ Yes ☐ No If not, explain: _______

____

(9) Activity of victim at time of accident: Cleaning bin/hopper _______

Loosening bridged material _______ Removing grating _______

Breaking or removing material on grizzly _______ Repair or maintenance _______

Other (specify) _______

(10) Tool or aid used (described): _______

(11) Was ladder involved in accident? ☐ Yes ☐ No

Condition: _______ Good _______ Poor If poor, explain: _______

____

(12) If victim was covered by material, how was he rescued? _______

____

(13) If person fell into material, complete Accident Data Form, Stepping or Kneeling on Object.

ACCIDENT DATA, POWER TOOLS AND WELDING

(1) Power tool or welder involved: Type: ____________________

(2) Age: __________ Manufacturer: __________ Model: __________

Power type: ____________________

(3) Was tool/bit/tip/rod the correct size for job? ☐ Yes ☐ No

____

(4) Was tool/welder the one normally used for task being performed?

☐ Yes ☐ No If not, why was this one being used? ____________________

____

(5) Was tool/welder being used according to safe job procedures at the operation? ☐ Yes ☐ No
If not, explain: ____________________

____

(6) Was tool/welder maintained properly? ☐ Yes ☐ No If not, explain (with particular attention to electrical cords, air hoses, gas hoses):

____

(7) What specifically produced injury? ____________________

____

____

(8) If electrical, complete Accident Data, Electrical Form; pressure vessel explosion, complete Accident Data, Exploding Vessels Under Pressure Form; fire, complete Accident Data, Fire Form.

ACCIDENT DATA, OTHER

This form is used only for "last resort" accident classification. It relates to very unusual accident phenomena not covered by one or more of the others.

(1) Is this accident unique in experience of investigator, witnesses, and the operations' officials?
☐ Yes ☐ No

If not, state other known occurrences: ______________________________

(2) Explain objects and conditions involved; how accident happened and what inflicted injury(ies):

____

(3) Was task being performed, a normal task in the operation? ☐ Yes ☐ No

Normal for job of victim(s)? ☐ Yes ☐ No

Was there an established safe job for this task in this mine? ☐ Yes ☐ No

Was victim(s) trained to do task? ☐ Yes ☐ No

(4) Other pertinent data: ______________________________

INVESTIGATOR'S RECOMMENDATIONS

(1) Recommendations

A. Implementation of countermeasures, and explain how, when, and by whom countermeasure is to be implemented: ____________________

B. If there appears to be a need for a company rule or regulation not now in existence, which would be useful as a countermeasure against future accidents, formulate the rule or regulation:

____

(2) Post-investigation follow-up

A. Countermeasure (identify from (1) A, above): ____________________

B. Date implemented: ____________________

C. Effectiveness (state how determined): ____________________

Appendix C: Causal Analysis Worksheets*

* Courtesy of the United States Department of Energy.

Equipment/Material Worksheet

☐ **Applicable** ☐ **Not Applicable**

Why was "Equipment/Material" a Cause?

Rate each subcategory cause:
D = Direct Cause
C = Contributing Cause
R = Root Cause

Equipment/Material Problem Subcategories	I	II	III	IV
1A = Defective or Failed Part				
1B = Defective or Failed Material				
1C = Defective Weld, Braze, or Soldered Joint				
1D = Error by Manufacturer in Shipping or Marking				
1E = Electrical or Instrument Noise				
1F = Contamination				

Cause Descriptions:

Recommended Corrective Actions:

Procedure Worksheet

☐ **Applicable** ☐ **Not Applicable**

Why was "Procedures" a Cause?

Rate each subcategory cause:
D = Direct Cause
C = Contributing Cause
R = Root Cause

Procedure Problem Subcategory	I	II	III	IV
2A = Defective or inadequate Procedure				
2B = Lack of Procedure				

Cause Descriptions:

Recommended Corrective Actions:

Personnel Error Worksheet

☐ **Applicable** ☐ **Not Applicable**

Why was "Personnel Error" a Cause?

Rate each subcategory cause:
D = Direct Cause
C = Contributing Cause
R = Root Cause

Personnel Error Subcategory	I	II	III	IV
3A = Inadequate Work Environment				
3B = Inattention to Detail				
3C = Violation of Requirement or Procedure				
3D = Verbal Communication Problem				
3E = Other Human Error				

Cause Descriptions:

Recommended Corrective Actions:

Design Problem Worksheet

☐ **Applicable** ☐ **Not Applicable**

Why was "Design" a Cause?

Rate each subcategory cause:
D = Direct Cause
C = Contributing Cause
R = Root Cause

Design Problem Subcategories	**I**	**II**	**III**	**IV**
4A = Inadequate Man-Machine Interface				
4B = Inadequate or Defective Design				
4C = Error in Equipment or Malarial Selection				
4D = Drawing, Specification, or Data Errors				

Cause Descriptions:

Recommended Corrective Actions:

Training Deficiency Worksheet

☐ **Applicable** ☐ **Not Applicable**

Why was "Training Deficiency" a Cause?

Rate each subcategory cause:
D = Direct Cause
C = Contributing Cause
R = Root Cause

Training Deficiency Subcategories	**I**	**II**	**III**	**IV**
5A = No Training Provided				
5B = Insufficient Practice or Hands-On Experience				
5C = Inadequate Content				
5D = Insufficient Refresher Training				
5E = Inadequate Presentation or Materials				

Cause Descriptions:

Recommended Corrective Actions:

Management Problem Worksheet

☐ **Applicable** ☐ **Not Applicable**

Why was "Management Problem" a Cause?

Rate each subcategory cause:

D = Direct Cause

C = Contributing Cause

R = Root Cause

Management Problem Subcategories	I	II	III	IV
6A = Inadequate Administrative Control				
6B = Work Organization/Planning Deficiency				
6C = Inadequate Supervision				
6D = Improper Resource Allocation				
6E = Policy Not Adequately Defined, Disseminated, or Enforced				
6D = Other				

Cause Descriptions:

Recommended Corrective Actions:

External Phenomena Worksheet

☐ **Applicable** ☐ **Not Applicable**

Why was "External Phenomena" a Cause?

Rate each subcategory cause:

D = Direct Cause

C = Contributing Cause

R = Root Cause

External Phenomena Subcategories	I	II	III	IV
7A = Weather or Ambient Condition				
7B = Power Failure or Transient				
7C = External Fire or Explosion				
7D = Theft, Tampering, Sabotage, Vandalism				

Cause Descriptions:

Recommended Corrective Actions:

Worksheet Summary

Problem/Deficiency Category		Direct Cause	Root Cause	Contributing Cause
Operational Readiness Problem	Equipment/Material Problem			
	Procedure Problem			
	Personnel Error			
Management/Field Bridge Problem	Design Problem			
	Training Deficiency			
Management Problem				
External Phenomenon				

Cause Descriptions:

Corrective Actions:

Appendix D: OSHA Safety and Health Training Requirements

GENERAL INDUSTRY TRAINING REQUIREMENTS (29 CFR PART 1910)

SUBPART E — MEANS OF EGRESS

Employee Emergency Plans and Fire Prevention Plans—1910.38(a)(5)(i), (ii) (a)–(c), (iii), (b)(4)(i) & (ii)

SUBPART F — POWERED PLATFORMS, MANLIFTS, AND VEHICLE-MOUNTED WORK PLATFORMS

Powered Platforms for Building Maintenance—Operations-Training—1910.66(i), (ii)(A)–(E) & (iii)–(v)
Care and use Appendix C, Section 1—1910.66 (e)(9)

SUBPART G — OCCUPATIONAL HEALTH AND ENVIRONMENTAL CONTROLS

Dip Tanks—Personal Protection—1910.94 (d)(9)(I)
Respirators—1910.94 (d)(9)(vi)
Inspection, Maintenance, and Installation—1910.94(d)(11)(v)
Hearing Protection—1910.95 (i)(4)
Training Program—1910.95 (k)(1)–(3)(i)–(iii)

SUBPART H — HAZARDOUS MATERIALS

Flammable and Combustible Liquids—1910.106(b)(5)(v)(2) & (3)
Explosives and Blasting Agents—1910.109(d)(3)(i) & (iii)
Bulk Delivery and Mixing Vehicles—1910.109(h)(3)(d)(iii)
Storage and Handling of Liquefied Petroleum Gases—1910.110(b)(16) & 1910.110(d)(12)(i)
Process Safety Management of Highly Hazardous Chemicals—1910.119(g)(1) (i) & (ii)
Contract Employer Responsibilities—1910.119(h)(3)(i) through (iv)
Mechanical Integrity—1910.119(j)(3)
Hazardous Waste Operations and Emergency Response—1910.120(e)(1)(i) & (ii); (2)(i)–(vii); (3)(i)–(iv) & (4)–(9)
Hazardous Waste Cleanup Workers—1910.120 Appendix C

New Technology Programs—1910.120(o)(i)
Hazardous Waste—Emergency Responders—1910.120(p)(8)(iii)(A)–(C)

Subpart I — Personal Protective Equipment

Personal Protective Equipment—1910.132(f)(1)(i)–(v); (2), (3)(i)–(iii) & (4)
Respiratory Protection—1910.134(k)(1)(i)–(vii); (2), (3) & (5)(i)–(iii)
Respiratory Protection for M. Tuberculosis—1910.139(a)(3); 1910.139(b)(3)

Subpart J — General Environmental Controls

Temporary Labor Camps—1910.142(k)(1) & (2)
Specifications for Accident Prevention Signs and Tags—1910.145(c)(1)(ii), (2)(ii) & (3)
Permit Required Confined Spaces—1910.146(g)(1) & (2)(i)–(iv)(3) & (4) & (k)(1)(i)–(iv)
The Control of Hazardous Energy (Lockout/Tagout) Lockout or Tagout Devices Removed—1910.147(a)(3)(ii); (4)(i)(D); (7)(i)(A)–(c); (ii)(A)–(F); (iii (A)–(C)(iv) & (8)
Outside Personnel—1910.147(f)(2)(i)

Subpart K — Medical Services and First Aid

Medical Services and First Aid—1910.151(a) & (b)

Subpart L — Fire Protection

Fire Protection—1910.155(c)(iv)(41)
Fire Brigades—1910.156(b)(1)
Training and Education—1910.156(c)(1)–(4)
Portable Fire Extinguishers—1910.157(g)(1), (2) & (4)
Fixed Extinguishing Systems—1910.160(b)(10)
Fire Detection Systems—1910.164(c)(4)
Employee Alarm Systems—1910.165(d)(5)

Subpart N — Materials Handling and Storage

Servicing of Multi-Piece and Single-Piece Rim Wheels—1910.177(c)(1)(i)–(iii); (2)(i)–(viii) & (3)
Powered Industrial Trucks—1910.178(1)
Moving the Load—1910.179(n)(3)(ix)
Crawler Locomotives and Truck Cranes—1910.180(i)(5)(ii)

Subpart O — Machinery and Machine Guarding

Mechanical Power Presses—1910.217(e)(3)—1910.217(f)(2)

Mechanical Power Presses—Instructions to Operators—1910.217(e)(2)
Training of Maintenance Personnel—1910.217(e)(3)
Operator Training—1910.217(H)(13)(i)(A)–(E) & (ii)
Forging Machines—1910.218(a)(2)(iii)

Subpart Q — Welding, Cutting, and Brazing

General Requirements—1910.252(a)(2)(xiii)(c)
Oxygen—Fuel Gas Welding and Cutting—1910.253(a)(4)
Arc Welding and Cutting—1910.254(a)(3)
Resistance Welding—1910.255(a)(3)

Subpart R — Special Industries

Pulp, Paper, and Paperboard Mills—1910.261(h)(3)(ii)
Laundry Machinery and Operating Rules—1910.264(d)(1)(v)
Sawmills—1910.265(c)(3)(x)
Logging—1910.266(i) & (2)(i)–(iv); (3)(i)–(vi); (4) & (5)(i)–(iv); (6) & (7)(i)–(iii); (8) & (9)
Telecommunications—1910.268(b)(2)(i)
Derrick Trucks—1910.268(j)(4)(iv)(D)
Cable Fault Locating—1910.268(l)(1)
Guarding Manholes—1910.268(o)(1)(ii)
Joint Power and Telecommunication Manholes—1910.268(o)(3)
Tree Trimming—Electrical Hazards—1910.268(q)(1)(ii)(A)–(D)
Electric Power Generation, Transmission, and Distribution—1910.269(b)(1)(i) & (ii); (d)(vi)(A)–(C); (vii); (viii)(A)–(C); & (ix)
Grain Handling Facilities—1910.272(e)(1)(i) & (ii) & (2)
Entry Into Bins, Silos, and Tanks—1910.272(g)(5)
Contractors—1910.272(h)(2)

Subpart S — Electrical Safety-Related Work Practices

Content of Training—1910.332(b)(1)

Subpart T — Commercial Diving Operations

Qualifications of Dive Team—1910.410(a)(1); (2)(i)–(iii); (3) & (4)

Subpart Z — Toxic Substances

Asbestos—1910.1001(j)(7)(i)–(iii)(A)–(H)
4-Nitrobiphenyl—1910.1003(e)(5)(i)(a)–(h)(i) & (ii)
Alpha-Naphthylamine—1910.1004(e)(5)(i)(a)–(h)(i) & (ii)
Methyl Chloromethyl Ether—1910.1006(e)(5)(i)(a)–(h)(i) & (ii)
3,3-Dichlorobenzidine (and its salts)—1910.1007(e)(5)(i)(a)–(h)(i) & (ii)

Bis-Chloromethyl Ether—1910.1008(e)(5)(i)(a)–(h)(i) & (ii)
Beta-Naphthylamine—1910.1009(e)(5)(i)(a)–(h)(i) & (ii)
Benzidine—1910.1010(e)(5)(i)(a)–(h)(i) & (ii)
4-Aminodiphenyl—1910.1011(e)(5)(i)(a)–(h)(i) & (ii)
Ethyleneimine—1910.1012(e)(5)(i)(a)–(h)(i) & (ii)
2-Acetylaminofluorene—1910.1014(e)(5)(i)(a)–(h)(i) & (ii)
4-Dimethylaminoazobenzene—1910.1015(e)(5)(i)(a)–(h)(i) & (ii)
N-Nitrosodimethylamine—1910.1016(e)(5)(i)(a)–(h)(i) & (ii)
Vinyl Chloride—1910.1017(j)(1)(i)–(ix)
Inorganic Arsenic—1910.1018(o)(1)(i) & (ii)(A)–(F) & (2)(i) & (ii)
Cadmium—1910.1027(m)(4)(i)–(iii)(A)–(H) & (m)(4)(iv)(A) & (B)
Benzene—1910.1028(j)(3)(i)–(iii)(A) & (B)
Coke Oven Emissions—1910.1029(k)(1)(i)–(iv)(a)–(e) & (k)(2)(i) & (ii)
Bloodborne Pathogens—1910.1030(g)(2)(i); (ii)(A)–(C); (iii)–(vii)(A)–(N); (viii) & (ix)(A)–(C)
Cotton Dust—1910.1043(i)(1)(i)(A)–(F) & (2)(i) & (ii)
1,2-Dibromo-3-Chloropropane—1910.1044(n)(1)(i) & (ii)(a)–(e) & (n)(2)(i) & (ii)
Acrylonitrile (Vinyl Cyanide)—1910.1045(o)(1) & (iii)(A)–(G)&(2)(i) & (ii)
Ethylene Oxide—1910.1047(j)(3)(i); (ii)(A)–(D) & (iii)(A)–(D)
Formaldehyde—1910.1048(n)(1)–(3)(i) & (ii)(A) & (B)(iii)–(vii)
4,4 Methylenedianiline—1910.1050(k)(3)(i) & (ii)(A) & (4)(i)(ii)
Ionizing Radiation Testing—1910.1096(f)(3)(viii)
Posting—1910.1096(i)(2)
Hazard Communication—1910.1200(h)(1), (2)(i)–(iii) & (3)(i)–(iv)
Occupational Exposure to Hazardous Chemicals in Laboratories—1910.1450(f)(1)(2) & (f)(4)(i)(A)–(C) & (ii)

SHIPYARD EMPLOYMENT TRAINING REQUIREMENTS (29 CFR PART 1915)

Subpart A — General Provisions

Commercial Diving Operations
Competent Person—1915.7(b)(1)(i)–(iv); (2)(i)–(iii)(A)–(C); & (c)(1)–(7)

Subpart B — Explosive and Other Dangerous Atmospheres

Confined and Enclosed Spaces—1915.12(d)(1) & (2)(i)–(iii), (3)(i)–(iii), (4)(i) & (ii), (5)(i) & (ii)
Precautions before Entering—1915.12(a)(1)(i)–(v)
Cleaning and Other Cold Work—1915.13(b)(2) & (4)
Certification before Hot Work is Begun—1915.14(b)(1)(i)–(v)
Maintaining Gas Free Conditions, Ship Repairing—1910.15(c)

Subpart C — Surface Preparation and Preservation

Painting—1915.35(b)(1) & (8)
Flammable Liquids—1915.36(a)(2) & (5)

Subpart D — Welding, Cutting, and Heating

Fire Prevention—1915.52(b)(3) & (c)
Welding, Cutting, and Heating in Way of Preservative Coatings —1915.53(b)
Welding, Cutting and Heating of Hollow Metal Containers and Structures Not Covered by 1915.12
Gas Welding and Cutting—1915.55(d)(1)–(6)
Arc Welding and Cutting—1915.56(d)(1)–(4)
Uses of Fissionable Material—1915.57(b)

Subpart E — Scaffolds, Ladders, and Other Working Surfaces

Scaffolds or Staging—1915.71(b)(7)

Subpart F — General Working Conditions

Work on or in the Vicinity of Radar and Radio—1915.95(a)
First Aid—1915.98(a)

Subpart G — Gear and Equipment for Rigging and Materials Handling

Ropes, Chains, and Slings—1915.112(c)(5)
Use of Gear—1915.116(1)
Qualifications of Operators—1915.117(a) & (b)

Subpart H — Tools and Related Equipment

Powder Actuated Fastening Tools—1915.135(a) & (c)(1)–(6)
Internal Combustion Engines, Other Than Ships' Equipment—1915.136(c)

Subpart I — Personal Protective Equipment

General Requirements—1915.152(e)(1)(i)–(v); (2), (3)(i)–(iii); & (4)
Respiratory Protection—1915.152(a)(4)
Personal Fall Arrest Systems—1915.159(d)
Positioning Device Systems—1915.160(d)

Subpart K — Portable, Unfired Pressure Vessels, Drums, and Containers, Other than Ships' Equipment

Portable Air Receivers and Other Unfired Pressure Vessels—1915.172(b)

Subpart Z —Toxic and Hazardous Substances

Asbestos—1915.1001(k)(9)(i)–(vi)(A)–(J)
Carcinogens—1915.1003
Vinyl Chloride—1915.1017
Inorganic Arsenic—1915.1018
Lead—1915.1025
Cadmium—1915.1027
Benzene—1915.1028
Bloodborne Pathogens—1915.1030
1,2-Dibromo-3-Chloropropane—1915.1044
Acrylonitrile—1915.1045
Ethylene Oxide—1915.1047
Formaldehyde—1915.1048
Methylenedianiline—1915.1050
Ionizing Radiation—1915.1096
Hazard Communication—1915.1200
Occupational Exposure to Hazardous Chemicals in Laboratories—1915.145

MARINE TERMINAL TRAINING REQUIREMENTS (29 CFR PART 1917)

Subpart A — Scope and Definitions

Commercial Diving Operations—1917.1(a)(2)(iii)
Electrical Safety-Related Work Practices—1917.1(a)(2)(iv)
Grain Handling Facilities—1917.1(a)(2)(v)
Hazard Communication—1917.1(a)(2)(vi)
Ionizing Radiation—1917.1(a)(2)(vii)
Hearing Protection—1917.1(a)(2)(viii)
Respiratory Protection—1917.1(a)(2)(x)
Servicing Multi-Piece and Single-Piece Rim Wheels—1917.1(a)(2)(xii)
Toxic and Hazardous Substances—1917.1(a)(2)(xiii)

Subpart B — Marine Terminal Operations

Hazardous Atmospheres and Substances—1917.23(b)(1)
Fumigants, Pesticides, Insecticides, and Hazardous Preservatives —1917.25(e)(2) & (3)
Personnel—1917.27(a)(1) & (b)(1) & (2)
Hazard Communication—1917.28
Emergency Action Plans—1917.30(a)(5)(i) & (ii)(A)–(C)(iii)

Subpart C — Cargo Handling Gear and Equipment

General Rules Applicable to Vehicles—1917.44(i) & (ii)(A)–(G)

Subpart D — Specialized Terminals

Terminal facilities—Handling Menhaden and Similar Species of Fish—1917.73(d)
Related Terminal Operations and Equipment
Welding, Cutting, and Heating (Hot Work)—1917.152(c)(4)

LONGSHORING TRAINING REQUIREMENTS (29 CFR PART 1918)

Subpart A —Scope and Definitions

Commercial Diving Operations —1918.1(b)(2)
Electrical Safety-Related Work Practices—1918.1(b)(3)
Hazard Communication—1918.1(b)(4)
Ionizing Radiation—1918.1(b)(5)
Hearing Protection—1918.1(b)(6)
Respiratory Protection—1918.1(b)(8)
Toxic and Hazardous Substances—1918.1(b)(9)

Subpart H — Handling Cargo

Containerized Cargo Operations—Fall Protection Systems—1918.85(k)(12)

Subpart I — General Working Conditions

Hazardous Atmospheres and Substances—1918.93(d)(3)
Ventilation and Atmospheric Conditions and Fumigants—1918.94(b)(v)
First-Aid and Life Saving Facilities—1918.97(b)
Qualifications of Machinery Operators—1918.98(a)(1)

CONSTRUCTION INDUSTRY TRAINING REQUIREMENTS (29 CFR PART 1926)

Subpart C — General Safety and Health Provisions

General Safety and Health Provisions—1926.20(b(2) & (4))
Safety Training and Education—1926.21(a)
Employee Emergency Action Plans—1926.26.35(e)(1) & (2)(i)–(iii) & (3)

Subpart D — Occupational Health and Environmental Controls

Medical Services and First Aid—1926.50©
Ionizing Radiation—1926.53(b)
Nonionizing Radiation—1926.54(a) & (b)
Gases, Vapors, Fumes, Dusts, and Mists—1926.55(b)
Hazard Communication—1926.59
Methylenedianiline—1926.60(l)(3)(i) & (ii)(A)–(C)

Lead in Construction—1926.62(l)(1)(i)–(iv); (2)(i)–(viii) & (3)(i) & (ii)
Process Safety Management of Highly Hazardous Chemicals—1926.64
Hazardous Waste Operations and Emergency Response—1926.65

Subpart E — Personal Protective and Life Saving Equipment

Hearing Protection—1926.101(b)
Respiratory Protection Subpart F Fire Protection and Prevention—1926.103(c)(1)
Fire Protection Subpart G Signs, Signals, and Barricades—1926.150(a)(5)
Signaling—1926.201(a)(2)

Subpart I — Tools—Hand and Power

Power-Operated Hand Tools—1926.302(e)(1) & (12)
Woodworking Tools—1926.304(f)

Subpart J — Welding and Cutting

Gas Welding and Cutting—1926.350(d)(1)–(6)
Arc Welding and Cutting—1926.351(d)(1)–(5)
Fire Prevention—1926.352(e)
Welding, Cutting, and Heating in Way of Preservative Coatings—1926.354(a)

Subpart K — Electrical

Ground Fault Protection—1926.404(b)(iii)(B)

Subpart L — Scaffolding

Scaffolding—Training Requirements—1926.454(a)(1)–(5) & (b)(1)–(4) & (c)(1)–(3)

Subpart M — Fall Protection

Fall Protection—Training Requirements—1926.503(a)(1) & (2)(ii)–(vii)

Subpart N — Cranes, Derricks, Hoists, Elevators, and Conveyors

Cranes and Derricks—1926.550(a)(1), (5) & (6)
Material Hoists, Personnel Hoists, and Elevators—1926.552(a)(1)

Subpart O — Motor Vehicles, Mechanized Equipment, and Marine Operations

Material Handling Equipment—1926.602(c)(1)(vi)

Powered Industrial Trucks (Forklifts)—1926.602(d)
Site Clearing—1926.604(a)(1)

Subpart P — Excavations

General Protection—1926.651(c)(1)(i)

Subpart Q — Concrete and Masonry Construction

Concrete and Masonry Construction—1926.701(a)

Subpart R — Steel Erection

Bolting, Riveting, Fitting-Up, and Plumbing-Up—1926.752(d)(4)

Subpart S — Underground Construction, Caissons, Cofferdams, and Compressed Air

Underground Construction—1926.800(d)
Compressed Air —1926.803(a)(1) & (2)

Subpart T — Demolition

Preparatory Operations—1926.850(a)
Chutes—1926.852(c)
Mechanical Demolition—1926.859(g)

Subpart U — Blasting and Use of Explosives

General Provisions—1926.900(a)
Blaster Qualifications—1926.901(c), (d), & (e)
Surface Transportation of Explosives—1926.902(b) & (i)
Firing the Blast—1926.909(a)

Subpart V — Power Transmission and Distribution

General Requirements—1926.950(d)(1)(ii)(a), (c), (vi) & (vii)
Overhead Lines—1926.955(b)(3)(i)
Underground Lines—1926.956(b)(1)
Construction in Energized Substations—1926.957(a)(1)

Subpart X — Stairways and Ladders

Ladders—1926.1053(b)(15)
Training Requirements—1926.1060(a)(i)–(v) & (b)

Subpart Y — Diving

Commercial Diving Operations

Subpart Z — Toxic and Hazardous Substances—1926.1076

Asbestos—1926.1101(9)(i)–(viii)(A)–(e)(10)
Carcinogens—1926.1103
Vinyl Chloride—1926.1117
Inorganic Arsenic—1926.1118
Cadmium—1926.1127(m)(4)(i)–(iii)(A)–(E)
Benzene—1926.1128
Coke Oven Emissions—1926.1129
1,2-Dibromo-3-Chloropropane—1926.1144
Acrylonitrile—1926.1145
Ethylene Oxide—1926.1147
Formaldehyde—1926.1148
Methylene Chloride—1926.1152

AGRICULTURE TRAINING REQUIREMENTS (29 CFR 1938)

Subpart B — Applicability of Standards

Temporary Labor Camps—1928.142
Logging—1928.266
Hazard Communication—1928.1200
Cadmium—1928.1027

Subpart C — Roll-Over Protective Structures

Roll-Over Protective Structures (ROPS) for Tractors Used in Agricultural Operations—1928.51(d)

Subpart D — Safety for Agricultural Equipment

Guarding of Farm Field Equipment, Farmstead Equipment, and Cotton Gins—1928.57(a)(6)(i)–(v)

Subpart M — Occupational Health

Cadmium—1928.1027

FEDERAL EMPLOYEE TRAINING REQUIREMENTS (29 CFR PART 1960)

Subpart B — Financial Management

Subpart D — Inspection and Abatement

Qualifications of Safety and Health Inspectors and Agency Inspections—1960.25(a)

Subpart E — General Services Administration and Other Federal Agencies

Safety and Health Services—1960.34(e)(1)

Subpart F — Occupational Safety and Health Committees

Agency Responsibilities—1960.39(b)

Subpart H — Training

Top Management—1960.54
Supervisors—1960.55(a) & (b)
Safety and Health Specialists—1960.56(a) & (b)
Safety and Health Inspectors—1960.57
Collateral Duty, Safety and Health Personnel, and Committee Members—1960.58
Employees and Employee Representatives—1960.59(a) & (b)
Training Assistance—1960.60(a)–(d)

Subpart K — Federal Safety and Health Councils

Role of the Secretary—1960.85(b)
Objectives of Field Councils—1960.87(d)

Appendix E: OSHA Regional Offices and State Plan Offices

OSHA REGIONAL OFFICES

Region I: Connecticut, Maine, Massachusetts, New Hampshire, Rhode Island, Vermont

JFK Federal Building Room E340
Boston, Massachusetts 02203
Phone: (617) 565-9860
FAX: (617) 565-9827

Region II: New Jersey, New York, Puerto Rico, Virgin Islands

201 Varick Street, Room 670
New York, New York 10014
Phone: (212) 337-2378
FAX: (212) 337-2371

Region III: Delaware, District of Columbia, Maryland, Pennsylvania, Virginia, West Virginia

The Curtis Center, Suite 740 West
170 S. Independence Mall West
Philadelphia, Pennsylvania 19106-3309
Phone: (215) 861-4900
FAX: (215) 861-4904

Region IV: Alabama, Florida, Georgia, Kentucky, Mississippi, North Carolina, South Carolina, Tennessee

61 Forsyth Street, SW
Atlanta, Georgia 30303
Phone: (404) 562-2300
FAX: (404) 562-2295

Region V: Illinois, Indiana, Michigan, Minnesota, Ohio, Wisconsin

230 South Dearborn Street, Rm. 3244
Chicago, Illinois 60604
Phone: (312) 353-2220
FAX: (312) 353-7774

Region VI: Arkansas, Louisiana, New Mexico, Oklahoma, Texas

525 Griffin Square Bldg., Rm. 602
Dallas, Texas 75202
Phone: (972) 850-4145
FAX: (972) 850-4149

Region VII: Iowa, Kansas, Missouri, Nebraska

Two Pershing Square
2300 Main Street, Suite 1010
Kansas City, Missouri 64108
Phone: (816) 283-8745
FAX: (816) 283-0547

Region VIII: Colorado, Montana, North Dakota, South Dakota, Utah, Wyoming

1999 Broadway, Suite 1690
Denver, Colorado 80202
Phone: (720) 264-6550
FAX: (720) 264-6585

Region IX: Arizona, California, Hawaii, Nevada, American Samoa

Guam, Trust Territory of the Pacific Islands
90 7th Street, Suite 18100
San Francisco, California 94103
Phone: (415) 625-2547
FAX: (415) 625-2534

Region X: Alaska, Idaho, Oregon, Washington

1111 Third Avenue, Suite 715
Seattle, Washington 98101-3212
Phone: (206) 553-5930
FAX: (206) 553-6499

STATE PLAN OFFICES

Alaska:

Alaska Department of Labor and Workforce Development
P.O. Box 111149
1111 W. 8th Street, Room 304
Juneau, Alaska 99801-1149
(907) 465-2700 Fax: (907) 465-2784

Arizona:

Industrial Commission of Arizona
800 W. Washington
Phoenix, Arizona 85007-2922
(602) 542-5795 Fax: (602) 542-1614

California:

California Department of Industrial Relations
455 Golden Gate Avenue
San Francisco, California 94102
(415) 703-5050 Fax: (415) 703-5059

Connecticut:

Connecticut Department of Labor
Conn-OSHA
200 Folly Brook Blvd.
Wethersfield, Connecticut 06109
(860) 263-6900 Fax: (860) 263-6940

Hawaii:

Hawaii Department of Labor and Industrial Relations
830 Punchbowl Street
Honolulu, Hawaii 96813
(808) 586-8844 Fax: (808) 586-9099

Illinois:

Illinois Department of Labor
1 W. Old State Capitol Plaza, Room 300
Springfield, Illinois 62701
217-782-6206 Fax: 217-782-0596

Indiana:

Indiana Department of Labor
State Office Building
402 West Washington Street, Room W195
Indianapolis, Indiana 46204-2751
(317) 232-2655 Fax: (317) 233-3790

Iowa:

Iowa Division of Labor Services
1000 E. Grand Avenue
Des Moines, Iowa 50319-0209
(515) 242-5870 Fax: (515) 281-7995

Kentucky:

Kentucky Department of Labor Cabinet
1047 U.S. Highway 127 South
Frankfort, Kentucky 40601
(502) 564-3070 Fax: (502) 564-5387

Maryland:

Maryland Occupational Safety and Health (MOSH)
1100 North Eutaw Street, Room 613
Baltimore, Maryland 21201-2206
(410) 767-2190 Fax: (301) 333-7909

Michigan:

Michigan Occupational Safety and Health Administration
7150 Harris Drive
Lansing, Michigan 48909-8143
(517) 322-1817 Fax: (517)322-1775

Minnesota:

Minnesota Department of Labor and Industry
443 Lafayette Road, North
St. Paul, Minnesota 55155-4307
(651) 284-5050 Fax: (651) 284-5741

Nevada:

Nevada Division of Industrial Relations
400 West King Street, Suite 200
Carson City, Nevada 89073
(702) 486-9020 Fax: (702) 990-0365

New Jersey:

New Jersey Department of Labor and Workforce Development
One John Fitch Plaza – State Office Building

Trenton, New Jersey 08625-0110
(609) 633-3896 Fax: (609) 292-3749

New Mexico:

New Mexico Environment Department
1190 St. Francis Drive, Suite N4050
Santa Fe, New Mexico 87502
(505) 827-2855 Fax: (505) 827-2836

New York:

New York Department of Labor
Governor W. Averell Harriman State Office Campus – Building 12, Room 158
Albany, New York 12240
(518) 457-1263 Fax: (518) 457-5545

North Carolina

North Carolina Department of Labor
111 Hillsborough Street
Raleigh, North Carolina 27601-1092
(919) 733-7166 Fax: (919) 733-6197

Oregon:

Oregon Occupational Safety and Health Division
Department of Consumer & Business Services
350 Winter Street, NE, Room 430
Salem, Oregon 97309-0405
(503) 378-3272 Fax: (503) 947-7461

Puerto Rico:

Puerto Rico Department of Labor and Human Resources
Prudencio Rivera Martínez Building, 21st Floor
505 Muñoz Rivera Avenue
Hato Rey, Puerto Rico 00918
(787) 754-2119 Fax: (787) 753-7670

South Carolina:

South Carolina Department of Labor, Licensing, and Regulation
Koger Office Park, Kingstree Building
110 Centerview Drive
Columbia, South Carolina 29211-1329
(803) 896-7665 Fax: (803) 896-7670

Tennessee:

Tennessee Department of Labor and Workforce Development
200 French Landing Drive
Nashville, Tennessee 37243-1002
(615) 741-2793 Fax: (615) 741-5078

Utah:

Utah Labor Commission
160 East 300 South
P.O. Box 146650
Salt Lake City, Utah 84114-6650
(801) 530-6901 Fax: (801) 530-7606

Vermont:

Vermont Department of Labor and Industry
5 Green Mountain Drive
Montpelier, Vermont 05601-0458
(802) 828-4000 Fax: (802) 828-4022

Virgin Islands:

Virgin Islands Department of Labor
3012 Golden Rock
Christiansted, St. Croix, Virgin Islands 00890
(340) 772-1315 Fax: (340) 772-4323

Virginia:

Virginia Department of Labor and Industry
Powers-Taylor Building
13 South 13th Street
Richmond, Virginia 23219
(804) 786-2377 Fax: (804) 371-6524

Washington:

Washington Department of Labor and Industries
7273 Linderson Way SW
Tumwater, Washington 98501-5414
(360) 902-4805 Fax: (360) 902-5619

Wyoming:

Wyoming Worker's Safety and Compensation Division
1510 East Pershing Building- West Wing
Cheyenne, Wyoming 82002
(307) 777-7786 Fax: (307) 777-3646

Appendix F: Sample Glove Selection Charts

Glove Chart

Type	Advantages	Disadvantages	Use Against
Natural rubber	Low cost, good physical properties, dexterity	Poor vs. oils, greases, organics Frequently imported; may be poor quality	Bases, alcohols, dilute water solutions; fair vs. aldehydes, ketones
Natural rubber blends	Low cost, dexterity, better chemical resistance than natural rubber vs. some chemicals	Physical properties frequently inferior to natural rubber	Same as natural rubber
Polyvinyl chloride (PVC)	Low cost, very good physical properties, medium cost, medium chemical resistance	Plasticizers can be stripped, frequently imported, may be poor quality	Strong acids and bases, salts, other water solutions, alcohols
Neoprene	Medium cost, medium chemical resistance, medium physical properties	NA	Oxidizing acids, anilines, phenol, glycol ethers
Nitrile	Low cost, excellent physical properties, dexterity	Poor vs. benzene, methylene chloride, trichloroethylene, many ketones	Oils, greases, aliphatic chemicals, xylene, perchloroethylene, trichloroethane; fair vs. toluene
Butyl	Specialty glove, polar organics	Expensive, poor vs. hydrocarbons, chlorinated solvents	Glycol ethers, ketones, esters
Polyvinyl alcohol (PVA)	Specialty glove, resists a very broad range of organics, good physical properties	Very expensive, water sensitive, poor vs. light alcohols	Aliphatics, aromatics, chlorinated solvents, ketones (except acetone), esters, ethers
Fluoro-elastomer (Viton)	Specialty glove, organic solvents	Extremely expensive, poor physical properties, poor vs. some ketones, esters, amines	Aromatics, chlorinated solvents, also aliphatics and alcohols
Norfoil (Silver Shield)	Excellent chemical resistance	Poor fit, easily punctures, poor grip, stiff	Use for Hazmat work

Glove Type and Chemical Use

Chemical	Neoprene	Natural Latex or Rubber	Butyl	Nitrile Latex
*Acetaldehyde	VG	G	VG	G
Acetic acid	VG	VG	VG	VG
*Acetone	G	VG	VG	P
Ammonium hydroxide	VG	VG	VG	VG
*Amyl acetate	F	P	F	P
Aniline	G	F	F	P
*Benzaldehyde	F	F	G	G
*Benzene	F	F	F	P
Butyl acetate	G	F	F	P
Butyl alcohol	VG	VG	VG	VG
Carbon disulfide	F	F	F	F
*Carbon tetrachloride	F	P	P	G
Castor oil	F	P	F	VG
*Chlorobenzene	F	P	F	P
*Chloroform	G	P	P	E
Chloronaphthalene	F	P	F	F
Chromic acid (50%)	F	P	F	F
Citric acid (10%)	VG	VG	VG	VG
Cyclohexanol	G	F	G	VG
*Dibutyl phthalate	G	P	G	G
Diesel fuel	G	P	P	VG
Diisobutyl ketone	P	F	G	P
Dimethylformamide	F	F	G	G
Dioctyl phthalate	G	P	F	VG
Dioxane	VG	G	G	G
Epoxy resins, dry	VG	VG	VG	VG
*Ethyl acetate	G	F	G	F
Ethyl alcohol	VG	VG	VG	VG
Ethyl ether	VG	G	VG	G
*Ethylene dichloride	F	P	F	P
Ethylene glycol	VG	VG	VG	VG
Formaldehyde	VG	VG	VG	VG
Formic acid	VG	VG	VG	VG
Freon 11	G	P	F	G
Freon 12	G	P	F	G
Freon 21	G	P	F	G
Freon 22	G	P	F	G
*Furfural	G	G	G	G
Gasoline, leaded	G	P	F	VG
Gasoline, unleaded	G	P	F	VG
Glycerine	VG	VG	VG	VG

continued

Glove Type and Chemical Use (continued)

Chemical	Neoprene	Natural Latex or Rubber	Butyl	Nitrile Latex
Hexane	F	P	P	G
Hydrochloric acid	VG	G	G	G
Hydrofluoric acid (48%)	VG	G	G	G
Hydrogen peroxide (30%)	G	G	G	G
Hydroquinone	G	G	G	F
Isooctane	F	P	P	VG
Isopropyl alcohol	VG	VG	VG	VG
Kerosene	VG	F	F	VG
Ketones	G	VG	VG	P
Lacquer thinners	G	F	F	P
Lactic acid (85%)	VG	VG	VG	VG
Laurie acid (36%)	VG	F	VG	VG
Lineoleic acid	VG	P	F	G
Linseed oil	VG	P	F	VG
Maleic acid	VG	VG	VG	VG
Methyl alcohol	VG	VG	VG	VG
Methylamine	F	F	G	G
Methyl bromide	G	F	G	F
*Methyl chloride	P	P	P	P
*Methyl ethyl ketone	G	G	VG	P
*Methyl isobutyl ketone	F	F	VG	P
Methyl methacrylate	G	G	VG	F
Monoethanolamine	VG	G	VG	VG
Morpholine	VG	VG	VG	G
Naphthalene	G	F	F	G
Naphthas, aromatic	G	P	P	G
*Nitric acid	G	F	F	F
Nitromethane (95.5%)	F	P	F	F
Nitropropane (95.5%)	F	P	F	F
Octyl alcohol	VG	VG	VG	VG
Oleic acid	VG	F	G	VG
Oxalic acid	VG	VG	VG	VG
Palmitic acid	VG	VG	VG	VG
Perchloric acid (60%)	VG	F	G	G
Perchloroethylene	F	P	P	G
Petroleum distillates (naphtha)	G	P	P	VG
Phenol	VG	F	G	F
Phosphoric acid	VG	G	VG	VG
Potassium hydroxide	VG	VG	VG	VG
Propyl acetate	G	F	G	F
Propyl alcohol	VG	VG	VG	VG
Propyl alcohol (iso)	VG	VG	VG	VG

Glove Type and Chemical Use (continued)

Chemical	Neoprene	Natural Latex or Rubber	Butyl	Nitrile Latex
Sodium hydroxide	VG	VG	VG	VG
Styrene	P	P	P	F
Stryene (100%)	P	P	P	F
Sulfuric acid	G	G	G	G
Tannic acid (65%)	VG	VG	VG	VG
Tetrahydrofuran	P	F	F	F
*Toluene	F	P	P	F
Toluene diisocyanate	F	G	G	F
*Trichloroethylene	F	F	P	G
Triethanolamine	VG	G	G	VG
Tung oil	VG	P	F	VG
Turpentine	G	F	F	VG
*Xylene	P	P	P	F

*Limited service
VG = Very Good
G = Good
F = Fair
P = Poor (not recommended)

Appendix G: Occupational Safety and Health Resources and Information Sources

All potential sources of information on occupational safety and health could not possibly be provided in one appendix. The listings here are some of the most recent and ones that have been useful.

BOOKS AND DOCUMENTS

Accident Hazard Analysis

Reese, C. D. *Accident/incident prevention techniques*. New York: Taylor & Francis, Inc., 2001.

Reese, C. D. and J. V. Eidson. *Handbook of OSHA construction safety & health (second edition)*. Boca Raton, FL: CRC Press/Taylor & Francis, 2006.

Accident Investigation

Reese, C. D. *Accident/incident prevention techniques*. New York: Taylor & Francis, Inc., 2001.

Reese, C. D. and J. V. Eidson. *Handbook of OSHA construction safety & health (second edition)*. Boca Raton, FL: CRC Press/Taylor & Francis, 2006.

United States Department of Energy. Accident/Incident Investigation Manual (SSDC 27, DOE/SSDC 76-45/27), second edition. Washington, D.C.: Systems Safety Development Center, November 1985.

Vincoli, J. W. *Basic guide to accident investigation and loss control*. New York: John Wiley & Sons, Inc., 1994.

Accident Prevention

Michaud, P. A. *Accident prevention and OSHA compliance*. Boca Raton, FL: CRC Press/Lewis Publishers, 1995.

Reese, C. D. *Accident/incident prevention techniques*. New York: Taylor & Francis, Inc., 2001.

Reese, C. D. and J. V. Eidson. *Handbook of OSHA construction safety & health (second edition)*. Boca Raton, FL: CRC Press/Taylor & Francis, 2006.

United States Department of Labor, National Mine Health and Safety Academy. *Accident Prevention Techniques*. Beckley, WV: U.S. Department of Labor, 1984.

United States Department of Labor, Mine Safety and Health Administration. Accident Prevention, Safety Manual No. 4, Beckley, WV: U.S. Department of Labor, revised 1990.

Construction Safety and Health

Hess, K. *Construction safety auditing made easy: A checklist approach to OSHA compliance*. Rockville, MD: Government Institutes, Inc., 1998.

Moran, M. M. *Construction safety handbook: A practical guide to OSHA compliance and injury prevention.* Rockville, MD: Government Institutes, Inc., 1996.

Reese, C. D. *Annotated dictionary of construction safety and health.* Boca Raton, FL: CRC Press/Lewis Publishers, 2000.

Reese, C. D. and J. V. Eidson. *Handbook of OSHA construction safety & health (second edition).* Boca Raton, FL: CRC Press/Taylor & Francis, 2006.

Consultants

Reese, C. D. and J. V. Eidson. *Handbook of OSHA construction safety & health.* Boca Raton, FL: CRC Press/Lewis Publishers, 1999.

Reese, C. D. *Accident/incident prevention techniques*. New York: Taylor & Francis, Inc., 2001.

Environmental

Arms, K. *Environmental science (second edition).* Upper Saddle Brook, NJ: HBJ College and School Division, 1994.

Henry, J. G. and G. W. Heinke. *Environmental science and engineering (second edition).* New York: Prentice Hall, 1995.

Jackson, A. R. and J. M. Jackson. *Environmental science: The natural environment and human impact.* New York: Longman, 1996.

Koren, H. and M. Bisesi. *Handbook of environmental health and safety (3rd edition), Principles and practices*, Volumes I and II. Boca Raton, FL: Lewis Publishers, 1996.

Lynn, L. *Environmental biology.* Northport, NY: Kendall-Hunt, 1995.

Manahan, S. E. *Fundamentals of environmental chemistry*. Boca Raton, FL: CRC Press/Lewis Publishers, 1993.

Moron, J. M. et al. *Introduction to environmental science*. New York: W. H. Freeman and Company, 1986.

Que Hee, S. S. *Hazardous waste analysis*. Rockville, MD: Government Institutes, 1999.

Schell, David J. *What environmental managers really need to know.* Rockville, MD: Government Institutes, 1999.

Spellman, F. R. and N. E. Whiting. *Environmental science and technology: Concepts and applications*. Rockville, MD: Government Institutes, 1999.

Sullivan, T. F. P. *Environmental law book (16th edition).* Rockville, MD: Government Institutes, 2001.

Wentz, C. A. *Hazardous waste management*. New York: McGraw-Hill, 1990.

Ergonomics

Erdil, M. and O. B. Dickerson. *Cumulative trauma disorders: Prevention, evaluation, and treatment*. New York: Van Nostrand Reinhold, 1997.

Eastman Kodak Company, *Ergonomic design for people at work: Volumes 1 and* 2. New York: Van Nostrand Reinhold, 1983.

Kromer, K. H. E. *Ergonomics design of material handling systems*. Boca Raton, FL: CRC Press/Lewis Publishers, 1997.

Kromer, K. H. E. and E. Grandjean. *Fitting the task to the human*. New York: Taylor & Francis, Inc., 1997.

Kromer, K., H. Kromer, and K. Kromer-Elbert. *Ergonomics: How to design for ease and efficiency*. Englewood Cliffs, NJ: Prentice Hall, 1994.

Laing, P. M. *Ergonomics: A practical guide (second edition)*. Itasca, IL: National Safety Council, 1993.
MacLeod, D. *The ergonomics edge*. New York: Van Nostrand Reinhold, 1995.
Putz-Anderson, V. *Cumulative trauma disorders: A manual for musculoskeletal disease of the upper limbs*. New York: Taylor & Francis, Inc., 1994.
Reese, C. D. and J. V. Eidson. *Handbook of OSHA construction safety & health*. Boca Raton, FL: CRC Press/Lewis Publishers, 1999.

Fleet Safety

National Safety Council. *Motor fleet safety manual. Third edition*. Itasca, IL: National Safety Council, 1986.
Reese, C. D. *Accident/incident prevention techniques*. New York: Taylor & Francis, Inc., 2001.
Reese, C. D. and J. V. Eidson. *Handbook of OSHA construction safety & health (second edition)*. Boca Raton, FL: CRC Press/Taylor & Francis, 2006.

Hazard Identification

National Safety Council, *Supervisors' safety manual (ninth edition)*. Itasca, IL: National Safety Council, 1997.
Reese, C. D. *Accident/incident prevention techniques*. New York: Taylor & Francis, Inc., 2001.
Reese, C. D. and J. V. Eidson. *Handbook of OSHA construction safety & health*. Boca Raton, FL: CRC Press/Lewis Publishers, 1999.
United States Department of Labor, Mine Health and Safety Administration. *Hazard Recognition and Avoidance: Training Manual (MSHA 0105)*. Beckley, WV: U.S. Department of Labor, revised May 1996.

Health Hazards

Levy, B. S. and D. H. Wegman. *Occupational health: Recognizing and preventing work-related disease (third edition)*. Boston, MA: Little, Brown and Company, 1995.
Reese, C. D. *Accident/incident prevention techniques*. New York: Taylor & Francis, Inc., 2001.

Industrial Hygiene

Hathway, G. J., N. H. Proctor, and J. P. Hughes. *Proctor & Hughes' chemical hazards of the workplace (fourth edition)*. New York: John Wiley & Sons, Inc., 1996.
Kamrin, M. *Toxicology*. Boca Raton, FL: CRC PRess/Lewis Publishers, 1988.
Plog, B. A. *Fundamentals of industrial hygiene (fifth edition)*. Itasca, IL: National Safety Council, 2001.
Reese, C. D. and J. V. Eidson. *Handbook of OSHA construction safety & health (second edition)*. Boca Raton, FL: CRC Press/Taylor & Francis, 2006.
Scott, R. *Basic concepts of industrial hygiene*. Boca Raton, FL: CRC Press/Lewis Publishers, 1997.

Job Hazard Analysis

Reese, C. D. *Accident/incident prevention techniques*. New York: Taylor & Francis, Inc., 2001.
Reese, C. D. and J. V. Eidson. *Handbook of OSHA construction safety & health (second edition)*. Boca Raton, FL: CRC Press/Taylor & Francis, 2006.

United States Department of Labor, Occupational Safety and Health Administration. Job Hazard Analysis (OSHA 3071). Washington, D.C.: U.S. Department of Labor, 1992

United States Department of Labor, Mine Safety and Health Administration. Job Safety Analysis: A Practical Approach (Instruction Guide No. 83). Beckley, WV: U.S. Department of Labor, 1990.

United States Department of Labor, Mine Safety and Health Administration. Job Safety Analysis (Safety Manual No. 5). Beckley, WV: U.S. Department of Labor, revised 1990.

United States Department of Labor. National Mine Health and Safety Academy. *Accident prevention techniques: Job safety analysis*. Beckley, WV: U.S. Department of Labor, 1984.

Job Safety Observation

Reese, C. D. *Accident/incident prevention techniques*. New York: Taylor & Francis, Inc., 2001.

Reese, C. D. and J. V. Eidson. *Handbook of OSHA construction safety & health (second edition)*. Boca Raton, FL: CRC Press/Taylor & Francis, 2006.

United States Department of Labor, Mine Safety and Health Administration, Safety Observation (MSHA IG 84). Beckley, WV: U.S. Department of Labor, revised 1991.

Office Safety and Health

Reese, C. D. *Office building safety and health*. Boca Raton, FL: CRC Press, 2004

OSHA Compliance

Blosser, F. *Primer on occupational safety and health*. Washington, D.C.: The Bureau of National Affairs, Inc., 1992.

Reese, C. D. *Accident/incident prevention techniques*. New York: Taylor & Francis, Inc., 2001.

Reese, C. D. and J. V. Eidson. *Handbook of OSHA construction safety & health (second edition)*. Boca Raton, FL: CRC Press/Taylor & Francis, 2006.

United States Department of Labor, Occupational Safety and Health Administration. Field Inspection Reference Manual (FIRM) (OSHA Instruction CPL 2.103). Washington, D.C.: U.S. Department of Labor, September 26, 1994.

United States Department of Labor, Occupational Safety and Health Administration, Office of Training and Education. *OSHA Voluntary Compliance Outreach Program: Instructors Reference Manual*. Des Plaines, IL: U.S. Department of Labor, 1993.

United States Department of Labor, Occupational Safety and Health Administration. OSHA 10- and 30-Hour Construction Safety and Health Outreach Training Manual. Washington, D.C.: U.S. Department of Labor, 1991.

United States Department of Labor, Occupational Safety and Heath Administration, General Industry. Code of Federal Regulations (Title 29, Part 1910). Washington, D.C.: Government Printing Office, 1998.

United States Department of Labor, Occupational Safety and Heath Administration, Construction. Code of Federal Regulations (Title 29, Part 1926). Washington, D.C.: Government Printing Office, 1998.

Psychology of Safety

Brown, P. L. and R. J. Presbie. *Behavior modification in business, industry and government*. Paltz, NY: Behavior Improvement Associates, Inc., 1976.

Geller, E. S. *The psychology of safety handbook*. Boca Raton, FL: CRC/Lewis Publishers, 2001.

Herzberg, F. "One more time: How do you motivate employees? *Harvard Business Review* (January-February, 1968): pp. 53–62.

Mager, R. F. *Analyzing performance problems*. Belmont, CA: Fearson Publishers, Inc., 1970.

Riggio, R. E. *Introduction to industrial/organizational psychology (third edition).* Upper Saddle River, NJ: Prentice Hall, 2000.

Reese, C. D. *Accident/incident prevention techniques*. New York: Taylor & Francis, Inc., 2001.

Reese, C. D. and J. V. Eidson. *Handbook of OSHA construction safety & health (second edition).* Boca Raton, FL: CRC Press/Taylor & Francis, 2006.

Regulations

Reese, C. D. *Accident/incident prevention rechniques*. New York: Taylor & Francis, Inc., 2001.

Reese, C. D. and J. V. Eidson. *Handbook of OSHA construction safety & health (second edition).* Boca Raton, FL: CRC Press/Taylor & Francis, 2006.

United States Department of Labor, Occupational Safety and Health Administration. General Industry Digest (OSHA 2201). Washington, D.C.: Government Printing Office, 1995.

United States Department of Labor, Occupational Safety and Health Administration. 29 Code of Federal Regulations 1910. Washington, D.C.: Government Printing Office, 1999.

United States Department of Labor, Occupational Safety and Health Administration. 29 Code of Federal Regulations 1926. Washington, D.C.: Government Printing Office, 1999.

Safety Hazards

Reese, C. D. *Accident/incident prevention techniques*. New York: Taylor & Francis, Inc., 2001.

Reese, C. D. and J. V. Eidson. *Handbook of OSHA construction safety & health.* Boca Raton, FL: CRC Press/Lewis Publishers, 1999.

United States Department of Energy. OSHA Technical Reference Manual. Washington, D.C.: U.S. Department of Energy, 1993.

Safety and Health Hazards

Goetsch, D. L. *Occupational safety and health for technologists, engineers, and managers (third edition).* Upper Saddle River, NJ: Prentice Hall, 1999.

Hagan, P. E., J. F. Montgomery, and J. T. O'Reilly, *Accident prevention manual for business and industry: Engineering and technology (12th edition).* Itasca, IL: National Safety Council, 2001.

Spellman, F. R. and, N. E. Whiting. *Safety engineering: Principles and practices.* Rockville, MD: Government Institutes, 1999.

Safety and Health Management

Kohn, J. P. and T. S. Ferry. *Safety and health management planning*. Rockville, MD: Government Institutes, 1999.

Dougherty, J. E. *Industrial safety management: A practical approach.* Rockville, MD: Government Institutes, 1999.

Hagan, P. E., J. F. Montgomery, and J. T. O'Reilly. *Accident prevention manual for business and industry: Administration & programs (12th edition).* Itasca, IL: National Safety Council, 2001.

Lack, R. W. *Essentials of safety and health management*. Boca Raton, FL: CRC Press/ Lewis Publishers, 1996.

Lack, R. W. *Safety, health, and asset protection: Management essentials (second edition).* Boca Raton, FL: Lewis Publishers, 2002.

Petersen, D. *Human error reduction and safety management (3rd edition).* New York: Van Nostrand Reinhold, 1996.

Petersen, D. *Safety management: A human approach (2nd edition).* Goshen, NY: Aloray, Inc., 1988.

Petersen, D. *Techniques of safety management: A systems approach (3rd edition).* Goshen, NY: Aloray Inc., 1989.

Reese, C. D. *Occupational health and safety management: A practical approach.* Boca Raton, FL: Lewis Publishers, 2003.

United States Department of Labor, Occupational Safety and Health Administration. *Federal Register: Safety and Health Program Management Guidelines (Vol. 54, No. 16).* pp. 3904–3916. Washington, D.C.: U.S. Department of Labor, January 26, 1989.

Service Industry Safety and Health

Reese, C. D., *Handbook of safety and health for the service industry, Volume 1: Industrial safety and health for goods and material services.* Boca Raton, FL: CRC Press, Taylor & Francis, 2008.

Reese, C. D., *Handbook of safety and health for the service industry, Volume 2: Industrial safety and health for infrastructure services.* Boca Raton, FL: CRC Press, Taylor & Francis, 2008.

Reese, C. D., *Handbook of safety and health for the service industry, Volume 3: Industrial safety and health for administrative services.* Boca Raton, FL: CRC Press, Taylor & Francis, 2008.

Reese, C. D., *Handbook of safety and health for the service industry, Volume 4: Industrial safety and health for people oriented services.* Boca Raton, FL: CRC Press, Taylor & Francis, 2008.

System Safety

ABS Group. *Root cause analysis handbook: A guide to effective incident investigation.* Rockville, MD: Government Institutes, 1999.

Bahr, N. J. *System safety engineering, and risk assessment: A practical approach.* New York: Taylor & Francis, Inc., 1997.

Reese, C. D. *Accident/incident prevention techniques.* New York: Taylor & Francis, Inc., 2001.

Kavianian, H. R. and C. A. Wentz, Jr. *Occupational and environmental safety engineering and management.* New York: Van Nostrand Reinhold, 1990.

United States Department of Energy, Office of Nuclear Energy. Root Cause Analysis Guidance Document. Washington, D.C.: U.S. Department of Energy, February 1992.

Training

Reese, C. D. *Accident/incident prevention techniques.* New York: Taylor & Francis, Inc., 2001.

Reese, C. D. and J. V. Eidson. *Handbook of OSHA construction safety & health (second edition).* Boca Raton, FL: CRC Press/Taylor & Francis, 2006.

Reese, C. D. *Occupational health and safety management: A practical approach.* Boca Raton, FL: Lewis Publishers, 2003.

United States Department of Labor. Training Requirements in OSHA Standards and Training Guidelines (OSHA 2254). Washington, D.C.: U.S. Department of Labor, 1998.

Workplace Violence

United States Department of Health and Human Services, National Institute for Occupational Safety and Health. NIOSH Current Intelligence Bulletin 57. Violence in the Workplace: Risk Factors and Prevention Strategies. Washington, D.C.: U.S. Department of Health and Human Services, 1996.
United States Department of Labor, Bureau of Labor Statistics. *National census of fatal occupational injuries*. Washington, D.C.: U.S. Department of Labor, 1998.
Warchol, G. Workplace Violence, 1992–96. National Crime Victimization Survey (Report No. NCJ-168634). Washington, D.C.: 1998.

PROFESSIONAL ORGANIZATIONS AND AGENCIES

These are national organizations that specialize in the many aspects of occupational safety and health. They have a wide range of resources as well as unique materials that have been developed by individuals and organizations with special expertise in occupational safety and health. Some key organizations and agencies are as follows:

Health and Environmental Assistance

ABIH (American Board of Industrial Hygiene)
4600 West Saginaw, Suite 101
Lansing, Michigan 48917
(517) 321-2638

ACGIH (American Conference of Governmental Industrial Hygienists)
Building D-7
6500 Glenway Avenue
Cincinnati, Ohio 45211
(513) 661-7881

AIHA (American Industrial Hygiene Association)
P.O. Box 8390
475 White Pond Drive
Akron, Ohio 44311
(216) 873-3300

Safety and Engineering Consensus Standards

ANSI (American National Standards Institute)
11 West 42nd Street
New York, New York 10038
(212) 354-3300

ASME (American Society of Mechanical Engineers)
345 East 47th Street
New York, New York 10017
(212) 705-7722

ASTM (American Society for Testing and Materials)
655 15th Street NW
Washington, District of Columbia 20005
(202) 639-4025

NSMS (National Safety Management Society
12 Pickens Lane
Weaverville, North Carolina 28787
(800) 321-2910

Professional Safety Organizations

ASSE (American Society for Safety Engineers)
1800 East Oakton Street
Des Plaines, Illinois 60016
(847) 699-2929

BCSP (Board of Certified Safety Professionals)
208 Burwash Ave.
Savoy, Illinois 61874
(312) 359-9263

ISEA (Industrial Safety Equipment Association)
1901 North Moore Street
Arlington, Virginia 22209
(703) 525-1695
FAX: (703) 528-2148

NSC (National Safety Council)
1121 Spring Lake Drive
Itasca, Illinois 60143-3201
(708) 285-1121

HFS (Human Factors Society)
P.O. Box 1369
Santa Monica, California 90406
(310) 394-1811

Specialty Associations (with Specific Expertise)

AWS (American Welding Society)
P.O. Box 351040
550 LeJeune Road, NW
Miami, Florida 33135
(305) 443-9353

AGA (American Gas Association)
1515 Wilson Blvd.
Arlington, Virginia 22209
(703) 841-8400

API (American Petroleum Institute)
1220 L Street, NW
Washington, District of Columbia 20005
(202) 682-8000
FAX: (202) 682-8159

ASHRAE (American Society of Heating, Refrigerating, and Air Conditioning Engineers)
1791 Tullie Circle, NE
Atlanta, Georgia 30329
(404) 636-8400

ASTD (American Society for Training and Development)
1640 King Street
P.O. Box 1443
Alexandria, Virginia 22313-2043
(703) 683-8129

CGA (Compressed Gas Association)
1235 Jefferson Davis Highway
Arlington, Virginia 22202
(703) 979-0900

Illuminating Engineering Society of North America
120 Wall Street, 17th Floor
New York, New York 10005
(212) 248-5000

Institute of Makers of Explosives
1120 19th Street, NW
Washington, District of Columbia 20036
(202) 429-9280

Laser Institute of America
12424 Research Parkway, Suite 130
Orlando, Florida 32826
(407) 380-1553

NFPA (National Fire Protection Association)
1 Batterymarch Park
Quincy, Massachusetts 02269
(800) 344-3555

National Propane Gas Association
1150 176th Street, NW
Washington, District of Columbia 20036
(202) 466-7200

The Chlorine Institute
2001 L Street
Washington, District of Columbia 20036
(202) 775-2790
FAX: (202) 223-7225

FEDERAL GOVERNMENT SOURCES

The federal government is not the enemy, as many individuals surmise. It is a great resource for all types of information such as publications, training materials, compliance assistance, audio-visuals, access to experts, and other assorted occupational safety and health aids. In most cases, resources offered by the federal government are current and the response time is very reasonable. Asking for information does not act as a trigger for your company to become a target for inspections or audits. The federal government would prefer to assist you in solving your safety and health issues before they become problems. You will be pleasantly surprised by the help that you receive. All you need to do is ask. A listing of government agencies that have information regarding occupational safety and health is as follows:

BLS (Bureau of Labor Statistics)
U.S. Department of Labor
Occupational Safety and Health Statistics
441 G Street, NW
Washington, District of Columbia 20212
(202) 523-1382

CDC (Center for Disease Control)
U.S. Department of Health and Human Services
1600 Clifton Avenue, NE
Atlanta, Georgia 30333
(404) 329-3311

EPA (Environmental Protection Agency)
410 M Street, SW
Washington, District of Columbia 20460
(202) 382-4361

GPO (U.S. Government Printing Office)
Superintendent of Documents
732 N. Capitol Street, NW
Washington, District of Columbia 20402
(202) 512-1800

MSHA (Mine Safety and Health Administration)
U.S. Department of Labor
4015 Wilson Blvd.
Arlington, Virginia 22203
(703) 235-1452

NAC (National Audio Visual Center)
National Archives and Records Administration
Customer Services Section CL
8700 Edgewood Drive
Capitol Heights, Maryland 20743-3701
(301) 763-1896

National Institute of Standards and Technology
U.S. Department of Commerce
National Engineering Laboratory
Route I-270 and Quince Orchard Road
Gaithersburg, Maryland 20899
(310) 921-3434

NIH (National Institutes of Health)
U.S. Department of Health and Human Services
9000 Rockville Pike
Bethesda, Maryland 20205
(310) 496-5787

NIOSH (National Institute for Occupational Safety and Health)
U.S. Department of Health and Human Services
Publications Dissemination
4676 Columbia Parkway
Cincinnati, Ohio 45226
(513) 533-8287 or (800) 35-NIOSH

NTIS (National Technical Information Services)
U.S. Department of Commerce
5285 Port Royal Road
Springfield, Virginia 22161
(703) 487-4636

OSHA (Occupational Safety and Health Administration—National Office)
U.S. Department of Labor
200 Constitution Avenue, NW
Washington, District of Columbia 20210
(202) 523-8151
OSHA (after Hours), National Hotline—(800) 321-OSHA

OSHA's Training Institute
1555 Times Drive
Des Plaines, Illinois 60018
(708) 297-4913

OSHA Publications Office
Room N3101
Washington, District of Columbia 20210
(202) 219-9667

OSHRC (Occupational Safety and Health Review Commission)
1825 K Street, NW
Washington, District of Columbia 20006
(202) 643-7943

ELECTRONIC SOURCES (INTERNET)

You must be connected to the Internet; this means you must select an Internet provider. This provider may be your local or long-distance phone company, your cable television, or it may be commercial services such as AOL, Yahoo!, etc. (just to name a few). Of course, it goes without saying that you will need a computer with a reasonably fast modem (the faster, the better), a telephone line, and some software such as Microsoft Explorer or Netscape Navigator (your Internet provider usually provides this), which allows you to browse the Internet.

Once you have access to the Internet, there are several good search engines that are helpful in finding information (Internet sites). These have names, Google being the most common. These search engines allow you to find the sites or locations of the information that you are interested in (i.e., machine guarding, fire safety). Internet addresses are constantly changing so the ability to search is critical.

The Internet sites have names that help you understand what they are. Some of the most common abbreviated names are as follows:

- http—this is a transfer protocol, a standard web programming language
- www—means World Wide Web and is a connective or networking component of the Internet
- com—means commercial
- edu—stands for education
- gov—means government
- org—stands for organization

Most Internet sites start with http://www. followed by an abbreviation for the entity (company or institution) and other numbers, symbols, or abbreviations that seem to make no sense. The ending is usually .com, .gov, .edu, or .org. You can use established site addresses, such as the ones that follow to access these specific locations:

Government

- **Addresses of Government Agencies**
  http://www.fedworld.gov
- **Agency for Toxic Substances and Disease Registry**
  http://atsdrl.atsdr.cdc.gov.8080/atsdrhome.html
- **ATSDR Hazardous Substance Release/Health Effects Database**
  http://atsdr1.atsdr.cdc.gov.8080/hazdat.html
- **Building and Fire Research Laboratory**
  http://www.bfrl.nist.gov/
- **Bureau of Labor Statistics**
  http://stats.bls.gov/
- **California Department of Industrial Relations**
  http://www.dir.ca.gov/
- **Centers for Disease Control and Prevention**
  http://www.cdc.gov/
- **Consumer Product Safety Commission**
  http://www.cpsc.gov/
- **Emerging Infectious Diseases Home Page**
  http://www.cdc.gov/ncidod/EID/eid.html
- **Federal Emergency Management Agency**
  http://www.fema.gov/
- **Mine Safety and Health Administration**
  http://www.msha.gov/
- **Mining Accident and Injury Information**
  http://www.msha.gov/STATINFO.HTML
- **National Agriculture Safety Database**
  http://www.cdc.gov/niosh/nasd/nasdhome.html
- **National Highway Traffic Safety Administration**
  http://www.nhtsa.dot.gov/
- **National Institute of Environmental Health Sciences**
  http://heww.niehs.nih.gov/
- **National Institutes of Health**
  http://www.nih.gov/
- **National Institute for Occupational Safety and Health**
  http://www.cdc.gov/niosh/homepage.html
- **Occupational Safety and Health Administration**
  http:/www.osha.gov/and http:www.osha.gov/STLC
- **OSHA Ergonomics**
  http://www.osha.gov/ergo
- **U.S. Department of Transportation**
  http://www.dot.gov/
- **U.S. Department of Energy Chemical Safety Program**
  http://tis-hq.eh.doe.gov/web/chem_safety/

- **U.S. Environmental Protection Agency**
  http://www.epa.gov/
- **U.S. Department of Health and Human Services**
  http://www.dhhs.gov/
- **U.S. Department of Labor Office of Inspector General**
  http://gatekeeper.dol.gov/dol/oig/

Associations and Societies

- **American Association of Occupational Health Nurses**
  http://www.aaohn.org/
- **American Chemical Society**
  http://www.acs.org/
- **American College of Occupational and Environmental Medicine**
  http://www.acoem.org
- **American Conference of Governmental Industrial Hygienists**
  http://www.acgih.org/
- **American Industrial Hygiene Association**
  http://www.aiha.org/
- **American National Standards Institute**
  http://web.ansi.org/default.htm
- **American Society of Heating, Refrigerating and Air-Conditioning Engineers**
  http://www.ashrae.org/
- **American Society of Safety Engineers**
  http://www.asse.org/
- **American Society for Testing and Materials**
  http://www.astm.org/
- **American Speech-Language-Hearing Association**
  http://www.asha.org/
- **Board of Certified Safety Professionals**
  http://www.bcsp.com/
- **British Safety Council**
  http://www.britishsafetycouncil.co.uk/
- **Building Owners and Managers Association International**
  http://www.boma.org/
- **Canada Safety Council**
  http://www.safety-council.org/english/index.htm
- **Canadian Society of Safety Engineering**
  http://www.csse.org/
- **Chemical Manufacturers Association**
  http://www.cmahq.corn/index.html
- **Industrial Safety Equipment Association**
  http://www.safetycentral.org/isea/
- **National Association of Demolition Contractors**
  http:/Iwww.voicenet.corn/-NAOC

- **National Association of Tower Erectors**
  http://www.daknet.corn/nate/
- **National Fire Protection Association**
  http://www.wpi.edu/-fpe/nfpa.html
- **National Hearing Conservation Association**
  http://www.globaldialag.corn/-nhca/
- **National Safety Council**
  http://www.nsc.org/

Many other electronic sites for information are available, and more are being posted at what seems like lightning speed. The electronic information system is very fluid, and new avenues of information are constantly emerging.

Index

B

C

D

J

K

L

M

N

O

R

T

Printed in the United States
by Baker & Taylor Publisher Services

Printed in the United States
by Baker & Taylor Publisher Services